Sociobiology of Communication: an interdisciplinary perspective

Sociobiology of Communication

An interdisciplinary perspective

EDITED BY

Patrizia d'Ettorre and David P. Hughes

OXFORD
UNIVERSITY PRESS

This book has been printed digitally and produced in a standard specification

in order to ensure its continuing availability

OXFORD
UNIVERSITY PRESS

Great Clarendon Street, Oxford OX2 6DP

Oxford University Press is a department of the University of Oxford.
It furthers the University's objective of excellence in research, scholarship,
and education by publishing worldwide in

Oxford New York

Auckland Cape Town Dar es Salaam Hong Kong Karachi
Kuala Lumpur Madrid Melbourne Mexico City Nairobi
New Delhi Shanghai Taipei Toronto

With offices in

Argentina Austria Brazil Chile Czech Republic France Greece
Guatemala Hungary Italy Japan Poland Portugal Singapore
South Korea Switzerland Thailand Turkey Ukraine Vietnam

Oxford is a registered trade mark of Oxford University Press
in the UK and in certain other countries

Published in the United States
by Oxford University Press Inc., New York

Reprinted 2011

British Library Cataloguing in Publication Data
Data available

Library of Congress Cataloging in Publication Data
Data available

ISBN 978-0-19-921683-3

Printed and bound by CPI Antony Rowe , Eastbourne

Preface

Communication bridges biology disciplines, and beyond

As we first designed this book, the title we had in mind was *Communication among social organisms*, and its aim was to make the most of an integrated and interdisciplinary approach in order to seek commonalties across a diversity of taxa expressing social behaviour. The ultimate hope was trying to identify the underlying general principles of communication. However, when thinking about a possible table of contents and list of contributors, it became obvious to us that the general principles would also apply to communication *within organisms* and perhaps even to non-organisms. Communication is the essence of any interaction, without communication social interactions are simply impossible. We wanted to present communication as a ubiquitous and unifying biological principle but our title didn't quite take us there.

Having bothered several colleagues and the OUP team with what was becoming a pressing issue, we were pleased to accept the suggestion of our friend Kevin Foster who—while enjoying a beer at the evening-pub during the 2007 congress of the European Society for Evolutionary Biology in Uppsala—came up with *Sociobiology of Communication*. We believe this title is duly qualifying for the plethora of communication issues that are addressed in this book, since *Sociobiology* is nothing less than the study of the biological bases of social behaviour, in particular its ecological and evolutionary basis.

The book is not intended to encyclopaedically encompass all aspects of social communication, but rather to offer a broad and novel perspective. We believe that, with our esteemed contributors, we have achieved this goal at least at three different levels. Firstly, we present a wide range of systems, from intra-genomic conflict to metazoan and bacterial cells, to insect, vertebrate and human societies. Secondly, we address conceptual, theoretical and empirical research in order to unveil both proximate and ultimate mechanisms shaping communication among and within organisms. And finally, we cross some historically defined frontiers between disciplines. The effort in understanding the general principles of communication not only bridges biology disciplines but may act as a joker when playing the cards of knowledge. The study of social communication is undoubtedly a common ground of interest, a *deus ex machina* that can resolve a long-lasting situation of incommunicability between the natural sciences, the social sciences and the humanities.

A journey through the chapters

In selecting the contributors to this book, we aimed to cover a broad array of model systems and levels of analysis, and to have both well-known established scholars and young researchers that are just beginning to influence the way we think about paradigms. The rational being that communication across academic generations would also help to achieve a high degree of interdisciplinary synthesis.

Amotz Zahavi has had a lasting influence on the way in which we interpret biological signals. In Chapter 1, he summarizes the essence of the Handicap Principle, introduces us to the fascinating world of Arabian babblers and their "selfish altruism", and argues that altruism in slime-molds can also be explained by individual selection. The chapter ends with his most recent research project on the evolution of chemical signals within multi-cellular organisms. Chemical signals are also the focus of Chapter 2 by Stephen Diggle and collaborators, but this time an explicit kin selection perspective is applied to bacteria and their quorum sensing, including communication between cells of the same species and of different microbial kingdoms and with interpretations ranging from altruism to coercion. Communication goes networking in Chapter 3 by Giuliano Matessi and co-workers, which shows that signalling and receiving strategies can be accurately explained with models based on social networks, particularly when studying bird communication in the field. The authors present us with a cautionary tale regarding the complexity of such networks when examined in full detail.

In Chapter 4, David Nash and Koos Boomsma highlight how even the extremely efficient communication systems of insect societies are vulnerable to social parasites that exploit the host communication system for their own ends. The prospects for coevolutionary arms races are reviewed and illustrated with key examples from long-term studies of *Maculinea* butterflies. Chapter 5, by Allen Moore and one of us, uses insects as model systems to explore the complexity of multi-component chemical communication and the nested levels of variation that characterize pheromones. Here, social selection and indirect genetic effects provide the framework for understanding the fine-tuned coordination of messages from senders and receivers. Chapter 6, by Jane Hurst and Robert Beynon, gives an overview of the power of scent in mammalian societies, with a comparative analysis of the role of the Major Histocompatibility Complex and Major Urinary Proteins in transmitting information about identity and status both in laboratory and wild rodents. In Chapter 7, Gabriela de Brito-Sanchez and collaborators disentangle the neurobiology of pheromone processing from peripheral to central brain units in the honey bee, arguing that advances in our understanding of the architecture of a mini-brain may soon reveal the neural basis of social olfactory communication in this model system.

Social communication and the powerful role of signals in rapid evolutionary change are highlighted by Marlene Zuk and Robin Tinghitella in Chapter 8, with a review on sexual signals and an example in which behavioural plasticity facilitated the elimination of a courtship acoustic signal in response to parasite pressure. We stay with sexual signals in Chapter 9, but this time we are our own models when Craig Roberts shows how physical characteristics can be cues for good genes in humans and suggests that the reliability of facial, bodily, vocal and olfactory traits in communicating mate quality might be extrapolated to understand the role of non-physical traits, such as 'body language' in our mate choice.

Another of us (David Hughes) introduces us to the world of extended phenotypes in Chapter 10, where we can see how parasites manipulate host behaviour and obfuscate communication in the advanced insect societies, and gain insight into the evolution of communication. Collective behaviour is the focus of Chapter 11 where David Sumpter and Åke Brännström argue that communication is key to make a group more than the sum of its parts, owing to synergy between cooperative signalling and thus resolving social dilemmas. The jump from group signalling to signalling within an individual body might seem insurmountable, but is in fact possible when taking an explicit cooperation and conflict angle. This is what David Haig offers in Chapter 12, where genomic imprinting exemplifies the role of internal conflicts in communication between and within organisms. Genomic imprinting is also the focus of Chapter 13. Here, Bernard Crespi considers the role that language and disordered social communication might have played in the evolution of autism and schizophrenia, mediated through genomic conflict.

In Chapter 14, the linguist James Hurford unveils the key features of human communication that have made us exclusively different from all the other animals: our language, our willingness to altruistically impart information by teaching. In Chapter 15, Livio Riboli-Sasco and collaborators propose that the answer is to be found in the autocatalytic nature of information transfer typical of teaching. Information copy number increases with teaching but not with other forms of altruism, and this dynamic process is likely to have contributed to our evolutionary success.

We end our journey through the eyes of a philosopher, Ronald de Sousa, who makes sense of the sociobiology of communication with a synthetic essay underlining what is *not* communication in the interactions of cells, organs, or individuals. Here we may find the way towards a conceptual unit: "What exactly, then, do all those phenomena have in common which may legitimately fall under the concept of 'communication'?"

We hope that this integrated and interdisciplinary perspective will successfully address both graduate students interested in social communication and professionals in evolutionary biology and behavioural ecology seeking novel inspiration. However, we will achieve our intimate goal only if a wider academic audience, including social and medical scientists, would be tempted to explore what evolutionary approaches can offer to their fields.

Is terminology an issue?

"**Hermogenes:** I have often talked over this matter, both with Cratylus and others, and cannot convince myself that there is any principle of correctness in names other than convention and agreement; any name which you give, in my opinion, is the right one, and if you change that and give another, the new name is as correct as the old [...].

Socrates: I dare say that you be right, Hermogenes: let us see—Your meaning is, that the name of each thing is only that which anybody agrees to call it?"

Plato, *Cratylus (dialogue)*

The thirty-one authors of this book, if asked individually to describe terms such as 'communication', 'social interaction' or 'signal', would each give a slightly different definition, perhaps emphasizing those features of a particular biological phenomenon that were most useful to develop their own research approaches. In general, is the plurality of definitions an authentic problem for the progress of science? Or is it an intellectual richness, which is enhancing the advancement of science? We certainly need agreement to progress, but sometimes controversy could be the driving force of new and unexpected discoveries.

We have tried to overcome possible semantic problems by asking all the authors to define specific terms in text boxes and we provide a general glossary at the end of the book (glossary entries are bold in the text of the Chapters). We hope to have succeeded in our goal of making understandable what we mean with a term in a specific context. There is probably no universal recipe on how to achieve agreement on terminology, and the terminology issue will thus continue to entertain students of any discipline. So our last word on this issue will mirror Socrates as he moves to close the dialogue "And when you have found the truth, come and tell me."

Patrizia d'Ettorre and David P. Hughes

Acknowledgements

We wish to sincerely thank all the authors. This book, and its impact, exists because of their palpable curiosity for a myriad of phenomena in our cultural and biological world. It has been our great pleasure to coax their thoughts onto the pages of this book. We are grateful to all of them for their willingness to communicate with us, and now you, the reader.

We are fortunate to be part of the *Centre for Social Evolution* in Copenhagen, a highly stimulating working environment. David R. Nash provided valuable help and suggestions. Koos Boomsma constantly encouraged us during this project and his enthusiasm erased our doubts. We very much appreciate his excellent advice throughout.

This volume would not exist without the *Marie Curie Action*, since this EU program made our scientific careers possible.

We are very grateful to Anna M. Schmidt, whose critical eye and efficiency have been essential in the final editing of this volume.

It has been a pleasure to work with the OUP staff, thanks to the enthusiasm of Ian Sherman and the proficient kindness of Helen Eaton.

And finally, to the special people in our lives that must experience the pain of this project through our moaning but none of its rewards. Inadequate compensation though it may be we are extremely grateful to them for their continued support—David thanks Alba and Jacopo and Patrizia thanks Mauro for continuing to love us despite the fact we have been married to this book for a while.

Par ici et ver là is an acrylic painting on sanded canvas, by François Géhan. The painting is from a collection representing a dream-like journey through a colourful *bestiaire improbable* (fantastic bestiary), inspired by the work of Jérôme Bosch and the Les Shadoks cartoons. The title of the painting is a play on words: *vers* is a direction (towards) but *ver* is a worm (as painted on the sign).

François Géhan graduated from L'Ecole Des Beaux Arts, Tours, France, and has exhibited his paintings since the early nineties. For further information please visit www.art-gehan.fr.

Contents

Preface v
Patrizia d'Ettorre and David P. Hughes

Acknowledgements ix
List of contributors xiii

1 **The handicap principle and signalling in collaborative systems** 1
Amotz Zahavi

2 **Communication in bacteria** 11
Stephen P. Diggle, Stuart A. West, Andy Gardner, and Ashleigh S. Griffin

3 **Communication in social networks of territorial animals: networking at different levels in birds and other systems** 33
Giuliano Matessi, Ricardo J. Matos, and Torben Dabelsteen

4 **Communication between hosts and social parasites** 55
David R. Nash and Jacobus J. Boomsma

5 **Chemical communication and the coordination of social interactions in insects** 81
Patrizia d'Ettorre and Allen J. Moore

6 **Chemical communication in societies of rodents** 97
Jane L. Hurst and Robert J. Beynon

7 **Neurobiology of olfactory communication in the honeybee** 119
Maria Gabriela de Brito-Sanchez, Nina Deisig, Jean-Christophe Sandoz, and Martin Giurfa

8 **Rapid evolution and sexual signals** 139
Marlene Zuk and Robin M. Tinghitella

9 **Communication of mate quality in humans** 157
S. Craig Roberts

10 **The extended phenotype within the colony and how it obscures social communication** 171
David P. Hughes

11 **Synergy in social communication** 191
David J.T. Sumpter and Åke Brännström

12 **Conflicting messages: genomic imprinting and internal communication** 209
David Haig

13 Language unbound: genomic conflict and psychosis in the origin of modern humans 225
Bernard J. Crespi

14 The evolution of human communication and language 249
James R. Hurford

15 Why teach? The evolutionary origins and ecological consequences of costly information transfer 265
Livio Riboli-Sasco, Sam Brown, and François Taddei

16 Grades of communication 275
Ronald de Sousa

Concluding remarks 289
David P. Hughes and Patrizia d'Ettorre

Glossary 291
Index 295

List of contributors

Beynon, Robert J. Protein Function Group, Faculty of Veterinary Science, University of Liverpool, Crown Street, Liverpool L69 7ZJ, UK.
r.beynon@liv.ac.uk

Boomsma, Jacobus J. Centre for Social Evolution, Department of Biology, University of Copenhagen, Universitetsparken 15, 2100 Copenhagen, Denmark.
jjboomsma@bio.ku.dk

Brown, Sam. Department of Zoology, University of Oxford, South Parks Rd, Oxford 0X1 3PS, UK.
ssaamm_uk@yahoo.com

Brännström, Åke. Mathematics Department, University of Uppsala, Box 480, 751 06 Uppsala, Sweden.
ake@brannstrom.org

Crespi, Bernard J. Department of Biosciences, Simon Fraser University, Burnaby BC V5A1S6, Canada.
crespi@sfu.ca

Dabelsteen, Torben. Animal Behaviour Group, Department of Biology, University of Copenhagen, Universitetsparken 15, 2100 Copenhagen, Denmark.
tdabelsteen@bio.ku.dk

de Brito-Sanchez, Maria Gabriela. Centre de Recherches sur la Cognition Animale, CNRS UMR 5169, Université Paul Sabatier—Toulouse III, 118 Route de Narbonne, 31062 Toulouse cedex 9, France.
debrito@cict.fr

Deisig, Nina. Centre de Recherches sur la Cognition Animale, CNRS UMR 5169, Université Paul Sabatier—Toulouse III, 118 Route de Narbonne, 31062 Toulouse cedex 9, France.
deisig@cict.fr

d'Ettorre, Patrizia. Centre for Social Evolution, Department of Biology, University of Copenhagen, Universitetsparken 15, 2100 Copenhagen, Denmark.
pdettorre@bio.ku.dk

Diggle, Stephen P. Institute of Infection, Immunity & Inflammation, Centre for Biomolecular Sciences University Park, University of Nottingham, Nottingham NG7 2RD, UK.
steve.diggle@nottingham.ac.uk

Gardner, Andy. Institute of Evolutionary Biology, School of Biological Sciences, University of Edinburgh, King's Buildings, Edinburgh, EH9 3JT, UK.
andy.gardner@ed.ac.uk

Giurfa, Martin. Centre de Recherches sur la Cognition Animale, CNRS UMR 5169, Université Paul Sabatier—Toulouse III , 118 Route de Narbonne, 31062 Toulouse cedex 9, France.
giurfa@cict.fr

Griffin, Ashleigh S. Institute of Evolutionary Biology, School of Biological Sciences, University of Edinburgh, King's Buildings, Edinburgh, EH9 3JT, UK.
a.griffin@ed.ac.uk

Haig, David. Department of Organismic and Evolutionary Biology, Harvard University, 26 Oxford Street, 02138 Cambridge, MA.
dhaig@oeb.harvard.edu

Hughes, David P. Centre for Social Evolution, Department of Biology, University of Copenhagen, Universitetsparken 15, 2100 Copenhagen, Denmark.
dphughes@bio.dk.uk

Hurford, James R. Language Evolution and Computation Research Unit, School of Philosophy, Psychology and Language Sciences, University of Edinburgh, Adam Ferguson Building, Edinburgh EH8 9LL, Scotland, UK.
jim@ling.ed.ac.uk

Hurst, Jane. Mammalian Behaviour & Evolution Group, Faculty of Veterinary Science, University of Liverpool, Leahurst Veterinary Field Station, Neston, South Wirral CH64 7TE, UK.
jane.hurst@liverpool.ac.uk

Matessi, Giuliano. Animal Behaviour Group, Department of Biology, University of Copenhagen, Universitetsparken 15, 2100 Copenhagen, Denmark.
gmatessi@bio.ku.dk

Matos, Ricardo. Animal Behaviour Group, Department of Biology, University of Copenhagen, Universitetsparken 15, 2100 Copenhagen, Denmark.
rmatos@bio.ku.dk

Moore, Allen J. Centre for Ecology & Conservation, School of Biosciences, University of Exeter, Cornwall Campus, Penryn TR10 9EZ, UK.
a.j.moore@exeter.ac.uk

Nash, David R. Centre for Social Evolution, Department of Biology, University of Copenhagen, Universitetsparken 15, 2100 Copenhagen, Denmark.
drnash@bio.ku.dk

Roberts, S. Craig. School of Biological Sciences, University of Liverpool, Liverpool L69 7ZB, UK.
craig.roberts@liverpool.ac.uk

Riboli-Sasco, Livio. Centre de Recherches Interdisciplinaires, Université Paris Descartes, 25, rue du Faubourg Saint Jacques 75014 Paris, France.
livio.riboli-Sasco@ens.fr

Sandoz, Jean-Christophe. Centre de Recherches sur la Cognition Animale, CNRS UMR 5169, Université Paul Sabatier—Toulouse III , 118 Route de Narbonne, 31062 Toulouse cedex 9, France.
sandoz@cict.fr

de Sousa, Ronald. Department of Philosophy, 170 St George Street #424, University of Toronto, Toronto, ON M5R 2M8, Canada.
sousa@chass.utoronto.ca

Sumpter, David J.T. Mathematics Department, University of Uppsala, Box 480, 751 06 Uppsala, Sweden.
david@math.uu.se

Taddei, François. Laboratoire de Génétique Moléculaire Évolutive et Médicale, INSERM U 571, Faculté de Médecine Necker, 156 rue de Vaugirard, 75730 Paris Cedex 15, France.
taddei@necker.fr

Tinghitella, Robin M. Department of Biology, University of California, Riverside CA 92521, USA.
rting001@student.ucr.edu

West, Stuart A. Institute of Evolutionary Biology, School of Biological Sciences, University of Edinburgh, King's Buildings, Edinburgh, EH9 3JT, UK.
stu.west@ed.ac.uk

Zahavi, Amotz. Department of Zoology, Tel-Aviv University, Tel-Aviv 69978, Israel.
zahavi@post.tau.ac.il

Zuk, Marlene. Department of Biology, University of California, Riverside CA 92521, USA.
marlene.zuk@ucr.edu

CHAPTER 1

The handicap principle and signalling in collaborative systems

Amotz Zahavi

1.1 Introduction: what is a signal?

Signals are cooperative systems: at the bare minimum, signalling involves one signaller and one receiver, because unless there is a potential receiver there is no point in signalling. More often additional individuals are involved: several signallers compete for the attention of one or more receivers and there might be eavesdropping (see Chapter 3). Signals evolve and persist over time when both signallers and receivers gain from their interaction.

I define signals as characters that evolve in a signaller in order to provide information to a receiver, aiming to change the behaviour of the receiver to the benefit of the signaller. Receivers benefit from a signal when the information encoded in the signal informs them that it is to their benefit to change their behaviour. Responding to a message that is not reliable is obviously non-adaptive. Hence, it is the receivers of the signal that select the signallers to invest in the reliability of the signal by responding to reliable signals and ignoring non-reliable ones. A signal is reliable when the investment in it is worthwhile to an honest signaller and not worthwhile to a cheater. In order to cooperate, signallers invest in producing reliable signals, and receivers benefit from responding to reliable information.

1.1.1 The evolution of reliable signals

All signals evolve from characters that were not signals to begin with. The evolution of the peacock's tail may illustrate the sequence of events that transformed a character first developed to function as a rudder to function also as a signal of mate choice (Fisher 1958). Clearly, the tail was originally functioning as a rudder for steering. Heavier peacocks require a longer tail as a rudder. Females that benefited from mating with heavier males were able to pick them by preferring males with longer tails; such males were likely to be heavier overall than males with shorter tails. At that time, although females benefited from considering the length of the peacock's tail in their preferences, the tail was not yet a signal.

Once many females started preferring males with long tails, it became beneficial for a male to increase the length of its tail beyond the length optimal for steering, in spite of the extra burden involved in carrying a long and less efficient tail. That extra investment in the length of the tail is the investment in the tail as a signal. Any exaggeration, however slight, means that the trait from which the signal is derived is no longer at the optimum selected by **natural selection** to serve its initial function—the new selection pressure for a longer tail as a signal is 'handicapping' the signaller (adding an extra burden).

Individuals differ in the extent to which they can invest in reducing the efficiency of a character. It is this differential investment that provides reliability to the signal. The extra investment (the handicap) provides more detailed and accurate information about the particular quality of the signaller that was originally of interest to the receivers.

The selection for a handicap creates a logical connection between the message encoded in a signal and its pattern; in other words, signals are not random patterns that code for particular messages. They are optimal patterns that have been selected to convey reliable and more accurate information concerning a certain quality. For example, a rich person can signal the degree of his wealth by wasting money. His signal is reliable since a poorer man cannot waste as much money. A courageous man can display the degree of his courage by taking a risk which a less courageous individual would not dare to take. On the other hand , taking a risk of bodily harm does not display wealth, and spending money does not display how brave a person is. The connection between the pattern of a signal and its message content is a powerful tool for understanding the messages encoded in signals.

Exploring the special investment (the handicap) required by a signal provides a better understanding of its message than the common practice of deducing the message encoded in the signal from the reaction of the receiver to it. The same information, displayed by the same signal, may cause different receivers to respond to it differently, according to their specific interests (see for example multi-purpose chemical signals, Chapter 5). A display of strength may deter a rival but may attract a mate or a potential collaborator. If we judge the function of a signal by the reaction of the receiver to it, the message of a signal that results in the retreat of the receiver would be considered a threat, and when the same signal attracts a mate, it would be considered a signal of courtship. I suggest that the signal encodes neither threat nor invitation, but rather dimensions of a quality, i.e. strength, which produces different reactions in different receivers. Thus, a study of the handicaps involved in a signal may provide better insights to the message encoded by it.

I first suggested the handicap principle in 1973 (Zahavi 1975) to resolve the evolution of signals of mate choice like the peacock's tail, but it soon became apparent to me that the handicap principle is a basic component of all signalling (Zahavi 1977). The handicap principle is an essential component in all signals and shows why signals take the form they do. It indicates the message encoded in the signal, helps to clarify to whom the signal is directed, and often helps one understanding the adaptive component of characters that otherwise seem to be maladaptive, such as **altruism** (Zahavi 1977; Zahavi and Zahavi 1997). In a way, the word 'handicap' is misleading because it has the connotation of a loss. Signallers are not losing—they invest in order to gain: an individual that takes on a reasonable handicap in order to signal is like a businessman investing in an advertisement. In our book (Zahavi and Zahavi 1997) we provide many examples, in various signalling **modalities**, that show the logical relationship between the patterns of signals and the messages encoded in them.

It is important to note that since signals evolve from characters that were not signals to begin with, but that were already used as a source of information, it is not always easy to determine whether a particular trait is just used by observers as a source of information (a **cue**) or whether it has already evolved to function as a signal. Many of the traits that serve as signals have a mixed value: they retain their original function, but in a handicapped manner, in order to convey more reliable information. For example, the peacock's tail still serves as a rudder, even though it clearly signals the quality of the male.

1.2 Altruism in babblers

One of the major problems faced by evolutionary biologists over the last 50 years has been how biological cooperations are able to persist. Why don't members of cooperations exploit the cooperation and use false signals in the interactions among their members? Models of indirect selection were constructed to explain these paradoxes by suggesting that the individual is compensated for its efforts by the fact that its group (**group selection**) or kinship (in the case of **kin selection**) benefits, and the altruist gains indirectly, as a member of the group. Other models (**reciprocal altruism**) suggest that the altruist stands a chance to benefit from reciprocation by the receiver of the altruistic act (Trivers 1971) or indirectly from other individuals (Alexander 1987).

The study of the social life of Arabian babblers (Fig. 1.1), song birds living in cooperative territorial groups, reveals the power of the handicap principle in explaining the evolution and patterns of their signals. Our observations also suggested that their apparent altruistic activities are in fact signals that advertise the claim of the 'altruist' for social prestige (Zahavi 1977, 1990). Babblers are seemingly altruistic: they act as sentinels when the rest of the group is feeding; they endanger themselves when they are exposed as sentinels and by giving warning calls; they help at the nest to feed nestlings that are not their offspring, and risk their lives to save a group member from predators or when fighting other groups. They also donate food to other adult members in the group (allo-feeding; Fig. 1.2).

Figure 1.1 An Arabian babbler (*Turdoides squamiceps*) acting as sentinel for the group.

We found that babblers compete to act as altruists. Dominants invest in the welfare of the group more than lower-ranking group members do, and they often interfere with the altruistic acts of lower-ranking group members while mobbing predators (Anava 1992), or during border fights (Berger 2002). Interference of dominants with the sentinel activities of lower-ranking individuals is common, especially during courtship periods, when the competition over copulation with the breeding females is most extreme (Carlisle and Zahavi 1986; Zahavi and Zahavi 1997; Dattner 2005; Kalishov *et al.* 2005). Such interference cannot be easily explained by models of group selection, kin selection, or reciprocal altruism. According to the handicap principle, it is possible to suggest that, for the altruist, the investment in the group is an investment in the reliability of its claim to social prestige. This is a suggestion based on **individual selection**, and does not require any model of kin

Figure 1.2 Allo-feeding between two Arabian babblers.

selection, group selection, or reciprocity. A similar interpretation of altruism may be applied to many other species, from humans to social insects, whose apparent altruistic behaviours are currently explained by models of indirect selection (Zahavi 1995, Zahavi and Zahavi 1997). The most difficult form of altruism to explain by indirect models like group or kin selection or reciprocal altruism is the unconditional altruism in which the altruist helps non-relatives that do not belong to its social group and from whom the altruist cannot expect any benefit in the future. Seeing altruism as a handicap signalling the quality of the altruist bypasses all these problems.

Lotem *et al.* (2003) developed a model showing that, in a population composed of reciprocating individuals, unconditional altruistic activity may evolve to function as a signal, supporting my claim that the eventual unconditional altruism is a **selfish** trait by which the altruist displays its quality. However, there is no need to start the model with the evolution of altruism from a reciprocating population. Altruistic activity like standing sentry can start to evolve as a signal from a trait that was not a signal to begin with. Babblers, for example, scout an area before they traverse open ground where they are vulnerable to predators. In the presence of predators, and also in the semidarkness of the morning, sentinels stay inside thickets. They scout the area from the safety of the canopy rather than from its top. However, scouting from the top is more efficient. Older and more experienced babblers that can better assess the degree of risk they can take dare to perch at the top of trees more than young ones do. Once group members are attentive to these differences between the more confident babblers and the fearful ones, it becomes beneficial for a babbler to take a greater risk and spend longer periods in scouting an area as a display (a signal) of its quality. The group benefits from the investment of the sentinel. But the benefit to the group is not the selection pressure that causes sentinel activity to evolve: the sentinel is acting in its own selfish interest, displaying its claim for social prestige. In this case, reciprocation is not expected; in fact, it is often actively rejected (Zahavi 1990).

According to this interpretation, the donor benefits directly from an increase in its social prestige. High social prestige functions for the altruist like an invisible peacock's tail: it deters rivals (who are often members of the same cooperation) and attracts collaborators. The collaborators may be potential mates, or individuals that join the cooperation for other benefits such as joint hunting, joint defence and so on. The deterrence of rivals is often much more important than the attraction of collaborators. In general, a high social prestige provides the individual with a greater share of the common resources of the cooperation, which in biological terms eventually translates into **fitness**. Thus, complex phenomena such as altruistic behaviours may serve as signals. The signalling component of the altruistic behaviour is a handicap that displays quality.

In 1990, Alan Grafen constructed a formal, mathematical model of the handicap principle that convinced those who are wary of verbal models that my verbal model of the evolution of the peacock's tail may work (Grafen 1990a,b). Grafen commented that "The handicap principle is a strategic principle, properly elucidated by game theory, but actually simple enough that no formal elucidation is really required" (Grafen 1990b, p.541). I think that the similar verbal model of the evolution of altruism is also simple and that no formal elucidation is really required in this case either.

1.3 Altruism in slime moulds

The interpretation of altruism in babblers as a selfish investment in advertisement, an interpretation that does not consider indirect benefits, tempted me to study the apparent altruism in slime moulds, a phenomenon that is interpreted by researchers using models of indirect selection, mostly group selection models. Slime moulds cooperate to the extent that some individuals undergo active cell death (condensation and fragmentation of cytoplasm and chromatin) in response to a chemical produced by other members of the cooperation. Similar phenomena occur among many bacteria (Shapiro 1998).

The following discussion on the function of DIF ('differentiation inducing factor', the technical name of a morphogenic chemical produced by slime moulds) can demonstrate the difficulties in determining whether or not a chemical is a signal. It also explains why we interpret the active cell death of slime moulds as a selfish act (Atzmoni *et al.* 1997; Zahavi 2005). Slime moulds are amoebae that under conditions of food shortage or other stress congregate to form a 'slug' that is composed of thousands and even many thousands of individuals. In the wild the slug migrates, looking for new grazing grounds. If food is not found, a fruiting body is produced. The fruiting body comprises live spores carried on a stalk composed of dead amoebae. The stalk is formed by about 30% of the population, most of them originally from the front of the advancing slug, named 'pre-stalk cells'. The chemical mechanism that induces these amoebae to become pre-stalk cells is well known: DIF that is secreted by the cells in the centre and rear of the slug binds to receptors on the membranes of the pre-stalk cells and is believed to serve as a signal. It creates a signal transduction in the pre-stalk cells, culminating in their migration to form a stalk in which they commit active cell death. The stalk lifts the spores above the ground and thus improves the chances of survival of the spores, an action that benefits the spores and therefore has been described as altruism. When the population in the front of the slug that was destined to die is experimentally removed, other cells that would otherwise have survived take their place and die. It is also well established that slime moulds also undergo active cell death when cooperating with unrelated individuals (Kaushik and Nanjundiah 2003). The slime moulds, therefore, are one of the cases that supposedly support group selection theory (Werfel and Bar-Yam 2004).

In my discussion with microbiologists it appears to me that most, if not all, believe that group selection plays a role in evolution. Consequently, they have no problem in interpreting the development of slime moulds by group selection models, explaining traits harmful to individuals by their benefit to the group. However, since I firmly believe that evolution is a consequence of individual selection only, I decided to take on the challenge of exploring what could be the advantage to the individual pre-stalk amoeba in undergoing active cell death. Together with my student Daniella Atzmony and in cooperation with Vidianand Nanjundiah from Bangalore, India, we developed a model that interprets the life history of slime moulds on the basis of individual selection (Atzmony *et al.* 1997).

There are phenotypic differences between the amoebae that form the front of the slug and those at the rear: when well-nourished individuals are mixed with undernourished ones, the latter are more likely to be in the front of the slug and consequently become the stalk cells that perish. One of the phenotypic curiosities in pre-stalk cells is their secretion of an enzyme that removes DIF from its membrane receptor. The phenomenon is traditionally interpreted as improving the response of the cells to the DIF signal itself. Our simple assumption, based on individual selection, was that when one individual provides another with a chemical that kills the other, that chemical is a poison. At the same conference in which we proposed our model, Shaulsky provided evidence that DIF is a noxious chemical that reduces the efficiency of mitochondria in synthesizing ATP. The sporulating cells survive the effect of DIF by producing additional mitochondria, while the dying pre-stalk cells do not, possibly because they do not have enough resources to do it (Shaulsky and Loomis 1995). But the pre-stalk cells do not simply perish, they undergo active cell death. What could be the advantage of active cell death for a unicellular organism? Our speculation is that by active cell death, in the vicinity of surviving cells, the stalk cells create a chance for some of their genes to transfect the germinating spores. Although it is a small chance, it is better than nothing. Hence as soon as an undernourished cell gets to the point where it has no chance of surviving, or of developing a spore, its best remaining chance is to take the path of becoming a stalk cell and undergo active cell death, with the expectation that one or more of its genes would survive (Zahavi 2005; Koren 2006). Indeed, Arnoult *et al.* (2001) found that during active cell death the DNA of pre-stalk cells is cut into fragments of around 5000 base pairs, which I interpret as pieces that could include whole genes. It is interesting to note that in the process of 'apoptosis' (active cell death in multicellular organisms), the DNA pieces are only around 200 base pairs, too small to include a gene. Obviously, the evolution of active cell death in slime moulds and other unicellular organisms preceded the evolution of apoptosis in multicellular organisms. It seems that a mechanism that enabled some unicellular organisms to have a chance of passing some of their genes to the next generation was later utilized by multicellular organisms, with a slight modification, to protect them from the damage that the DNA of dying cells in the body might inflict on the rest of the organism.

If one views the slime-mould life cycle through the lens of individual selection, there are still two more questions: why should every sporulating cell invest in secreting DIF, rather than letting others secrete it and exploiting their efforts? And why should stalk cells produce the enzyme that cleaves DIF from their receptor? Obviously, my answers to these questions are speculative. It may be that DIF, which is harmful to mitochondria, protects the spores from predation. If so, an amoeba that does not secrete DIF is more vulnerable to predation. As to the pre-stalk cells, DIF is a chemical that can go through membranes without the help of membrane receptors. Stalk cells that cannot survive the effect of DIF use membrane receptors to keep it outside the cell. The enzyme that removes DIF from the receptor, and most probably degrades it, prevents the entrance of more DIF molecules into the pre-stalk cells.

According to our speculations, the behaviour of the slime moulds is not altruistic. DIF, which is considered a signal in group selection models, may not be a signal at all. It probably functions as a poison produced by the sporulating cells, each of which is secreting it for its own sake, in order to defend itself from predators. Stalk cells try to defend themselves against this poison by producing membrane receptors and enzymes that prevent DIF from entering their cytoplasm.

The trigger that causes pre-stalk cells to undergo active cell death may not be a signal sent by other cells (that is, a character produced by an individual in order to change the behaviour of others), but rather a poisonous chemical secreted by the sporulating cells to defend themselves from predation. Pre-stalk cells cannot sporulate in the presence of the poison (DIF). Thus, they make the best of a bad situation by trying to help some of their genes survive by undergoing active cell death.

1.4 The handicap principle in chemical signals

Chemical signals are not different from signals in any other modality, such as visual and acoustic (see Chapter 5 for a discussion of chemical signals as composite traits). Like other signals, they too require investment in reliability. The investment may be in the ability of the signaller to bear damage caused by the signalling chemical; or it may be the difficulty of producing a particular chemical. An example of signals that cause damage may be the use of carotenoids as signals of quality by birds (Hill 1990): although small amounts of carotenoids may be beneficial—since carotenoids quench radicals—larger amounts cause damage since they increase the lifetime of radicals (Haila 1999). Hence only high-quality individuals that can bear the damage can assume intense carotenoid coloration (Zahavi 2007).

An example of signals that are difficult to produce may be the mating **pheromones** of yeast cells, complex molecules such as glycoproteins that require special investment for their synthesis. The alpha mating peptide of yeasts is produced from a complex glycoprotein pro-peptide. Nahon *et al.* (1995) suggested that the handicap by which yeast cells choose a mate is in the complex glycoprotein pro-peptide rather than in the short alpha peptide. The synthesis of the pro-peptide requires oligosaccharides that may represent phenotypic quality. It may be that only individuals of a particular quality are able to synthesize it with the complete set of sugar units (Nahon *et al.* 1995). A short peptide, on the other hand, may not be a good medium for advertising phenotypic quality. It is very likely that in other cases in which short peptides are assumed to be signals it is in fact the complex pro-peptides that are responsible for the reliability of the information (messages) encoded in them.

1.5 Signals within the multicellular organism

All the somatic cells within a multicellular body (except for the germ line) share completely the same interests. It may seem, then, that there is no need to invest in evolving costly signals to ensure the reliability of signals within the multicellular body. However, even a superficial survey of signals within the body reveals that many of them are loaded with heavy investments, just like signals between organisms (Zahavi 1993; Zahavi and Zahavi 1997). Snyder and Bredt (1992), in a review of the biological function of nitric oxide (NO) as a signal, remark that it is surprising that evolution uses such a noxious chemical as a signal. Many common signals are noxious small molecules, such as steroids and dihydroxyphenylalanine (DOPA; a precursor of dopamine) or complex glycoproteins, such as follicle-stimulating **hormone** (FSH) and luteinizing hormone (LH). Often the same chemicals used as signals within the body are also used as signals among organisms, where reliability is obviously necessary, e.g. c-AMP and glycoproteins. I suggest therefore, that signals within the body require special investment in reliability, like signals among organisms. The reason for that requirement of reliability may be to avoid signalling by cell phenotypes that should not signal, or to inhibit the signalling cells from producing too much of the signal. Using handicaps fulfils these requirements. The investment (the handicap) ensures that the quantity of the signal is correlated to a certain quality or a certain physiological state of the signalling cell (whatever that quality or state may be). Like signals among organisms, the pattern of the signal—the chemical properties of the signal—is therefore related to the message encoded in the signal.

It is reasonable to assume that a chemical signal within the body, like signals among organisms, is not a molecule selected to instruct the receiver to take certain actions. Rather, it appears to function as an indication of the state of the signalling cells. Like signals among organisms, a signalling cell provides information by a chemical molecule that is an analogue of a particular quality or state of the signalling cell. The information influences a decision in the receiving cell. Just as in signalling threat or courtship between individuals, different cells may respond in different ways to the same information. The response to the same signal depends on the phenotypic quality of the receiving cell: some cells enhance their development, others arrest it; some do not respond at all, while still others undergo apoptotic cell death.

As in chemical signals among organisms, the handicap in chemical signals within the body may involve the cost to the signalling cell of assembling chemical structures that low-quality cells may not be able to produce, such as a complex glycoprotein; or it may show the ability of the signalling cell to withstand the noxious nature of a chemical it produces such as steroids, NO and carbon monoxide (CO).

Physiologists and endocrinologists typically study the effects of a particular signal on other cells; they usually do not ask what is the objective information transferred by the signal. We are presenting here the theory that signals have their effect because they carry reliable information on particular qualities of the signalling cells. The type of investment required to produce the signal within the signalling cell may therefore point to the message encoded in the signal—whether the signal reflects the energy potential in the signalling cell, its reduction/oxidation potential, or the availability of certain chemicals to it.

Within the body, even more than in chemical signals acting among organisms, it is important to distinguish between true signals that evolved in order to transfer information and chemicals that produce an effect in other cells but have not evolved in order to carry such information. There are clearly enzymes and membrane proteins that serve the cells for other reasons then for passing information, but which other cells react to.

In conclusion, signals are characters that evolve in a signaller in order to provide information to a receiver. The signaller benefits if by signalling it may change the behaviour of the receiver in a way that benefits the signaller. It is to the benefit of receivers of signals to react only to reliable signals. The signaller invests in the reliability of its signals by handicapping itself in something that is directly related to the information provided by the signal. Understanding the handicap in a signal points to that information, and provides a better understanding of the interactions among cooperating individuals, based on models of strict individual selection.

Summary

Signalling systems are by nature collaborations, since for a signal to be effective, the receiver has to cooperate with the signaller. The handicap principle ensures the reliability of signals, and is an essential component in all signals. The handicap principle explains why signals evolve their particular patterns, and the relationship of the patterns to the messages encoded in them. We use the handicap principle to understand signalling among Arabian Babblers—the patterns by which they advertise their qualities to mates, rivals, and predators. The handicap principle also explains the altruism of babblers as a selfish investment in advertising prestige. Recent theoretical studies have used the handicap principle to interpret the evolution of chemical signalling among organisms (pheromones) and within multicellular organisms (hormones), and the messages encoded in such chemical signals.

Acknowledgements

Avishag Zahavi has been a partner in the development of this chapter. Naama Zahavi-Ely edited it and markedly improved the English presentation. Thanks also to Patrizia and David who invited me to write this chapter.

References

Alexander, R.D. (1987). *The Biology of Moral Systems.* Aldine de Gruyter, New York.

Anava, A. (1992). The value of mobbing behaviour for the individual babbler (*Turdoides squamiceps*). MSc Thesis, Ben-Gurion University of the Negev, Israel.

Arnoult, D., Tatischeff, I., Estaquier, J. *et al.* **(2001).** On the evolutionary conservation of the cell death pathway: mitochondrial release of an apoptosis-inducing factor during *Dictyostelium discoideum* cell death. Molecular *Biology of the Cell*, **12**, 3016–3030.

Atzmoni, D., Zahavi, A., and Nanjundiah, V. (1997). Altruistic behaviour in *Dictyostelium discoideum* explained on the basis of individual selection. *Current Science*, **72**, 142–145.

Berger, H. (2002). Interference and competition while attacking intruder in groups of Arabian Babblers (*Turdoides squamiceps*). MSc Thesis, Tel-Aviv University, Israel.

Carlisle, T.R. and Zahavi, A. (1986). Helping at the nest, allofeeding and social status in immature Arabian babblers. *Behavioral Ecology and Sociobiology*, **18**, 339–351.

Dattner, A. (2005). Allopreening in the Arabian babbler (*Turdoides squamiceps*). MSc Thesis, Tel-Aviv University, Israel.

Fisher, R.A. (1958). *The Genetical Theory of Natural Selection.* Dover Publications, New York.

Hill, G. (1990). Female house finches prefer colourful males: sexual selection for a condition-dependent trait. *Animal Behaviour*, **40**, 563–572.

Haila, K.K. (1999). Effects of carotenoids and carotenoid-tocopherol interaction on lipid oxidation *in vitro*. PhD Thesis, University of Helsinki, Finland.

Grafen, A. (1990a). Sexual selection unhandicapped by the Fisher process. *Journal of Theoretical Biology*, **144**, 473–516.

Grafen, A. (1990b). Biological signals as handicaps. *Journal of Theoretical Biology*, **144**, 517–546.

Kalishov, A., Zavahi, A., and Zahavi, A. (2005). Allofeeding in Arabian Babblers (*Turdoides squamiceps*). *Journal of Ornithology*, **146**, 141–150.

Kaushik, S. and Nanjundiah, V. (2003). Evolutionary questions raised by cellular slime mould development. *Proceedings of the Indian National Science Academy*, **B69**, 825–852.

Koren, A. (2006). Is *Saccharomyces cerevisiae* apoptotic cell death associated with gene transfer? *IUBMB Life*, **58**, 203–207.

Lotem, A., Fishman, M., and Stone, L. (2003). From reciprocity to unconditional altruism through signaling benefits. *Proceedings of the Royal Society B: Biological Sciences*, **270**, 199–205.

Mcgraw, K.J., Hill, G.E., and Parker, R.S. (2005). The physiological costs of being colourful: nutritional control of carotenoid utilization in American goldfinch, *Carduelis tristis*. *Animal Behaviour*, **69**, 653–660.

Nahon, E., Atzmony, D., Zahavi, A., and Granot, D. (1995). Mate selection in yeast: a reconsideration of the signals and the message encoded by them. *Journal of Theoretical Biology*, **172**, 315–322.

Shapiro, J.A. (1998). Thinking about bacterial populations as multicellular organisms. *Annual Review of Microbiology*, **52**, 81–104.

Shaulsky, G. and Loomis, W.F. (1995). Mitochondrial DNA replication but no nuclear DNA replication during development of *Dictyostelium. Proceedings of the National Academy of Sciences of the USA*, **92**, 5660–5663.

Snyder, H.S. and Bredt, D.S. (1992). Biological roles of nitric oxide. *Scientific American*, **266**, 28–35.

Trivers, R. (1971). The evolution of reciprocal altruism. *Quarterly Review of Biology*, **46**, 35–57.

Werfel, J, and Bar-Yam, Y. (2004). The evolution of reproductive restraint through social communication.

Proceedings of the National Academy of Sciences of the USA, **101**, 11019–11024.
Wynne-Edwards, V.C. (1962). *Animal Dispersion in Relation to Social Behaviour*. Oliver and Boyd, Edinburgh.
Zahavi, A. (1975). Mate selection a selection for a handicap. *Journal of Theoretical Biology*, **53**, 205–214.
Zahavi, A. (1977). Reliability in communication systems and the evolution of altruism. In: B. Stonehouse and C. M. Perrins (eds), *Evolutionary Ecology*, pp. 253–259. Macmillan, London.
Zahavi, A. (1990). Arabian babblers: the quest for social status in a cooperative breeder. In: P.B. Stacey and W.D. Koenig (eds), *Cooperative Breeding in Birds. Long-term Studies of Ecology and Behaviour*, pp. 103–130. Cambridge University Press, Cambridge.
Zahavi, A. (1993). The fallacy of conventional signalling. *Philosophical Transactions. of the Royal Society B: Biological Sciences*, **338**, 227–230.
Zahavi A. (1995). Altruism as a handicap—the limitation of kin selection and reciprocity. *Avian Biology*, **26**, 1–3.
Zahavi, A. (2003). Indirect selection and individual selection in sociobiology: my personal views on theories of social behaviour. *Animal Behaviour*, **65**, 859–863.
Zahavi, A. (2005). Is group selection necessary to explain social adaptations in microorganisms? *Heredity*, **94**, 143–144.
Zahavi, A. (2007). Sexual selection, signal selection and the handicap principle. In: B.G.M. Jamieson (ed.), *Reproductive Biology and Phylogeny of Birds. Part B*, pp 143–159. Science Publishers, Plymouth.
Zahavi, A. and Zahavi, A. (1997). *The Handicap Principle: a Missing Piece of Darwin's Puzzle*. Oxford University Press, New York.

CHAPTER 2

Communication in bacteria

Stephen P. Diggle, Stuart A. West, Andy Gardner, and Ashleigh S. Griffin

2.1 Introduction: communication in a unicellular world

In 1905, the pioneering plant pathologist E. F. Smith suggested that 'a multiple of bacteria are stronger than a few and thus by union are able to overcome obstacles too great for the few' (Smith 1905). This was for the time a remarkable statement, because until recently it was considered by most microbiologists that bacterial cells were unicellular organisms that existed in isolation from each other. It is now well established that bacteria are highly interactive and possess an extraordinary repertoire for intercellular communication and social behaviours such as group migration, conjugal plasmid transfer (sexual transfer of genetic material between cells), antibiotic resistance, biofilm maturation (development of 'slime cities'), and virulence which, although not a social trait, can be a consequence of social behaviour (Williams *et al.* 2007).

Indeed, some workers have suggested that these behaviours are similar to those observed in social insects, vertebrates, and humans. For example, *Myxococcus xanthus* cells exhibit socially dependent swarming across surfaces (Velicer and Yu 2003) which allows the population to seek out bacterial prey in a manner reminiscent of hunting packs of wolves (Dworkin 1996; Crespi 2001). In a similar fashion, biofilms (a collection of bacterial cells enclosed in a polysaccharide matrix) have been likened to ant nests and beehives (Crespi 2001; Diggle *et al.* 2007b). Furthermore, bacteria such as *Pseudomonas aeruginosa* can modulate the immune response, reminiscent of helminth parasites, and antibiotic resistance due to the production of extracellular enzymes (e.g. β-lactamase) could be considered to be a group defence mechanism (Diggle *et al.* 2007b).

Perhaps the paradigm for bacterial cooperation and communication can be seen in the diverse **quorum sensing** (QS) systems found in both Gram-negative and Gram-positive bacteria (Diggle *et al.* 2007a; Williams *et al.* 2007). QS describes the phenomenon whereby the accumulation of 'signalling' molecules in the surrounding environment enables a single cell to sense the number of bacteria (cell density), and therefore the population as a whole can make a coordinated response. The signal produced regulates its own production (autoinduction) and so this leads to a positive-feedback response and greatly increased signal production. At critical cell densities, the binding of a regulator protein to the signal leads to the switch on of genes controlled by QS and a coordinated population response.

It is important to note that many studies on QS in bacteria have been performed under laboratory conditions, and it needs to be determined whether QS is an artefact of laboratory growth (Redfield 2002; Hense *et al.* 2007). It is possible that this may be the case for certain organisms, but it has been shown, for example, that *P. aeruginosa* makes QS signal molecules in the lungs of cystic fibrosis patients (Collier *et al.* 2002; Middleton *et al.* 2002). Despite this, it is still not known whether QS is important in the development and establishment of chronic infections in this population. Therefore, the idea that QS is for the 'common good' of the bacterial population has yet to be significantly

Table 2.1 Bacterial cooperative behaviours known to be regulated by QS systems

QS-controlled behaviour	Bacterial species
Biofilms	*Aeromonas hydrophila, Burkholderia cenocepacia, Pseudomonas aeruginosa, Pseudomonas putida, Serratia liquefaciens*
Exoproteases	*Aeromonas hydrophila, Aeromonas salmonicida, Burkholderia pseudomallei, Pseudomonas aureofaciens, Serratia liquefaciens*
Plasmid conjugation	*Agrobacterium tumefaciens, Rhizobium leguminosarum*
Exoenzymes	*Burkholderia cenocepacia, Erwinia carotovora, Chromobacterium violaceum, Pseudomonas aeruginosa, Serratia* spp. ATCC 39006, *Serratia proteamaculans*
Swarming motility	*Burkholderia cenocepacia, Pseudomonas aeruginosa, Serratia liquefaciens, Yersina enterocolitica, Yersinia pseudotuberculosis*
Siderophore production	*Burkholderia cenocepacia*
Virulence	*Agrobacterium vitiae, Burkholderia cenocepacia, Burkholderia pseudomallei, Burkholderia mallei, Erwinia carotovora, Pseudomonas syringae, Pseudomonas aeruginosa*
Pigment production	*Chromobacterium violaceum, Pseudomonas aureofaciens, Pseudomonas chlororaphis, Serratia* spp. ATCC 39006, *Serratia marcescens*
Antibiotics	*Erwinia carotovora, Serratia* spp. ATCC 39006
Exopolysaccharides	*Pantoea stewartii, Pseudomonas syringae*
Aggregation	*Rhodobacter sphaeroides, Yersinia pseudotuberculosis*
Swimming motility	*Yersinia enterocolitica, Pseudomonas syringae*
Root nodulation/symbiosis	*Rhizobium leguminosarum, Sinorhizobium meliloti*
Biosurfactant production	*Pseudomonas aeruginosa, Serratia liquefaciens, Serratia marcescens*
Sliding motility	*Serratia marcescens*
Bioluminescence	*Vibrio fischeri*

proven. That aside, many of the behaviours regulated by QS appear to be cooperative and could be described as public goods, for example exoenzymes, biosurfactants, antibiotics, and exopolysaccharides (Table 2.1).

The importance of QS to a bacterium can be seen when studying the opportunistic pathogen *P. aeruginosa*. In this organism, a hierarchical QS system has been estimated to regulate at least 6% of the genome (Hentzer *et al.* 2003; Schuster *et al.* 2003; Wagner *et al.* 2003) which is a possible reason why *P. aeruginosa* is so highly adaptable and able to inhabit a wide range of diverse environmental niches.

It is often assumed in the microbiology literature that QS behaviour is cooperative and is for the good of the population as whole (Shapiro 1998; Henke and Bassler 2004) and little attention has been given to the evolutionary implications of QS. Understanding cooperative behaviour is one of the greatest challenges faced by evolutionary biologists, and the dictum of the survival of the fittest makes it unclear why one organism should behave for the good of another (Hamilton 1964). This chapter will review QS in bacteria and integrate this with the literature on animal signalling. We will discuss the nature of QS signals and signalling between single species and mixed species (bacterial cross-talk) and whether QS is truly cooperative. We will also explore whether QS in bacteria can be used to answer fundamental questions, such as how social behaviours can be maintained in natural populations.

2.2 When is a signal not a signal?

As will be described later, many diverse compounds have been identified as bacterial cell-to-cell QS signal molecules. Furthermore, interactions between different species of bacteria, and even between prokaryotes and eukaryotes, have also been widely described. There are several characteristics that a typical QS signal should display: (1) the production of the QS signal takes place during specific stages of growth or in response to particular environmental changes; (2) the QS signal accumulates in the extracellular environment and is recognized by a specific bacterial receptor; (3) the accumulation of a critical threshold concentration of the QS signal generates a concerted response; and (4) the cellular response extends beyond the physiological changes required to metabolize or detoxify the molecule (Winzer *et al.* 2002). Even taking these factors into consideration, it is also important to define what a **signal** is using terminology that is accepted amongst evolutionary biologists when discussing signalling between higher organisms (Keller and Surette 2006; Diggle *et al.* 2007b) (see also Chapter 1).

In a seemingly simple scenario, when we see cell A produce a substance X that elicits a response in cell B it is tempting to conclude that the substance produced is a signal, i.e. cell A is trying to tell cell B something. The word 'signal' is widely used to define such substances in the context of QS, or communication between bacterial cells. However, broad use of this term can be misleading and obscure the details of the interaction between cells that it attempts to describe. This has been well illustrated by research on communication and signalling in animals, where considerable confusion has arisen through different researchers using the same term to mean different things, or different terms to mean the same things (Maynard Smith and Harper 2003).

Confusion over terminology can be avoided if the different kinds of interactions that we observe when cell A elicits a response in cell B are differentiated, depending upon their consequences for cell A and cell B (Table 2.2) (Maynard Smith and Harper 2003). Specifically, a signal is defined as 'characters that evolve in a signaller in order to provide information to a receiver, aiming to change the behaviour of the receiver to the benefit of the signaller' (see Chapter 1). This definition distinguishes a signal from a **cue**, where the production of substance X by cell A has not evolved *because* of its effect on cell B. For example, substance X may be a waste product produced by cell A that is detected by cell B. To demonstrate that substance X is a signal and not a cue it is necessary to show that it evolved because of the response it elicits. If the production of substance X by cell A forces a costly response from cell B we differentiate this from signalling and term it coercion or chemical manipulation.

Table 2.2 Different types of communication identified by their fitness consequences on the sender and receiver

	Evolved because of effect on sender (Cell A)	Benefits receiver to respond (Cell B)
Signal	+	+
Cue	−	+
Coercion	+	−

Beneficial (+), Costly (−).

Do these semantic points really matter? The answer is yes, for two reasons. Firstly, it is important for general understanding if there is a consensus on the use of terms. This is a lesson hard-learned by biologists working on signalling in higher organisms (Maynard Smith and Harper 2003), as well as more generally in the field of social evolution (West *et al.* 2007b). Secondly, and more importantly, we can make very different predictions about the behaviour of bacterial cells depending on whether they are communicating by a signal, a cue, or coercion (Table 2.2). For example, if a molecule is a signal, then we can say several things:

1. It is beneficial to cell B to respond.
2. The response of cell B benefits cell A.
3. It might be possible for a signaller to cheat in the amount of signal that it produces either to: (a) free-ride on the back of other signallers (avoiding the cost of producing substance X, i.e. signal negative), (b) manipulate responders (signal can become coercive), or (c) not respond to the presence of signal and therefore not produce signal-controlled public goods (signal blind).
4. There must be some mechanism that provides a shared interest to cells A and B, otherwise cheats would invade and make the signalling unstable—later we discuss how **kin selection** provides a solution to this problem.
5. A signalling system is likely to be more complex than a system involving a cue, to remain stable in the face of evolution for individuals to make less substance X or for individuals to respond less.

2.3 The discovery of cell-to-cell communication in bacteria

Whilst the term 'quorum sensing' has only been in use since 1994 (Fuqua *et al.* 1994), cell-to-cell communication in bacteria has an experimental history that dates back to the early 1960s. Early work on fruiting body formation in *M. xanthus* (Mcvittie *et al.* 1962) and on streptomycin production in *Streptomyces griseus* (Khokolov *et al.* 1967) challenged the common view that bacteria behaved as isolated single cells.

One of the earliest reports of a classical cell density-dependent phenotype was by Nealson *et al.* (1970) who showed that the addition of spent culture supernatants of the marine luminescent bacterium *Vibrio fischeri* (formally *Photobacterium fischeri*) to low-density cultures of the same organism induced the production of bioluminescence due to the presence of a substance they termed an autoinducer (Nealson *et al.* 1970). When in a confined area such as a flask, or in symbiosis in a light organ found in certain species of squid, the autoinducer molecules accumulate to a critical concentration (usually at high bacterial cell densities) which, in turn, induces expression of the genes responsible for bioluminescence.

The autoinducer responsible for the regulation of bioluminescence was later identified as *N*-(3-oxohexanoyl) homoserine lactone (3-oxo-C6-HSL) (Eberhard *et al.* 1981). The structural and regulatory genes necessary for bioluminescence and 3-oxo-C6-HSL production were identified and

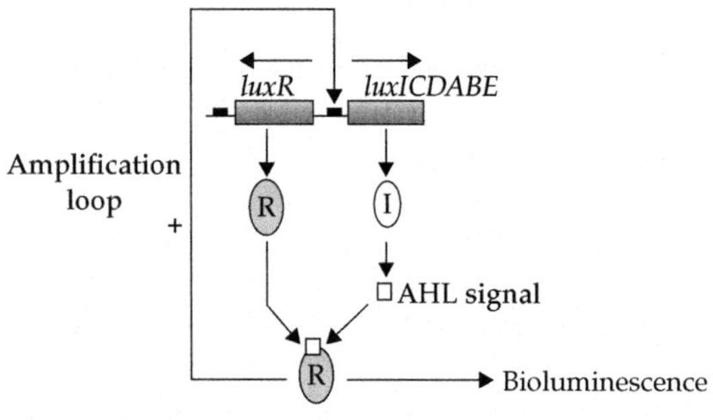

Figure 2.1 The LuxR /AHL-driven quorum sensing module of *V. fischeri*. LuxR is the AHL receptor and LuxI is the AHL signal synthase. Many bacteria possess multiple LuxR /LuxI /AHL modules which work in a similar manner.

termed the *lux* regulon (Engebrecht *et al.* 1983). This regulon is organized into two divergently transcribed operons (operons are units of coordinated gene activity which regulate protein synthesis in prokaryotes). The leftward operon comprises the *luxR* gene which encodes the transcriptional regulator protein LuxR. The rightward operon consists of six genes arranged as *luxICDABE*. The *luxI* gene encodes an autoinducer synthase responsible for the synthesis of 3-oxo-C6-HSL. The *luxCDABE* genes are involved in generating the products required for the luciferase reaction and the induction of bioluminescence. The genetic regulation of bioluminescence in *V. fischeri* is illustrated in Fig. 2.1. This elegant mechanism of gene regulation was thought to be a phenomenon restricted to bioluminescence production in a few marine *Vibrio* species; however, it is now known that this type of system is widespread in Gram-negative bacteria.

In the early 1990s it was discovered that the production of the β-lactam antibiotic, 1-carbapen-2-em-3-carboxylic acid (carbapenem) by the terrestrial plant pathogen *Erwinia carotovora* was also regulated by 3-oxo-C6-HSL (Bainton *et al.* 1992a,b). This finding led to the intriguing possibility that many bacteria may use *N*-acylhomoserine lactones (AHLs) in order to regulate specific phenotypes. This was confirmed when Bainton *et al.* (1992a) used plasmid-based AHL-biosensors to detect AHL molecules from spent culture supernatants from *P. aeruginosa, Serratia marcescens, Erwinia herbicola, Citrobacter freundii, Enterobacter agglomerans,* and *Proteus mirabilis* (Bainton *et al.* 1992a). Since this work, many other Gram-negative bacteria have been shown to produce different types of AHL molecules and all have homologues of LuxI and LuxR proteins of *V. fischeri* (Table 2.3). AHL-mediated QS is responsible for the regulation of a wide variety of different phenotypes in these organisms.

Although the distribution of Gram-negative bacteria that produce AHLs is widespread, there are some Gram-negative species that have failed to exhibit any activity in any of the AHL biosensor assays available, for example *Escherichia coli* and *Salmonella* species. However, this does not mean that they are incapable of producing and sensing a signal, and Gram-negative bacteria often utilize alternative QS signal molecules. The

Table 2.3 LuxR/AHL-dependent QS systems in Gram-negative bacteria

Organism	Major AHL(s)	LuxR	LuxI	Phenotypes
Aeromonas hydrophila	C4-HSL, C6-HSL	AhyR	AhyI	Biofilms, exoproteases
Aeromonas salmonicida	C4-HSL, C6-HSL	AsaR	AsaI	Exoprotease
Agrobacterium tumefaciens	3-oxo-C8-HSL	TraR	TraI	Plasmid conjugation
Agrobacterium vitiae	C14:1-HSL, 3-oxo-C16:1-HSL	AvsR	AvsI	Virulence
Burkholderia cenocepacia	C6-HSL, C8-HSL	CepR, CciR	CepI, CciI	Exoenzymes, biofilm formation, swarming motility, siderophore, virulence
Burkholderia pseudomallei	C8-HSL, C10-HSL, 3-hydroxy-C8-HSL, 3-hydroxy-C10-HSL, 3-hydroxy-C14-HSL	PmlIR1, BpmR2, BpmR3	PmlI1, PmlI2, PmlI3	Virulence, exoprotease
Burkholderia mallei	C8-HSL, C10-HSL	BmaR1, BmaR3, BmaR4, BmaR5	BmaI1, BmaI3	Virulence
Chromobacterium violaceum	C6-HSL	CviR	CviI	Exoenzymes, cyanide, pigment
Erwinia carotovora subsp. carotovora	3-oxo-C6-HSL	ExpR, CarR	CarI (ExpI)	Carbapenem, exoenzymes, virulence
Pantoea (Erwinia) stewartii	3-oxo-C6-HSL	EsaR	EsaI	Exopolysaccharide
Pseudomonas aeruginosa	C4-HSL; C6-HSL, 3-oxo-C12-HSL	LasR, RhlR, QscR, VqsR	LasI, RhlI	Exoenzymes, exotoxins, protein secretion, biofilms, swarming motility, secondary metabolites, 4-quinolone signalling, virulence
Pseudomonas aureofaciens	C6-HSL	PhzR, CsaR	PhzI, CsaI	Phenazines, protease, colony morphology, aggregation, root colonization
Pseudomonas chlororaphis	C6-HSL	PhzR	PhzI	Phenazine-1-carboxamide
Pseudomonas putida	3-oxo-C10-HSL, 3-oxo-C12-HSL	PpuR	PpuI	Biofilm formation
Pseudomonas syringae	3-oxo-C6-HSL	AhlR	AhlI	Exopolysaccharide, swimming motility, virulence
Rhizobium leguminosarum bv. *viciae*	C14:1-HSL, C6-HSL, C7-HSL, C8-HSL, 3-oxo-C8-HSL, 3-hydroxy-C8-HSL	CinR, RhiR, RaiR, TraR, BisR, TriR	CinI, RhiI, RaiI	Root nodulation/symbiosis, plasmid transfer, growth inhibition; stationary phase adaptation
Rhodobacter sphaeroides	7-cis-C14-HSL	CerR	CerI	Aggregation
Serratia spp. ATCC 39006	C4-HSL, C6-HSL	SmaR	SmaI	Antibiotic, pigment, exoenzymes
Serratia liquefaciens MG1	C4-HSL, C6-HSL	SwrR	SwrI	Swarming motility, exoprotease, biofilm development, biosurfactant
Serratia marcescens SS-1	C6-HSL, 3-oxo-C6-HSL, C7-HSL, C8-HSL	SpnR	SpnI	Sliding motility, biosurfactant, pigment, nuclease, transposition frequency
Serratia proteamaculans B5a	3-oxo-C6-HSL	SprR	SprI	Exoenzymes
Sinorhizobium meliloti	C8-HSL, C12-HSL, 3-oxo-C14-HSL, 3-oxo-C16:1-HSL, C16:1-HSL, C18-HSL	SinR, ExpR, TraR	SinI	Nodulation efficiency, symbiosis, exopolysaccharide
Vibrio fischeri	3-oxo-C6-HSL	LuxR	LuxI	Bioluminescence
Yersinia enterocolitica	C6-HSL, 3-oxo-C6-HSL, 3-oxo-C10-HSL, 3-oxo-C12-HSL, 3-oxo-C14-HSL	YenR, YenR2	YenI	Swimming and swarming motility
Yersinia pseudotuberculosis	C6-HSL, 3-oxo-C6-HSL, C8-HSL	YpsR, YtbR	YpsI, YtbI	Motility, Aggregation

cabbage pathogen *Xanthomonas campestris* employs a low-molecular-weight diffusible factor unrelated to AHLs to regulate expression of virulence determinants such as extracellular enzymes and exopolysaccharide (Barber *et al.* 1997). Furthermore, another plant pathogen, *Ralstonia solanacearum*, uses a 3-hydroxypalmitic acid methyl ester as a volatile signal molecule (Clough *et al.* 1997). *Myxococcus xanthus* also produces non-AHL signals. This Gram-negative bacterium is capable of forming complex multicellular structures that play a role in starvation survival. In order to coordinate this, *M. xanthus* produces two different signals, the A-signal and the C-signal. The A-signal, produced under nutrient limitation and at high cell densities, is the first signal that triggers multicellular behaviour. Analysis has revealed that the A-signal is a mixture of amino acids and small peptides (Kuspa *et al.* 1992). Following the formation of a layer of cells triggered by the A-signal, the production of the C-signal gives rise to the next stages in the development process, cell aggregation and sporulation.

The molecules identified and the processes controlled in *M. xanthus* are very different from those associated with AHLs and there have now been multiple signalling systems described, using different chemical signals, in the same organism. For example, *P. aeruginosa* has been shown to produce two AHL-distinct classes of molecules (2-alkyl-4-quinolones and cyclic dipeptides) with signalling activity in addition to AHLs (Holden *et al.* 1999; Pesci *et al.* 1999; Diggle *et al.* 2006a). This suggests that the signal may be tailored to particular physiological or environmental conditions depending upon its physical properties. Some examples of bacterial QS signals can be seen in Fig. 2.2.

Signalling is not restricted to Gram-negative bacteria: a number of Gram-positive bacteria have been shown to employ small, modified oligopeptides

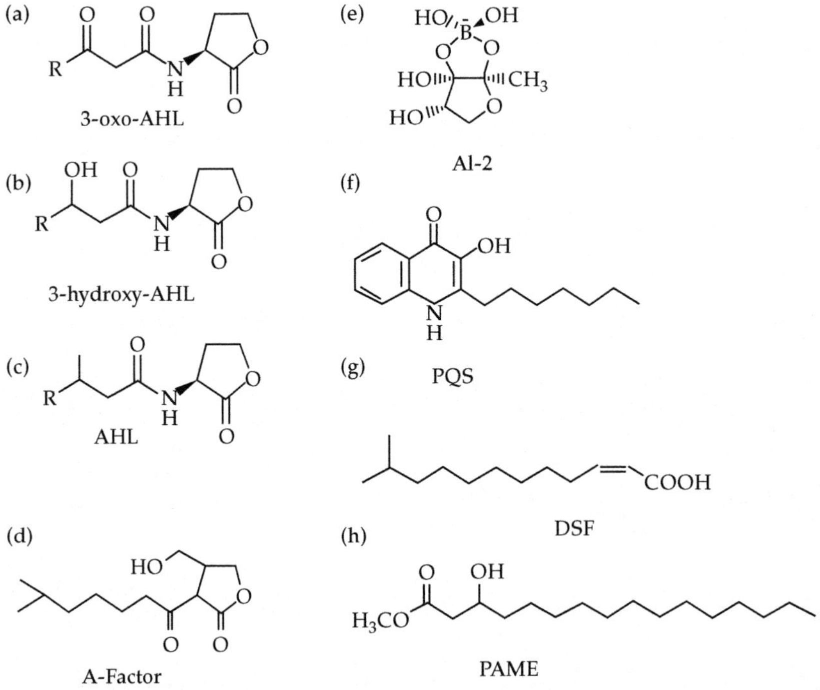

Figure 2.2 Structures of quorum sensing signal molecules found in bacteria. (a) 3-oxo-AHL; (b) 3-hydroxy-AHL; (c) *N*-acyl homoserine lactone (R ranges from C1-C15); (d) A-factor (2-isocapryloyl-3-hydroxy-methyl-γ-butyrolactone; (e) AI-2 (autoinducer-2); (f) The Pseudomonas quinolone signal (PQS, 2-heptyl-3-hydroxy-4(1*H*)-quinolone); (g) DSF (diffusible factor, methyl dodecenoic acid); (h) PAME (hydroxyl-palmitic acid methyl ester).

as extracellular signalling molecules. These peptides activate gene expression by interacting with two-component histidine protein kinase signal transduction systems (Kleerebezem *et al.* 1997). For example, in *Staphylococcus aureus* the expression of a number of cell density-dependent virulence factors is regulated by the global regulatory locus *agr* (accessory gene regulator) (Williams *et al.* 2007).

2.4 Evolutionary problems of signalling and cooperation

2.4.1 The problems of communication and cooperation

Two problems that have received much attention in the field of evolutionary biology are cooperation and communication (Hamilton 1964; Maynard Smith and Harper 2003), and these two issues come together in QS (Brown and Johnstone 2001; Redfield 2002; Keller and Surette 2006; Diggle *et al.* 2007b). In this section we consider the conditions under which QS to coordinate cooperation can be evolutionarily stable. We base our review of the relevant theory on Diggle *et al.* (2007b).

The problem of cooperation is why should an individual carry out a cooperative behaviour that is costly to perform, but benefits other individuals or the local group (Hamilton 1964). Such cooperation is vulnerable to invasion by cheaters who do not cooperate, but gain the benefit from the cooperation of others. This problem is well known in the fields of economics and human morality, where it is termed the tragedy of the commons (Hardin 1968). The tragedy is that, as a group, individuals would do better with cooperation, but this is not stable because each individual gains by selfishly pursuing its own short-term interests.

We have recently reviewed this problem in a microbial context elsewhere (West *et al.* 2006, 2007a). An obvious case in which it arises is when cells produce extracellular products for nutrient acquisition (Dinges *et al.* 2000; Greig and Travisano 2004; Griffin *et al.* 2004), antibiotics (Riley and Wertz 2002), immune modulation molecules (Brown 1999; Tateda *et al.* 2003; Hooi *et al.* 2004), antibiotic degradation compounds (e.g. β-lactamases) (Ciofu *et al.* 2000), and bio-surfactants (e.g. rhamnolipids) for motility (Velicer and Yu 2003; Daniels *et al.* 2004). These products are costly to an individual to produce, but provide a benefit to the individuals in the local group or population. Economic and evolutionary theory refers to such things as public goods (Dionisio and Gordo 2006). Many bacterial products termed 'virulence factors' are likely to be public goods—their coordinated production leading to damage to the host. The problem in these cases is that cheaters who do not pay the cost of producing such goods can still gain the benefit from neighbouring cooperators who do (for an experimental demonstration see Griffin *et al.* (2004) and Diggle *et al.* (2007c). This makes the cooperative production of public goods unstable, unless a mechanism such as kin selection operates (see below) (West and Buckling 2003).

The problem of communication is how can communication be reliable (Maynard Smith and Harper 2003)? Why do individuals convey honest information about themselves, to the benefit of other individuals? Why would they not give a false signal to their selfish advantage? If communication isn't reliable, then why should the receiver listen to it? The problem is reviewed for communication in general by Maynard Smith and Harper (2003) and within the specific context of bacteria by Keller and Surette (2006) (see also Chapter 1).

2.4.2 The problem of quorum sensing

Quorum sensing is generally assumed to coordinate cooperative behaviours in bacteria. Specifically, QS appears to provide a means for individual bacteria to assess local cell density and to engage in cooperation once a threshold density has been reached. Many cooperative ventures will not be worthwhile until a sufficient number of cells are present, so one would expect facultative cooperation based on the presence of cues such as QS molecules that act as a proxy for cell density. The idea is that signalling molecules are released, and that this rate of release is further increased by signalling molecules. This leads to positive feedback at high cell densities, and a dramatic increase in cooperative effort (Diggle *et al.* 2007a; Williams *et al.* 2007). (See Chapter11 for a related discussion on collective behaviours in other taxa.)

However, this communication may potentially be invaded by cheats that exploit this system (Brown and Johnstone 2001; Redfield 2002; Keller and Surette 2006). One possibility is a cheat that does not produce QS molecules (signal negative), and so benefits from monitoring the local cell density without investing effort into the dissemination of this information. An alternative possibility would be for a cheat to neither make the costly signal nor to respond to it (signal blind). A further possibility is for a signal blind cheat to make a signal but not respond. The crucial point here is that both signalling and responding to a signal with the production of public goods are costly. Consequently, there must be benefits that outweigh these—otherwise the system could be invaded by cheats that did not signal or cooperate.

As has previously been discussed, there are many species of bacteria that use QS to regulate the production of public goods and are therefore exploitable by cheats. It is important to note that many *P. aeruginosa* clinical isolates are 'signal blind' (i.e. they may or may not make minimal amounts of signal but, importantly, do not respond to a signal) (Denervaud *et al.* 2004; Smith *et al.* 2006), and so it is desirable to understand the costs and benefits of QS from an empirical perspective. A fundamental first step is to determine the **fitness** consequences of producing and responding to a signal. Calculating the number of ATP molecules required to make signal, Keller and Surette (2006) suggested that the cost of production of QS molecules varies from low to high depending on the type of signal molecule produced (Keller and Surette 2006).

Whilst there is undoubtedly a cost in making a signal, it is likely that the cost of responding is more metabolically expensive, especially when you consider that 6% of the *P. aeruginosa* genome changes in response to the addition of QS molecules. Given high costs, QS signalling or response could be potentially exploitable by QS cheats (Keller and Surette 2006; Diggle *et al.* 2007b). In theory, QS cheats could take the form of either: (1) a 'signal negative' strain which does not make the molecule but can respond to it, or (2) a 'signal blind' strain

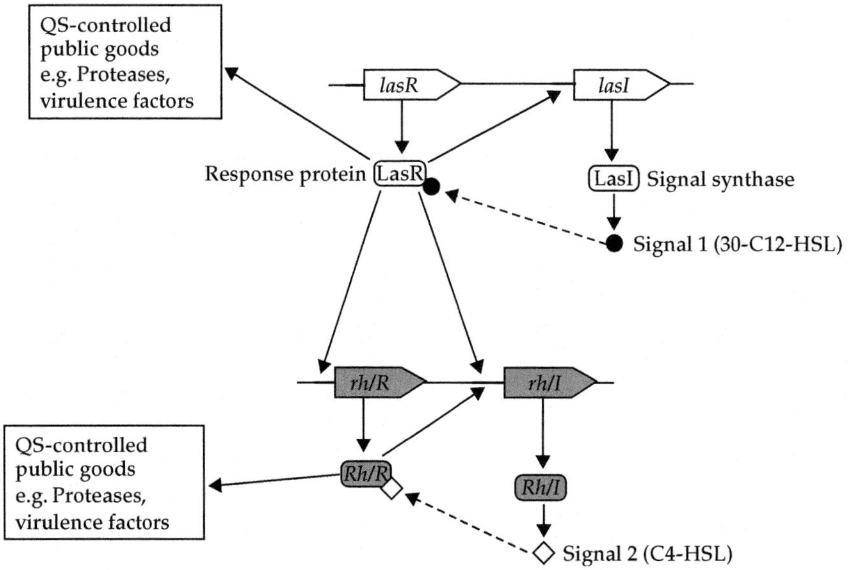

Figure 2.3 The hierarchical quorum sensing (QS) system of *P. aeruginosa*. The QS cascade is induced at high population cell densities when within the cell, the LasR response protein binds to a critical concentration of 3O-C12-HSL signal which has been produced by neighboring cells and taken up from the surrounding environment. This results in activation of the *las* QS system and the production of a number of QS-regulated public goods such as the proteases. Activation of the *las* system is also important in the induction of the *rhl* QS system which is also required for the production of proteases and a number of other *rhl*-controlled public goods.

which may (or may not) make signal but, more importantly, does not respond to it.

Recently we have been addressing empirically (using *P. aeruginosa*) whether QS is costly and subject to cheating behaviour (Diggle *et al.* 2007c). In *P. aeruginosa*, QS is controlled by two pathways (homologous to the *V. fischeri luxIR* system) which regulate the production of AHL signalling molecules (Fig. 2.3). These two systems are termed *las* and *rhl*, and use different AHL signal molecules, synthesized via LasI [*N*-(3-oxododecanoyl)-homoserine lactone (3O-C12-HSL)], and RhlI [*N*-butanoylhomoserine lactone (C4-HSL)], respectively (Latifi *et al.* 1995, 1996; Winson *et al.* 1995). Importantly, in *P. aeruginosa* QS regulates many potential social traits such as virulence, biofilm formation, and swarming motility. To examine the consequences of QS for social fitness, we focused on the *las* QS pathway because this system is top of the QS hierarchy (Fig. 2.3) (Latifi *et al.* 1996; Pesci *et al.* 1997), and a mutation in the *las* system results in the general abolition of QS.

We constructed both signal negative (*lasI-*) and signal blind (*lasR-*) mutants. Importantly, in the laboratory we can experimentally alter the level of signal perceived by either the wild type or the signal negative mutant by adding synthetic signal, which is chemically identical to that produced by *P. aeruginosa*, to cultures (Chhabra *et al.* 2003). We first examined the fitness consequences of QS in a situation where cooperation is favoured. A group of exoproducts whose production is controlled by QS in *P. aeruginosa* are the proteases. We examined the growth of the wild type and the signal negative and signal blind mutants in a medium where the ability to make proteases is required for growth. We found that: (1) both the signal negative and signal blind mutants grew very poorly in this medium when compared with the parental wild-type strain; (b) addition of synthetic signal to the signal negative strain significantly improved growth, as would be expected, because this will stimulate the production of proteases; (c) addition of signal to the signal blind strain resulted in no improvement in growth, as would be expected because the cells do not respond to the signal (Diggle *et al.* 2007c). This shows that QS can provide a benefit at the population level, by increasing the production of cooperative exoproducts, that can aid growth in certain environmental conditions.

We then determined whether the production of the QS signal molecules and cooperative QS-dependent exoproducts (public goods) is costly. We did this by comparing the growth rate of the mutants and the wild type in nutrient-rich Luria–Bertani (LB) broth, where the exoproducts produced by QS are not needed for growth. In these conditions, the QS mutants were able to grow to a higher density than the wild type. Addition of synthetic signal molecule to the signal negative mutant resulted in growth profiles similar to those seen for the wild type, suggesting that the response to QS signal molecules is costly as similar results were not seen when signal was added to the signal blind strain. These results suggests that upon entry to the stationary phase, QS signalling and the production of QS-dependent public goods place a heavy metabolic load on the cell (Diggle *et al.* 2007c).

Thus, it can be shown experimentally that QS is a social trait susceptible to exploitation and invasion by cheats. Given this, how is QS maintained in natural populations? The most likely explanation is kin selection, with cooperation being favoured because it is between close relatives.

2.4.3 A kin selection model of quorum sensing

Kin selection theory provides an explanation for cooperation or communication between relatives (Hamilton 1964). By helping a close relative reproduce, an individual is still passing on its own genes to the next generation, albeit indirectly. This theory is formalized by **Hamilton's rule** (Hamilton 1964), which states that **altruistic** cooperation is favoured when $rb - c > 0$; where c is the fitness cost to the altruist, b is the fitness cost to the beneficiary, and r is their genetic relatedness. This predicts that individuals should be more likely to cooperate when social partners are more closely related (higher r). For example, high levels of production of public goods are predicted when relatedness is higher among interacting bacteria (West and Buckling 2003). Relatedness can often be extremely high in bacteria because limited dispersal and clonal reproduction can lead to the individuals interacting over a small area being predominantly clone-mates (West *et al.* 2006).

Brown and Johnstone (2001) developed a kin selection model of QS. They assumed:

1. Signalling is costly to the individual. The fitness of an individual cell is negatively correlated to the amount of signalling by that individual.
2. The production of public goods, in response to QS, is costly to the individual. The fitness of an individual cell is negatively correlated to the amount of public goods produced by that individual.
3. The production of public goods provides a benefit to the local group of interacting cells (the group). The fitness of an individual cell is positively correlated to the average amount of public goods produced by the local individuals.
4. The benefit of producing public goods is greater at higher population densities. The fitness benefit to an individual cell of a certain level of local production of public goods is positively correlated with cell density.

Brown and Johnstone (2001) then made predictions for the evolutionarily stable level of signalling (production of signalling molecule) and public goods production (cooperation). A behaviour is described as an **evolutionarily stable strategy** (ESS) if it cannot be invaded or beaten by a mutant performing any other strategy once it has been adopted by the majority of individuals (Maynard Smith and Price 1973). In particular, they examined the consequences of variation in mean population density and relatedness (r). They found that:

1. *Result 1.* The ESS level of signalling and public goods production both increased with greater population densities. At low densities there is little to be gained from the cooperative production of public goods.
2. *Result 2.* The ESS level of production of public goods increased with higher relatedness between interacting bacteria (Fig. 2.4a). This is expected because greater levels of cooperation are favoured with a higher relatedness. However, appreciable levels of cooperation can be predicted even when relatedness is relatively low.
3. *Result 3.* The ESS level of signalling showed a domed relationship with relatedness (Fig. 2.4b). At high relatedness there is a shared interest in cooperation, and in cheap signalling. At low relatedness, there is no selection for cooperation, and hence no selection for signalling to coordinate this. With intermediate relatedness, there can still be selection to produce public goods, but it is in the individual's interest to produce fewer public goods than the other local cells (because $r < 1$). This favours higher levels of signalling in an attempt to manipulate competitors to cooperate more (which in turn leads to the signal being increasingly ignored). This is termed 'competitive devaluation of signal strength' (Brown and Johnstone 2001).

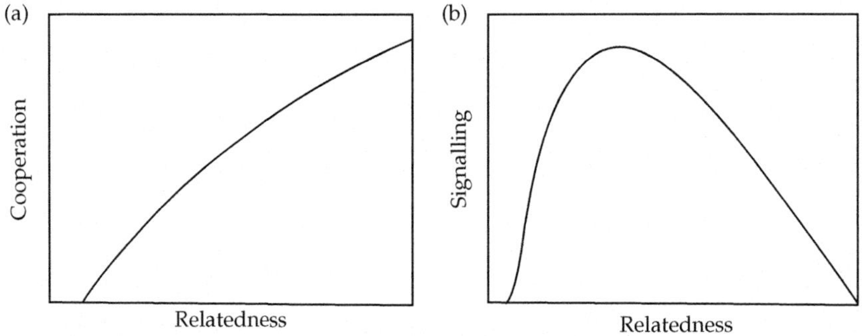

Figure 2.4 Brown and Johnstone's theoretical model of quorum signalling. (a) Cooperation effort increases with increasing relatedness, because the inclusive fitness benefits of cooperation are maximal at high relatedness and minimal at low relatedness. (b) Signalling effort is a dome-shaped function of relatedness, because at low relatedness there is little inclusive fitness benefit to be accrued from organizing a cooperative venture, and at high relatedness there is little conflict so that a cheap signal is favoured, whereas at intermediate relatedness cooperation is worthwhile yet there is also scope for conflict so a costly signal is required to initiate competition.

Experimentally we tested Brown and Johnstone's theory that QS can be maintained by kin selection. Using a QS-positive wild type (QS positive) and a signal blind cheat, mixed together (1:1) in a medium where the ability to quorum sense is essential for survival, we found that QS was favoured at a relatively high relatedness. This is in agreement with Brown and Johnstone's prediction that cooperation would increase with higher relatedness (Fig. 2.4a). Under conditions of high relatedness, and a number of rounds of selection, the wild-type cells constituted 100% of the total population. In contrast, in conditions of low relatedness, the cheats increased in frequency to approximately 60% after a number of rounds of selection. Therefore, low relatedness within a population allows cheats who do not quorum sense to exploit the individuals who do (Diggle *et al.* 2007c).

2.4.4 Other models of quorum sensing

Brown and Johnstone's (2001) model provides a clear and elegant application of kin selection theory to QS. However, as they stress, it makes many simplifications, the relaxing of which may have important consequences. Furthermore, much more has been learnt about QS since, and we should also consider alternative possible explanations for QS.

Brown and Johnstone's (2001) model could be extended to investigate the consequences of several biological complexities. It has been found that signalling molecules can have multiple functions, and this would alter the relative cost and benefit of their production, as well as how this would vary with the social context. For example, they can also function as antibiotics (Stein 2005), potentially as public goods such as iron-scavenging molecules (Kaufmann *et al.* 2005; Diggle *et al.* 2007d), and as potent immune modulators (Tateda *et al.* 2003; Hooi *et al.* 2004). Production and secretion of signal molecules may also be linked to the production of other molecules through excretion in membrane vesicles (Mashburn and Whiteley 2005). Another possibility is that different types of signal need to be considered, with different costs or specificities. It appears that specificity and cost vary across signals, with cheap-to-produce signals being utilized very generally across species, and more expensive signals being more specific, within species, possibly even within lineages (Keller and Surette 2006).

Kin selection is not the only possible explanation for cooperation (Sachs *et al.* 2004; see an individual-level hypothesis by Zahavi in Chapter1) An alternative explanation for cooperation is that it provides a direct benefit to the individual performing the behaviour, which outweighs the cost of performing the behaviour (i.e. it is mutualistic not altruistic). An example of this would be if the waste product of one species provided a benefit to individuals of a second species (by-product benefit), and hence the second species could be selected to cooperatively help individuals of the first species, in order to increase the by-product benefits (Sachs *et al.* 2004). It would be extremely interesting to see whether communication between species can be evolutionarily stable in such cases. There are several other forms of direct benefit to cooperation that could be examined from a QS and communication perspective—for example, when cooperation is stabilized between non-relatives by policing or punishment of non-cooperators (Frank 2003).

2.5 Defining signalling in bacteria

As discussed earlier, the fact that a compound produced by cell A elicits a response in cell B does not necessarily mean that there is true signalling between the cells and may represent cell B using the molecule as a 'cue' or cell A coercing cell B into a certain action. In this section we discuss examples of QS between single populations and mixed populations of bacteria and suggest whether this can be considered signalling, a response to a cue, or a coercion (see also Keller and Surette 2006).

In general, communication in bacteria can be divided into three main areas:

1. Intraspecies: communication arising or occurring within a single bacterial species.
2. Interspecies: communication arising between two or more distinct species of bacteria.
3. Interkingdom: communication arising between a bacterial species and a higher organism.

2.5.1 Intraspecies communication

In Gram-negative bacteria, the most intensely studied QS systems rely upon the interaction of AHL signal molecules synthesized by LuxI-type AHL synthases, with LuxR-type transcriptional regulator proteins (see Section 2.3). A simple example of this can be seen in the marine bacterium *V. fischeri* (Nealson *et al.* 1970). This organism forms a symbiotic relationship with the squid *Euprymna scolopes* where it colonizes the light organ (McFall-Ngai and Ruby 2000). At low cell densities the bacterial population does not luminesce but at high densities there is a coordinated switch on of bioluminescence. This production of light has been shown to be mediated by a diffusible AHL molecule (3O-C6-HSL) synthesized by the LuxI protein. At a critical concentration, 3O-C6-HSL binds to LuxR and the complex activates expression of the *luxCDABE* operon resulting in coordinated production of bioluminescence. Under laboratory conditions, it is possible to stimulate early induction of bioluminescence simply by providing the cells with exogenous 3O-C6-HSL. It is not entirely clear why *V. fischeri* cells have a shared interest that favours signalling and cooperation to produce light. Possibilities are a high relatedness between the cells within a light organ, or the avoidance of punishment from the host squid if light is not produced (analogous to why rhizobia fix nitrogen for their host plants (West *et al.* 2002a; Kiers *et al.* 2003)). Indeed, it appears to be the case that the squid can enforce bioluminescence by altering the environment such that *lux*-deficient strains are defected in light organ colonization. It was hypothesized that a diminished level of oxygen consumption by *lux*-deficient strains is responsible for the reduced fitness (Visick *et al.* 2000).

As many species of Gram-negative bacteria have been shown to produce AHL signalling molecules, then similar examples can be seen in other species (Diggle *et al.* 2007a; Williams *et al.* 2007). Some bacteria have been shown to regulate the production of virulence determinants in a cell density-dependent manner. For example, *Erwinia carotovora* subsp. *carotovora* coordinately produces both exoenzymes, which destroy plant tissue, and the antibiotic carbapenem in response to critical concentrations of 3O-C6-HSL (Jones *et al.* 1993). Similarly the opportunistic pathogen *P. aeruginosa* regulates an arsenal of extracellular virulence factors using a complex hierarchical QS cascade involving two major AHL molecules, namely 3O-C12-HSL and C4-HSL (Venturi 2006). In such cases it is likely that these are examples where QS molecules can be classed as 'signals' between cells as the production by cell A has evolved due to its effects on cell B which in turn has evolved a response to the signal (Maynard Smith and Harper 2003). We suspect that kin selection is the mechanism to explain the evolutionary stability of such signalling, as discussed in Section 2.4. Although the AHL family of QS molecules have been described in a wide variety of Gram-negative bacterial species (Lazdunski *et al.* 2004), crucially they tend to differ between bacterial species. AHLs consist of a conserved homoserine lactone ring connected via an amide bond to an acyl side chain which can vary in length from 4 to 18 carbons. In addition, these side chains may or may not be modified with a 3-hydroxy or a 3-oxo group, potentially providing a large variety of AHL molecules. Many species of bacteria will only respond to their cognate molecule(s) providing a certain degree of specificity, and therefore AHL signalling is generally of an intraspecies nature. Some bacteria, however, are able to 'exploit' AHLs produced by another species, and this will be discussed later.

Whilst it is plausible to view AHLs as signals between cells of the same species, the situation is often more complicated as some AHLs have been shown to have multiple functions. For example 3O-C12-HSL produced by *P. aeruginosa* has been reported to have immunomodulatory properties (Telford *et al.* 1998; Tateda *et al.* 2003). It is unlikely that this involves signalling between the host and bacteria. More likely, this represents 3O-C12-HSL 'chemically manipulating' or 'coercing' the host immune response to the benefit of the bacterial population.

The world of microbial communication is not limited to Gram-negative bacteria. Gram-positive bacteria also produce QS molecules but tend to utilize post-translationally modified autoinducing peptides (AIPs). For example, *S. aureus* uses AIPs to regulate the production of exotoxins in response to a critical concentration of peptide (Novick 2003).

Explaining within-species cooperative signalling requires some kind of mechanism (see also Chapter11). The production of a costly signal for the common good makes this type of communication exploitable by cheats who do not contribute to signal production but reap the benefits of QS-mediated behaviour, for example acquisition of nutrients provided by QS-dependent exoenzyme production. In fact, recent work has shown that many *P. aeruginosa* clinical isolates are QS defective and make very few virulence factors when grown in the laboratory (Denervaud *et al.* 2004; Schaber *et al.* 2004; Lee *et al.* 2005) suggesting that it may be beneficial not to signal under certain environmental conditions, or that cheats can invade in long-term infections (West *et al.* 2006). As local populations of cells are likely to be closely related, then one way that cooperation can be maintained is via kin selection, which requires a sufficiently high relatedness between cooperating individuals (West *et al.* 2006). Limited dispersal (population viscosity) would tend to keep relatives together (Hamilton 1964). In this case, indiscriminate altruism may be favoured because neighbours will tend to be relatives (Hamilton 1964; Queller 1992; West *et al.* 2002b). This type of mechanism is likely to be of huge importance in microorganisms where asexual reproduction means that single cells colonize and grow in a local area. In this case, the individuals interacting over a small area will be clonal, which can be very conducive to the evolution of cooperation.

2.5.2 Interspecies communication—bacterial 'cross-talk'

A third class of QS signal molecule has been described in the marine bacterium *Vibrio harveyi*. Bioluminescence in this organism is cooperatively regulated by AHLs and a molecule termed autoinducer-2 (AI-2) which is a furanosyl borate diester produced by the enzyme LuxS (Chen *et al.* 2002). The identification of the *luxS* gene required for the production of AI-2 production (Surette *et al.* 1999) sparked an exponential increase in AI-2 signalling research. The reason being that the *luxS* gene can be found in a wide variety of bacterial genera (Winzer *et al.* 2002, 2003).

Importantly, representatives of both Gram-negative and Gram-positive bacteria carry this particular gene, and consequently AI-2 production has been demonstrated in many species of bacteria. This has led to the hypothesis that AI-2 is employed as a means of interspecific communication or 'bacterial Esperanto' (Winans 2002). This idea is difficult to explain from an evolutionary point of view, as cooperation between species is even harder to explain than within species. The major difference is that kin selection, as discussed in Section 2.3, will not be important across species. There are mechanisms by which cooperation can be favoured between species, such as by-product benefit (Sachs *et al.* 2004), or to avoid punishment (West *et al.* 2002a; Kiers *et al.* 2003), but these are expected to be rarer (West *et al.* 2006).

It must therefore be questioned whether AI-2 can be defined as a true signal. For this to be the case AI-2 must: (1) be diffused from the cell, (2) be taken up by a neighbouring cell, (3) elicit a response from that cell because the receiver's response has evolved, (4) benefit both producer and receiver. Clearly points 1 and 2 are met with respect to AI-2 but there are major doubts about points 3 and 4. Despite AI-2 being produced by many genera, there is very little evidence linking it with direct activation of any specific genes. Studies in many different bacteria have shown that *luxS* mutants differ phenotypically from wild-type strains; however, this can often be explained because of a defect in a metabolic pathway. It is now well known that LuxS plays an important role in bacterial metabolism, contributing to the recycling of *S*-adenosyl-L-methionine (SAM), of which AI-2 is a metabolic by-product (Winzer *et al.* 2003). To date only bioluminescence in *V. harveyi* (Surette *et al.* 1999), and an ABC transporter in *Salmonella typhimurium* (termed Lsr) (Taga *et al.* 2001) have been shown to be regulated by AI-2. In these species, we can speculate that AI-2 may be used as a cooperative signal in an intraspecies context. Theoretically, these species could also use AI-2 from other organisms to regulate these respective traits. In this case, however, it is inaccurate to use the term interspecies signalling as the receiver's response has not evolved in parallel with the producing bacterial species. In this scenario we can say that both *V. harveyi* and

S. typhimurium use the metabolic by-product AI-2 as an environmental 'cue' to regulate gene expression. Interspecies signalling has also been suggested between avirulent oropharyngeal flora (OF) (AI-2 +ve) and *P. aeruginosa* (*luxS* and AI-2 –ve) within the cystic fibrosis (CF) lung (Duan *et al.* 2003). Co-incubation of *P. aeruginosa* with OF bacteria resulted in an increase in virulence gene expression which was attributed, at least in part, to AI-2. The mechanism for this is unknown as *P. aeruginosa* does not make AI-2 but we suggest that this is not an example of interspecies signalling. It is more likely that *P. aeruginosa* is able to use AI-2 as a cue, perhaps to assess its surroundings, or it may be that OF bacteria 'coerce' or manipulate *P. aeruginosa* into increased virulence which may provide them with more nutrients.

Interspecies signalling between bacterial species using AHL molecules has also been suggested. *Pseudomonas aeruginosa* and *Burkholderia cepacia* often occur together in the lungs of people with cystic fibrosis, where they are associated with high morbidity and mortality (Eberl and Tummler 2004; Govan and Deretic 1996). *Burkholderia cepacia* has been shown to up-regulate the production of virulence determinants in response to AHLs produced by *P. aeruginosa*, although this does not appear to happen the other way round. This type of behaviour has also been termed 'bacterial cross-talk' which is suggestive of a cooperative venture between two or more species. In this case, it suggests that *B. cepacia* uses *P. aeruginosa* AHLs as a cue to alter its behaviour rather than there being signalling between the two bacterial species. *Pseudomonas aeruginosa* pays the cost of producing AHLs, possibly for within-species signalling, but appears to gain no benefit from *B. cepacia* in return.

2.5.3 Interkingdom communication across the prokaryote/eukaryote divide

Several recent reports have demonstrated that bacterial QS molecules (specifically AHLs) can affect gene expression in eukaryotes as many eukaryotic **hormones** structurally resemble AHLs. Generally this has been termed interkingdom signalling or global sensing (Shiner *et al.* 2005). AHL molecules have been experimentally demonstrated to affect a number of animal cell types including murine and human primary cells (Telford *et al.* 1998), breast cancer cells (Li *et al.* 2004), bone marrow macrophages (Tateda *et al.* 2003), and primary porcine arterial smooth muscle cells (Lawrence *et al.* 1999). In addition, plant behaviour has also been shown to be modified by AHLs. The zoospores of the seaweed *Enteromorpha* have been shown to settle preferentially on AHL-producing biofilms of the marine bacterium *Vibrio anguillarum* (Joint *et al.* 2002). Furthermore, higher organisms have mechanisms that appear to downregulate QS in microorganisms. For example, the marine red alga *Delisea pulchra* produces a halogenated furanone that disrupts QS in several species of bacteria including the swarming motility of *Serratia liquefaciens* (Givskov *et al.* 1996). This furanone has also been shown to disrupt *P. aeruginosa* biofilms (Hentzer *et al.* 2002). These AHL 'mimics' attract interest as possible alternatives to antibiotic therapy. Whether these examples demonstrate signalling using small molecules between prokaryotes and eukaroytes is open to debate. In general, studies performed to date appear to show that either (1) the signalling bacterium manipulates or coerces its host into a certain action rather than there being a truly evolved signalling system between the two (cf. coercion strategies, Chapters 4 and 10) or (2) as in the example of the zoospore settlement, the eukaryote utilizes bacterial AHLs as an environmental cue as a guide to future action.

2.6 Complexities of bacterial communication

In agreement with behavioural studies on organisms such as birds, mammals, and insects, signalling in bacteria has a number of complexities that offer problems from an evolutionary perspective.

First, the signal can be degraded (as also occurs for other modalities such as sound and **pheromones**). This degradation can be environmental in nature or due to the action of certain enzymes. This signal interference has often been suggested as a possible way of controlling the virulence of pathogenic bacterial species (i.e. breaking the lines of communication) and thus leading to novel therapies. AHL signals are rendered biologically inactive in alkaline environments (Yates *et al.* 2002) and therefore, in certain environmental niches, signalling may be ineffective. In theory, the levels of QS signalling may be greatly influenced by environmental conditions but whether this alters the cost and benefit of either making a signal, or responding, has not been explored. AHLs can also be degraded by enzymes produced by bacteria, a process known as quorum quenching (Dong and Zhang 2005). Examples include AiiA, an AHL lactonase produced by a *Bacillus* spp. (Dong *et al.* 2001), and PvdQ, an AHL-acylase produced by *P. aeruginosa* (Sio *et al.* 2006). This raises many interesting questions, which could be empirically tested. What effect can an AHL-degrading species have on an AHL producer? For instance, does degradation interfere with key social behaviours such as population swarming or result in the reduction of a number of harmful AHL-dependent exoproducts which is ultimately beneficial to the degrading organism? Can this behaviour be considered coercive or spiteful, and are there indirect or direct fitness benefits for the AHL degrader? Is AHL degradation evolutionary stable or is it subject to invasion by cheats who do not make the degrading enzymes?

Second, the genes required for signal generation (*luxI* homologues) and response (*luxR* homologues) are not always found on the bacterial chromosome. A number of these homologues have been identified on plasmids such as the *Agrobacterium* Ti plasmid (Zhang *et al.* 1993) and *Rhizobium* symbiotic plasmids (Smith 2001; Wisniewski-Dye and Downie 2002). While this may just represent an easy way to obtain QS mechanisms, could it also be a mechanism by which signalling is forced onto a cell that doesn't contain the QS machinery, coercing it into cooperative behaviour? An important point here is the **conflicting** interests of the bacteria involved, and the plasmids themselves. Third, QS molecules are not just signals. A number of other roles have been assigned to QS molecules which suggests they can also function as public goods, for example iron chelators (Diggle *et al.* 2007d), immunomodulatory compounds(Pritchard 2006), and biosurfactants (Daniels *et al.* 2006). QS compounds can also be harmful or spiteful, for example the AIP lantibiotics typified by lactococcal nisin and produced by *Lactococcus lactis* are potent bacteriocides against many Gram-positive organisms (Stein 2005). The consequences of QS signals having multiple functions needs to be explored theoretically (Brown and Johnstone 2001; Diggle *et al.* 2007b).

Another complexity of studying signalling in bacteria is that most bacterial species are capable of forming structured multicellular communities known as biofilms (Kolter and Greenberg 2006). Biofilms are ubiquitous, being found in such diverse environments as dental plaques, wounds, rock surfaces, and at the bottom of rivers. They have a definite structure, including water channels, which may involve a number of different 'specialist' cells and they are often enclosed by a exopolysaccharide matrix which can make them difficult to eradicate. It is also comparatively harder to empirically study cells growing in a biofilm compared with planktonic cells. However, biofilms are of particular interest from an evolutionary perspective, because the close proximity of individuals in a biofilm can make cooperation and communication particularly important.

Many forms of cooperation can be involved in the establishment and growth of a biofilm, such as the cooperative production of an extracellular matrix which surrounds the biofilm, and may be important in maintaining structure (Davies and Geesey 1995; Nivens *et al.* 2001; Friedman and Kolter 2004; Matsukawa and Greenberg 2004; Diggle *et al.* 2006b). In addition, numerous other public goods can be important in biofilms, such as rhamnolipid, a biosurfactant which aids in biofilm detachment (Boles *et al.* 2005), and microvesicles which are a component of the extracellular matrix and can contain signal molecules and proteases (Schooling and Beveridge 2006). Quorum sensing may play an important role in the development and structuring of biofilms produced by certain bacterial species, as suggested by the poor biofilm formation of some QS mutants (Davies *et al.* 1998), although, perhaps surprisingly, not a great deal is known generally about QS and biofilm development which may stem from the fact that biofilms are difficult to study experimentally. However, it has been shown in *P. aeruginosa* that QS plays a role in biofilm differentiation (Fig. 2.5).

The evolutionary implications of QS in biofilms are also uncertain. It could be expected that kin

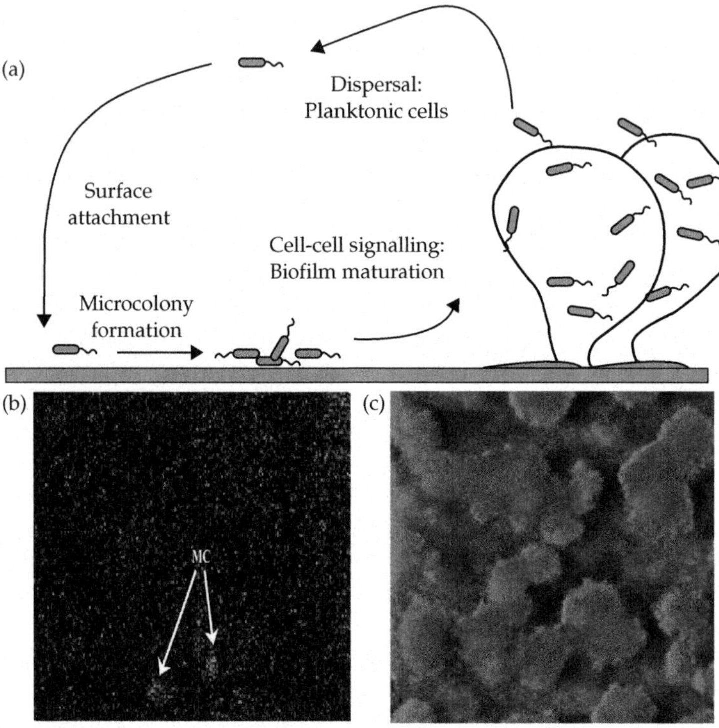

Figure 2.5 Lifecycle of a bacterial biofilm. (a) Planktonic cells are released from mature biofilms, and via motility mechanisms they settle on a new surface. Cells become irreversibly attached and begin to form microcolonies. Mechanisms such as cell-cell signalling systems lead to the differentiation of mature biofilm structures. Diagram adapted from (Kolter and Losick 1998). (b) Scanning electron microscopy image of *P. aeruginosa* attachment to stainless steel coupon. The formation of microcolonies can be observed (MC), image taken from (Diggle *et al.* 2006b). (c) Scanning confocal microscopy image of mature 5 day old *P. aeruginosa* biofilms grown in flow cell chambers. Image courtesy of S. Crusz.

selection is important in biofilms initiated by one or a small number of clonal lineages. However, naturally formed biofilms very rarely contain just one species of bacteria, let alone a single clonal lineage. For example, the colonization of human teeth and the oral mucosa can involve up to 500 species of bacteria (Kolenbrander *et al.* 2002). Nonetheless, kin selection may still be important in such cases if social interactions take place on a local scale. For example, if the benefit of producing the materials that structure the biofilm, such as exopolysaccharides, or other public goods (perhaps regulated by QS), are shared primarily with neighbouring cells, then the clonal growth of bacteria means that these benefits can still be shared with close relatives (Xavier and Foster 2007). In this case it might be useful to think of biofilms as consisting of a number of clonal lineages (groups of lineages), with cooperation primarily within lineages but competition primarily between lineages.

Some workers suggest that QS is not true cell-to-cell signalling and that is an artefact of laboratory conditions. Redfield (2002), has argued that autoinducer molecules are not released to signal to other cells. Redfield suggests that autoinducer secretion and response may have a more direct benefit, by allowing individual cells to determine how rapidly secreted molecules move away from the cell. This diffusion sensing (DS) could allow cells to regulate secretion of costly public goods to minimize losses owing to extracellular diffusion and mixing. This is an alternative explanation for the evolution of QS, and diffusion effects could also be incorporated into kin selection models. It is important to consider how QS and DS may have evolved. In the case of QS, it may be assumed that it evolved because of group benefits, but when you consider DS this is likely to have evolved because of benefits to the individual. It should also be considered that production of these molecules may have initially evolved for one reason (e.g. DS), but is now maintained for another (e.g. QS). Hense *et al.* (2007) recently introduced a new term: efficiency sensing (ES) which unifies the QS and DS theories. Efficiency sensing argues that sensing will have been favoured by both individual and group benefits as the cells measure a combination of cell density, mass-transfer properties, and spatial cell distribution (Hense *et al.* 2007). The hypotheses need not be alternatives, as it may be the case that benefits of DS are crucial for the maintenance of this trait, yet they are still monitored for QS purposes. It is likely that both functions will be of importance in understanding when and why these molecules are produced.

2.7 Conclusions and future perspectives

Quorum sensing systems are widespread amongst Gram-negative and Gram-positive bacterial species. However, when one compares the microbiological literature on QS with the animal literature it is sometimes questionable as to whether QS in bacteria should always be regarded as signalling. It is likely that QS within species represents signalling, because the natural history of many microorganisms means that the interactions will be between close relatives and therefore cooperation and signalling can be explained by kin selection. However, QS is often described between species or even across kingdoms, and in these cases QS signals may be used as chemical cues or as coercion molecules. It is possible that these interactions may be due to signalling, but cooperation between species requires special conditions that are only rarely met. There is therefore increasing uncertainty as to what form of communication QS represents in bacteria and the challenges for the future will rely on experimental studies that examine the costs and benefits of communication to both the sender and responder. Furthermore, cooperation and communication need to be expanded empirically into biofilms as this is the natural state of growth for many species of bacteria. It is important to note that in one particular environment where biofilms are formed (the cystic fibrosis lung), *P. aeruginosa* QS signal blind mutants are often isolated; the reasons for this are poorly understood. Understanding the interactions between strains found within such environments will provide unique insights into eradicating problematic organisms such as *P. aeruginosa*.

Summary

The term quorum sensing (QS) is used to describe communication between bacterial cells, whereby a coordinated population response is controlled by diffusible signal molecules. Quorum sensing has not only been described between cells of the same species (intraspecies), but also between bacterial species (interspecies), and between bacteria and higher organisms (interkingdom). Here we compare the evolutionary literature on animal signalling and cooperation with the microbiological literature on QS, and discuss whether bacterial QS can be considered true signalling. From an evolutionary perspective, intraspecies signalling can be explained using models such as kin selection, but explanations become more difficult when communication is described between species. It is likely that this often involves QS molecules being used as 'cues' by other species as a guide to future action or as coercing molecules whereby one species will 'coerce' another into a response.

Acknowledgements

We would like to gratefully acknowledge the Royal Society as S.P.D., S.A.W., A.G., and A.S.G. are all supported by Royal Society Fellowships.

References

Bainton, N.J., Bycroft, B.W., Chhabra, S.R. *et al.* (1992a). A general role for the Lux autoinducer in bacterial-cell signaling—control of antibiotic biosynthesis in *Erwinia*. *Gene*, **116**, 87–91.

Bainton, N.J, Stead, P., Chhabra, S.R. *et al.* (1992b). N-(3-oxohexanoyl)-L-homoserine lactone regulates carbapenem antibiotic production in *Erwinia carotovora*. *Biochemical Journal*, **288**, 997–1004.

Barber, C.E, Tang, J.L, Feng, J.X. *et al.* (1997). A novel regulatory system required for pathogenicity of *Xanthomonas campestris* is mediated by a small diffusible signal molecule. *Molecular Microbiology*, **24**, 555–566.

Boles, B.R, Thoendel, M., and Singh, P.K. (2005). Rhamnolipids mediate detachment of *Pseudomonas aeruginosa* from biofilms. *Molecular Microbiology*, **57**, 1210–1223.

Brown, S.P. (1999). Cooperation and conflict in host-manipulating parasites. *Proceedings of the Royal Society Series B: Biological Sciences*, **266**, 1899–1904.

Brown, S.P. and Johnstone, R.A. (2001). Cooperation in the dark: signalling and collective action in quorum-sensing bacteria. *Proceedings of the Royal Society Series B; Biological Sciences*, **268**, 961–965.

Chen, X, Schauder, S, Potier, N. *et al.* (2002). Structural identification of a bacterial quorum-sensing signal containing boron. *Nature*, **415**, 545–549.

Chhabra, S.R, Harty, C., Hooi, D.S.W. *et al.* (2003). Synthetic analogues of the bacterial signal (quorum sensing) molecule N-(3-oxododecanoyl)-L-homoserine lactone as immune modulators. *Journal of Medicinal Chemistry*, **46**, 97–104.

Ciofu, O., Beveridge, T.J., Kadurugamuwa, J, Walther-Rasmussen, J., and Hoiby, N. (2000). Chromosomal beta-lactamase is packaged into membrane vesicles and secreted from *Pseudomonas aeruginosa*. *Journal of Antimicrobial Chemotherapy*, **45**, 9–13.

Clough, S.J., Flavier, A.B., Schell, M.A., and Denny, T.P. (1997). Differential expression of virulence genes and motility in *Ralstonia (Pseudomonas) solanacearum* during exponential growth. *Applied and Environmental Microbiology*, **63**, 844–850.

Collier, D.N., Anderson, L., McKnight, S.L. *et al.* (2002). A bacterial cell to cell signal in the lungs of cystic fibrosis patients. *FEMS Microbiology Letters*, **215**, 41–46.

Crespi, B.J. (2001). The evolution of social behavior in microorganisms. *Trends in Ecology and Evolution*, **16**, 178–183.

Daniels, R, Vanderleyden, J., and Michiels, J. (2004). Quorum sensing and swarming migration in bacteria. *FEMS Microbiology Reviews*, **28**, 261–289.

Daniels, R, Reynaert, S, Hoekstra, H. *et al.* (2006). Quorum signal molecules as biosurfactants affecting swarming in *Rhizobium etli*. *Proceedings of the National Academy of Sciences of the USA*, **103**, 14965–14970.

Davies, D.G. and Geesey, G.G. (1995). Regulation of the alginate biosynthesis gene *algC* in *Pseudomonas aeruginosa* during biofilm development in continuous culture. *Applied and Environmental Microbiology*, **61**, 860–867.

Davies, D.G., Parsek, M.R, Pearson, J.P., Iglewski, B.H., Costerton, J.W., and Greenberg, E.P. (1998). The involvement of cell-to-cell signals in the development of a bacterial biofilm. *Science*, **280**, 295–298.

Denervaud, V., TuQuoc, P, Blanc, D. *et al.* (2004). Characterization of cell-to-cell signaling-deficient *Pseudomonas aeruginosa* strains colonizing intubated patients. *Journal of Clinical Microbiology*, **42**, 554–562.

Diggle, S.P., Cornelis, P., Williams, P., and Camara, M. (2006a). 4-Quinolone signalling in *Pseudomonas aeruginosa*: Old molecules, new perspectives. *International Journal of Medical Microbiology*, **296**, 83–91.

Diggle, S.P., Stacey, R.E., Dodd, C., Camara, M., Williams, P., and Winzer, K. (2006b). The galactophilic lectin, LecA, contributes to biofilm development in *Pseudomonas aeruginosa*. *Environmental Microbiology*, **8**, 1095–1104.

Diggle, S.P., Crusz, S.A., and Camara, M. (2007a). Quorum Sensing. *Current Biology*, **17**, R907–R910.

Diggle, S.P., Gardner, A., West, S.A., and Griffin, A.S. (2007b). Evolutionary theory of bacterial quorum sensing: when is a signal not a signal? *Philosophical Transactions of the Royal Society B: Biological Sciences*, **362**, 1241–1249.

Diggle, S.P., Griffin, A.S., Campbell, G.S., and West, S.A. (2007c). Cooperation and conflict in quorum sensing bacterial populations. *Nature*, **450**, 411–414.

Diggle, S.P., Matthijs, S., Wright, V.J. *et al.* (2007d). The *Pseudomonas aeruginosa* 4-quinolone signal molecules HHQ and PQS play multifunctional roles in quorum sensing and iron entrapment. *Chemistry and Biology*, **14**, 87–94.

Dinges, M.M., Orwin, P.M., and Schlievert, P.M. (2000). Exotoxins of *Staphylococcus aureus*. *Clinical Microbiology Reviews*, **13**, 16–34.

Dionisio, F. and Gordo, I. (2006). The tragedy of the commons, the public goods dilemma, and the meaning of rivalry and excludability in evolutionary biology. *Evolutionary Ecology Research*, **8**, 321–332.

Dong, Y.H. and Zhang, L.H. (2005). Quorum sensing and quorum-quenching enzymes. *Journal of Microbiology*, **43**, 101–109.

Dong, Y.H., Wang, L.H., Xu, J.L., Zhang, H.B., Zhang, X.F., and Zhang, L.H. (2001). Quenching quorum-sensing-dependent bacterial infection by an N-acyl homoserine lactonase. *Nature*, **411**, 813–817.

Duan, K.M., Dammel, C., Stein, J., Rabin, H., and Surette, M.G. (2003). Modulation of *Pseudomonas aeruginosa* gene expression by host microflora through interspecies communication. *Molecular Microbiology*, **50**, 1477–1491.

Dworkin, M. (1996). Recent advances in the social and developmental biology of the myxobacteria. *Microbiological Reviews*, **60**, 70–102.

Eberhard, A., Burlingame, A.L., Eberhard, C., Kenyon, G.L., Nealson, K.H., and Oppenheimer, N.J. (1981). Structural identification of autoinducer of *Photobacterium fischeri* luciferase. *Biochemistry*, **20**, 2444–2449.

Eberl, L. and Tummler, B. (2004). *Pseudomonas aeruginosa* and *Burkholderia cepacia* in cystic fibrosis: genome evolution, interactions and adaptation. *International Journal of Medical Microbiology*, **294**, 123–131.

Engebrecht, J., Nealson, K., and Silverman, M. (1983). Bacterial bioluminescence—isolation and genetic analysis of functions from *Vibrio fischeri*. *Cell*, **32**, 773–781.

Frank, S.A. (2003). Perspective: repression of competition and the evolution of cooperation. *Evolution*, **57**, 693–705.

Friedman, L. and Kolter, R. (2004). Two genetic loci produce distinct carbohydrate-rich structural components of the *Pseudomonas aeruginosa* biofilm matrix. *Journal of Bacteriology*, **186**, 4457–4465.

Fuqua, W.C., Winans, S.C., and Greenberg, E.P. (1994). Quorum sensing in bacteria—the LuxR-LuxI family of cell density-responsive transcriptional regulators. *Journal of Bacteriology*, **176**, 269–275.

Givskov, M., DeNys, R., Manefield, M. *et al.* (1996). Eukaryotic interference with homoserine lactone-mediated prokaryotic signaling. *Journal of Bacteriology*, **178**, 6618–6622.

Govan, J.R.W. and Deretic, V. (1996). Microbial pathogenesis in cystic fibrosis: mucoid *Pseudomonas aeruginosa* and *Burkholderia cepacia*. *Microbiological Reviews*, **60**, 539–574.

Greig, D. and Travisano, M. (2004). The prisoner's dilemma and polymorphism in yeast SUC genes. *Proceedings of the Royal Society Series B: Biological Sciences*, **271**, S25–S26.

Griffin, A.S, West, S.A., and Buckling, A. (2004). Cooperation and competition in pathogenic bacteria. *Nature*, **430**, 1024–1027.

Hamilton, W.D. (1964). Genetical evolution of social behaviour, I & II. *Journal of Theoretical Biology*, **7**, 1–52.

Hardin, G. (1968). Tragedy of the commons. *Science*, **162**, 1243–1248.

Henke, J.M. and Bassler, B.L. (2004). Bacterial social engagements. *Trends in Cell Biology*, **14**, 648–656.

Hense, B.A, Kuttler, C, Muller, J, Rothballer, M, Hartmann, A., and Kreft, J.U. (2007). Does efficiency sensing unify diffusion and quorum sensing? *Nature Reviews Microbiology*, **5**, 230–239.

Hentzer, M, Riedel, K, Rasmussen, T.B. *et al.* (2002). Inhibition of quorum sensing in *Pseudomonas aeruginosa* biofilm bacteria by a halogenated furanone compound. *Microbiology*, **148**, 87–102.

Hentzer, M, Wu, H, Andersen, J.B. *et al.* (2003). Attenuation of *Pseudomonas aeruginosa* virulence by quorum sensing inhibitors. *EMBO Journal*, **22**, 3803–3815.

Holden, M.T.G, Chhabra, S.R, de Nys, R. *et al.* (1999). Quorum-sensing cross talk: isolation and chemical characterization of cyclic dipeptides from *Pseudomonas aeruginosa* and other Gram-negative bacteria. *Molecular Microbiology*, **33**, 1254–1266.

Hooi, D.S.W, Bycroft, B.W, Chhabra, S.R, Williams, P., and Pritchard, D.I. (2004). Differential immune modulatory activity of *Pseudomonas aeruginosa* quorum-sensing signal molecules. *Infection and Immunity*, **72**, 6463–6470.

Joint, I., Tait, K., Callow, M.E. *et al.* (2002). Cell-to-cell communication across the prokaryote-eukaryote boundary. *Science*, **298**, 1207–1207.

Jones, S., Yu, B., Bainton, N.J. *et al.* (1993). The Lux autoinducer regulates the production of exoenzyme virulence determinants in *Erwinia carotovora* and *Pseudomonas aeruginosa*. *EMBO Journal*, **12**, 2477–2482.

Kaufmann, G.F., Sartorio, R, Lee., S.H. *et al.* (2005). Revisiting quorum sensing: Discovery of additional chemical and biological functions for 3-oxo-N-acylhomoserine lactones. *Proceedings of the National Academy of Sciences of the USA*, **102**, 309–314.

Keller, L. and Surette, M.G. (2006). Communication in bacteria: an ecological and evolutionary perspective. *Nature Reviews Microbiology*, **4**, 249–258.

Khokolov, A.S., Tovarova, I.I., Borisova, L.N. *et al.* (1967). A-factor assuring the biosynthesis of streptomycin by a mutant strain of *Actinomyces streptomycini*. *Dokladi Na Bolgarskata Akademiya Na Naukite*, **1777**, 232–235.

Kiers, E.T., Rousseau, R.A., West, S.A., and Denison, R.F. (2003). Host sanctions and the legume-rhizobium mutualism. *Nature*, **425**, 78–81.

Kleerebezem, M., Quadri, L.E.N., Kuipers, O.P., and deVos, W.M. (1997). Quorum sensing by peptide pheromones and two-component signal-transduction systems in Gram-positive bacteria. *Molecular Microbiology*, **24**, 895–904.

Kolenbrander, P.E., Andersen, R.N, Blehert, D.S, Egland, P.G, Foster, J.S., and Palmer, R.J. (2002). Communication among oral bacteria. *Microbiology and Molecular Biology Reviews*, **66**, 486–505.

Kolter, R. and Losick, R. (1998). One for all and all for one. *Science*, **280**, 226–227.

Kolter, R. and Greenberg, E.P. (2006). Microbial sciences—the superficial life of microbes. *Nature*, **441**, 300–302.

Kuspa, A., Plamann, L., and Kaiser, D. (1992). A-signaling and the cell-density requirement for *Myxococcus xanthus* development. *Journal of Bacteriology*, **174**, 7360–7369.

Latifi, A., Winson, M.K., Foglino, M. *et al.* (1995). Multiple homologs of LuxR and LuxI control expression of virulence determinants and secondary metabolites through quorum sensing in *Pseudomonas aeruginosa* PAO1. *Molecular Microbiology*, **17**, 333–343.

Latifi, A., Foglino, M., Tanaka, K., Williams, P., and Lazdunski, A. (1996). A hierarchical quorum-sensing cascade in *Pseudomonas aeruginosa* links the transcriptional activators LasR and RhIR (VsmR) to expression of the stationary-phase sigma factor RpoS. *Molecular Microbiology*, **21**, 1137–1146.

Lawrence, R.N., Dunn, W.R., Bycroft, B. *et al.* (1999). The *Pseudomonas aeruginosa* quorum-sensing signal molecule, N-(3-oxododecanoyl)-L-homoserine lactone, inhibits porcine arterial smooth muscle contraction. *British Journal of Pharmacology*, **128**, 845–848.

Lazdunski, A.M., Ventre, I., and Sturgis, J.N. (2004). Regulatory circuits and communication in gram-negative bacteria. *Nature Reviews Microbiology*, **2**, 581–592.

Lee, B., Haagensen, J.A.J., Ciofu, O., Andersen, J.B., Hoiby, N., and Molin, S. (2005). Heterogeneity of biofilms formed by nonmucoid *Pseudomonas aeruginosa* isolates from patients with cystic fibrosis. *Journal of Clinical Microbiology*, **43**, 5247–5255.

Li, L., Hooi, D, Chhabra, S.R, Pritchard, D. and Shaw, P.E. (2004). Bacterial N-acylhomoserine lactone-induced apoptosis in breast carcinoma cells correlated with down-modulation of STAT3. *Oncogene*, **23**, 4894–4902.

McFall-Ngai, M.J. and Ruby, E.G. (2000). Developmental biology in marine invertebrate symbioses. *Current Opinion in Microbiology*, **3**, 603–607.

McVittie, A., Zahler, S.A., and Messik, F. (1962). Developmental biology of *Myxococcus*. *Journal of Bacteriology*, **84**, 546–551.

Mashburn, L.M. and Whiteley, M. (2005). Membrane vesicles traffic signals and facilitate group activities in a prokaryote. *Nature*, **437**, 422–425.

Matsukawa, M. and Greenberg, E.P. (2004). Putative exopolysaccharide synthesis genes influence *Pseudomonas aeruginosa* biofilm development. *Journal of Bacteriology*, **186**, 4449–4456.

Maynard Smith, J. and Harper, D. (2003). *Animal Signals*. Oxford: Oxford University Press.

Maynard Smith, J. and Price, G.R. (1973). The logic of animal conflict. *Nature*, **246**, 15–18.

Middleton, B., Rodgers, H.C., Camara, M., Knox, A.J., Williams, P., and Hardman, A. (2002). Direct detection of N-acylhomoserine lactones in cystic fibrosis sputum. *FEMS Microbiology Letters*, **207**, 1–7.

Nealson, K.H., Platt, T., and Hastings, J.W. (1970). Cellular control of synthesis and activity of bacterial luminescent system. *Journal of Bacteriology*, **104**, 313–322.

Nivens, D.E., Ohman, D.E., Williams, J., and Franklin, M.J. (2001). Role of alginate and its O acetylation in formation of *Pseudomonas aeruginosa* microcolonies and biofilms. *Journal of Bacteriology*, **183**, 1047–1057.

Novick, R.P. (2003). Autoinduction and signal transduction in the regulation of staphylococcal virulence. *Molecular Microbiology*, **48**, 1429–1449.

Pesci, E.C., Pearson, J.P., Seed, P.C., and Iglewski, B.H. (1997). Regulation of *las* and *rhl* quorum sensing in *Pseudomonas aeruginosa*. *Journal of Bacteriology*, **179**, 3127–3132.

Pesci, E.C., Milbank, J.B.J., Pearson, J.P. *et al.* (1999). Quinolone signaling in the cell-to-cell communication system of *Pseudomonas aeruginosa*. *Proceedings of the National Academy of Sciences of the USA*, **96**, 11229–11234.

Pritchard, D.I. (2006). Immune modulation by *Pseudomonas aeruginosa* quorum-sensing signal molecules. *International Journal of Medical Microbiology*, **296**, 111–116.

Queller, D.C. (1992). Does population viscosity promote kin selection. *Trends in Ecology and Evolution*, **7**, 322–324.

Redfield, R.J. (2002). Is quorum sensing a side effect of diffusion sensing? *Trends in Microbiology*, **10**, 365–370.

Riley, M.A. and Wertz, J.E. (2002). Bacteriocin diversity: ecological and evolutionary perspectives. *Biochimie*, **84**, 357–364.

Sachs, J.L., Mueller, U.G., Wilcox, T.P., and Bull, J.J. (2004). The evolution of cooperation. *Quarterly Review of Biology*, **79**, 135–160.

Schaber, J.A., Carty, N.L., McDonald, N.A. *et al.* (2004). Analysis of quorum sensing-deficient clinical isolates of *Pseudomonas aeruginosa*. *Journal of Medical Microbiology*, **53**, 841–853.

Schooling, S.R. and Beveridge, T.J. (2006). Membrane vesicles: an overlooked component of the matrices of biofilms. *Journal of Bacteriology*, **188**, 5945–5957.

Schuster, M., Lostroh, C.P., Ogi, T., and Greenberg, E.P. (2003). Identification, timing, and signal specificity of *Pseudomonas aeruginosa* quorum-controlled genes: a transcriptome analysis. *Journal of Bacteriology*, **185**, 2066–2079.

Shapiro, J.A. (1998). Thinking about bacterial populations as multicellular organisms. *Annual Review of Microbiology*, **52**, 81–104.

Shiner, E.K., Rumbaugh, K.P., and Williams, S.C. (2005). Interkingdom signaling: Deciphering the language of acyl homoserine lactones. *FEMS Microbiology Reviews*, **29**, 935–947.

Sio, C.F., Otten, L.G., Cool, R.H. *et al.* (2006). Quorum quenching by an N-acyl-homoserine lactone acylase from *Pseudomonas aeruginosa* PAO1. *Infection and Immunity*, **74**, 1673–1682.

Smith, E.E., Buckley, D.G., Wu, Z.N. *et al.* (2006). Genetic adaptation by *Pseudomonas aeruginosa* to the airways of cystic fibrosis patients. *Proceedings of the National Academy of Sciences of the USA*, **103**, 8487–8492.

Smith, E.F. (1905). Bacteria in relation to plant disease. *Carnegie Institution Report*. Carnegie Institute of Washington, Washington, DC.

Smith, J. (2001). The social evolution of bacterial pathogenesis. *Proceedings of the Royal Society Series B: Biological Sciences*, **268**, 61–69.

Stein, T. (2005). *Bacillus subtilis* antibiotics: structures, syntheses and specific functions. *Molecular Microbiology*, **56**, 845–857.

Surette, M.G., Miller, M.B., and Bassler, B.L. (1999). Quorum sensing in *Escherichia coli*, *Salmonella typhimurium*, and *Vibrio harveyi*: a new family of genes responsible for autoinducer production. *Proceedings of the National Academy of Sciences of the USA*, **96**, 1639–1644.

Taga, M.E., Semmelhack, J.L., and Bassler, B.L. (2001). The LuxS-dependent autoinducer Al-2 controls the expression of an ABC transporter that functions in Al-2 uptake in *Salmonella typhimurium*. *Molecular Microbiology*, **42**, 777–793.

Tateda, K., Ishii, Y., Horikawa, M. *et al.* (2003). The *Pseudomonas aeruginosa* autoinducer N-3-oxododecanoyl homoserine lactone accelerates apoptosis in macrophages and neutrophils. *Infection and Immunity*, **71**, 5785–5793.

Telford, G., Wheeler, D., Williams, P. *et al.* (1998). The *Pseudomonas aeruginosa* quorum-sensing signal molecule N-(3-oxododecanoyl)-1-homoserine lactone has immunomodulatory activity. *Infection and Immunity*, **66**, 36–42.

Velicer, G.J. and Yu, Y.T.N. (2003). Evolution of novel cooperative swarming in the bacterium *Myxococcus xanthus*. *Nature*, **425**, 75–78.

Venturi, V. (2006). Regulation of quorum sensing in *Pseudomonas*. *FEMS Microbiology Reviews*, **30**, 274–291.

Visick, K.L., Foster, J. Doino, J., McFall-Ngai, M., and Ruby, E.G. (2000). *Vibrio fischeri lux* genes play an important role in colonization and development of the host light organ. *Journal of Bacteriology*, **182**, 4578–4586.

Wagner, V.E., Bushnell, D., Passador, L., Brooks, A.I., and Iglewski, B.H. (2003). Microarray analysis of *Pseudomonas aeruginosa* quorum-sensing regulons: effects of growth phase and environment. *Journal of Bacteriology*, **185**, 2080–2095.

West, S.A. and Buckling, A. (2003). Cooperation, virulence and siderophore production in bacterial parasites. *Proceedings of the Royal Society Series B: Biological Sciences*, **270**, 37–44.

West, S.A., Kiers, E.T., Simms, E.L., and Denison, R.F. (2002a). Sanctions and mutualism stability: why do rhizobia fix nitrogen? *Proceedings of the Royal Society Series B: Biological Sciences*, **269**, 685–694.

West, S.A., Pen, I., and Griffin, A.S. (2002b). Conflict and cooperation—cooperation and competition between relatives. *Science*, **296**, 72–75.

West, S.A., Griffin, A.S., Gardner, A., and Diggle, S.P. (2006). Social evolution theory for microorganisms. *Nature Reviews Microbiology*, **4**, 597–607.

West, S.A., Diggle, S.P., Buckling, A., Gardner, A., and Griffin, A.S. (2007a). The social lives of microbes. *Annual Review of Ecology, Evolution and Systematics*, **38**, 53–77.

West, S.A., Griffin, A.S., and Gardner, A. (2007b). Social semantics: altruism, cooperation, mutualism, strong reciprocity and group selection. *Journal of Evolutionary Biology*, **20**, 415–432.

Williams, P., Winzer, K., Chan, W.C., and Camara, M. (2007). Look who's talking: communication and quorum sensing in the bacterial world. *Philosophical Transactions of the Royal Society B: Biological Sciences*, **362**, 1119–1134.

Winans, S.C. (2002). Bacterial Esperanto. *Nature Structural Biology*, **9**, 83–84.

Winson, M.K., Camara, M., Latifi, A. *et al.* (1995). Multiple N-acyl-L-homoserine lactone signal molecules regulate production of virulence determinants and secondary metabolites in *Pseudomonas aeruginosa*. *Proceedings of the National Academy of Sciences of the USA*, **92**, 9427–9431.

Winzer, K., Hardie, K.R., and Williams, P. (2002). Bacterial cell-to-cell communication: Sorry, can't talk now—gone to lunch! *Current Opinion in Microbiology*, **5**, 216–222.

Winzer, K., Hardie, K.R., and Williams, P. (2003). LuxS and autoinducer-2: their contribution to quorum sensing and metabolism in bacteria. *Advances in Applied Microbiology*, **53**, 291–296.

Wisniewski-Dye, F. and Downie, J.A. (2002). Quorum-sensing in *Rhizobium*. *Antonie Van Leeuwenhoek International Journal of General and Molecular Microbiology*, **81**, 397–407.

Xavier, J.B. and Foster, K.R. (2007). Cooperation and conflict in microbial biofilms. *Proceedings of the National Academy of Sciences of the USA*, **104**, 876–881.

Yates, E.A., Philipp, B., Buckley, C. *et al.* (2002). N-acylhomoserine lactones undergo lactonolysis in a pH-, temperature-, and acyl chain length-dependent manner during growth of *Yersinia pseudotuberculosis* and *Pseudomonas aeruginosa*. *Infection and Immunity*, **70**, 5635–5646.

Zhang, L.H., Murphy, P.J., Kerr, A., and Tate, M.E. (1993). Agrobacterium conjugation and gene-regulation by N-acyl-L-homoserine lactones. *Nature*, **362**, 446–448.

CHAPTER 3

Communication in social networks of territorial animals: networking at different levels in birds and other systems

Giuliano Matessi, Ricardo J. Matos, and Torben Dabelsteen

3.1 Introduction

An individual animal will, during its lifetime, interact with a number of different conspecifics. The type and frequency of interactions and the life stage at which they take place will determine the social structure of the population. This structure often assumes the form of a network (see Box 3.1), with multiple individuals connected by multiple associations, and is referred to as a **social network** (Newman 2003). Communication allows individuals to share information and plays a central role in determining their social behaviour. Signalling interactions are embedded in a social network and the combination of signal, habitat, population, and individual properties will determine the number of conspecifics within signalling range of the individual and forming a network of receivers and signallers—the **communication network** (McGregor 2005). Knowledge of the properties of communication networks is important for understanding information flow and its effects on the social behaviour of individuals, because signals are specifically designed to transmit information, and communicating in networks introduces special opportunities and special challenges for information exchange. What is the relationship between social and communication networks? How does their interaction affect social behaviour and the flow of information among their members? These are some of the questions we will address in this chapter. We begin with broad definitions of the concepts we will be discussing, laying the framework on which we will gradually add detail and focus, finally suggesting answers to those questions and making predictions.

We define an animal as social when individuals interact repeatedly and have the opportunity of recognizing each other, at least at the level of neighbour vs. stranger. We exclude only large anonymous aggregations (e.g. flocks of waders, shoals of fish, or herds of ungulates). The social network of an individual is the set of all conspecifics with which the individual interacts, either directly or indirectly (e.g. by witnessing interactions among them). In exceptional cases a network may include heterospecifics, as in cleaner fish and their clients (Bshary and D'Souza 2005). A **signal** is any morphological trait or behaviour which has evolved with the purpose of transmitting information, as opposed to **cues**, which also contain information but are not designed to transmit it. Communication is therefore the transfer of information between conspecifics through the use of signals (see also Chapter 1 and 2). The communication network of an individual is the set of all conspecifics within signalling range at a given time. Communication networks are thus often contained within social networks and both types of networks vary over time. In the next sections we will see how space,

Box 3.1 Definitions

Network: a set of individual animals of the same species with connections between them deriving from interactions of any kind.

Social individuals: animals of the same species which interact repeatedly and therefore can potentially recognize each other.

Information: a property of a source (e.g. behaviour) eliciting a change in the state of a receiver in a biologically functional manner.

Eavesdropping or social eavesdropping: gathering absolute and relative information on other individuals by attending to their signalling interactions.

Audience or bystander: individual that is present during, but does not take part in, signalling interactions between others; the individual may or may not eavesdrop.

Audience effect: changes in the signalling behaviour of individuals during an interaction caused by the presence of an audience.

Apparent vs. evolutionary audiences: audiences that cause an effect only when present and detected vs. audiences that were historically common in the environment of the signaller and have generated selection for signal form and content.

(Jablonka 2002; Dabelsteen 2005; Peake 2005; Matos and Schlupp 2005)

time, behaviour, environment, and physiology add to the definitions of social and communication networks to improve our understanding of information exchange in social animals.

3.2 A model of social and communication networks

3.2.1 Networks in space and time

In its original definition, an individual's communication network includes all conspecifics within signalling range of that individual at any particular time (McGregor and Dabelsteen 1996). This concept can explain signalling and receiving behaviours that are not obvious if viewed under a dyadic perspective on communication. For example, in a number of taxa individuals are able to eavesdrop (Box 3.1) and use the acquired information in future encounters (e.g. McGregor *et al.* 2001; Peake *et al.* 2002; Amy and Leboucher 2007; Matessi *et al.* 2007). Chorus signalling, victory displays, and quiet song are also better understood when considering the presence of multiple receivers and signallers (McGregor 2005). Communication network studies focus on information collected by individuals (nodes) through the exchange of signals in various sensory modalities. Therefore, links between nodes represent signals (see Box 3.2). Social network studies focus on associations and their effects on population structure (e.g. group fission and fusion; Croft *et al.* 2004), with the most basic type of link between nodes being spatial association. Links may be based on other types of interaction between individuals, e.g. cooperation events, aggression episodes, or grooming (e.g. Lusseau 2003; Croft *et al.* 2006). All such links contain information, both for the parties involved and for bystanders, and have different and variable perception ranges. Networks based on different link types include different numbers of individuals. Social and communication networks as we define them are both centred on and relative to the individual, and refer to the evolutionarily relevant links the individual has with conspecifics, i.e. links which have an influence on the individual's **fitness**. The ability to collect and use the information exchanged in such a network will depend on the physical properties of signals or other information sources, the perceptive abilities of the individual, an individual's responsiveness to the information, its position in the environment, and the properties of the biophysical and social environments. A hypothetical crow foraging in a forest can collect information from the N_V individuals it can hold visual contact with (visual range), the N_Q individuals it can hear bustling through the leaves (cue range), the N_S individuals with whom it can exchange calls of different type (signal range), and so forth. If trees are in full leaf (environmental state e_1), all N will be relatively small (e.g. Blumenrath and Dabelsteen 2004); if the wind is blowing (environmental state e_2) N_Q and N_S may be small; if the foraging party is particularly large and noisy (social group state g_1; e.g. Freeberg 2006), the size of N_Q and N_S will depend on individuals' ability to separate, locate, and identify the source of each type of noise (e.g. leaves, cognitive state c_1) or vocalization (cognitive state c_2; e.g. Sturdy *et al.* 2001) and on the acoustic properties, like amplitude and frequency, of the sounds created (cue state q_1) or used (signal state s_1; e.g. Langemann and Klump 2005). The information available to the individual crow x at any particular time t through its communication network will be:

$$CN(x)_t = e_i g_i c_i s_i N_S(t) \tag{3.1}$$

while the information available through its social network will be:

$$SN(x)_t = CN(x)_t + e_i g_i c_i N_V(t) + e_i g_i c_i q_i N_Q(t). \tag{3.2}$$

Of course, each state contributes at differing degrees to all N, but the example is simplified for clarity. The composition of these social and communication networks changes almost instantaneously, over the duration of a single signal in the case of communication networks.

Using the original definition of networks implies looking only at the signal range in space and is thus limiting. How will a communication network be described when the sources of information are, for example, chemical? Some chemical signals and cues can have small active spaces (e.g. mammalian scent marks on fixed substrates or cuticular hydrocarbons in social insects; see Chapters 4, 5, and 6), but contain information which can be collected on different points of the time scale. Chemical signals are often used to provide long-lasting information about individual identity, sex, reproductive state, and territory ownership. In many cases the sender of the signal need not be present for the information to be passed on, to the sender's (and possibly the receiver's) advantage. Therefore an individual's communication network can include individuals which left useful information in the past but are not physically present when the information transfer takes place (e.g. Hurst 2005, Darden *et al.* 2008; Chapter 7).

The relevance of signals will depend on how fast the signals decay. The rate of decay may be due to the type of signal used and depends on how the signal-to-noise ratio changes over time. For example, volatile chemicals may decrease in concentration or be altered in structure by environmental agents relatively fast, whereas scratch markings on hard substrates may be resistant to environmental agents and hence long lasting. Information may also decay because the state of the information source changes, e.g. resources are depleted, hierarchy positions shift, oestrous starts or ends (see Chapters 5 and 6). A signal with a slow decay rate may significantly extend the communication networks of individuals.

Memory of past interactions or information must be taken into account when defining social networks. Many species are able to retain information about different individuals, at least for some time, and apply logical operations like transitive inference to such information (e.g. Peake *et al.* 2001). Knowledge of the position of a competitor in the dominance hierarchy, for example, is based on both current information (e.g. **badge of status**) and information gathered during past interactions involving the individual. Memory becomes crucial in social interactions based on reciprocal **altruism** and is a necessary condition for the evolutionary stability of some forms of cooperation (Aktipis 2006; Alonso-Sanz and Martin 2006). Continuing therefore with the above example, the social network experienced by the crow will include individuals it remembers having seen find food, has vocally competed with, or has witnessed competing at a food source on previous occasions, i.e. information from previous networks. This will be weighted by the ability of individual crows to retain and reuse information (a memory state m_1; e.g. Stevens and Hauser 2004) and by the rate at which information decays in the environment (e.g. a property of cues d_1 and signals d_2) or within the individual (a property of memory, d_3). The overall information available to crow x through the social network will be:

$$OSN(x)_t = SN(x)_t + d_i m_i SN(x)_{t-1}. \tag{3.3}$$

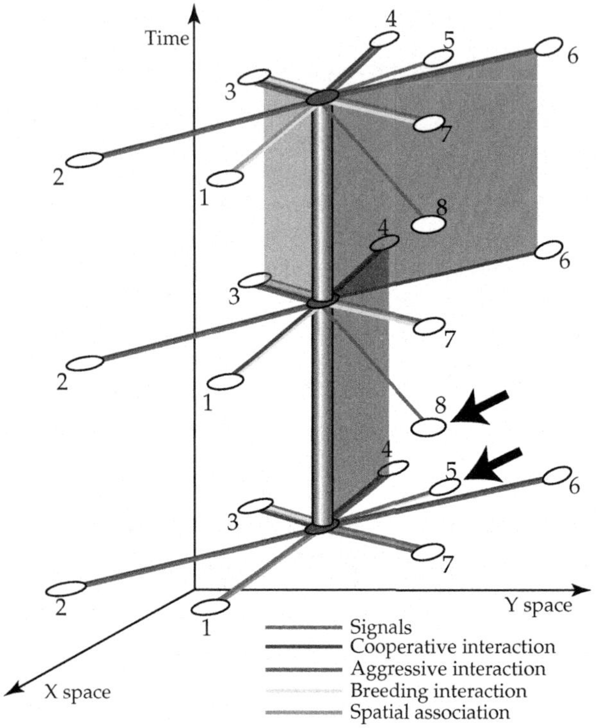

Figure 3.1 Graphic representation of the Rainbow Networks Model. At each moment in time the focal individual (red circle) is connected to a number of other individuals (open numbered circles) through different types of link (coloured lines, see legend in Plate 1). Some link types are more persistent (shaded colour areas) than others. Communication networks, which are based on signals (purple lines), change rapidly because individuals (n. 5 and 8, marked by arrows) join and leave the network according to their signalling and receiving status. Only bi-dimensional space (x and y) and time coordinate axes are represented for graphical simplicity. (see Plate 1)

We have defined communication networks and shown how information obtained through them may form the basis for social networks.

The complexity of network phenomena and their interaction, as described above, combined with the lack of a general consensus on their definitions induced us to propose the Rainbow Networks Model, which describes a general and integrated structure for social and communication networks (Fig. 3.1). In this model a focal individual is like a rod standing parallel to the time axis in multi-dimensional space. At any instant in time, this individual will be connected to others by links of different type (signals, cooperations, aggressions, associations). Each N in equations (3.1)–(3.3) therefore indicates the individuals connected to the focal individual through a particular type of link, which can be represented by a line of a particular colour. Some individuals will be linked by many different colours (i.e. a rainbow), while others will be linked by only a few. The communication network of the focal individual is defined by the line colour representing signal links (e.g. purple in Fig. 3.1). Signal links can be broken down further by signalling modality (e.g. red, blue, and yellow in different proportions). Different link types (i.e. colours) may have different durations or rates of decay, and therefore some colours will persist for different times, for example through memory and persistent signals or cues. Links that persist over time may be represented by vertical planes (Fig. 3.1).

We can express the theoretical lifetime social network (λ) of an individual x as:

$$\lambda_{sn}(x) = \int_0^T OSN(x)_t \, dt \tag{3.4}$$

where T is the lifetime and $OSN(x)_t$ the overall network of the individual at time t. To quantify an individual's social network would involve a sampling procedure where the network is determined at regular intervals (Δt) over the chosen part of the individual's lifetime. The samples thus form a time series of networks representing that individual's social network over the sampling period. We can then estimate the social network of individual x as:

$$L_{sn}(x) = \sum OSN(x)_k \tag{3.5}$$

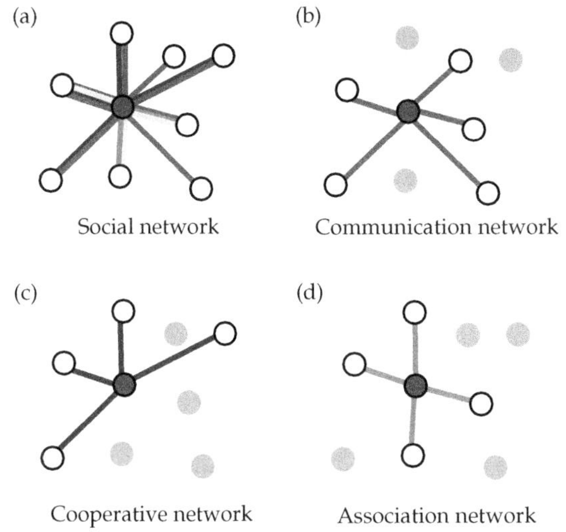

Figure 3.2 The social network (a) of the focal individual (red circle) can be filtered according to the different link types (b. signals; c. cooperation events; d. spatial association) to produce different types of network. Some individuals will be included (open circles) or excluded (shaded circles) from each network according to the type of interaction they have with the focal individual. The communication network (b) of an individual is particularly relevant for information flow since information exchange is the specific function of signals. (see Plate 2)

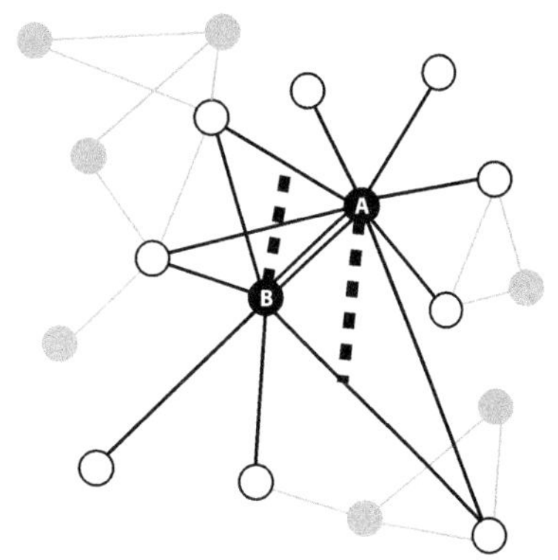

Figure 3.3 The communication networks of different individuals (A and B, solid circles) intersect to produce the community level network. In communication networks information can flow directly (solid lines) or indirectly (e.g. eavesdropping, dashed lines) between individuals.

where k is the number of sampling intervals. To analyse the effects of specific social relationships or types of information on the behaviour and fitness of the focal individual the network can be filtered by a specific link type or strength (Fig. 3.2). The ecology and life history of an individual will determine the extension of its lifetime network in space and time. The social network of a community will be given by the intersecting social networks of each individual (Fig. 3.3). At this point the static and dynamic properties of a social network can be analysed using social network theory (Newman 2003).

3.2.2 Networks and the behavioural ecology of individuals

Behaviour, ecology, and physiology influence the specific structure and extension of an individual animal's network environments, as we illustrated above. The most relevant aspect of an individual's behavioural ecology in determining its social network is its level of sociality. Individuals living in non-territorial communities with no or little social structure may thus have more or less continuously shifting communication networks and broad, unstable, and relatively short-lived social networks. Winter flocks of passerines, foraging parties in social Corvidae, and bachelor groups in both birds and mammals are examples of non-territorial communities of individuals with temporarily overlapping social networks and dominance links (e.g. Schubert *et al.* 2007).

Many birds and other vertebrates live in pair-territorial communities, in which a male and a female defend a portion of the habitat containing one or more resources that allow them to breed. The pair and their dependant offspring are within signalling range of each other most of the time and thus form a cluster. Each individual is also member of a broader social network composed of networks of neighbouring pairs and their offspring, plus floaters and erratic individuals occasionally encountered. Social and genetic **polygamy** will add links and nodes to these networks according to the specific breeding strategies adopted. Pair-territorial communities are typical of many small passerines like great tits (*Parus major*), blue tits (*Cyanistes caeruleus*), whitethroats (*Sylvia communis*), and blackbirds (*Turdus merula*), just to mention

Box 3.2 Networking glossary

Node: the fundamental unit of a network, an individual animal in the present case.

Link: the line connecting two nodes representing a social bond (e.g. members of the same group) or a signalling interaction.

Degree or connectivity: number of links which converge on the node.

Link strength: frequency or duration of an association between two nodes (e.g. frequency of co-membership in a group or number of signalling interactions within a time interval).

Cluster: set of nodes each of which is linked to each of the others. Three nodes are in a cluster if the links between them form a triangle. The level of clustering in a network is essentially the proportion of nodes connected in such triangles.

Structure: groups of nodes that have a high density of links within them, with a lower density of links between groups. It is connected to network clustering.

Roles: sets of nodes which share either some phenotypic trait or which perform similar functions within a community (e.g. male **vs.** female; parent, offspring, and helper; producer **vs.** scrounger).

Keystone individuals or roles: nodes or roles the removal of which profoundly changes the shape and dynamics of the network. They may be nodes showing the highest degree or the only nodes linking two clusters. Removal of the possibility to assume a keystone role produces the same result as removal of keystone individuals.

(Newman 2003; Sih and Watters 2005)

a few for which there is abundant information on breeding and signalling strategies (see also Bennett and Owens 2002; Cockburn 2006). A few examples can also be found among mammals, for example the swift fox (*Vulpes velox*) and other canidae, some ungulates, and some primates.

Some species live in extended family groups in which a breeding pair shares a territory with adult offspring and often other unrelated adults, which cooperate in territorial defence and care of the current dependant offspring (family-territorial communities). Each individual has networks that overlap those of other family members most of the time, and the family forms a cluster which may be stable over more than one reproductive event. Neighbouring family groups and floaters will be part of the network of some family members, representing competitors or potential social and/or genetic breeding partners. Because of the presence of related and unrelated individuals sharing a particular habitat and its resources the dynamics and information flow within the family group are potentially more complex than in pair-territorial communities. Among birds some classic examples are Florida scrub-jays (*Aphelocoma coerulescens*), Arabian babblers (*Turdoides squamiceps*; see Chapter 1), and Australian fairy-wrens. There are many examples among mammals such as wolves (*Canis lupus*) and other carnivores, meerkats (*Suricata suricatta*), rodents (see Chapter 6) and many apes and monkeys. Some species of fish, e.g. cichlids like *Neolamprologus pulcher*, also show a family-territorial social structure.

Seabird species like gulls, terns, auks, penguins, petrels, and albatrosses typically breed in large colonies. Their network structure may resemble that of non-territorial species in the sense of individuals being within signalling range of many others with whom they form only weak social associations. However, because of colony stability and site fidelity each individual may be part of a more or less stable cluster. A pair and their dependant offspring form a primary family social network, and close proximity and colony-level cooperation (e.g. mobbing and adoption) may extend that network. Information available at the colony level may contribute to the lifetime social network of an individual. Besides seabirds, colonies are found among herons, many shorebirds and some sparrows, swallows, and corvids. Lapwings (*Vanellus vanellus*), rock sparrows (*Petronia petronia*), and house sparrows (*Passer domesticus*) are examples of species that share features of pair-territorial and colonial systems. Bats, seals, and rodents are examples of colonial mammals.

Species where males gather in **leks** to attract females represent an extreme in which each male is within signalling range of most other males on the lek most of the time. Males experience tight and relatively stable male social networks that are overlapping to a great extent during lekking. The networks of the signalling males are, however, rapidly changing because of transiting females. Each female may experience a separate set of

Box 3.3 Levels of sociality and networks

We describe and illustrate with network graphs the effects of the level of sociality experienced by an animal on the structure and properties of its social network and on the network at the community level. The illustrated networks represent a snapshot of the community network. Real social networks, and communication networks in particular, show a high link density and complexity (many nodes in receiving range of each other), so we illustrate only a subset of them (e.g. the strongest or most relevant) for clarity.

- **Non-territorial community (Fig. 3.4a):** Circles represent individual nodes with different roles (e.g. solid = producers, open = scroungers). Lines of differing thickness and style represent links of different strength (e.g. increasing frequency of interaction). Some nodes may have a higher degree than others (arrow; e.g. dominant individuals). Movement and interactions of individual nodes are not spatially constrained; therefore any structure in the network does not reflect geography. Clustering is usually unclear or weak. The graph on the right shows a predicted unimodal distribution of link strengths.

(a)

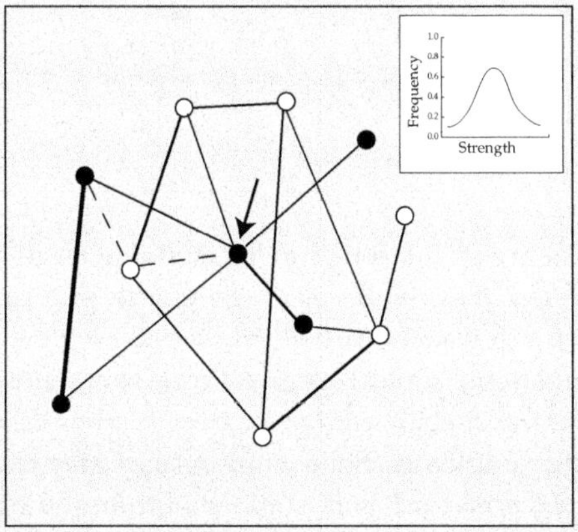

- **Pair-territorial community (Fig. 3.4b):** Nodes of different shapes and shading represent different roles (e.g. solid circles = males, open circles = females, open squares = offspring). Lines as above. Nodes are clustered in a structure which reflects territory geography. The graph shows a predicted bimodal distribution of link strengths, with the higher peak corresponding to links within territories. Extra-pair relationships and intra-sexual competition link nodes across territory boundaries (shaded dotted lines) and may affect link strength within the territory (e.g. differential parental investment).

(b)

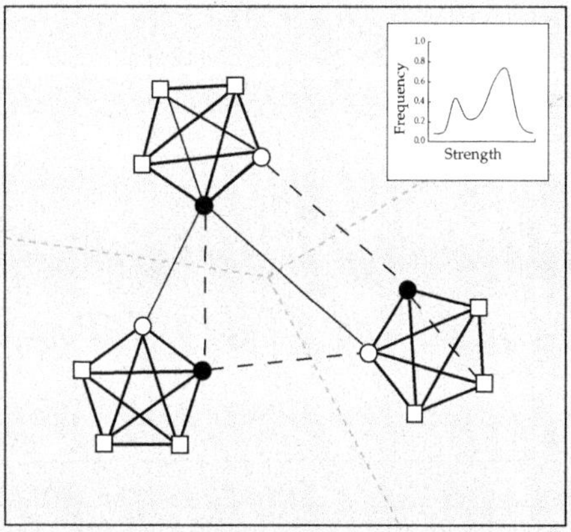

- **Family-territorial community (Fig. 3.4c):** Nodes and lines as above (shaded circles = helpers). Nodes are clustered in a structure which reflects territory geography, but some roles (e.g. helpers) may increase connectivity across boundaries (shaded dotted lines). Links across boundaries may also reflect past connections in persistent networks (e.g. helpers inheriting a neighbour's mate). The graph shows a distribution of link strengths derived from the previous example

continues

Box 3.3 *continued*

but skewed to the high end, as some roles fill out the gap between the two peaks (e.g. helpers may show intermediate link strengths).

(c)

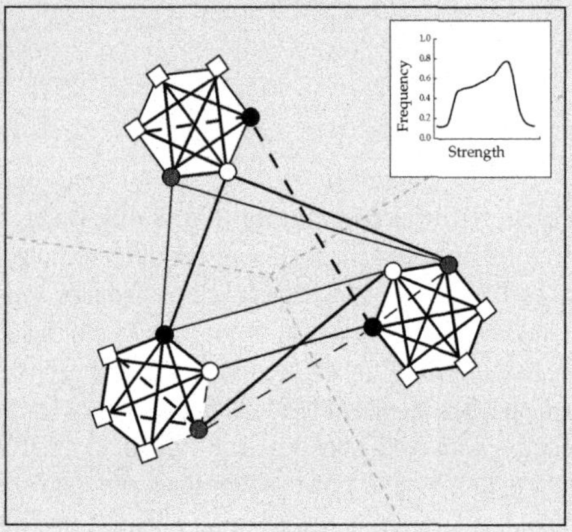

(d)

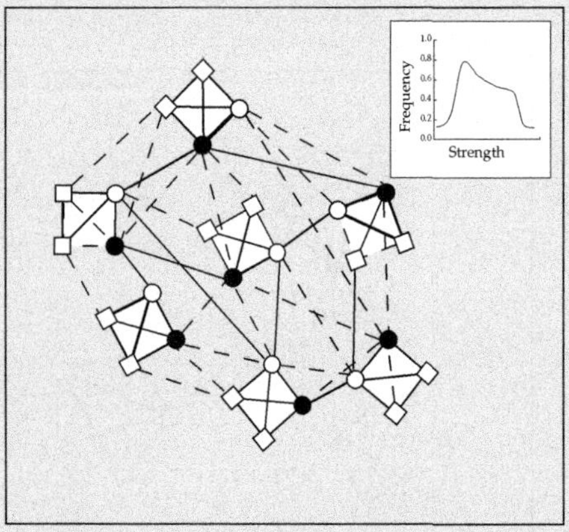

- **Colonial community (Fig. 3.4d):** Nodes show clustering around breeding pairs. Each node is within receiving distance of many other nodes, but only a small subset of links and nodes is illustrated. Roles are also more uniformly linked due to node spatial density and colony-level phenomena (e.g. cooperative mobbing, resource stealing, or offspring adoption). The graph shows a more uniform distribution of link strengths for similar reasons.
- **Lekking community (Fig. 3.4e)**: Nodes show a high level of clustering within roles (e.g. solid circles = displaying males). As in colonial communities, only a subset of links is illustrated for clarity. Links may not be uniformly distributed across roles, some nodes linking preferentially to nodes with roles differing from their own (e.g. open circles and open squares = females alone care for offspring, few female–female links). Some nodes may have a very high degree (arrow; e.g. hot-shot central males). The graph shows a strongly skewed distribution of link strengths, with the strongest and most frequent links among nodes displaying in the lek.

(e)

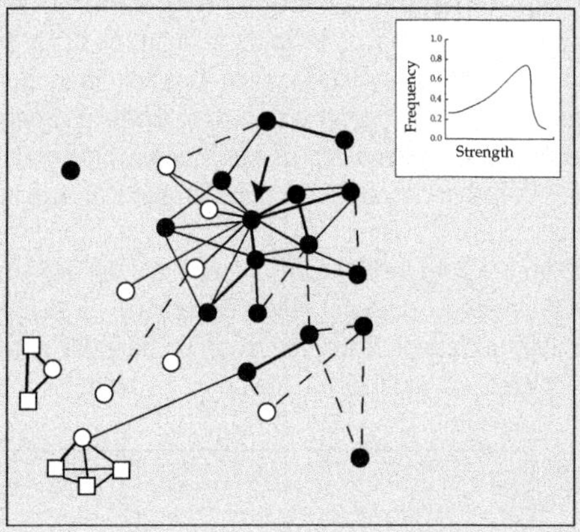

networks, and be only temporarily part of a specific male's network. Nonetheless, females may represent the nodes holding together the population-level network (Höglund and Alatalo 1995). The stability of the lek network will depend on the site fidelity of individuals and on the turnover rate of display site owners. Lekking mating systems are present in many different taxonomic groups from arthropods to vertebrates. Among birds, leks are well studied in Tetraonidae (sage grouse, *Centrocercus urophasianus*, and black grouse, *Tetrao terix*), in waders (ruffs, *Philomachus pugnax*), and in numerous tropical species. Prominent mammalian lekking species are fallow deer (*Cervus dama*) and other ungulates. Some anurans (e.g. grey treefrogs, *Hyla versicolor*) and some mouth-brooding cichlids have been described as lekking (Höglund and Alatalo 1995).

A single individual may go through different levels of sociality through its lifetime, and therefore experience different social network structures (for a bird: non-territorial or helper as juvenile, pair-territorial at first reproduction, and family-territorial later; for a mammal: from bachelor floater to territory owner, then dominant in a lek and back to territory owner, e.g. topi, *Damaliscus lunatus*; Bro-Jørgensen and Durant 2003). Seasonality, movement patterns, density, and habitat (e.g. winter flocks vs. spring territories; uniform and stable vs. variable habitats; specialists vs. generalists etc.) will also affect the size and stability of both communication and social networks experienced by an individual through the number and frequency of encounters and the value of the information concerning them and its decay (states e_i, g_i, d_i in Eqn 3.2). The sensitivity of sensory modalities and information storage and retrieving ability are traits which can vary among individuals and even within an individual according to its state (age, nutrition, circulating **hormones**, parasite level, etc.). These will extend or reduce the range of information gathering strategies available to different individuals in different contexts.

3.3 Information flow and information-gathering strategies in networks

Living in a network environment is essentially an exercise in balancing advertisement, privatization, and the gathering of information. The patterns of information flow in networks defined by the different levels of sociality will be conditioned by the degree of overlap of fitness interests among the individual nodes in the network, which is determined only in part by the genetic relatedness among nodes. In other words, information flow will be different at different points of a cooperation–competition gradient, and the most profitable information-exchange strategies will change accordingly (Searcy and Nowicki 2005; see also Chapter 2). In all types of communities (see Section 3.2.2) there will be some basic costs and constraints to information flow dictated by the trade-off between exchanging information and all other activities required for self-maintenance and reproduction (foraging, avoiding predators, thermoregulation, etc.), while other costs and constraints will be specific to the type of community. We describe for each community how its network structure affects patterns of information flow, privatizing and advertising strategies, and the costs and constraints of information exchange.

3.3.1 Non-territorial communities (Box 3.3, Fig. 3.4a)

We expect information to flow more or less uniformly and unhindered among all nodes in a non-territorial community, though flow between distant nodes may be constrained by abiotic factors. Link strength will therefore in general decrease with increasing node distance and the distribution of strength values will be unimodal, with mean strength possibly being low and determined by the size and density of the network. Horizontal transmission of information (within generations or age classes) will be approximately as frequent as vertical transmission (between generations or age classes). We expect the stability of social networks to depend on community stability, but to be generally low. Networks may show some level of clustering by node role or following geographical constraints. Most individuals will play similar roles and may have equal interests in information exchange with all other individuals, though of course sexual and parental roles will still have an influence during the breeding season. Foraging or roosting groups may show differences in information flow in the presence of producer/scrounger roles (e.g. Coolen *et al.* 2001), with a high rate of information exchange between the two roles, but little within them. A possible exception is the presence of cooperation, e.g. producers sharing resources and information about resources, or scroungers joining in exploiting or stealing from producers. In the presence of strong dominance hierarchies information flow may be directional, e.g. some information may go from high to low ranks and other vice versa. Most networks in this type of community will be made of non-related individuals, for whom fitness interests will not overlap, therefore most information exchanges will be competition based and with strict honesty requirements. The degree of divergence of interests may depend on the level of cooperation among individuals combined with community membership stability, which affects the frequency of repeated interactions.

Opportunities for privatizing information will be uniformly distributed among the nodes in social networks within non-territorial communities, and they will possibly be limited, distance dependent, and environmentally determined, even though the selective pressures for privatization may be strong due to the competitive nature of the networks. Likewise, opportunities for eavesdropping (Box 3.1) should be frequent, and the benefits relatively high. Audience effects (Box 3.1) may be evolutionarily encoded in signal properties or as specific fixed strategies of signallers, but not be detectable within particular interactions occurring in the presence or absence of bystanders (Dabelsteen 2005; Matos and Schlupp 2005). The majority of signals may be public, i.e. designed for advertising and broadcasting. We would expect to find a larger proportion of generalized signals (intended for different types of receivers) than of specialized signals (intended for specific types of receivers), though we might find special strategies for addressing a specific individual in the network (e.g. threats using signal matching and directional signals).

In non-territorial communities the lack of spatial structure in the distribution of individuals may give a high risk of interference during interactions and of redirected aggression for bystanders. In dense communities these costs will be high, and so will the constraint of identifying and discriminating information sources. There will also be costs due to competition for receiver attention and masking by noise or other information sources. Dense communities may also be more detectable by predators, increasing the trade-offs between information exchange and anti-predator behaviour. Individuals in loose communities will incur information search costs but experience a lower risk of interference and redirected aggression. Habitat structure and community density will interact, so that individuals in dense communities living in closed habitats may have high costs of information source discrimination but low risk of predation, and vice versa for dense communities in open habitats.

3.3.2 Pair-territorial communities (Box 3.3, Fig. 3.4b)

We expect most of the information in pair-territorial communities to flow between pair members and between parents and offspring (i.e. nodes within a territory) and less between nodes across territory boundaries in the breeding season as a whole. At the start of the breeding season during the establishment of territories and before mating, information flow may be quite intense across territory boundaries. After mating most links between nodes within territories will be stronger than links across territories, and the distribution of link strengths will be bimodal. The social networks will be structured, with clustering corresponding to the geographical distribution of territories. The type and complexity of the information transmitted, together with the signals used to transmit it, could differ within and across territory boundaries. For example, different signalling modalities will be used due to varying distances among nodes in the network (larger across territories than within territories). Therefore, signals will mainly be acoustic across territories. However, signals used within territories and for direct boundary disputes may be multimodal (see Chapter 6). Some roles within the network will be relatively stable (e.g. male vs. female, parent vs. offspring, neighbour vs. stranger), while others may be more unstable (floater vs. territorial, mated vs. unmated). Information may flow with different intensities among or between roles and may often be directional. Fitness interests will overlap to some degree among nodes within a territory, but not at all across boundaries. Therefore we expect information exchange to be at least partly cooperative among nodes within territories and mostly competitive across territories.

In networks of pair-territorial individuals, privatizing information from others outside the territory should be easier than in other communities, especially in closed habitats. Territory boundaries will by definition space individuals out, therefore the breeding pair and their offspring can simply adopt signals which have a smaller range than the territory size. Signallers may also choose more anonymous signals, thus reducing the utility of information for others (Dabelsteen 2005). Mating status, mate or territory quality, and breeding stage are some of the types of information which individuals may want to keep private from rivals or potential extra-pair partners outside the territory (Vignal *et al.* 2004). On the other hand, the potential usefulness of such information will add pressure to eavesdrop, and individuals will adopt behaviours that facilitate eavesdropping, such as patrolling borders furtively, choosing optimal positions for receiving from both parts, or intruding into territories without signalling (e.g. Mennill *et al.* 2004; Balsby and Dabelsteen 2005). The frequency of eavesdropping will depend on environmental and signal states (e_i and s_i in Eqns 3.1–3.3). Who eavesdrops will depend among other things on the breeding strategies available to the sexes (e.g. Matessi *et al.* 2005, 2007). The partly cooperative nature of information exchange within the territory may reduce the scope for privatizing of information among nodes within territories and close physical proximity may also make it more difficult to eavesdrop. An exception may be when both sexes have an interest in concealing attempts at attracting an extra partner. Pied flycatcher (*Ficedula hypoleuca*) males, for instance, sing to attract a second female at a nest box that is outside the hearing range of the first female, probably to avoid interference from her (Lampe *et al.* 2007). Another exception may be on matters of parental investment. If an individual's level of parental investment is negatively correlated to that of its partner, individuals are essentially scrounging on the partner's effort (Hinde and Kilner 2007), and the pressure for privatization and eavesdropping may be significant. Apparent audiences should in theory be more relevant than evolutionary audiences, because territoriality spaces individuals out and makes absence of competitors the standard social context. Therefore in the presence of detectable bystanders that are strangers to the breeding pair and their offspring we expect signallers to behave differently from in their absence, i.e. audience effects should be measurable by variations in individual behaviour. However, evolutionary audiences may have significantly shaped courtship signalling in many pair-territorial species. In whitethroats, neighbouring males will interfere with courtship if they detect it (Balsby and Dabelsteen 2005). The same has been observed in blackbirds and rock sparrows, and probably therefore courtship signalling is often quiet and inconspicuous, as expected for privatized signalling (Balsby and Dabelsteen 2003; Dabelsteen 2005; Matessi *et al.* 2007). Advertising and broadcasting may be more typical of information exchanges across territory boundaries. A possible exception is if an individual's parental investment is positively correlated to that of its partner, in which case it would pay the partner to advertise its contribution to parental care. The higher stability of roles and the pressure of eavesdropping will promote the use of signals specialized for addressing particular types of receivers (mate, offspring, neighbour) and generalized signals will be mainly used in advertising across territory boundaries.

Individuals in pair-territorial communities face costs and constraints due to the need to defend a territory and guard the mate, and may still be unable to prevent neighbour interference and extra-pair copulations. An individual may need to leave its territory to collect information about neighbours, facing the risks of entering unknown habitat (e.g. unknown location of shelter), intrusions in its own territory, extra-pair mating by its partner, and brood predation. Information value will have to be weighed against these costs. The spatial separation of some individuals implies costs for transmitting information. Signals have to be strong to reach intended receivers in other territories, and therefore may require more energy to produce, and may be exploited by bystanders and predators. Individuals may be forced to choose receiving and transmitting positions which are exposed or otherwise suboptimal. Detecting the direction and distance of other members of the network may require special behaviours, e.g. perching (Holland *et al.* 1998, 2001), which are in **conflict** with other activities. The transmission properties of the environment will set the basic constraint level, and will produce differences among species and populations.

3.3.3 Family-territorial communities (Box 3.3, Fig. 3.4c)

Information flow among nodes in networks within family-territorial communities will resemble that in pair-territorial networks, with the added complexity of some nodes (i.e. adult helpers) which may have strong links to a number of nodes outside the territory boundary. Information flow should therefore show a shallower decreasing gradient from the territory centre and outwards. Helpers may be seen as filling the network space between within-territory and across-territory information exchanges through links joining them to the breeding pair, but also to adults in other territories as an expression of the conflict between staying and helping vs. dispersing and breeding. These effects should be even more evident when considering long-term social networks, as breeders may retain and use information collected at the time in which they were helpers. Helpers who take over the territory and mate from the breeder they have helped may have access to information about mate and territory quality unavailable to strangers (e.g. Cockburn *et al.* 2003). The distribution of link strengths will again be unimodal, but with a peak skewed towards the high range. The networks will again be structured, but the clustering of nodes may be weaker due to the effects of helpers. We expect intense information transfer within the territory, but also relatively intense transfer across territories, and information should travel horizontally, vertically, and diagonally (if helpers are unrelated to the young). Some signals may be specialized for interacting with helpers (e.g. Sharp and Hatchwell 2006). Some node roles will possibly be more stable than others, again through the action of helpers. We may also expect the network structure to show keystone individuals (Lusseau 2003; Sih and Watters 2005; Box 3.2). We expect to see a gradient in overlap of fitness interests, given the more continuous variation in relatedness provided by helpers, and therefore there should be a clear gradient in the frequency of cooperative vs. competitive interactions.

The main challenge to privatization of information in networks within family-territorial communities is the presence of nodes within the territory boundary (e.g. non-reproducing unrelated helpers) the fitness interests of which diverge clearly from those of other family members. The spatial structure of networks will still allow isolation of the groups of nodes within a territory from nodes outside. From the point of view of information exchange, nodes with helper roles are somewhat intermediate between the position of neighbours and floaters on one side, and the breeding pair with their offspring on the other. Helper nodes will be interested in privatizing some information, e.g. attempts to mate with one of the breeders (like neighbours) while advertising others, e.g. their contribution to offspring care (like breeders; e.g. Doutrelant and Covas 2007). They will at the same time be under strong **selection** to collect information at the expense of the breeders, e.g. to eavesdrop on chances of breeding or inheriting the territory (like neighbours and floaters), but also to collect information useful for territorial defence, e.g. to eavesdrop on risks of intrusions (like breeders). Breeders will mostly be under pressure to privatize information concerning themselves and their partner from helpers (e.g. fertility; Crockford *et al.* 2007), while collecting information concerning them (e.g. contribution to offspring care). These complex information exchange dynamics should be revealed by measurable differences in behaviour and signalling of breeders in the presence and absence of helpers during interactions (audience effects). The degree of **reproductive skew** (Emlen 1995) in the breeding system will enhance or mitigate conflicts about information exchange. Signal and environmental states, s_i and e_i in Eqns (3.1)–(3.3), will play a significant role in these interactions, but the environmental properties within the territory will be crucial in allowing privacy. Cognitive states, c_i, like divided attention and sensory ability, may additionally be important in determining the opportunities for advertising, privatizing, and collecting information (e.g. Voigt *et al.* 2007).

In family-territorial communities individuals have to take into account the presence of other adults that regularly use the territory. Therefore, if the costs of brood predation and intrusion during information collecting may be lower in these communities due to the presence of other defenders of the territory, the risks of extra-pair mating or lost mating opportunities are higher. If there is negotiation through signalling between brood and caretaker on how much to provide, then there are costs for helpers that risk being expelled or exploited by eavesdropping parents, and costs for parents that risk being exploited by eavesdropping helpers (e.g. Mulder and Langmore 1993). Helpers may face the strongest constraints and trade-offs of information collection if they need information on their neighbours to improve their chances of future breeding.

3.3.4 Colonial communities (Box 3.3, Fig. 3.4d)

The networks shown by colonial communities should be the result of a combination of non-territorial and pair-territorial networks. Thus information flow should show properties of both community types. We expect high information exchange between mates and between pairs and their offspring, but colony-level phenomena such as conflicts over space, adoption, communal defence, and resource information centres will promote interactions among all nodes. The networks show a clear clustering based on family groups. Link strengths will decrease with distance between nodes and the distribution of link strength will be bimodal with the smallest peak in the high range representing family interactions and the highest peak in the low range representing colony-level phenomena. Information, e.g. about predators, may travel very fast in these networks due to high node density. Information will probably travel as much horizontally as vertically and diagonally. Given the potentially high levels of noise there will be selection for signalling strategies and signal properties which favour targeting of specific nodes or roles (e.g. Bonadonna and Nevitt 2004). We expect to observe a wide range of generalized signals, shared among many nodes. The division of roles will be similar to pair-territorial systems, but may additionally include producer/scrounger-type roles. The level of cooperation in information exchange will be dependent only partly on the average relatedness within a colony, since there will be ample scope for the action of reciprocal altruism and mutualism.

The most relevant challenge to privatizing and advertising information in networks of colonial communities will be posed by the high density of nodes and the noisiness of the social and physical environments (e.g. Poesel *et al.* 2007). High density may make privatizing difficult but favour anonymity beyond the nearest neighbours. Eavesdropping may still be a useful way to collect information, but its adoption may be difficult to detect since no specific behaviours to facilitate it may be necessary. Likewise, audience effects may be difficult to prove because bystanders in colonies are essentially always present and are therefore likely to represent evolutionary audiences (Matos and Schlupp 2005). High node density will promote network-specific advertising strategies. Therefore victory displays and winner/loser effects, among others, could be relatively common (Dugatkin and Druen 2004; Bower 2005; Kazem and Aureli 2005; Oliveira 2005). Advertising, privatising, and information collection will be mostly under the influence of social, cognitive, signal, and cue states (g_i, c_i, s_i, and q_i, Eqns 3.1 and 3.2). Memory and information decay (m_i and d_i, Eqn 3.3) also play a role, because individuals have to deal with large quantities of rapidly changing information originating from many unknown individuals.

Individuals in colonial communities have two main sources of costs: the close proximity of many other individuals and the large amount of information available simultaneously. Close proximity among individuals implies a high risk of redirected aggression while collecting information and of exploitation or scrounging while distracted by exchange of information. The large amount of information available simultaneously may impose physiological costs for filtering and selecting information sources, and may require specialized behaviours and mechanisms for directing signals and detecting the direction and position of an information source. The trade-offs with predator detection and resource gathering may, on the other hand, be relaxed due to colony-level defence and information centres.

3.3.5 Lekking communities (Box 3.3, Fig. 3.4e)

Networks within lekking communities will probably diverge the most from those described above. Information flow will be high among the displaying individuals within the lek itself (mostly males), and also among the latter and the choosing individuals outside the lek (mostly females). Some nodes may preferentially link to others with roles differing from their own (e.g. females interacting mainly with males on the lek or with offspring outside the lek) while others may link mainly to nodes with similar roles (e.g. males competing with other males). In the presence of mate choice copying, females will also use information from, and therefore link to, other females (e.g. Höglund *et al.* 1995). The distribution of link strengths will be skewed to the high range, with the strongest links being among neighbouring nodes within leks. The networks will be structured with a clustering of displaying males and more scattered and less connected females. These networks have the highest chance of showing keystone individuals or roles, probably those individuals occupying a dominant or central position in the lek among males, and possibly dominant or older females if mate choice copying is relevant (Coleman *et al.* 2004; Hill and Ryan 2006). Information, for instance about the arrival of females, will travel fast and mostly horizontally. Communication networks in a lek will be dynamic because male nodes may shift during the lekking season. Fidelity of nodes to a lek may make some information available in networks across seasons. As the main constraints will be noise and getting attention, we expect to find high signal complexity but possibly low information complexity. Signals will possibly be tailored to address the two main roles within these networks, i.e. competitors and mating partners. There is no overlap of fitness interests within leks, therefore almost all interactions will be competitive, including those between males and females.

Leks are all about advertising. Territories are defended with the sole purpose of displaying. Privatizing interactions is impossible. Even the interactions most commonly privatized in other contexts, courtship and mating, will be public. This implies that eavesdropping will be widespread and bystanders always present. In fact, leks may result from selection pressure on females to reduce information gathering costs through the combined action of eavesdropping on male interactions and mate choice copying in networks. In these communities one may expect winner and loser effects to be most widespread. In leks, selection may favour low-ranking individuals who adopt 'good loser' strategies, i.e. signalling strategies to minimize the costs of being perceived as a loser by audiences (Peake and McGregor 2004). These may include the use of overt aggression during signalling interactions (e.g. Matos 2002; Ophir and Galef 2004) or redirected aggression (e.g. Kazem and Aureli 2005). Collecting and transmitting information in leks will obviously be under the influence of signal and cue states (states s_i and q_i, Eqns 3.1 and 3.2) but the social and cognitive states will be most relevant (g_i and c_i). Decay states (d_i, Eqn 3.3) will affect information about fighting ability (e.g. through exhaustion), fertility and sperm competition (e.g. through sperm depletion), or dominance rank (e.g. when signalled by plumage).

In lekking communities the costs of information exchange lie mostly in the competitive nature of the social environment. Risks linked to redirected aggression, harassment, and social punishment are probably stronger in these communities than in others. The costs linked to predation risk and lost resource gathering may have been significant on the evolutionary time scale by determining general lek dynamics, but are probably less relevant to the individual, since by being part of a lek it has already prioritized information exchange over other needs. In a lek there will also be constraints due to the amount of information exchanged simultaneously, requiring selection and filtering.

3.4 Case studies of communication networks in different communities

We present the current available knowledge on communication and information flow in a selection of bird species representative of different social structures and illustrate how the concepts presented above can provide useful insights to their behaviour. We chose the great tit, a species which has been studied from the perspective of communication networks, as representative of a pair-territorial community structure. We chose Australian malurid wrens to represent family-territorial communities, and illustrate cases for which there is information on communication behaviour but which have not been viewed from a communication networks perspective.

3.4.1 A pair-territorial species: the great tit

Great tit males defend all-purpose territories during the breeding season and join flocks of different composition and size during the remaining part of the year. However, they may defend the territory throughout the year when food conditions allow it. A pair raises one or two clutches per year on the territory. At fledging the family breaks up and juveniles create or join summer foraging flocks of continuously changing composition due to emigration and mortality. In autumn and winter adults may join these flocks, which are now more permanent in composition and may fuse and form larger aggregations around rich food sources. Flock members frequently communicate with each other using different agonistic, contact, and alarm calls and, at close range, also agonistic displays (Hinde 1952). Dominant males have wider black belly stripes, a badge of status, and males are dominant over females (Järvi and Bakken 1984; Järvi *et al.* 1987). The process of establishment of territories at the end of winter usually takes 1–2 months. During this period forthcoming male territory owners may engage in intense territorial disputes, sometimes throughout the day (Hinde 1952). These may involve physical fights, visual displays, calling, and intense singing interactions during which males attempt to overlap each other's songs. Song overlapping is one way of expressing dominance over the opponent (Dabelsteen *et al.* 1996; Langemann *et al.* 2000). Once territories are established, territorial activity decreases and most of the singing takes place at dawn and early morning. Matching of song lengths and song types during singing interactions is probably common at this time where neighbouring territory owners share some of their song types (McGregor and Krebs 1982). The timing of pair formation may vary with respect to territory establishment, but males and females associate more closely when prospecting and choosing nest holes. During the fertile period females will gradually spend more time inside the nest hole before starting full incubation and are called out by their males to be fed and/or to copulate (Lind *et al.* 1996). Males vocalize from a position close to the nest entrance using quiet songs or calls. Females recognize their mates even if they use songs shared with neighbours and answer with quiet calls before leaving the nest (Blumenrath *et al.* 2007). Females may leave the nest independently to forage or possibly engage in extra-pair copulations (EPC). Female mate choice of both social and extra-pair partner is probably based on both male and territory quality, although it is unclear how the different measures are weighted (Otter *et al.* 2001).

The scenario emerging from the biology of great tits indicates that information is central to decision-making, both during and outside the breeding season. The ownership of a territory gives males a base for advertising themselves to surrounding males and females, and females a reliable and safe receiving station for collecting information crucial to mate choice. Song transmission experiments before and after foliation using different song types and sender and receiver positions indicate that songs may range over about three consecutive territories after foliation and during incubation, but often about six territories before foliation and during territory establishment and pair formation (Blumenrath and Dabelsteen 2004). Communication networks therefore include more nodes when individuals need most information. Being inside nest boxes complicates the receiving conditions of females (Blumenrath *et al.* 2004). Both experiments showed that pure notes transmit better than buzzes and therefore would be the optimal choice for public signalling. Females need to compare territory quality, resource holding potential (RHP), and willingness to contribute to parental care of potential social partners. They also need information about the genetic quality (relative to other males and compatible with themselves) of males they will choose to fertilize their eggs. Relative RHP is likely to be indicated by dominance relationships between males. Mated females may gather this information by eavesdropping on male singing interactions (Otter *et al.* 1999), although a single eavesdropping event seems insufficient for females to make decisions about EPCs (Otter *et al.* 2001). Males need information about the position and quality (relative to their own) of trespassing strangers and neighbours to assess potential risks to their territory and mate. They need to address these specific males at a distance if necessary. Song degradation measures seem to provide cues that allow males to range singing rivals (McGregor and Krebs 1984). Like females, males are able to gather relative information about male dominance relationships by eavesdropping on singing interactions between others and use it in later interactions (Peake *et al.* 2001, 2005). They seem to be able to combine information gathered indirectly by eavesdropping with information from direct experience with one of the contestants in the eavesdropped interaction (Peake *et al.* 2002). Males are also able to address a particular male at a distance by matching its song types and their durations (McGregor *et al.* 1992). Males need information about the breeding status of extra-territorial females to decide on EPC opportunities. One option would be to eavesdrop on courtship interactions between neighbouring pairs. However, this may be difficult at a distance because the events that lead to courtship, e.g. calling out, and courtship itself are usually very quiet (Gompertz 1961; Lind *et al.* 1996). Probably, this privatizing has evolved as a strategy to counter interference from eavesdropping neighbouring males (Dabelsteen *et al.* 1998; Balsby and Dabelsteen 2005; Dabelsteen 2005). Both sexes need information later on about the needs of the offspring and the contribution of the mate to meeting these needs. The latter information may be obtained by watching closely the mate's feeding activity, whereas the former can be obtained from the begging signals of the young, and sometimes by an interaction of the two (Hinde and Kilner 2007).

A social network representation of these relationships would unveil two different network structures: one in which clustering is weak or absent and one which shows a clear clustering of nodes, corresponding to the two types of communities experienced by great tits, non-territorial winter flocks, and pair-territorial communities. There may be a transition phase with some clustering if pairs form before territories are established. During and after territory establishment a territory holder links to all other territory holders within signalling range, although most strongly to neighbours, as well as to females who have not yet chosen a social partner and trespassing males in search for a vacant territory. After pairing a clear clustering appears, with each cluster consisting of a pair, later expanded to a family with dependant young. Gradually the links between the members of each cluster grow stronger than those to nodes outside the territory. The male may still have strong links to rivals during the fertile period of his mate and to extra-pair females, while the female may link to extra-pair males in her fertile period. When the young become independent and join summer flocks the links to the parents may disappear or be weakened, but may be strengthened again if young and parents rejoin in autumn or winter flocks.

Although great tits have already been studied from the communication networks perspective there are still many unanswered questions. For example, do females gather information about potential social partners by eavesdropping on posturing and calling interactions in winter flocks or on the intense singing interactions during territory establishment? How do links and clustering develop during this very dynamic period? How important is eavesdropping for extra-pair mate choice? Is memory of information gathered by eavesdropping constrained by time passed and number of eavesdropping events? Do great tits use strategies to counter or facilitate eavesdropping?

3.4.2 A family-territorial group of species: Australian fairy-wrens

Australian fairy-wrens (family Maluridae, genus *Malurus*) are a group of cooperatively breeding passerines with stable year-round territories defended by a resident group of two to eight adults. Breeding pairs may occupy the same territory for several years. Most species are sexually dimorphic, with males brightly coloured in blue, purple, and black and drab camouflaged females. Visual communication accounts for a significant portion of the signalling repertoire, and both males and females sing. In splendid (*Malurus splendens*) and superb (*Malurus cyaneus*) fairy-wrens social groups are composed of a dominant pair and a variable number of related and unrelated helpers, mostly males. All adults contribute to territory defence and feeding the young. Genetically, these birds present among the most extreme cases of extra-pair paternity recorded, where more than 70% of offspring are not fathered by the dominant male. Many young are fathered by extra-group males and some by helpers, so the dominant male frequently cares for unrelated young. The mating system seems a product of female choice rather than male–male competition. The within-group paternity achieved by a dominant male is negatively related to group size and extra-group paternity is positively related to male age (Rowley and Russell 1997; Dunn and Cockburn 1999; Webster *et al.* 2004).

Visual courtship signals are very conspicuous. The 'petal carrying' display, in which a male openly carries flower petals strongly contrasting with its plumage colours as nuptial gifts, is often associated with a conspicuous and inefficient form of undulating 'sea horse' flight. This is especially true for displays by extra-group males, who apparently do not attempt to conceal their forays. Males travel across as many as six territories during their forays (Rowley and Russell 1990). On the other hand, there seems to be no temporal association between courtship displays and copulations, which are secretive. Females encourage displays and solicit copulations from both helpers and extra-group males and seem to prefer distant males. The helper's contribution to brood care provides females with an alternative to paternal care, thus allowing more freedom in mate choice without the costs of retaliation by the dominant male. The dominant male collects information about helper contributions and punishes defecting helpers. Female choice is affected by a combination of genetic similarity and quality of the males. The larger the territory of a group the more helpers and neighbours are available for the female to choose from and the more opportunities the dominant male has to perform extra-group courtship. Dispersing females often use foreign territories as staging posts before settling and prefer territories with many supernumerary individuals (Mulder and Langmore 1993; Mulder *et al.* 1994; Mulder and Magrath 1994; Brooker and Rowley 1995; Green *et al.* 1995; Dunn and Cockburn 1996; Cockburn *et al.* 2003; Webster *et al.* 2004; Tarvin *et al.* 2005). Territories are defended vocally, and females are particularly active singers. They can discriminate classes of individuals by their song (dominants, helpers, neighbours, and strangers), but discrimination is negatively related to group and territory size and general vocal activity negatively related to length of territory tenancy (Payne *et al.* 1988, 1991; Cooney and Cockburn 1995; Rowley and Russell 1997).

Information is as central to fairy-wren reproductive behaviour as it was for great tits. Male and female behaviour is conditioned by information-gathering needs and territory ownership seems even more crucial, as suggested by a number of authors. Female mate choice in superb fairy-wrens '...will be limited by *imperfect information* about male quality and differing proximity to higher quality males' (Mulder *et al.* 1994, p. 227, our italics). Females need information on extra-group and helper male quality, on helper contribution to care, on productivity of available territories, and on neighbouring females' territorial activity. Dominant males need information on breeding status of extra-group females, especially distant ones, but also on the contribution of helpers to care, and need to spread information about themselves. Helpers need information on within-group and extra-group female status and on the presence and parental contribution of the dominant male. A network representation of these relationships would show dominant males with links to many females, and possibly stronger links to extra-group females than to their own. Females would in turn be linked to many different males, possibly with a more uniform distribution of link strengths. Helpers would also link to both within- and extra-group individuals. The spatial structure would possibly be less pronounced and the clustering weaker than other family-territorial species, given the high levels of extra-group linking. Do females eavesdrop on courtship interactions between males and extra-group females? Do they copy each other's mate choice? Do they observe interactions between dominant and helper males? Are dominant males affected by the presence of neighbouring females during their courtship of a resident female? Do some very successful dominant males act as keystone individuals in the network by conditioning information flow? Do they eavesdrop on interactions between extra-group females and other males? Do they eavesdrop on interactions between helpers and nestlings? Do helpers eavesdrop on male–female interactions within and outside the group? How does information travel through the territories, for example about the quality of a male or the status of a female? Why are courtship and copulation decoupled and why are copulations so discreet when courtship is not? These are just some of the questions that are stimulated by looking at the biology of cooperatively breeding fairy-wrens from a communication network perspective, and hopefully in the future these will help to better understand individual variation and general rules in social behaviour.

3.5 Concluding remarks

We brought together in a coherent form the concepts of communication and social network and illustrated how they can profitably be integrated to improve our understanding of the dynamics in animal societies. Our analysis is by necessity limited, and we are certain that the two perspectives can be integrated even further and produce even more interesting hypotheses and predictions, as is suggested by other recent work in the same direction (Terry and Lachlan 2005; Bonnie and Earley 2007; see also Chapter 4). Studies on communication networks are growing in number and taxonomic coverage, but we still know too little and much work is still needed. We hope to have pointed to some potential future directions of investigation. Many analytical tools and theoretical concepts have been developed for the study of social networks, but the application of social network theory to questions about animal behaviour is still in its infancy, and here too we hope to have contributed a step to its growth. We look forward to interesting new developments from the integrated study of communication in social networks of animals.

Summary

Communication allows individuals to share information and plays a central role in determining animal social behaviour. Animals live in social networks of multiple individuals connected by links representing different interaction types. Signalling interactions form the base of the communication network (i.e. all conspecifics within signalling range) experienced by an individual and are particularly important for information exchange. Looking at interactions within a network has helped identify and explain the diverse signalling and receiving strategies adopted by animals and may likewise help explain other social interactions. In this chapter we presented a network model which integrates the concepts of communication and social network. We illustrated how this model can affect information exchange in animal communities with different social structures and ecologies. We finally gave some concrete examples of the questions that arise and can be answered when looking at the behavioural ecology of birds from a network perspective.

Acknowledgements

We thank Peter McGregor, Tom Peake, Andy Terry, Safi Darden, Thorsten Balsby, Darren Croft, Ryan Earley, Andy Sih, and others for the many exciting and fruitful discussions about some of the ideas expressed here. We are grateful to Patrizia d'Ettorre and David Hughes, editors of this volume, for inviting this contribution. The work presented here is supported by Framework Grant no. 21–04–0403 from the Danish National Research Council to T.D.

References

Aktipis, C.A. (2006). Recognition memory and the evolution of cooperation: how simple strategies succeed in an agent-based world. *Adaptive Behavior*, **14**, 239–247.

Alonso-Sanz, R. and Martin, M. (2006). Memory boosts cooperation. *International Journal of Modern Physics C*, **17**, 841–852.

Amy, M. and Leboucher, G. (2007). Male canaries can visually eavesdrop on conspecific food interactions. *Animal Behaviour*, **74**, 57–62.

Balsby, T.J.S. and Dabelsteen, T. (2003). Male calling between courtship sequences in whitethroats: a way to counter intrusions from neighbouring rivals. *Behavioural Processes*, **63**, 149–157.

Balsby, T.J.S. and Dabelsteen, T. (2005). Simulated courtship interactions elicit neighbour intrusions in the whitethroat, *Sylvia communis*. *Animal Behaviour*, **69**, 161–168.

Bennett, P.M. and Owens, I.P.F. (2002). *Evolutionary Ecology of Birds*. Oxford University Press, Oxford.

Blumenrath, S.H. and Dabelsteen, T. (2004). Degradation of great tit (*Parus major*) song before and after foliation: implications for vocal communication in a deciduous forest. *Behaviour*, **141**, 935–958.

Blumenrath, S.H., Dabelsteen, T., and Pedersen, S.B. (2004). Being inside nest boxes: Does it complicate the receiving conditions for great tit *Parus major* females? *Bioacoustics*, **14**, 209–223.

Blumenrath, S.H., Dabelsteen, T., and Pedersen, S.B. (2007). Vocal neighbour-mate discrimination in female great tits despite high song similarity. *Animal Behaviour*, **73**, 789–796.

Bonadonna, F. and Nevitt, G.A. (2004). Partner-specific odor recognition in an Antarctic seabird. *Science*, **306**, 835.

Bonnie, K.E. and Earley, R.L. (2007). Expanding the scope for social information use. *Animal Behaviour*, **74**, 171–181.

Bower, J.L. (2005). The occurrence and function of victory displays within communication networks. In: P.K. McGregor (ed.) *Animal Communication Networks*, pp. 114–126. Cambridge University Press, Cambridge.

Bro-Jørgensen, J. and Durant, S.M. (2003). Mating strategies of topi bulls: getting in the centre of attention. *Animal Behaviour*, **65**, 585–594.

Brooker, M. and Rowley, I. (1995). The significance of territory size and quality in the mating strategy of the splendid fairy-wren. *Journal of Animal Ecology* ,**64**, 614–627.

Bshary, R. and D'Souza, A. (2005). Cooperation in communication networks: indirect reciprocity in interactions between cleaner fish and client reef fish. In P.K. McGregor (ed.), *Animal Communication Networks*, pp. 521–539. Cambridge University Press, Cambridge.

Cockburn, A. (2006). Prevalence of different modes of parental care in birds. *Proceedings of the Royal Society B: Biological Sciences*, **273**, 1375–1383.

Cockburn, A., Osmond, H.L., Mulder, R.A., Green, D.J., and Double, M.C. (2003). Divorce, dispersal and incest avoidance in the cooperatively breeding superb fairy-wren *Malurus cyaneus*. *Journal of Animal Ecology*, **72**, 189–202.

Coleman, S.W., Patricelli, G.L., and Borgia, G. (2004). Variable female preferences drive complex male displays. *Nature*, **428**, 742–745.

Coolen, I., Giraldeau, L.A., and Lavoie, M. (2001). Head position as an indicator of producer and scrounger tactics in a ground-feeding bird. *Animal Behaviour*, **61**, 895–903.

Cooney, R. and Cockburn, A. (1995). Territorial defense is the major function of female song in the superb fairy-wren, *Malurus cyaneus*. *Animal Behaviour*, **49**, 1635–1647.

Crockford, C., Wittig, R.M., Seyfarth, R.M., and Cheney, D.L. (2007). Baboons eavesdrop to deduce mating opportunities. *Animal Behaviour*, **73**, 885–890.

Croft, D.P., Krause, J., and James, R. (2004). Social networks in the guppy (*Poecilia reticulata*). *Proceedings of the Royal Society B: Biological Sciences*, **271**, S516–S519.

Croft, D.P., James, R., Thomas, P.O.R., Hathaway, C., Mawdsley, D., Laland, K.N., and Krause, J. (2006). Social structure and co-operative interactions in a wild population of guppies (*Poecilia reticulata*). *Behavioral Ecology and Sociobiology*, **59**, 644–650.

Dabelsteen, T. (2005). Public, private or anonymous? Facilitating and countering eavesdropping. In: P.K. McGregor (ed.), *Animal Communication Networks*, pp. 38–62. Cambridge University Press, Cambridge.

Dabelsteen, T., McGregor, P.K., Shepherd, M., Whittaker, X., and Pedersen, S.B. (1996). Is the signal value of overlapping different from that of alternating during matched singing in Great Tits? *Journal of Avian Biology*, **27**, 189–194.

Dabelsteen, T., McGregor, P.K., Lampe, H.M., Langmore, N.E., and Holland, J. (1998). Quiet song in song birds: An overlooked phenomenon. *Bioacoustics*, **9**, 89–105.

Darden, S.K., Steffensen, L.K., and Dabelsteen, T. (2008) Information transfer among widely spaced individuals: latrines as a basis for communication networks in the swift fox. *Animal Behaviour*, **75**, 425–432.

Doutrelant, C. and Covas, R. (2007). Helping has signalling characteristics in a cooperatively breeding bird. *Animal Behaviour*, **74**, 739–747.

Dugatkin, L.A. and Druen, M. (2004). The social implications of winner and loser effects. *Proceedings of the Royal Society B: Biological Sciences*, **271**, S488–S489.

Dunn, P.O. and Cockburn, A. (1996). Evolution of male parental care in a bird with almost complete cuckoldry. *Evolution*, **50**, 2542–2548.

Dunn, P.O. and Cockburn, A. (1999). Extrapair mate choice and honest signaling in cooperatively breeding superb fairy-wrens. *Evolution*, **53**, 938–946.

Emlen, S.T. (1995). An evolutionary theory of the family. *Proceedings of the National Academy of Sciences of the USA*, **92**, 8092–8099.

Freeberg, T.M. (2006). Social complexity can drive vocal complexity: group size influences vocal information in Carolina chickadees. *Psychological Science*, **17**, 557–561.

Gompertz, T. (1961). The vocabulary of the great tit. *British Birds*, **54**, 369–394, 409–418.

Green, D.J., Cockburn, A., Hall, M.L., Osmond, H., and Dunn, P.O. (1995). Increased opportunities for cuckoldry may be why dominant male fairy-wrens tolerate helpers. *Proceedings of the Royal Society B: Biological Sciences*, **262**, 297–303.

Hill, S.E. and Ryan, M.J. (2006). The role of model female quality in the mate choice copying behaviour of sailfin mollies. *Biology Letters*, **2**, 203–205.

Hinde, C.A. and Kilner, R.M. (2007). Negotiations within the family over the supply of parental care. *Proceedings of the Royal Society B: Biological Sciences*, **274**, 53–60.

Hinde, R.A. (1952). The behaviour of the great tit (*Parus major*) and some other related species. *Behaviour* (Supplement), **2**, 1–201.

Höglund, J. and Alatalo, R.V. (1995). *Leks*. Princeton University Press, Princeton, NJ.

Höglund, J., Alatalo, R.V., Gibson, R.M., and Lundberg, A. (1995). Mate-choice copying in black grouse. *Animal Behaviour*, **49**, 1627–1633.

Holland, J., Dabelsteen, T., Pedersen, S.B. and Larsen, O.N. (1998). Degradation of wren *Troglodytes troglodytes* song: implications for information transfer and ranging. *Journal of the Acoustical Society of America*, **103**, 2154–2166.

Holland, J., Dabelsteen, T., Bjørn, C.P., and Pedersen, S.B. (2001). The location of ranging cues in wren song: Evidence from calibrated interactive playback experiments. *Behaviour*, **138**, 189–206.

Hurst, J.L. (2005). Scent marking and social communication. In: P.K. McGregor (ed.), *Animal Communication Networks*, pp. 219–243. Cambridge University Press, Cambridge.

Jablonka, E. (2002). Information: its interpretation, its inheritance, and its sharing. *Philosophy of Science*, **69**, 578–605.

Järvi, T. and Bakken, M. (1984). The function of the variation in the breast stripe of the great tit (*Parus major*). *Animal Behaviour*, **32**, 590–596.

Järvi, T., Walso, O., and Bakken, M. (1987). Status signaling by *Parus major*—an experiment in deception. *Ethology*, **76**, 334–342.

Kazem, A.J.N. and Aureli, F. (2005). Redirection of aggression: multiparty signalling within a network? In: P.K. McGregor (ed.), *Animal Communication Networks*, pp. 191–218. Cambridge University Press, Cambridge.

Lampe, H.M., Larsen, O.N., Pedersen, S.B., and Dabelsteen, T. (2007). Song degradation in the hole-nesting pied flycatcher *Ficedula hypoleuca*: Implications for polyterritorial behaviour in contrasting habitat-types. *Behaviour*, **144**, 1161–1178.

Langemann, U. and Klump, G.M. (2005). Perception and acoustic communication networks. In: P.K. McGregor (ed.) *Animal Communication Networks*, pp. 451–480. Cambridge University Press, Cambridge.

Langemann, U., Tavares, J.P., Peake, T.M., and McGregor, P.K. (2000). Response of great tits to escalating patterns of playback. *Behaviour* **137**, 451–471.

Lind, H., Dabelsteen, T., and McGregor, P.K. (1996). Female great tits can identify mates by song. *Animal Behaviour*, **52**, 667–671.

Lusseau, D. (2003). The emergent properties of a dolphin social network. *Proceedings of the Royal Society B: Biological Sciences*, **270**, S186–S188.

McGregor, P.K. (ed.) (2005). *Animal Communication Networks*. Cambridge University Press, Cambridge.

McGregor, P.K. and Dabelsteen, T. (1996). Communication networks. In: D.E. Kroodsma and E.H. Miller (eds), *Ecology and Evolution of Acoustic Communication in Birds*, pp. 409–425. Cornell University Press, Ithaca, NY.

McGregor, P.K. and Krebs, J.R. (1982). Song types in a population of great tits (*Parus major*)—their distribution, abundance and acquisition by individuals. *Behaviour*, **79**, 126–152.

McGregor, P.K. and Krebs, J.R. (1984). Sound degradation as a distance cue in great tit (*Parus major*) song. *Behavioral Ecology and Sociobiology*, **16**, 49–56.

McGregor, P.K., Dabelsteen, T., Shepherd, M., and Pedersen, S.B. (1992). The signal value of matched singing in great tits—evidence from interactive playback experiments. *Animal Behaviour*, **43**, 987–998.

McGregor, P.K., Peake, T.M., and Lampe, H.M. (2001). Fighting fish *Betta splendens* extract relative information from apparent interactions: what happens when what you see is not what you get. *Animal Behaviour*, **62**, 1059–1065.

Matessi, G., McGregor, P.K., Peake, T.M., and Dabelsteen, T. (2005). Do male birds intercept and use rival courtship calls to adjust paternity protection behaviours? *Behaviour*, **142**, 507–524.

Matessi, G., McGregor, P.K., Peake, T.M., and Dabelsteen, T. (2007). Female rock sparrows (*Petronia petronia*), not males, respond differently to simulations of different courtship interaction outcomes. *Behaviour*, **144**, 735–752.

Matos, R.J. (2002). Social influences on signalling: audience effects and communication networks. PhD Thesis, University of Copenhagen.

Matos, R.J. and Schlupp, I. (2005). Performing in front of an audience: signallers and the social environment. In: P.K. McGregor (ed.), *Animal Communication Networks*, pp. 63–83. Cambridge University Press, Cambridge.

Mennill, D.J., Ramsay, S.M., Boag, P.T., and Ratcliffe, L.M. (2004). Patterns of extrapair mating in relation to male dominance status and female nest placement in black-capped chickadees. *Behavioral Ecology*, **15**, 757–765.

Mulder, R.A. and Langmore, N.E. (1993). Dominant males punish helpers for temporary defection in superb fairy-wrens. *Animal Behaviour*, **45**, 830–833.

Mulder, R.A. and Magrath, M.J.L. (1994). Timing of prenuptial molt as a sexually selected indicator of male quality in superb fairy-wrens (*Malurus cyaneus*). *Behavioral Ecology*, **5**, 393–400.

Mulder, R.A., Dunn, P.O., Cockburn, A., Lazenby-Cohen, K.A., and Howell, M.J. (1994). Helpers liberate female fairy-wrens from constraints on extra-pair mate choice. *Proceedings of the Royal Society Series B: Biological Sciences*, **255**, 223–229.

Newman, M.E.J. (2003). The structure and function of complex networks. *SIAM Review*, **45**, 167–256.

Oliveira, R.F. (2005). Hormones, social context and animal communication. In: P.K. McGregor (ed.), *Animal Communication Networks*, pp. 481–520. Cambridge University Press, Cambridge.

Ophir, A.G. and Galef, B.G. (2004). Sexual experience can affect use of public information in mate choice. *Animal Behaviour*, **68**, 1221–1227.

Otter, K., McGregor, P.K., Terry, A.M.R., Burford, F.R.L., Peake, T.M., and Dabelsteen, T. (1999). Do female great tits (*Parus major*) assess males by eavesdropping? A field study using interactive song playback. *Proceedings of the Royal Society B: Biological Sciences*, **266**, 1305–1309.

Otter, K.A., Stewart, I.R K., McGregor, P.K., Terry, A.M.R., Dabelsteen, T., and Burke, T. (2001). Extra-pair paternity among great tits *Parus major* following manipulation of male signals. *Journal of Avian Biology*, **32**, 338–344.

Payne, R.B., Payne, L.L., and Rowley, I. (1988). Kin and social relationships in splendid fairy-wrens—recognition by song in a cooperative bird. *Animal Behaviour*, **36**, 1341–1351.

Payne, R.B., Payne, L.L., Rowley, I., and Russell, E.M. (1991). Social recognition and response to song in cooperative red-winged fairy-wrens. *Auk*, **108**, 811–819.

Peake, T.M. (2005). Eavesdropping in communication networks. In: P.K. McGregor (ed.), *Animal Communication Networks*, pp. 13–37. Cambridge University Press, Cambridge.

Peake, T.M. and McGregor, P.K. (2004). Information and aggression in fishes. *Learning and Behavior*, **32**, 114–121.

Peake, T.M., Terry, A.M.R., McGregor, P.K., and Dabelsteen, T. (2001). Male great tits eavesdrop on simulated male-to-male vocal interactions. *Proceedings of the Royal Society B: Biological Sciences*, **268**, 1183–1187.

Peake, T.M., Terry, A.M.R., McGregor, P.K., and Dabelsteen, T. (2002). Do great tits assess rivals by combining direct experience with information gathered by eavesdropping? *Proceedings of the Royal Society B: Biological Sciences*, **269**, 1925–1929.

Peake, T.M., Matessi, G., McGregor, P K., and Dabelsteen, T. (2005). Song type matching, song type switching and eavesdropping in male great tits. *Animal Behaviour*, **69**, 1063–1068.

Poesel, A., Dabelsteen, T., and Pedersen, S.B. (2007). Implications of conspecific background noise for features of blue tit, *Cyanistes caeruleus*, communication networks at dawn. *Journal of Ornithology*, **148**, 123–128.

Rowley, I. and Russell, E. (1990). Philandering—a mixed mating strategy in the splendid fairy-wren *Malurus splendens*. *Behavioral Ecology and Sociobiology*, **27**, 431–437.

Rowley, I. and Russell, E. (1997). *Fairy-Wrens and Grasswrens–* Maluridae. Oxford University Press, Oxford.

Schubert, K.A., Mennill, D.J., Ramsay, S.M., Otter, K.A., Boag, P.T., and Ratcliffe, L.M. (2007). Variation in social rank acquisition influences lifetime reproductive success in black-capped chickadees. *Biological Journal of the Linnean Society*, **90**, 85–95.

Searcy, W.A. and Nowicki, S. (2005). *The Evolution of Animal Communication: Reliability and Deception in Signaling Systems*. Princeton University Press, Princeton, NJ.

Sharp, S.P. and Hatchwell, B.J. (2006). Development of family specific contact calls in the long-tailed tit *Aegithalos caudatus*. *Ibis*, **148**, 649–656.

Sih, A. and Watters, J.V. (2005). The mix matters: behavioural types and group dynamics in water striders. *Behaviour*, **142**, 1417–1431.

Stevens, J.R. and Hauser, M.D. (2004). Why be nice? Psychological constraints on the evolution of cooperation. *Trends in Cognitive Sciences*, **8**, 60–65.

Sturdy, C.B., Phillmore, L.S., Sartor, J.J., and Weisman, R.G. (2001). Reduced social contact causes auditory perceptual deficits in zebra finches, *Taeniopygia guttata*. *Animal Behaviour*, **62**, 1207–1218.

Tarvin, K.A., Webster, M.S., Tuttle, E.M., and Pruett-Jones, S. (2005). Genetic similarity of social mates predicts the level of extrapair paternity in splendid fairy-wrens. *Animal Behaviour*, **70**, 945–955.

Terry, A.M.R. and Lachlan, R.F. (2005). Communication networks in a virtual world. In: P.K. McGregor (ed.), *Animal Communication Networks*, pp. 604–627. Cambridge University Press, Cambridge.

Vignal, C., Mathevon, N., and Mottin, S. (2004). Audience drives male songbird response to partner's voice. *Nature*, **430**, 448–451.

Voigt, C., Leitner, S., and Gahr, M. (2007). Socially induced brain differentiation in a cooperatively breeding songbird. *Proceedings of the Royal Society B: Biological Sciences*, **274**, 2645–2651.

Webster, M.S., Tarvin, K.A., Tuttle, E.M., and Pruett-Jones, S. (2004). Reproductive promiscuity in the splendid fairy-wren: effects of group size and auxiliary reproduction. *Behavioral Ecology*, **15**, 907–915.

CHAPTER 4

Communication between hosts and social parasites

David R. Nash and Jacobus J. Boomsma

4.1 Introduction: social parasitism

The evolution of societies by **natural** or **cultural selection** is normally driven by the benefits of collaboration disproportionately available to groups (Axelrod and Hamilton 1981). These advantages tend to increase with the number of society members, because larger societies tend to accumulate more substantial amounts of resources, both as biomass of their members and as infrastructures or stored products. However, social organization comes at a cost, because outsiders that are not part of a society can gain more predictable benefits when exploiting larger, rather than smaller societies, so that arms races of deceit and defence are inevitable (Diamond 1997). Given sufficient evolutionary time, we should thus expect all societies to become burdened by forms of **social parasitism**, either by members that take more than their fair share, or by strangers. Although there exists a substantial literature on dishonest social exploitation in human and other vertebrate societies (Davies 2000; Altizer *et al.* 2003; Summers *et al.* 2003), it is only the insect societies of ants, bees, wasps, and termites that have existed long enough to have evolved a complete spectrum of alternative modes of social parasitism. We will therefore focus in this chapter on social insects and review the current state of our understanding of social parasitism in the **eusocial** hymenoptera and termites. This understanding includes different ways of invasion, exploitation, and defence strategies, and the principles of communication that allow insect social parasites to persist.

The **fitness** of the members of a social insect colony is closely tied to the amount of resources that it can obtain, and ultimately convert into reproductive offspring (Oster and Wilson 1978). Social insects generally gather resources together, and store them in discreet places (the nest or hive; Fig. 4.1). Resources may be stored without much manipulation (e.g. by seed-harvesting ants), may be converted into a form that eases storage and subsequent distribution within the colony (e.g. honey), or may be stored within the tissues of colony mates (e.g. the repletes of honeypot ants, or the brood of many other ant species). Many ant species also enter into trophobiotic associations (associations in which they obtain food from living organisms) with Hemiptera (Way 1963; Stadler and Dixon 2005), the larvae of lycaenid butterflies (Pierce *et al.* 2002), or **symbiotic** fungi (North *et al.* 1997; Aanen and Boomsma 2006). In these cases, the associated trophobionts can also be regarded as a resource that is gathered, concentrated, and maintained by the colony. Finally, the workforce of the colony is a resource in its own right, which can potentially be exploited by social parasites.

While predators destroy a colony or exhaust its resources quickly and completely, social parasites exploit the colony resources over an extended period of time. They are thus fundamentally different from the normal micro- and macro-parasites that may infect individual colony members, which tend to have short generation times (Boomsma *et al.* 2005). To protect the resources of the colony against intruders, social insects have evolved many defensive strategies, often involving elaborate protective nest structures, and show adaptations in both behaviour and morphology. Social insect colonies have therefore been characterized as 'factories

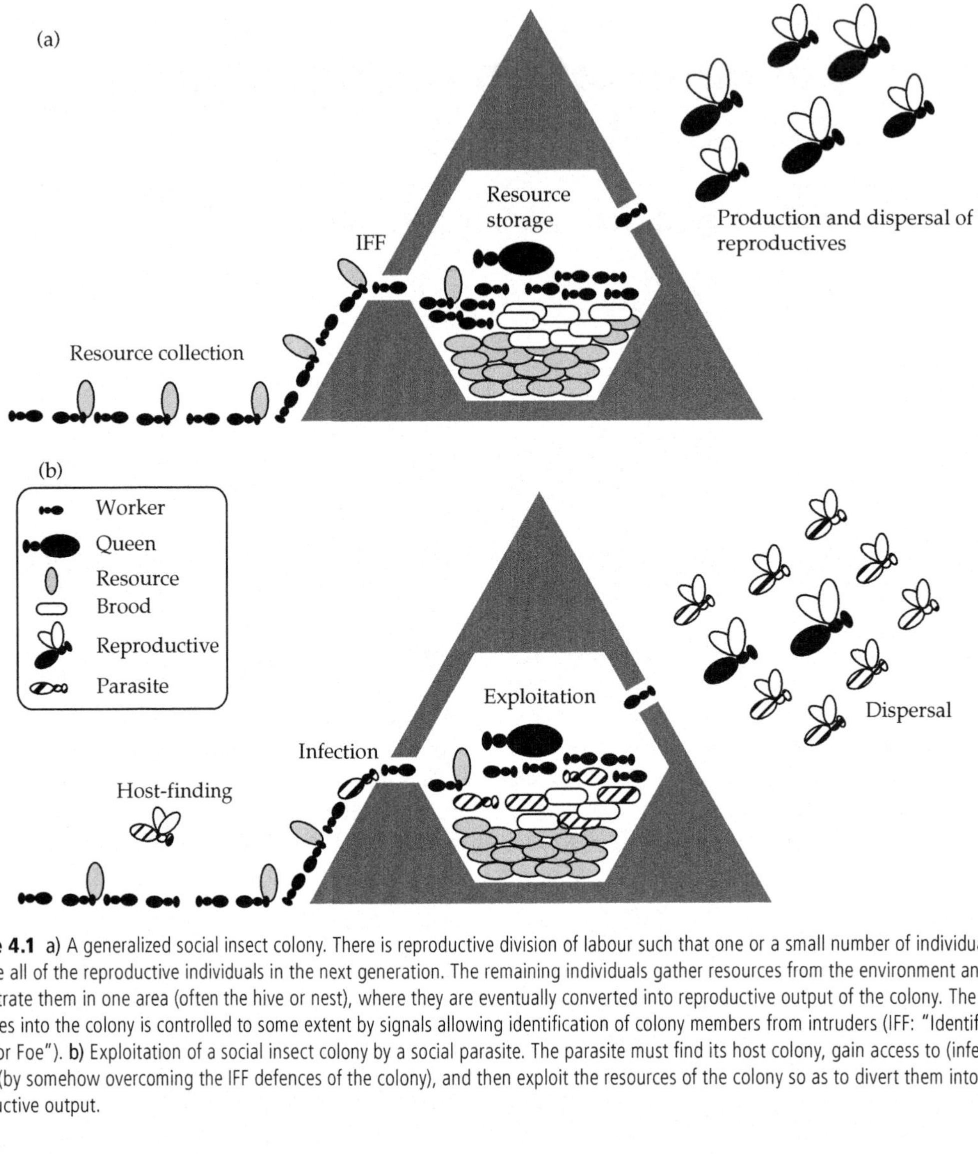

Figure 4.1 a) A generalized social insect colony. There is reproductive division of labour such that one or a small number of individuals produce all of the reproductive individuals in the next generation. The remaining individuals gather resources from the environment and concentrate them in one area (often the hive or nest), where they are eventually converted into reproductive output of the colony. The flow of resources into the colony is controlled to some extent by signals allowing identification of colony members from intruders (IFF: "Identification Friend or Foe"). b) Exploitation of a social insect colony by a social parasite. The parasite must find its host colony, gain access to (infect) the colony (by somehow overcoming the IFF defences of the colony), and then exploit the resources of the colony so as to divert them into its own reproductive output.

within fortresses' (Wilson 1968). Even more vital to the protection of the resources of the colony, however, is a complex set of **signals** that allow colony members to distinguish between colony members and intruders (Fig. 4.1)—essentially an immunological response to 'self' and 'non-self'.

Social parasites must overcome three barriers in order to successfully exploit their hosts (Fig. 4.1; cf. Schmid-Hempel and Ebert 2003). Firstly, they must successfully locate their host colony; secondly, they must successfully infect it by entering the colony or it resource store; and finally, they must acquire access to the social host's resources and convert them into reproductive output. While host-finding may involve searching for **cues** produced by the host, infection and exploitation are the stages in which the parasite must actively deceive the host into accepting it and allowing its resources to be exploited.

Pressure from social parasites may induce selection for host defences and counter measures of the parasites. Some social parasites may use force to infect a colony, and physical barriers to prevent subsequent eviction. For example, the larva of the lycaenid butterfly, *Lyphyra brassolis*, feeds on brood of the weaver ant, *Oecophylla smaragdina*, by barging its way into the ant nest, protected from the attacks of the workers by a heavily thickened cuticle, and grows and pupates inside the ant nest protected by this larval skin (Dodd 1902; Braby 2000). However, the majority of long-term social parasites gain access to colony resources by exploiting the signals used by the social insects themselves (Fig. 4.1), to avoid being recognized as non-self. This strategy is not unique to social parasites, as circumventing the host's self/non-self recognition system is also a common feature of most microbial parasites (e.g. Tsutsui 2004). However, the macroscopic scale of social parasites and their hosts means that signals are normally more complex, while at the same time being more amenable to experimental study.

4.2 Communication between social parasites and their hosts

Throughout this chapter we will use the term communication in a broad sense, to mean the transfer of information between two or more individuals. More specifically, this can be put into the framework information theory (Shannon 1948), where communication is based on a signal (or cue) provided by a transmitter which is passed through a particular **channel** to a receiver, and which reduces the uncertainty of the receiver. The distinction between a signal and a cue can be subtle, particularly in the context of communication between parasites and hosts, whose fundamental interests are at odds. In Chapter 1, signals were defined as 'characters that evolve in a transmitter in order to provide information to a receiver, aiming to change the behaviour of the receiver to the benefit of the transmitter'. Cues, on the other hand, can be regarded as characters that are a source of information, but which did not evolve to change the behaviour of the receiver to the benefit of the transmitter. When we are considering the hosts of social parasites as transmitters, most of the communication with their social parasite receivers involves the use of information sources that have not evolved specifically to communicate with parasites. Rather, they have evolved as signals for colony members or other receivers, but are used as cues by the parasites (this could also be regarded as a form of eavesdropping; see Chapter 3). It therefore depends on the type and context of the interaction as to whether these should be regarded as cues or signals, but we will follow the convention here of calling them cues. Communication between social parasites as transmitters and their hosts as receivers, on the other hand, usually appears to involve specifically designed signals.

4.2.1 Communication within social insect colonies

Social insects use a plethora of signals and cues via many different channels to communicate information to their nestmates and to other colonies (see Chapter 5). Visual cues may be used to distinguish between individuals or to assess dominance in polistine wasps (Tibbetts 2002; Tibbetts and Dale 2004), tactile and vibrational signals may be used to provide information on the distance and direction to foraging resources by honeybees (von Frisch and Chadwick 1967; Dyer 2002), and chemical cues and signals may be used to distinguish nestmates from non-nestmates (Gamboa *et al.* 1987; Vander Meer and Morel 1998), to provide information about the state of the colony (Vargo 1998; Nielsen *et al.* 1999) or the identity of different castes (Vander Meer and Morel 1988; Winston and Slessor 1992; d'Ettorre *et al.* 2004), and to mobilize defenders when a colony is threatened (Blum and Brand 1972). All forms of within-colony communication can potentially be exploited by social parasites.

How the communicatory signals of social insects are exploited by social parasites depends to some extent on the locations where information transfer between colony members occurs. Signals or cues that are produced outside the nest, including those associated with resource acquisition (e.g. foraging trails) and the organization of work or defence (e.g. alarm pheromones), may provide cues to social parasites for locating and identifying a host colony (e.g. Cammaerts *et al.* 1990; Dejean and Beugnon 1996; Pernal *et al.* 2005). However, subsequent host infection and exploitation will often involve within-colony signals and cues that hosts use for identifying colony members from outsiders. For further discussion of communication within social insect colonies see Chapter 5.

4.3 Types of social parasites

Various types of interactions have been classified as social parasitism, and differ in exploitative details. Although the division between defined forms of social parasitism is not always clear-cut, several broad categories can be distinguished.

4.3.1 Temporary social parasites

Several species of social insects do not found colonies independently, but enter already established colonies and kill or usurp the current queen. The social parasite will then lay eggs, which are initially raised by the workers of the host species. Over time the proportion of the worker force that are the offspring of the host will decrease, so that the colony gradually becomes composed of individuals of the temporary social parasite only (Wheeler 1904a; Hölldobler and Wilson 1990). From this point onwards, the colony continues to function in the same way as an independently founded colony.

4.3.2 Slave-makers

Slave-making only occurs in ants and involves raiding of colonies of other species to collect larvae and pupae and bring them back to the slave-making ant colony. Here, this stolen brood will continue development to eventually become an integrated alien worker force in the colony (Wasmann 1905; Hölldobler and Wilson 1990). The effect of slave-raiding on colonies of slave species may be devastating (Fischer-Blass *et al.* 2006), so that densities of slave-making social parasites are often low. Slave-makers often initiate their colonies through temporary social parasitism (see Section 4.3.1), but slave-raiding ensures that host workers remain available for the life of the slave-maker colony.

4.3.3 Inquilines

Inquiline parasites enter established colonies of social insects and remain in the colony while they reproduce (Hamilton 1888; Wheeler 1904b; Hölldobler and Wilson 1990). The parasite and host continue to reproduce side-by-side within the colony, but the host has a reduced fitness through diversion of resources that would normally be available for reproduction to the parasite (Bekkevold and Boomsma 2000; Passera *et al.* 2001). If the host queen is killed by the social parasite (which has been called queen-intolerant inquilinism), then this is similar to temporary social parasitism, but without the parasite colony having to go through an ergonomic phase of only worker production.

4.3.4 Brood parasites

Many social parasites can also be regarded as brood parasites, since they divert resources away from the larvae or pupae being reared by the colony, either by preying on the brood within the colony or by collecting resources from the workers that would otherwise have been delivered to the brood. Brood parasitism is normally a special case of inquilinism. Social brood parasites show many parallel adaptations to those known from more familiar vertebrate brood parasites such as cuckoos and cowbirds (Davies *et al.* 1989).

4.3.5 Anarchistic workers

Reproductive division of labour in social insects varies in degree between different species. In cases where workers have retained the ability to reproduce, some of these workers may become socially parasitic by using colony resources to produce mostly their own offspring rather than serve the collective by raising the queen's brood (Bourke 1988). In the Cape honeybee, these lineages are parthenogenetic and able to invade other colonies (Oldroyd 1999).

4.3.6 Egg-dumpers

Queens or workers of established colonies may visit neighbouring colonies and lay eggs, which will be raised by the workers of those colonies (Akre *et al.* 1976; Lopez-Vaamonde *et al.* 2004). This phenomenon is also known as *drift* between colonies, although not all drift may result in parasitism of neighbouring colonies (Sumner *et al.* 2007).

4.4 The parasitism–mutualism continuum

The distinction between parasitism, in which a host loses fitness as a result of its association with a symbiont, and mutualism, in which a host gains fitness, is not always clear-cut. Parasitism and mutualism are in reality opposite ends of the same spectrum of relative costs and benefits that symbionts obtain from their interaction (Ewald 1987).

Social insects engage in many interactions with other animals that are normally regarded as mutualistic (Hölldobler and Wilson 1990). These often take the form of protection mutualisms, in which the social insect colony provides refuge from parasites and predators in return for food rewards. However, there should always be selection on mutualists to gain ever more resources from their partners while giving up fewer of their own resources in return (Herre *et al.* 1999; Frank 2003). Only an exact knowledge of the costs and benefits of any particular interaction can allow us to classify it as mutualistic or parasitic, and such analyses have seldom been carried out. In the few documented cases available, costs and particularly benefits appear to be highly variable (Pierce *et al.* 1987; Cushman and Whitham 1989; Pierce and Nash 1999). This means that many apparently mutualistic interactions may in fact be parasitic, depending on ecological conditions (Cushman and Beattie 1991; Bronstein *et al.* 2003), or even entirely parasitic (Gomulkiewicz *et al.* 2003).

Conversely, it is also possible that interactions that are currently regarded as being socially parasitic may in fact be mutualistic when costs and benefits are examined in greater detail (c.f. Charlat and Mercot 2001).

4.5 The parasitism–predation divide

The division between parasitism and other forms of antagonistic interaction is even less clear-cut than that between parasitism and mutualism. A frequent definition of a parasite is an organism that gains resources from its host without killing it. This aspect of parasitism is far from useful when considering social parasitism, since the host colony is made up of many individuals and often has partitioned resources, so that most predators of social insects or their resources are unlikely to destroy the entire colony, while, conversely, many species commonly regarded as social parasites do annihilate the host colony. A more useful distinction between parasitism and other forms of antagonism is that parasitism is a long-term interaction with a single host. For social parasites, this distinction can also become rather blurred because the host is made up of numerous individuals and because many commonly recognized social parasites (e.g. slave-making ants) may associate with several host colonies. The criterion of long-term interaction, however, probably gives the best working definition of social parasitism.

Once again, however, there is a continuum between social parasitism and what could be termed *social predation*—the exploitation of the resources of a colony over a short period of time. Many of the same selection pressures apply across the range of interaction lengths, so it is not surprising that social parasites and social predators often show convergent adaptations to exploit their hosts, and exploit host communication in similar ways. A striking example of this is the death's-head hawk moth, *Acherontia atropos*, which frequently enters bee hives as an adult to feed on stored honey. It gains access to the hive and protection from the guard bees by mimicking the surface hydrocarbons of its hosts (Moritz *et al.* 1991). In addition, it has frequently been suggested that the characteristic squeaking sounds produced by the moth mimic the piping of queen bees, and also serve to reduce aggression by worker bees (Maeterlinck 1901; Moritz *et al.* 1991), but as far as we are aware, no formal analysis of this has been carried out.

Where social insects have trophobiotic associations with other organisms, exploitation of the trophobionts by predators can also be regarded as social predation, although the term *agro-predation* has also been coined for this particular form of social exploitation. Agro-predation has been described for several species of ant that exploit the fungus gardens of attine ants (Adams *et al.* 2000;

Dijkstra and Boomsma 2003), but it can equally well be applied to specialist predators of ant-tended hemipterans or lycaenid butterflies. Again, these specialist predators often show convergent adaptation to social parasites, including chemical camouflage and mimicry (Pierce 1995; Lohman *et al.* 2006; Pasteels 2007). In a few cases, agro-predators may have a long-term association with social insect colonies (Majerus *et al.* 2007), and so can be regarded as true social parasites.

4.6 Who are the social parasites?

4.6.1 Intraspecific social parasitism

Intraspecific social parasitism is probably a far more widespread phenomenon than has hitherto been acknowledged. Particularly in ants and wasps there are many species in which multiple queens breed in the same nest. Powerful molecular marker studies are increasingly showing that some of these queens obtain a higher share in the reproductives produced by a colony while they contribute less than average to the worker brood (Bargum and Sundström 2007; Kummerli and Keller 2007). Intraspecific social parasitism of this kind is generally assumed to be the incipient stage in the evolution of species that specialize on a socially parasitic lifestyle, either via sympatric (Buschinger 1986; Radchenko and Elmes 2003) or allopatric (Lowe *et al.* 2002) speciation. Some cases, such as the microgyne form of *Myrmica rubra*, appear to be on the border of speciation, as gene flow between host and parasitic morph may still be happening at a low frequency (Steiner *et al.* 2006).

There is recent evidence that some patrilines (workers in **polyandrous** colonies who share the same father) within colonies can also express socially parasitic traits and thus obtain a more than proportional share of the colony's reproductive output. Recent studies in honeybees (Moritz *et al.* 2005) and *Acromyrmex* leaf-cutting ants (Hughes and Boomsma 2008) have shown that such biases do exist. However, the mechanisms by which these genetic lineages achieve their biased fitness gains are only partially understood. In general, we should expect that the evolution of socially parasitic patrilines is more constrained than the evolution of socially parasitic matrilines (offspring of the same queen in **polygynous** colonies), because socially parasitic queens may be adopted when a colony has grown large and is reasonably well buffered against small fitness loads.

From a purely mechanistic point of view, the exploitation of communication signals by intraspecific brood parasites is expected to be relatively straightforward, since the parasites share the same set of cues and signals as their hosts. For selection to favour intraspecific social parasitism, there must be mechanisms that allow the parasites to recognize their own relatives, and there must also be sufficient genetic differentiation between hosts and parasites that any general effects on colony productivity do not also reduce the fitness of the parasites. Such conditions are likely to be countered by colony-level selection in most cases, which is expected to reduce the accuracy of the cue information required to make such discrimination (Hölldobler and Carlin 1987; Boomsma *et al.* 2003; Hannonen and Sundström 2003). The mixing of individual cuticular hydrocarbons within ant colonies to produce a single *gestalt* odour has probably arisen partly to avoid such exploitation (Ratnieks 1991). Although any 'green beard' recognition tags can also be expected to induce strong selection on non-parasitic nestmates to discover and eliminate this type of cheat, the use of such cues by polygynous *Solenopsis invicta* ants has been demonstrated (Keller and Ross 1998). This interaction, in which queens not bearing an odour cue that is associated with a particular haplotype possessed by workers are killed by those workers, can potentially lead to a form of temporary social parasitism, although whether the particular conditions for this ever occur in the field is unclear.

Cases in which social parasites exploit the resources of neighbouring colonies of the same species are less problematic, since there is strong selection to maintain colony recognition abilities. How such recognition is overcome by the parasitic individuals remains largely unknown, although the social parasites may provide particularly plastic cues (Breed *et al.* 1992), or show behaviours that minimize contacts with workers guarding the nest entrances (Neumann *et al.* 2003).

4.6.2 Sister-taxa social parasitism

Emery (1909) remarked that many social parasites of ants are closely related to their hosts, and subsequent confirmation of this general finding has led to this phenomenon being referred to as Emery's rule. This form of social parasitism is particularly clear in the slave-making ants, where the social parasites and victim hosts are often very closely related (Hölldobler and Wilson 1990). A recent study by Savolainen and Vepsalainen (2003) showed that the ant genus *Myrmica* has an entire series of independent sister lineages split into a social parasite and host, that morphological differentiation between host and parasite increases with the coalescence time of the split, and that strict associations according to Emery's rule tend to be lost by horizontal acquisition of secondary hosts in older socially parasitic lineages. Work by Foitzik and co-workers on slave-makers in leptothoracine ants shows similar patterns (Beibl *et al.* 2005; Brandt *et al.* 2007).

In sister taxa, common ancestry of signal-producing and -receiving apparatus should facilitate the exploitation of host signals by social parasites. However, as soon as the separation of host and parasite gene pools is complete, the possibility of evolutionary arms races between hosts and parasites in signal design emerges, and increases the sensitivity of selection to the frequency of parasites among the host population (see Section 4.10). That is, when gene flow is absent, selection for mimicry has to be actively maintained in a dynamic process of antagonistic co-evolution, rather than being present by default.

4.6.3 Distantly related social parasites

The most challenging type of social parasitism to explain from an evolutionary point of view is that of social parasites that are only distantly related to their hosts. Here, exploitation of social signals or cues must arise *de novo*, or at least from the modification of some other signalling system. Distantly related organisms are likely to differ in their physiological apparatus, such that the costs of signal production are also likely to differ. While it is often implicitly assumed that a signal or cue will be more costly to produce for a distantly related parasite, this may not always be the case. For example, many social parasites that feed on the brood or workers of a colony may be able to acquire the components needed for production of cuticular hydrocarbons relatively cheaply from their hosts.

The phylogenetic distinctness of distantly related social parasites means that they are unlikely to be overlooked, so that many examples of this type of social parasitism have long been recognized (e.g. Hamilton 1888). This contrasts with the sister-taxa forms of social parasitism, where many cryptic species probably remain to be discovered. An example case in point is two inquiline parasites of *Acromyrmex* leaf-cutting ants that were only discovered during the last decade (De Souza *et al.* 2007, Schultz *et al.* 1998).

4.7 How do social parasites exploit host communication?

4.7.1 Disguise

Most social parasites gain access to the resources of the host colony by producing signals that disguise their true nature from the hosts, allowing them to appear as members of the colony, as mutualists, or as unthreatening. Such disguises may use any of the communication channels used by the host colony (see Table 4.1), although the preponderant use of chemical channels reflects the prime importance of this mode of communication in social insects in general (Wilson 1965). Disguise by social parasites generally falls into three categories: insignificance, camouflage, and mimicry.

4.7.1.1 Insignificance

One possible way of avoiding detection is to lack any of the cues that can be detected by the host, so that discrimination is impossible (Fig. 4.2b). This has been termed 'chemical insignificance' in the chemical communication literature. Although several studies have shown that social parasites may indeed have remarkably few cuticular hydrocarbons (both number and quantity of compounds) relative to their host (Lenoir *et al.* 2001; Lambardi *et al.* 2007), the mechanism by which chemical insignificance evolves is far from clear. For example, in

Table 4.1 Selected examples of communication channels used by social insects, and of their exploitation by social parasites

Communication channel	Social insect examples	Social parasite examples
Visual signals	Recognition of face patterns of *Polistes* wasps.[1]	Mimicry of face patterns in vespine wasps[2]
Auditory signals	Quacking and piping in honeybees.[3] Bee dance language.[4]	Stridulation of *Maculinea* butterfly larvae.[5]
Substrate-borne vibrations	Ant vibrational alarm.[6]	Stridulation of *Maculinea* butterfly larvae.[7]
Tactile signals	Begging behaviour of ant larvae.[8]	Begging behaviour of *Atemeles*[9] and *Claviger*[10] beetles and *Maculinea* butterfly larvae.[11]
Volatile chemical signals	Alarm[12] and Queen[13] pheromones.	Secretions from the Tentacular Organs of Lycaenidae.[14] Propaganda substances of *Formica subintegra* ants[15]
Cuticular chemical signals	CHC colony recognition in ants and termites.[16]	*Maculinea* butterfly larvae.[17] Slave-making ants.[18]

References and notes: [1]Tibbetts 2002, Tibbetts and Dale 2004. [2]This is implied by Wilson 1971. [3]Michelsen *et al.* 1986a. [4]Michelsen *et al.* 1986b. [5]Schönrogge *et al.* 2005. [6]Hickling and Brown 2000. [7]DeVries *et al.* 1993. [8]Creemers *et al.* 2003. [9]Hölldobler and Wilson 1990. [10]Cammaerts 1995. [11]Elmes *et al.* 1991. [12]Blum 1992, Blum and Brand 1972. [13]Winston and Slessor 1992. [14]Henning 1983, Pierce *et al.* 2002. [15]Regnier and Wilson 1971. [16]Howard and Blomquist 2005. [17]Akino *et al.* 1999, Elmes *et al.* 2002, Schönrogge *et al.* 2004. [18]d'Ettorre et al. 2002.

cases of intraspecific social parasitism, why are breeders with moderately reduced quantities of cuticular hydrocarbons (a necessary intermediate stage towards total insignificance) ignored rather than attacked? It is possible that such individuals are perceived as subordinates by the dominants that express more abundant hydrocarbon profiles than average (Monnin *et al.* 1998; Heinze *et al.* 2002). However, we have very few data to substantiate such speculation in all but the smallest (usually queenless) insect societies, and are generally ignorant about which, if any, of the cuticular hydrocarbon peaks have such a quality signalling function (d'Ettorre *et al.* 2004). This is clearly an area where further work is required.

The apparent commonness of chemical insignificance as an infection method of interspecific social parasites of ants (see Section 4.9) may provide insight into the nestmate recognition systems used. The currently accepted model of ant nestmate recognition is the *'gestalt'* model where nestmate hydrocarbons are compared against a learned template (e.g. Queller and Strassmann 2002; Errard *et al.* 2006). With such a system it is difficult to see how a social parasite that lacks many of the required cues would not be perceived as an intruder, since it would have a very poor match to the colony template. However, there are alternative models of colony recognition based on sensory habituation (Wallis 1963; Kaib *et al.* 2002), under which chemical insignificance would have a clear advantage, since a lack of cues would provide the same set of stimuli as cues to which a guard worker has become habituated.

4.7.1.2 Camouflage

Camouflage has been used in many different ways in the literature, but here we follow Howard *et al.* (1990a) in using it to mean the acquisition of cues or signals by a social parasite from the environment or from other organisms, and their subsequent use to disguise itself from its host (Fig. 4.2c). It is thus a 'passive' process, in that the social parasite does not directly produce the cue or signal itself, although the acquisition of the cue may be very active. Transfer of recognition cues between colony members is common in social insects (Vienne *et al.* 1995; Dahbi *et al.* 1999; Lenoir *et al.* 2001), and is the basis of the so-called *gestalt* model of colony recognition (Crozier and Dix 1979). Hence, it is not surprising that many social parasites engage in interactions with their hosts that seem designed to transfer such cues, for example engaging in food exchange with their host ants (e.g. Cammaerts 1995; see also Fig. 4.3b) or capturing and feeding on some host individuals (e.g. Elgar and Allan 2004)

4.7.1.3 Mimicry

Mimicry involves the production of a cue or signal by a social parasite that is used to disguise itself

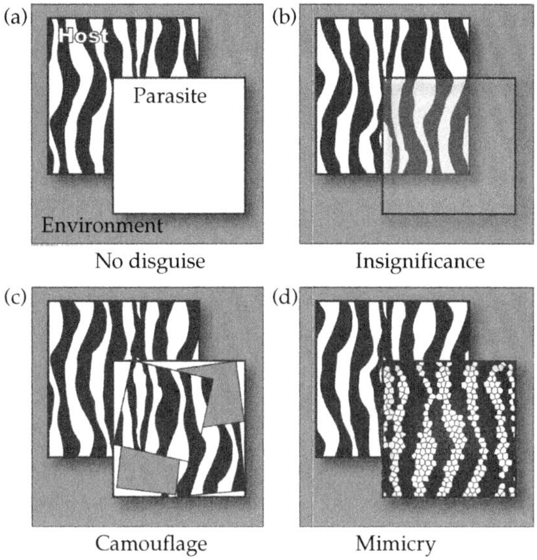

Figure 4.2 The three different methods of disguise used by social parasites. a) An undisguised parasite presents cues that easily enable a potential host to discriminate between itself, the parasite and the environment. b) An insignificant parasite provides few or no cues that are detectable to the host, and so is effectively indistinguishable from the host itself or the environment (in visual terms, it is transparent). c) A parasite may employ camouflage by collecting host or environmental cues or signals, and presenting them to the host. This is often the most accurate form of disguise, but needs specialized behaviours to obtain camouflage materials without being detected. d) A parasite may use mimicry by manufacturing its own signals that match cues from the host (or the environment). Such signals may be produced in very different ways from those produced by the host, and may not match all aspects of the cues that are mimicked.

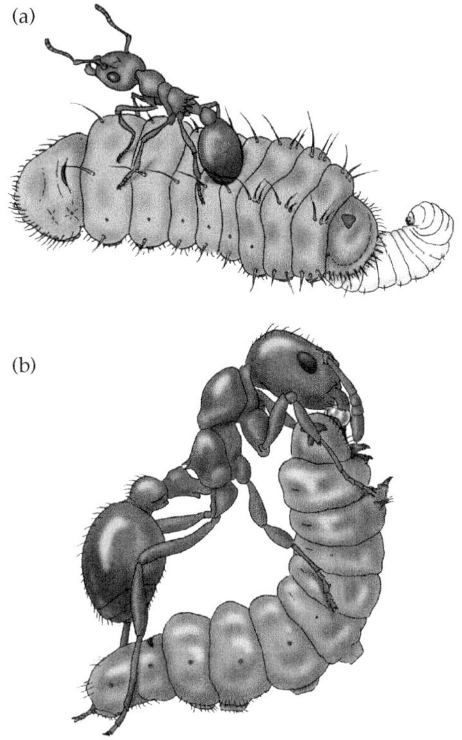

Figure 4.3 Exploitation of *Myrmica* ants by *Maculinea larvae* once inside the host nest. a) The predatory species *Maculinea arion* feeds on host ant larvae, and otherwise tries to avoid contact with worker ants. b) The cuckoo species *Maculinea alcon* both feeds directly on ant brood, and is fed trophallactically by workers of its host. Larvae stay among the brood and have frequent contact with host workers.

from its host by reproducing a cue or signal characteristic of another organism (the model—often the host itself) (Fig. 4.2d). This is an active process in which the signal is produced either during development or on demand. Distinguishing between camouflage and mimicry is often difficult, particularly in the case of chemical mimicry, but the costs and evolutionary pathways involved in these two strategies are quite distinct. We might expect, however, that mimics will match the overall suite of signals produced by their models less well than organisms employing camouflage, for the simple reason that not all aspects of the model will be equally important to the host. For example, while the cuticles of ants are covered in many hydrocarbons, it is likely that only a subset of these are used for species and colony recognition, so that a mimic can be effective just by producing these compounds, while a parasite employing camouflage is likely to use all of the compounds it acquires.

4.7.2 Other forms of exploitation via signals

Although disguise is the method of communicatory exploitation that most social parasites seem to employ, alternative manipulative signals have evolved in some cases. Two of these, appeasement and propaganda signals, have been examined in some detail and will be briefly reviewed here.

4.7.2.1 Appeasement signals

Social insects are normally aggressive toward intruders. However, several authors have suggested that social parasites may gain access to colony resources by producing signals that appease the hosts, so that aggression is reduced or eliminated (Wilson 1971; Wojcik 1989; Hölldobler and Wilson 1990). It is difficult to see how such signals can be evolutionarily stable if their only function is to reduce aggression without giving any reward to the host. Some social parasites do indeed give small rewards, which may provide short-term benefits to the host. For example, caterpillars of the inquiline butterfly *Maculinea arion* will provide a sugar-rich droplet from the dorsal nectary organ during their adoption into the host nest (Chapman 1916).

4.7.2.2 Propaganda signals

Rather than deceiving the host, another strategy for gaining entry to a colony to exploit its resources is to distract host workers away from their normal guard duties by producing signals that divert them into other activities. Such signals have been termed 'propaganda' signals (Regnier and Wilson 1971). They are particularly common among slave-making ants, and generally either mimic alarm pheromones, which cause general panic among the defenders (e.g. Regnier and Wilson 1971), or are substances that are smeared onto host workers which then make them appear hostile to the other host colony workers, resulting in their subsequent attack (e.g. Allies *et al.* 1986).

4.7.2.3 Supernormal stimuli, receiver biases, sensory traps, and addictive signals

For vertebrate brood parasites such as the cuckoo it has been suggested that the success of parasitism is enhanced by providing exaggerated signals that manipulate the host into caring for the brood parasite in preference to its own offspring (e.g. Kilner *et al.* 1999). Although this concept has caused much debate in the field of behavioural ecology (e.g. Noble *et al.* 1999), it should be equally applicable to many of the cases of sensory manipulation used by insect social parasites. For example, larvae of the socially parasitic butterfly *Maculinea rebeli* are fed in preference to larvae of their ant host, and achieve an apparently high social status within the colony, so that they are among the first members of the colony to be moved during nest emigration (Thomas *et al.* 1998). A currently popular alternative way of looking at such signals is in terms of receiver bias (Arak and Enquist 1993), where a receiver's response to signals (or cues) that are novel, but differ only slightly from those normally received, can be exaggerated. For social insects, receiver bias among honeybees for floral visual signals has recently been demonstrated (Naug and Arathi 2007). In order to successfully exploit supernormal stimuli or receiver biases, the signals produced by a social parasite must be outside the range of those normally produced by the host, and at the same time be sufficiently rarely encountered to prevent selection on hosts to reduce their response to exaggerated signals.

If hosts have sensory traps, i.e. responses to particular stimuli that are general and predictable, then these too are open to exploitation by social parasites, and indeed these have been suggested as the basis of the evolution of many symbiotic interactions (Edwards and Yu 2007). The putative general brood **pheromone** of some ants (e.g. Bigley and Vinson 1975) may provide an example of such a system that could be exploited by several social parasites, and this has been suggested to be the basis of exploitation of *Myrmica* ants by *Maculinea* butterflies (Fiedler *et al.* 1996). Other authors, however, have concluded that the susceptibility of such signals to exploitation will prevent their evolution, and that the behaviour of social insects that has been attributed to brood pheromones can be explained in other ways (e.g. Morel and Vander Meer 1988). The guest ant, *Formicoxenus nitidulus*, may also exploit a sensory trap to infect and possibly exploit its *Formica* host ants. Workers appears to have non-volatile chemicals (possibly alkadienes) on their cuticle that act as a repellent, causing any host ant that seizes a *Formicoxenus nitidulus* to immediately drop it (Martin *et al.* 2007). This deterrent signal is effective against several different species of *Formica* hosts, suggesting that it exploits a general, predictable response.

Another possible form of sensory trap exploitation is the production of addictive stimuli by social parasitism. While this intriguing possibility seems to have become entrenched in popular fiction (e.g. Werber 1998), its validity remains to be demonstrated. Many social parasites do, however, appear to modify the signals that they produce to make them more attractive to their hosts. For example, caterpillars of the Lycaenid butterfly *Niphanda fusca* have relatively high concentrations of the amino acid glycine in their dorsal nectary organ secretions, which apparently act as a taste enhancer for their *Camponotus* host ants (Wada *et al.* 2001).

Holen *et al.* (2001) modelled the use of appeasement substances by parasites, and concluded that they could be evolutionarily stable as long as parasites are rare. We would expect the same to apply for signals that exploit sensory traps or are addictive.

4.8 Interaction with the social structure of the host colony

The organization of social insect colonies has profound effects on the types of within- and between-colony communication systems that they use. For example, the number of times that a queen has mated is correlated with worker policing behaviour (Foster and Ratnieks 2000, 2001) and the signals that that involves (Endler *et al.* 2004), and information about the number of queens within a colony can be encoded in surface hydrocarbon profiles (Provost *et al.* 1992). It is not surprising, therefore, that the social structure of host colonies can have effects on the exploitation strategies of social parasites, and vice versa.

The vast majority of eusocial insect societies are headed by a single queen who has mated with a single male (Boomsma *et al.* in press), which appears to be the ancestral state (Boomsma *et al.* in press), and one of the defining features of social insects (Boomsma 2007). However, **polygyny** and **polyandry** have evolved multiple times (Boomsma *et al.* in press), both of which lead to an increase in the genetic diversity of social insect colonies (Boomsma and Ratnieks 1996). There is both theoretical (Sherman *et al.* 1988; Schmid-Hempel 1994) and empirical (Shykoff and Schmid-Hempel 1991; Liersch 1998; Palmer and Oldroyd 2003; Hughes and Boomsma 2004) support for the idea that high genetic diversity among colony members reduces colony susceptibility to pathogens and parasites. However, a greater genetic diversity among colony members also means that thresholds for self/non-self recognition are likely to be higher, and that there is less between-colony variation in recognition signals (Fig. 4.4). Both of these features mean that more genetically diverse colonies are likely to be more easily exploited by social parasites, a pattern that contrasts with their exploitation by other types of parasites and pathogens. Gardner *et al.* (2007) showed that indeed colonies of *Formica lemani* infested by the inquiline hoverfly *Microdon mutabilis* had lower within-colony relatedness than uninfected colonies - a pattern that also holds true for *Myrmica rubra* nests infected by *Maculinea alcon* (Fig. 4.4; Als 2001).

Social parasites often alter the social structure of their host colonies. This is seen most clearly in slave-makers, where the host worker force is reared under a completely different (unrelated) social structure, and in those temporary social parasites that kill the host queens. However, there may also be more subtle effects when social parasites alter the communication within host colonies. For example, it has been suggested that *Maculinea* butterflies mimic signals of queens of their host ants, both behaviourally (Elmes *et al.* 1991) and acoustically (K. Schönrogge *et al.* unpublished data) once they are inside the host colony.

It is also possible for social parasites to directly alter the signals produced by their host colony. For example, while the trophallactic exchange of material between hosts and social parasite individuals within the colony allows parasites to acquire host-derived cuticular hydrocarbons, it can also be a channel for parasite-derived hydrocarbons to be provided to hosts and distributed throughout the host colony (Lorenzi 2003).

4.9 Specific examples of communications in social parasitism

In this section we will briefly review some typical examples of social parasitism that have been studied in their natural context. The examples are biased towards social parasites of ants, both because this reflects our own backgrounds and expertise and because exploitative interactions with ants are far commoner than similar interactions with

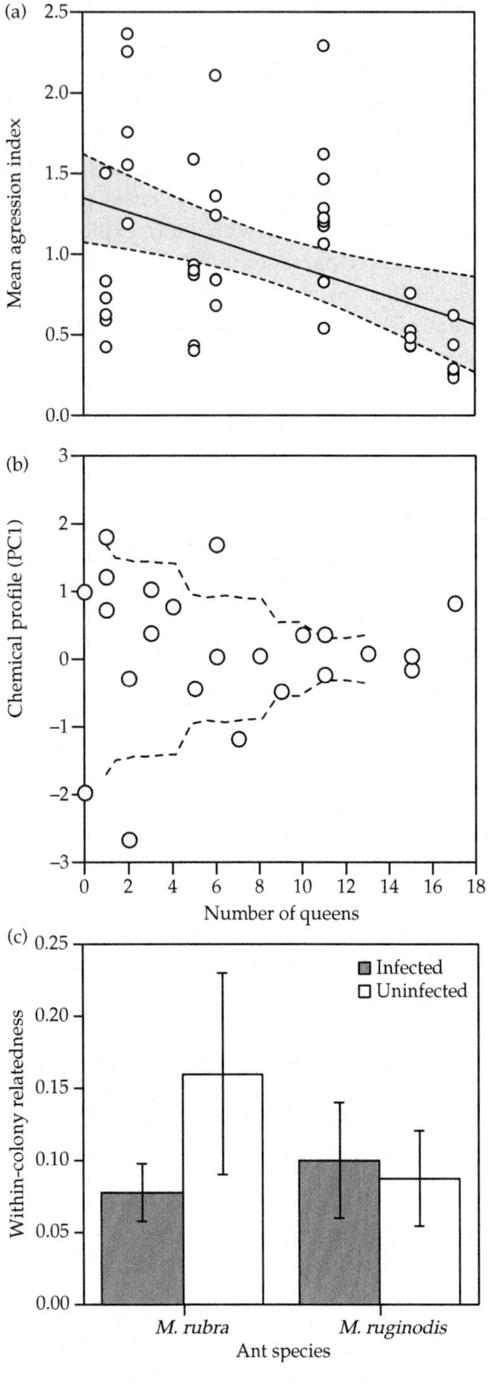

Figure 4.4 Relationship between social structure of *Myrmica rubra* colonies and their susceptibility to parasitism by *Maculinea alcon.* a) The relationship between inter-colony aggression in *Myrmica rubra* and the number of queens present. The aggression index is that used by (d'Ettorre *et al.* 2000). The reduction in aggression with increasing number of queens suggests that genetically more heterogeneous colonies accept a wider range of phenotypes as nestmates. **b)** Reduction in the variance in chemical profiles between colonies with increasing number of queens. Chemical profiles are summarized as the first principal component of a principal component analysis of 53 hydrocarbon peaks from a single ant worker from each of 23 colonies of *Myrmica rubra* not infected by *Maculinea alcon.* The dotted lines show ±1SD for a 7-point moving sample window **c)** Relatedness within colonies of *Myrmica rubra* and *Myrmica ruginodis* collected on the island of Læsø that are either infected with *Maculinea alcon* caterpillars or uninfected (Als 2001). Relatedness is significantly higher in uninfected than in infected *Myrmica rubra* colonies (unpaired relatedness-difference by nest = 0.0783, $p = 0.0064$), but not in *Myrmica ruginodis* colonies (unpaired relatedness-difference by nest = 0.0116, $p = 0.48$). Relatedness values and statistics are calculated from variation in the microsatellites *Myrt 4*, *MP-67* and *Trachy 11/12* among at least 10 workers from each of ten nests in each category, using the program Relatedness 5.08 (Queller and Goodnight 1989)

bees, wasps, and termites (Wilson 1971). We particularly concentrate on the interactions between *Maculinea* butterflies and ants, and provide some new data from our own research that demonstrate the complexity of communication strategies that only emerges when an interaction is studied exhaustively.

4.9.1 Microgyne *Myrmica rubra* (intraspecific inquiline)

Variability in queen size among *Myrmica* ants has been noted for some time, and a distinction was made between *microgyne* and *macrogyne* forms of *Myrmica ruginodis* (at that time referred to as *M. rubra*) by Brian and Brian (1955), followed by much pioneering ecological work on these forms (Elmes 1991). A microgyne form of *M. rubra* was also known to exist, but the differences between this and the microgyne form of *M. ruginodis* were not clearly apparent until the work of Elmes (1976), in part because of the nomenclatural confusion of *M. rubra* and *M. ruginodis* in the early 20th century. Seifert (1993) raised the microgyne form of *M. rubra* to the status of a species, *Myrmica microrubra,* which was considered to be an inquiline parasite of *M. rubra* colonies. Since then, genetic and other studies have revealed that there is still gene flow between *M. rubra* and *M. microrubra* leading to a recent suggestion to synonymize the species (Steiner *et al.* 2006). However, locally, there are high levels of genetic differentiation between *M. rubra* and *M. microrubra* suggesting that this 'species' may represent a syndrome of recurrently evolving intraspecific socially parasitic traits that are on the brink of becoming reproductively isolated.

4.9.2 *Acromyrmex insinuator* (related inquiline)

This inquiline leaf-cutter ant social parasite was only recognized as a species separate from its host relatively recently, following genetic studies on apparently polygynous colonies of its host *Acromyrmex echinatior* (Schultz *et al.* 1998). It provides an archetypal example of Emery's rule, being the sister taxa of its host (Sumner *et al.* 2004a), from which it may have only diverged recently, as it shows a combination of traits of its host (Sumner *et al.* 2003a,b) and derived traits shown by other intrageneric inquilines (Schultz *et al.* 1998; Bekkevold and Boomsma 2000; Sumner *et al.* 2004b). Mated queens of the parasite gain access to host colonies, where they utilize the resources in the fungus garden to produce their own offspring. While some *A. insinuator* workers are produced, the vast majority of the parasite's offspring develop as sexuals, and at the expense of the host, which normally produces no sexuals at all (Bekkevold and Boomsma 2000). Thus the social parasite effectively castrates the host, a common strategy of parasites for diverting host resources away from reproduction and to the parasite (Baudoin 1975). The host fungus garden is exhausted by the parasite in one or two reproductive seasons, and the host colony subsequently dies (Bekkevold and Boomsma 2000).

While it is still not clear how queens of *A. insinuator* gain access to colonies of *A. echinatior*, a study on the workers produced by *A. insinuator* has shown that they are very depauperate in cuticular hydrocarbons, and those that are found on the cuticle are unusually long-chain molecules outside the range normally found on ants (Lambardi *et al.* 2007). The use of chemical insignificance as a method of infecting the host colonies therefore seems likely.

One interesting twist in this system is that there is evidence that closely related free-living *Acromyrmex octospinosus* use not only their own cuticular hydrocarbons for nestmate recognition, but also compounds derived from their mutualistic fungus (Richard *et al.* 2007). *Acromyrmex octospinosus* is far less susceptible to parasitism by *A. insinuator* than *A. echinatior* (Schultz *et al.* 1998), although both potential hosts are sympatric and use the same genetic pool of fungal symbionts (Richard *et al.* 2007). It is possible, therefore, that the incorporation of fungus-derived chemicals into the recognition system of *A. octospinosus* has provided it with a degree of immunity from parasitism by *A. insinuator*.

Acromyrmex is so far the only genus of fungus-growing ants in which inquiline social parasites have been found. This is remarkable because there are several independent evolutionary developments towards inquilinism within this genus, including several species that were previously described under the genus *Pseudoatta*, which is now known to be a highly derived *Acromyrmex* (Sumner *et al.* 2004a). A recent study by De Souza *et al.* (2007) discovered yet another *Acromyrmex* inquiline, suggesting that *Acromyrmex* is at least as rich in social parasites as *Myrmica*. This underlines that the question as to why some clades have many and others have no inquilines is an important one to address in future comparative studies.

4.9.3 The Amazon ant *Polyergus rufescens* (related slave-maker)

This species was one of the first slave-making species to be recognized (Rennie 1834), and is probably the best-studied in terms of chemical communication with its hosts. Newly mated queens of this species must enter nests of their hosts (various *Formica* species), kill the queen, and take over reproduction using the resources of the colony. The initial infection of a host colony is achieved through a dual strategy of chemical insignificance (Lenoir *et al.* 2001) and the use of Dufour's gland substances that reduce aggression by the host ants. This was initially put forward as one of the few known examples of an appeasement substance (Mori *et al.* 2000b; see above), but subsequent bioassays suggest that it is a repellent that effectively reduces contact between the parasite queen and host workers (d'Ettorre *et al.* 2000). Once the colony is taken over, the queen rapidly acquires a cuticular chemical profile almost identical to that of the host queen (Lenoir *et al.* 2001). This is one of the clearest cases of chemical camouflage, since she will actively bite and rub the carcass of the dead queen to obtain cuticular hydrocarbons (Mori *et al.* 2000a), a behaviour that has also been described in *Polyergus breviceps* (Topoff and Zimmerli 1993; Johnson *et al.* 2001).

Shortly after establishment, the parasite workers begin to emerge into a worker force dominated by the host workers. Callow workers allowed to eclose in isolation show a cuticular hydrocarbon profile that closely matches that of workers of their most common host species (*Formica cunicularia*), suggesting true chemical mimicry (d'Ettorre *et al.* 2002), but in nests of other hosts they quickly match the hydrocarbon profiles of those species, probably through chemical camouflage following trophallaxis (d'Ettorre *et al.* 2002).

The cuticular chemical profiles of host workers remain somewhat distinct from those of *P. rufescens* workers, and the contents of their *post-pharyngeal gland* (believed to be the reservoir for the *gestalt* mixture of hydrocarbons) are even more distinct (Bonavita-Cougourdan *et al.* 1997), providing a possible insight into which components of the mixture on the cuticle are involved in nestmate recognition.

4.9.4 *Microdon* flies (unrelated inquilines)

Microdon is a very large, primitive and diverse genus of syrphid flies (Stahls *et al.* 2003), mostly found in Central and South America, but with representatives that are ant social parasites in northern temperate regions (e.g. Duffield 1981). The adult flies lay their eggs just outside the nests of the host ants, and the larvae enter the host nests and feed on the ant brood. They pupate within the nest, inside a particularly tough puparium. Each *Microdon* species seems to be highly host specific, not only depending on a single host species but on very local populations of that species (Elmes *et al.* 1999). The eggs of *Microdon mutabilis* appear to be protected by a chemical coating applied by the mother during oviposition, which is specific to local populations of its host ant, *Formica lemani*, and which is relatively volatile so that protection only lasts a few hours (Schönrogge *et al.* 2006). At least three Nearctic species of *Microdon* have been observed being carried by their host ants in the same manner as, and together with, their own cocoons, so that disguise as cocoons by chemical mimicry or camouflage has been suggested (Garnett *et al.* 1985). Further investigations of two of these species have confirmed similarity between the surface hydrocarbons of the *Microdon* and host brood (Howard *et al.* 1990a,b) and radioactive labelling studies suggest that *Microdon albicomatus* uses true chemical mimicry (Howard *et al.* 1990b).

Microdon mutabilis shows remarkably low dispersal, so that subsequent generations of parasites are most likely to parasitize the same host colony (Schönrogge *et al.* 2006). Such vertical (or strictly within-host) transmission of parasites to the same colony between generations is expected to provide strong selection for reduced (Read 1994) or intermediate (Antia *et al.* 1994) virulence. By preying selectively on eggs and young larvae, *M. mutabilis* appears to alter the social structure of its host colonies, so that reproduction is favoured over colony growth (c.f. Oster and Wilson 1978). It is unclear whether such a strategy may reduce the effect of the parasite on host fitness, or allow the persistence of the parasite by effectively enforcing dispersal in its host.

4.9.5 *Maculinea* butterflies (unrelated inquilines)

The large blue butterflies of the genus *Maculinea* have been more intensively studied in recent years than any other social parasite group. This is partly because these butterflies have become conservation icons among the invertebrates (e.g. Pyle *et al.* 1981; Settele *et al.* 2005), but also because butterflies in general tend to be better studied than other insect groups (e.g. Thomas and Clarke 2004).

Larvae of *Maculinea* butterflies spend their first three larval instars developing on or inside the flowers of specific host plants (Als *et al.* 2004), after which they moult one last time, leave the plant, and drop to the ground where they wait to be discovered by workers of *Myrmica* ants (Frohawk 1924). After discovery by a *Myrmica* worker, the *Maculinea* larva will be picked up and taken back to the ant nest. This 'adoption' procedure varies somewhat among the different species. In *Maculinea rebeli*, the larva is quickly picked up by any *Myrmica* worker that finds it, and carried back to the host colony (Elmes *et al.* 1991). In *Maculinea arion*, a complex and apparently ritualized series of interactions takes place between caterpillar and ant, involving the secretion of many droplets of a sugar-rich fluid from the caterpillar's dorsal nectary organ (DNO) (Frohawk 1924). Caterpillars of *Maculinea alcon* in Denmark are likely to encounter two different potential host ants and one non-host *Myrmica* species (Als *et al.* 2002), and interact somewhat differently with each of these ant species (Als *et al.* 2001). In a preliminary set of observations (Kiesbüy 1999) it was found that *M. alcon* caterpillars produced droplets of secretion from the DNOs in three out of four laboratory adoptions by *Myrmica ruginodis* colonies, whereas no droplets were produced in five adoptions by *Myrmica rubra* colonies (Fisher exact test; $P = 0.048$). The role of the contents of the DNO still remains to be investigated, but in other lycaenids that have specific interactions with ants, amino acids and possibly other compounds that are attractive to the host ants are added to this secretion (Pierce and Nash 1999; Wada *et al.* 2001).

Maculinea butterflies are generally quite specific as to which *Myrmica* species they parasitize (Thomas *et al.* 1989; Elmes *et al.* 1994, Stankiewicz and Sielezniew 2002; Als *et al.* 2004), and much of this variation can be explained in terms of how closely they mimic the cuticular chemistry of different potential host species (Akino *et al.* 1999; Elmes *et al.* 2002; Schönrogge *et al.* 2004; but see Schlick-Steiner *et al.* 2004; Pech *et al.* 2007). Caterpillars of *M. alcon* in Denmark take longer to be picked up in the field (mean ± SE, 37.6 ± 20.7 s) than caterpillars of *M. rebeli* in the Haute Alpes, France (2.1 ± 1.1 s; data from Elmes *et al.* 1991), although the basis for this difference is unclear, since these interactions involve both different ant species and substantially different habitats. It takes colonies of non-host *Myrmica scabrinodis* an order of magnitude longer to adopt caterpillars of *M. alcon* from Denmark

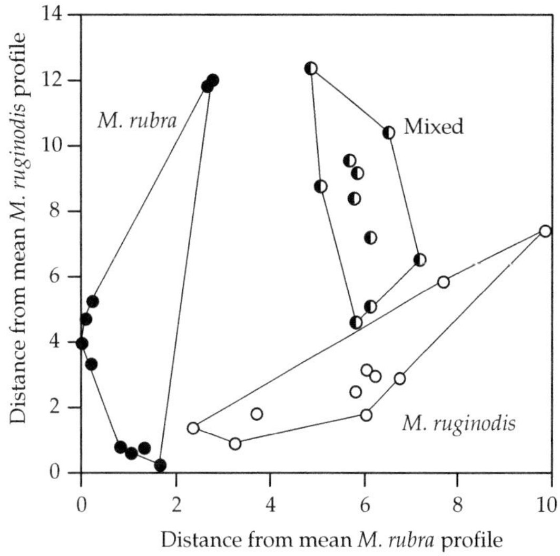

Figure 4.5 General matching of host ant hydrocarbon profiles in three populations of *Maculinea alcon* that exploit different *Myrmica* host ant species in Denmark. Each point represents the chemical distance (squared Mahalanobis distance) of an individual pre-adoption caterpillar of *Maculinea alcon* from the average chemical profile of *Myrmica rubra* and *Myrmica ruginodis*. Distances are based on fifteen compounds found consistently on the surface of all species. In those populations that use a singe host ant species, chemical distances from that species are consistently low, while those from the alternative species are much more variable. In the mixed population, caterpillars are relatively dissimilar to both host ants.

than colonies of *M. rubra* or *M. ruginodis*, which are suitable as hosts (Als *et al.* 2001, 2002). The difference in host use and adoption time by populations of *M. alcon* in Denmark can largely be explained by matching of host surface hydrocarbons (Fig. 4.5; Nash *et al.* 2008).

Once inside an appropriate *Myrmica* nest, the different *Maculinea* species follow one of two foraging strategies (Elmes *et al.* 1991; Als *et al.* 2004). *Maculinea arion*, *Maculinea teleius* and *Maculinea nausithous* gain all their nutrition from feeding directly on the ant brood (Fig. 4.3a), and tend to avoid direct contact with the worker ants whenever possible (Thomas and Elmes 1998). *Maculinea alcon* and *M. rebeli* feed on the ant brood, but are also fed by workers ants by trophallaxis (Fig. 4.3b; Thomas and Elmes 1998). This means that there is much more scope for taking advantage of chemical camouflage, and indeed caterpillars of *M. alcon* and *M. rebeli* collected from within ant nests very closely match the surface hydrocarbon profiles of theirs hosts, although there is also some evidence of active chemical mimicry (Schönrogge *et al.* 2004).

While chemical mimicry or camouflage appear to be the most important mechanisms by which *Maculinea* butterflies integrate into their host ant colonies, there are also other forms of communication with the host that appear to be important, while not being fully understood. For example, larvae of *M. alcon* and *M. rebeli* appear to mimic the behaviour of worker ants or ant larvae when soliciting trophallaxis (Fig. 4.3; Elmes *et al.* 1991), and both larvae and pupae of *Maculinea* butterflies produce sounds that may be both air- and substrate-borne (Devries *et al.* 1993; Schönrogge *et al.* 2005). The effects of the presence or absence of a queen on *Maculinea* larval survival (Fig. 4.6; Thomas and Wardlaw 1990), the high social rank that larvae of *M. rebeli* achieve in nests of their primary hosts (Thomas *et al.* 1998), and the similarity in sounds produced by *Maculinea* larvae and *Myrmica* queens (K. Schönrogge *et al.* unpublished data) have led to the suggestion that once integrated into the host nest, *Maculinea* larvae may produce an entire spectrum of signals characteristic of the queens of their host ants.

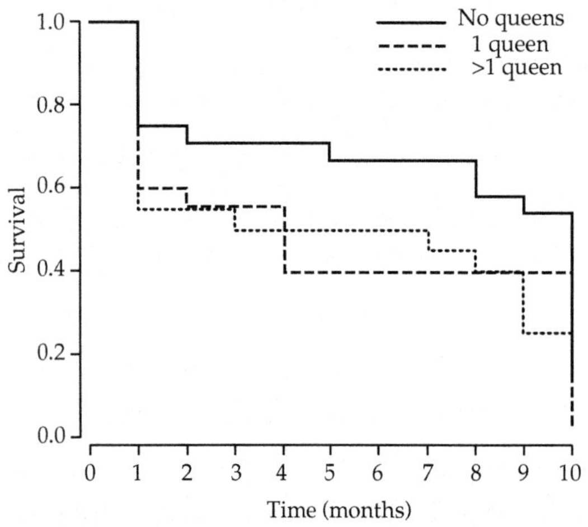

Figure 4.6 The effect of host ant queens on survivorship of *Maculinea alcon*. Survivorship of larvae of *Maculinea alcon* raised in laboratory nests of *Myrmica rubra* provided with different numbers of queens. All nests were created by splitting polygynous colonies into three subunits consisting of 200 worker ants, 0.5 g of brood and 0, 1 or more queens. The number of queens in the >1 treatment varied between colonies. Each nest was allowed to adopt four caterpillars of *Maculinea alcon*, and their survivorship was assessed on a monthly basis for 10 months (all surviving individuals eclosed as adults before month 11). A Cox nonparametric proportional hazard model of survivorship, with nests nested within numbers of queens, shows a difference in survivorship between the three treatments that is on the border of significance (LR $\chi^2 = 5.65$, d.f. = 2, p = 0.058). There is a significant difference in survivorship between caterpillars in nests without queens and those with queens (LR $\chi^2 = 5.41$, d.f. = 1, p = 0.020).

4.10 Concluding remarks

While there are many species of social parasites, they have a very uneven distribution between the different groups of social insects, with many more social parasites being found among the ants than among the bees, wasps, and termites. Wilson (1971) attempted to provide an overview of the diversity of social parasites within each group (although his nomenclature differs somewhat from ours). For social parasites that parasitize sister taxa he lists about 165 species within 46 genera for ants, about 34 species within 8 genera for bees, about 35 species within 5 genera for wasps, but no termite sister-taxa parasites. For unrelated inquilines, he lists 27 families of invertebrates that are social parasites of ants, 8 that parasitize termites, but only 3 and 2, respectively, that parasitize bees and wasps. It is also striking, however, that there is no information about the basic biology of more than 40% of the interactions between other invertebrates and social insects, something that has remained essentially unchanged over the last 35 years.

The dominance of ants as hosts for both sister taxa and unrelated social parasites (Thomas *et al.* (2005) estimate that there may be 10,000–20,000 species of ant social parasites) may well be related to their dominance in terrestrial ecosystems (Wilson 1971; Hölldobler and Wilson 1990; Folgarait 1998) and their efficiency at gathering resources (Carroll and Janzen 1973). The lack of unrelated inquilines among the bees and wasps is not very surprising, since they essentially have annual colonies, which makes exploitation by inquilines an unproductive strategy for all but the fastest developing social parasites. The dynamics of interactions between bees and wasps and their unrelated social parasites therefore come to resemble those between hosts and pathogens, as exemplified by the epidemic outbreaks of many social parasites of the honeybee such as the varroa mite (Martin 1998).

The absence of sister-taxa social parasites among the termites is the most puzzling pattern among the social parasites, and one that still awaits explanation. One possibility, however, is that it is the interaction between social systems and within-colony communication that provides a barrier to the evolution of social parasitism in termites and facilitates it in ants. Termites are strictly monogamous, which means that all workers within a colony are full siblings (see Boomsma 2007 for a recent review), so that the task of discriminating between colony mates and intruders is simplified. Many ant species, on the other hand, have polygynous colonies, that usually arise from secondary adoption of (often unrelated) queens (Reeve *et al.* 1993). This means that there are mechanisms present for the acceptance of individuals that do not provide exactly the same signals, and which are therefore open to exploitation by social parasites. In this context it is interesting to note that the genetic basis for hydrocarbon profiles used in colony recognition has proved to be relatively easy to demonstrate in termites (e.g. Adams 1991; Husseneder *et al.* 1998; Kaib *et al.* 2004), while there seems to be a large environmental component to nestmate recognition systems in (polygynous) ants (e.g. Stuart 1987; Liang and Silverman 2000).

In this chapter we have attempted to provide a brief overview of the main characteristics of communication between social parasites and their social insect hosts. It is clear that while many putative examples of social parasitism are known, there are very few for which the intimate details of how parasites communicate with and manipulate their hosts have been investigated in depth.

One pattern that seems to emerge from the discussion and examples that we have provided is that many social parasites employ multiple strategies to exploit their hosts, sometimes combining insignificance, camouflage, and mimicry at different stages in the process of exploitation (e.g. *Polyergus rufescens*), and often employing several different communication channels at once.

Different methods of host deception are likely to not only have different evolutionary origins but also different evolutionary consequences. True mimicry means that host and parasite are using a different apparatus to produce the signals that are acting as model and mimic, which allows the possibility of evolutionary arms races in signals and signal mimics, involving co-evolutionary changes in degree of mimicry by parasites and in discrimination ability of hosts. Camouflage, however, means that the signals of models and 'mimics' are produced in the same way, so that there is no possibility for co-evolution in the signals themselves (although there may still be selection on hosts to prevent parasites obtaining their camouflage materials). Insignificance is a strategy that is highly dependent on the ways that hosts can detect parasites, so that there is likely to be another type of ongoing evolutionary arms race—one in which hosts are selected to constantly evolve new channels for detecting parasites, and parasites are selected to reduce the cues available within these channels.

If discrimination is costly to hosts, then the frequency of parasitism is likely to be a key factor in determining whether hosts are likely to mount an evolutionary response to parasites. Social parasites are, in general, rare, which is another reason why so little is known about many of their life histories. However, there is great geographical variation in the prevalence of social parasites (Als *et al.* 2002; Foitzik *et al.* 2003), and many social parasites seem to exist in a geographical mosaic with their hosts, which may reflect geographical variation in co-evolutionary selection and host response (Fischer and Foitzik 2004; Thompson 2005; Nash *et al.* 2008). The geographical dimension, however, is missing from many other studies of social parasites, and this should prove a fruitful area of research in the future.

Although the concepts involved in social parasite studies are not new—indeed most of the foundations of the field were laid down at the end of the 19th century in the pioneering work of Wasmann (1886, 1888, 1889a,b)—modern techniques are allowing a much more detailed examination of the interactions between social parasites and hosts. Genetic markers now provide the possibility to examine the relative contribution of different lineages within social insect colonies to the maintenance and reproductive functions of the colony (e.g. Hughes and Boomsma 2008), as well as allowing the discovery of cryptic social parasites (e.g. Schultz *et al.* 1998; Schönrogge *et al.* 2002; Savolainen and Vepsalainen 2003). We predict that the examination of intraspecific social parasitism and its relationship to alternative reproductive tactics will become a growth area in the next decades.

The growth of genomics also has the potential to allow the genetic basis of communication between parasites and hosts to be examined. Those social parasites such as *Maculinea alcon* that show geographical variation in host use should be particularly useful model systems for tracking down the underlying genetics of host mimicry.

Finally, studies of social parasites have the potential to elucidate the importance of different types of communication within social insects, since they can provide examples of how communication channels are hijacked, and so allow the relative importance of different signal components to be examined. This is an area that has so far remained largely unexplored, but for which we see a bright future.

Summary

All parasites need to evade host defences to be successful. Social parasites, however, face unique challenges and opportunities. Their hosts are particularly well defended against intruders, but their social communication systems provide an alternative means of exploitation if social parasites can evolve ways to subvert this system for their own ends. Here we briefly review the range of tactics used by social parasites to exploit their hosts, and the communication channels and strategies used. Detailed analysis is presented of a few key systems that have been particularly well studied (*Maculinea* butterflies, *Microdon* flies, and slave-making and inquiline ants). Finally we examine general patterns of how social parasites use communication with their hosts to enhance their success, and the consequences that this has for the co-evolutionary interaction between social parasites and their hosts.

Acknowledgements

We thank Patrizia d'Ettorre and David Hughes for suggesting the idea behind this chapter, and for fruitful dialogue. Some of the ideas presented here developed from a workshop on communication organized by the Centre for Social Evolution (CSE), Copenhagen University. We are particularly indebted to Elizabeth Tibbetts, David Harper, and Craig Roberts for stimulating discussions at that workshop. We are also grateful to Thomas Als, Maëlle Durey, and Thomas Hansen for permission to present some of their unpublished data on *Maculinea alcon*, and to all members of the CSE for a stimulating research atmosphere. Funding to the authors was provided by a grant to the CSE from the Danish National Research Foundation.

References

Aanen, D.K. and Boomsma, J.J. (2006). Social-insect fungus farming. *Current Biology*, **16**, 1014–1016.

Adams, E.S. (1991). Nest-mate recognition based on heritable odors in the termite *Microcerotermes arboreus*. *Proceedings of the National Academy of Sciences of the USA*, **88**, 2031–2034.

Adams, R.M.M., Mueller, U.G., Schultz, T.R., and Norden, B. (2000). Agro-predation: usurpation of attine fungus gardens by *Megalomyrmex* ants. *Naturwissenschaften*, **87**, 549–554.

Akino, T., Knapp, J.J., Thomas, J.A., and Elmes, G.W. (1999). Chemical mimicry and host specificity in the butterfly *Maculinea rebeli*, a social parasite of *Myrmica* ant colonies. *Proceedings of the Royal Society B: Biological Sciences*, **266**, 1419–1426.

Akre, R.D., Garnett, W.B., Macdonald, J.F., Greene, A., and Landolt, P. (1976). Behavior and colony development of *Vespula pensylvanica* and *V. atropilosa* (Hymenoptera: Vespidae). *Journal of the Kansas Entomological Society*, **49**, 63–84.

Allies, A.B., Bourke, A.F.G., and Franks, N.R. (1986). Propaganda substances in the cuckoo ant *Leptothorax kutteri* and the slave-maker *Harpagoxenus sublaevis*. *Journal of Chemical Ecology*, **12**, 1285–1293.

Als, T.D.A. (2001). *Evolutionary and Ecological Interactions between* Maculinea *Parasites and their* Myrmica *Ant Hosts*. Department of Ecology and Genetics, University of Aarhus, Århus, Denmark.

Als, T.D., Nash, D.R., and Boomsma, J.J. (2001). Adoption of parasitic *Maculinea alcon* caterpillars

(Lepidoptera: Lycaenidae) by three *Myrmica* ant species. *Animal Behaviour*, **62**, 99–106.

Als, T.D., Nash, D.R., and Boomsma, J.J. (2002). Geographical variation in host-ant specificity of the parasitic butterfly *Maculinea alcon* in Denmark. *Ecological Entomology*, **27**, 403–411.

Als, T.D., Vila, R., Kandul, N.P. *et al.* (2004). The evolution of alternative parasitic life histories in large blue butterflies. *Nature*, **432**, 386–390.

Altizer, S., Nunn, C.L., Thrall, P.H. *et al.* (2003). Social organization and parasite risk in mammals: integrating theory and empirical studies. *Annual Review of Ecology, Evolution and Systematics*, **34**, 517–547.

Antia, R., Levin, B.R., and May, R.M. (1994). Within-host population dynamics and the evolution and maintenance of microparasite virulence. *The American Naturalist*, **144**, 457–472.

Arak, A. and Enquist, M. (1993). Hidden preferences and the evolution of signals. *Philosophical Transactions of the Royal Society B: Biological Sciences*, **340**, 207–213.

Axelrod, R. and Hamilton, W. (1981). The evolution of cooperation. *Science*, **211**, 1390–1396.

Bargum, K. and Sundström, L. (2007). Multiple breeders, breeder shifts and inclusive fitness returns in an ant. *Proceedings of the Royal Society B: Biological Sciences*, **274**, 1547–1551.

Baudoin, M. (1975). Host castration as a parasitic strategy. *Evolution*, **29**, 335–352.

Beibl, J., Stuart, R.J., Heinze, J., and Foitzik, S. (2005). Six origins of slavery in formicoxenine ants. *Insectes Sociaux*, **52**, 291–297.

Bekkevold, D. and Boomsma, J.J. (2000). Evolutionary transition to a semelparous life history in the socially parasitic ant *Acromyrmex insinuator*. *Journal of Evolutionary Biology*, **13**, 615–623.

Bigley, W.S. and Vinson, S.B. (1975). Characterization of a brood pheromone isolated from sexual brood of the imported fire ant, *Solenopsis invicta*. *Annals of the Entomological Society of America*, **68**, 301–304.

Blum, M.S. and Brand, J.M. (1972). Social insect pheromones: their chemistry and function. *Integrative and Comparative Biology*, **12**, 553–576.

Bonavita-Cougourdan, A., Bagneres, A.G., Provost, E., Dusticier, G., and Clement, J.L. (1997). Plasticity of the cuticular hydrocarbon profile of the slave-making ant *Polyergus rufescens* depending on the social environment. *Comparative Biochemistry and Physiology B: Biochemistry and Molecular Biology*, **116**, 287–302.

Boomsma, J.J. (2007). Kin selection versus sexual selection: why the ends do not meet. *Current Biology*, **17**, R673–R683.

Boomsma, J.J. and Ratnieks, F.L.W. (1996). Paternity in eusocial Hymenoptera. *Philosophical Transactions of the Royal Society B: Biological Sciences*, **351**, 947–975.

Boomsma, J.J., Nielsen, J., Sundström, L. *et al.* (2003). Informational constraints on optimal sex allocation in ants. *Proceedings of the National Academy of Sciences of the USA*, **100**, 8799–8804.

Boomsma, J.J., Schmid-Hempel, P., and Hughes, W.O.H. (2005). Life histories and parasite pressure across the major groups of social insects. In: M. Fellowes, G. Holloway, and J. Rolff (eds), *Insect Evolutionary Ecology*, pp. 139–175. CABI Publishing, Wallingford.

Boomsma, J.J., Kronauer, D.J.C., and Pedersen, J.S. (in press). The evolution of social insect mating systems. In: J. Gadau and J. Fewell (eds), *Organization of Insect Societies—Integrated Research Projects.* Harvard University Press, Cambridge, MA.

Bourke, A.F.G. (1988). Worker reproduction in the higher eusocial hymenoptera. *The Quarterly Review of Biology*, **63**, 291–311.

Braby, M.F. (2000). *Butterflies of Australia: their Identification, Biology and Distribution.* CSIRO, Canberra.

Brandt, M., Fischer-Blass, B., Heinze, J., and Foitzik, S. (2007). Population structure and the co-evolution between social parasites and their hosts. *Molecular Ecology*, **16**, 2063–2078.

Breed, M.D., Snyder, L.E., Lynn, T.L., and Morhart, J.A. (1992). Acquired chemical camouflage in a tropical ant. *Animal Behaviour*, **44**, 519–523.

Brian, M.V. and Brian, A.D. (1955). On the two forms macrogyna and microgyna of the ant *Myrmica rubra* L. *Evolution*, **9**, 280–290.

Bronstein, J.L., Wilson, W.G., and Morris, W.F. (2003) Ecological dynamics of mutualist/antagonist communities. *The American Naturalist*, **162**, S24–S39.

Buschinger, A. (1986). Evolution of social parasitism in ants. *Trends in Ecology and Evolution*, **1**, 155–160.

Cammaerts, R. (1995). Regurgitation behavior of the *Lasius flavus* worker (Formicidae) towards the myrmecophilous beetle *Claviger testaceus* (Pselaphidae) and other recipients. *Behavioral Processes*, **34**, 241–264.

Cammaerts, R., Detrain, C., and Cammaerts, M.C. (1990). Host trail following by the myrmecophilous beetle *Edaphopaussus favieri* (Fairmaire) (Carabidae, Paussinae). *Insectes Sociaux*, **37**, 200–211.

Carroll, C.R. and Janzen, D.H. (1973). Ecology of foraging by ants. *Annual Review of Ecology and Systematics*, **4**, 231–257.

Chapman, T.A. (1916). What the larva of *Lycaena arion* does during its last instar. *Transactions of the Royal Entomological Society of London*, **1915**, 291–297.

Charlat, S. and Mercot, H. (2001). Did *Wolbachia* cross the border? *Trends in Ecology and Evolution*, **16**, 540–541.

Creemers, B., Billen, J., and Gobin, B. (2003). Larval begging behaviour in the ant *Myrmica rubra*. *Ethology, Ecology and Evolution*, **15**, 261–272.

Crozier, R.H. and Dix, M.W. (1979). Analysis of two genetic models for the innate components of colony odor in social Hymenoptera. *Behavioral Ecology and Sociobiology*, **4**, 217–224.

Cushman, J.H. and Beattie, A.J. (1991). Mutualisms: assessing the benefits to hosts and visitors. *Trends in Ecology and Evolution*, **6**, 193–195.

Cushman, J.H. and Whitham, T.G. (1989). Conditional mutualism in a membracid–ant association: temporal, age-specific, and density-dependent effects. *Ecology*, **70**, 1040–1047.

Dahbi, A., Hefetz, A., Cerdá, X., and Lenoir, A. (1999). Trophallaxis mediates uniformity of colony odor in *Cataglyphis iberica* ants (Hymenoptera, Formicidae). *Journal of Insect Behavior*, **12**, 559–567.

Davies, N.B. (2000). *Cuckoos, Cowbirds and Other Cheats*, T. & A. D. Poyser, London.

Davies, N.B., Bourke, A.F.G., and Brooke, M.D.L. (1989). Cuckoos and parasitic ants: interspecific brood parasitism as an evolutionary arms race. *Trends in Ecology and Evolution*, **4**, 274–278.

Dejean, A. and Beugnon, G. (1996). Host ant trail following by myrmecophilous larvae of liphyrinae (lepidoptera, lycaenidae). *Oecologia*, **106**, 57–62.

Devries, P.J., Cocroft, R.B., and Thomas, J. (1993). Comparison of acoustical signals in *Maculinea* butterfly caterpillars and their obligate host *Myrmica* ants. *Biological Journal of the Linnean Society*, **49**, 229–238.

De Souza, D.J., Soares, I.M.F., and Della Lucia, T.M.C. (2007). *Acromyrmex ameliae* sp. n.(Hymenoptera: Formicidae): a new social parasite of leaf-cutting ants in Brazil. *Insect Science*, **14**, 251–257.

Diamond, J. (1997). *Guns, Germs and Steel: a Short History of Everybody for the Last 13,000 Years*. Jonathan Cape, London.

Dijkstra, M.B. and Boomsma, J.J. (2003). *Gnamptogenys hartmani* Wheeler (Ponerinae: Ectatommini): an agro-predator of *Trachymyrmex* and *Sericomyrmex* fungus-growing ants. *Naturwissenschaften*, **90**, 568–571.

Dodd, F.P. (1902). Contribution to the lifehistory of *Liphyra brassolis*, Westw. *The Entomologist*, **35**, 153–156.

Duffield, R.M. (1981). Biology of *Microdon fuscipennis* (Diptera: Syrphidae) with interpretations of the reproductive strategies of *Microdon* species found north of Mexico. *Proceedings of the Entomological Society of Washington*, **83**, 716–724.

Dyer, F.C. (2002). The biology of the dance language. *Annual Review of Entomology*, **47**, 917–949.

Edwards, D.P. and Yu, D.W. (2007). The roles of sensory traps in the origin, maintenance, and breakdown of mutualism. *Behavioral Ecology and Sociobiology*, **61**, 1321–1327.

Elgar, M.A. and Allan, R.A. (2004). Predatory spider mimics acquire colony-specific cuticular hydrocarbons from their ant model prey. *Naturwissenschaften*, **91**, 143–147.

Elmes, G.W. (1976). Some observations on the microgyne form of *Myrmica rubra* L.(Hymenoptera, Formicidae). *Insectes Sociaux*, **23**, 3–21.

Elmes, G.W. (1991). Mating strategy and isolation between the two forms, macrogyna and microgyna, of *Myrmica ruginodis* (Hym. Formicidae). *Ecological Entomology*, **16**, 411–423.

Elmes, G.W., Thomas, J.A., and Wardlaw, J.C. (1991). Larvae of *Maculinea rebeli*, a large-blue butterfly, and their *Myrmica* host ants: wild adoption and behaviour in ant-nests. *Journal of Zoology*, **223**, 447–460.

Elmes, G.W., Thomas, J.A., Hammarstedt, O., Munguira, M.L., Martin, J., and Van Der Made, J.G. (1994). Differences in host-ant specificity between Spanish, Dutch and Swedish populations of the endangered butterfly, *Maculinea alcon* (DENIS et SCHIFF.) (Lepidoptera). *Memorabilia Zoologica*, **48**, 55–68.

Elmes, G.W., Barr, B., Thomas, J.A., and Clarke, R.T. (1999). Extreme host specificity by *Microdon mutabilis* (Diptera: Syrphidae), a social parasite of ants. *Proceedings of the Royal Society B: Biological Sciences*, **266**, 447–453.

Elmes, G.W., Akino, T., Thomas, J., Clarke, R.T., and Knapp, J.J. (2002). Interspecific differences in cuticular hydrocarbon profiles of *Myrmica* ants are sufficiently consistent to explain host specificity by *Maculinea* (large blue) butterflies. *Oecologia*, **130**, 525–535.

Emery, C. (1909). Über den Ursprung der dulotischen, parasitischen und myrmekophilen Ameisen. *Biologisches Centralblatt (Berlin)*, **29**, 352–362.

Endler, A., Liebig, J., Schmitt, T. *et al.* (2004). Surface hydrocarbons of queen eggs regulate worker reproduction in a social insect. *Proceedings of the National Academy of Sciences of the USA*, **101**, 2945–2950.

Errard, C., Hefetz, A., and Jaisson, P. (2006). Social discrimination tuning in ants: template formation and chemical similarity. *Behavioral Ecology and Sociobiology*, **59**, 353–363.

d'Ettorre, P., Errard, C., Ibarra, F., Francke, W., and Hefetz, A. (2000). Sneak in or repel your enemy: Dufour's gland repellent as a strategy for successful usurpation in the slave-maker *Polyergus rufescens*. *Chemoecology*, **10**, 135–142.

d'Ettorre, P., Mondy, N., Lenoir, A., and Errard, C. (2002). Blending in with the crowd: social parasites integrate into their host colonies using a flexible chemical signature. *Proceedings of the Royal Society B: Biological Sciences*, **269**, 1911–1918.

d'Ettorre, P., Heinze, J., Schulz, C., Francke, W., and Ayasse, M. (2004). Does she smell like a queen? Chemoreception of a cuticular hydrocarbon signal in the ant *Pachycondyla inversa*. *Journal of Experimental Biology*, **207**, 1085–1091.

Ewald, P.W. (1987). Transmission modes and evolution of the parasitism–mutualism continuum. *Annals of the New York Academy of Sciences*, **503**, 295–306.

Fiedler, K., Hölldobler, B., and Seufert, P. (1996). Butterflies and ants: the communicative domain. *Cellular and Molecular Life Sciences*, **52**, 14–24.

Fischer, B. and Foitzik, S. (2004). Local co-adaptation leading to a geographical mosaic of coevolution in a social parasite system. *Journal of Evolutionary Biology*, **17**, 1026–1034.

Fischer-Blass, B., Heinze, J., and Foitzik, S. (2006). Microsatellite analysis reveals strong but differential impact of a social parasite on its two host species. *Molecular Ecology*, **15**, 863–872.

Foitzik, S., Fischer, B., and Heinze, J. (2003). Arms races between social parasites and their hosts: geographic patterns of manipulation and resistance. *Behavioral Ecology*, **14**, 80–88.

Folgarait, P.J. (1998) Ant biodiversity and its relationship to ecosystem functioning: a review. *Biodiversity and Conservation*, **7**, 1221–1244.

Foster, K.R. and Ratnieks, F.L. (2000). Facultative worker policing in a wasp. *Nature*, **407**, 692–693.

Foster, K.R. and Ratnieks, F.L.W. (2001). The effect of sex-allocation biasing on the evolution of worker policing in hymenopteran societies. *The American Naturalist*, **158**, 615–623.

Frank, S.A. (2003). Perspective: repression of competition and the evolution of cooperation. *Evolution*, **57**, 693–705.

von Frisch, K. and Chadwick, L.E. (1967). *The Dance Language and Orientation of Bees*. Belknap Press of Harvard University Press, Cambridge, MA.

Frohawk, F.W. (1924). *Natural History of British Butterflies: a Complete, Original, Descriptive Account of the Life-History of Every Species Occurring in the British Islands, Together With Their Habits, Times of Appearance, and Localities, and Illustrated by F. W. Frohawk*, 2 Volumes. Hutchinson, London.

Gamboa, G.J., Klahn, J.E., Parman, A.O., and Ryan, R.E. (1987). Discrimination between nestmate and non-nestmate kin by social wasps (*Polistes fuscatus*, Hymenoptera: Vespidae). *Behavioral Ecology and Sociobiology*, **21**, 125–128.

Gardner, M.G., Schönrogge, K., Elmes, G.W., and Thomas, J.A. (2007). Increased genetic diversity as a defence against parasites is undermined by social parasites: *Microdon mutabilis* hoverflies infesting *Formica lemani* ant colonies. *Proceedings of the Royal Society B: Biological Sciences*, **274**, 103–110.

Garnett, W.B., Akre, R.D., and Sehlke, G. (1985). Cocoon mimicry and predation by myrmecophilous diptera (Diptera: Syrphidae). *The Florida Entomologist*, **68**, 615–621.

Gomulkiewicz, R., Nuismer, S.L., and Thompson, J.N. (2003). Coevolution in variable mutualisms. *The American Naturalist*, **162**, S80–S93.

Hamilton, J. (1888). Catalogue of the myrmecophilous Coleoptera with bibliography and notes. *Canadian Entomologist*, **20**, 161–166.

Hannonen, M. and Sundström, L. (2003). Worker nepotism among polygynous ants. *Nature*, **421**, 910.

Heinze, J., Stengl, B., and Sledge, M.F. (2002). Worker rank, reproductive status and cuticular hydrocarbon signature in the ant, *Pachycondyla cf. inversa*. *Behavioral Ecology and Sociobiology*, **52**, 59–65.

Henning, S.F. (1983). Chemical communication between lycaenid larvae (Lepidoptera: Lycaenidae) and ants (Hymenoptera: Formicidae). *Journal of the Entomological Society of Southern Africa*, **46**, 341–366.

Herre, E.A., Knowlton, N., Mueller, U.G., and Rehner, S.A. (1999). The evolution of mutualisms: exploring the paths between conflict and cooperation. *Trends in Ecology and Evolution*, **14**, 49–53.

Hickling, R. and Brown, R.L. (2000). Analysis of acoustic communication by ants. *The Journal of the Acoustical Society of America*, **108**, 1920.

Holen, O.H., Saetre, G.P., Slagsvold, T., and Stenseth, N.C. (2001). Parasites and supernormal manipulation. *Proceedings of the Royal Society B: Biological Sciences*, **268**, 2551–2558.

Hölldobler, B. and Carlin, N.F. (1987). Anonymity and specificity in the chemical communication signals of social insects. *Journal of Comparative Physiology A: Sensory, Neural and Behavioral Physiology*, **161**, 567–581.

Hölldobler, B. and Wilson, E.O. (1990). *The ants*, Harvard University Press, Cambridge, MA.

Howard, R.W. and Blomquist, G.J. (2005). Ecological, behavioral, and biochemical aspects of insect hydrocarbons. *Annual Review of Entomology*, **50**, 371–393.

Howard, R.W., Akre, R.D., and Garnett, W.B. (1990a). Chemical mimicry in an obligate predator of carpenter ants (Hymenoptera, Formicidae). *Annals of the Entomological Society of America*, **83**, 607–616.

Howard, R.W., Stanley Samuelson, D.W., and Akre, R.D. (1990b). Biosynthesis and chemical mimicry of cuticular hydrocarbons from the obligate predator, *Microdon albicomatus* Novak (Diptera, Syrphidae) and its ant prey, *Myrmica incompleta* Provancher (Hymenoptera,

Formicidae). *Journal of the Kansas Entomological Society*, **63**, 437–443.

Hughes, W.O.H. and Boomsma, J.J. (2004). Genetic diversity and disease resistance in leaf-cutting ant societies. *Evolution*, **58**, 1251–1260.

Hughes, W.O.H. and Boomsma, J.J. (2008). Genetic royal cheats in leaf-cutting ant societies. *Proceedings of the National Academy of Sciences of the USA*, 0710262105.

Husseneder, C., Brandl, R., Epplen, C., Epplen, J.T., and Kaib, M. (1998). Variation between and within colonies in the termite: morphology, genomic DNA, and behaviour. *Molecular Ecology*, **7**, 983–990.

Johnson, C.A., Vander Meer, R.K., and Lavine, B. (2001). Changes in the cuticular hydrocarbon profile of the slave-maker ant queen, *Polyergus breviceps* Emery, after killing a *Formica* host queen (Hymenoptera: Formicidae). *Journal of Chemical Ecology*, **27**, 1787–1804.

Kaib, M., Franke, S., Francke, W., and Brandl, R. (2002). Cuticular hydrocarbons in a termite: phenotypes and a neighbour-stranger effect. *Physiological Entomology*, **27**, 189–198.

Kaib, M., Jmhasly, P., Wilfert, L. *et al.* (2004). Cuticular hydrocarbons and aggression in the termite *Macrotermes subhyalinus*. *Journal of Chemical Ecology*, **30**, 365–385.

Keller, L. and Ross, K.G. (1998). Selfish genes: a greenbeard in the red fire ant. *Nature*, **394**, 573.

Kiesbüy, H.C. (1999). *Adoption of* Maculinea alcon *by* Myrmica *Ants: Effects of Host Ant Species on Caterpillar and Ant Behaviour.* Department of Genetics and Ecology, University of Aarhus, Århus, Denmark.

Kilner, R.M., Noble, D.G., and Davies, N.B. (1999). Signals of need in parent-offspring communication and their exploitation by the common cuckoo. *Nature*, **397**, 667–672.

Kummerli, R. and Keller, L. (2007). Reproductive specialization in multiple-queen colonies of the ant *Formica exsecta*. *Behavioral Ecology*, **18**, 375.

Lambardi, D., Dani, F.R., Turillazzi, S., and Boomsma, J.J. (2007). Chemical mimicry in an incipient leaf-cutting ant social parasite. *Behavioral Ecology and Sociobiology*, **61**, 843–851.

Lenoir, A., d'Ettorre, P., Errard, C., and Hefetz, A. (2001). Chemical ecology and social parasitism in ants. *Annual Review of Entomology*, **46**, 573–599.

Liang, D. and Silverman, J. (2000). 'You are what you eat': diet modifies cuticular hydrocarbons and nestmate recognition in the Argentine ant, *Linepithema humile*. *Naturwissenschaften*, **87**, 412–416.

Liersch, S. (1998). Genetic variation within social insect colonies reduces parasite load. *Proceedings of the Royal Society B: Biological Sciences*, **265**, 221–225.

Lohman, D.J., Liao, Q., and Pierce, N.E. (2006). Convergence of chemical mimicry in a guild of aphid predators. *Ecological Entomology*, **31**, 41–51.

Lopez-Vaamonde, C., Koning, J.W., Brown, R.M., Jordan, W.C., and Bourke, A.F.G. (2004). Social parasitism by male-producing reproductive workers in a eusocial insect. *Nature*, **430**, 557–560.

Lorenzi, M.C. (2003). Social wasp parasites affect the nestmate recognition abilities of their hosts (*Polistes atrimandibularis* and *P. biglumis*, Hymenoptera, Vespidae). *Insectes Sociaux*, **50**, 82–87.

Lowe, R.M., Ward, S.A., and Crozier, R.H. (2002). The evolution of parasites from their hosts: intra- and interspecific parasitism and Emery's rule. *Proceedings of the Royal Society B: Biological Sciences*, **269**, 1301–1305.

Maeterlinck, M. (1901) *The Life of the Bee* (reprinted 2004). Kessinger Publishing, Whitefish, MT.

Majerus, M.E.N., Sloggett, J.J., Godeau, J.F., and Hemptinne, J.L. (2007). Interactions between ants and aphidophagous and coccidophagous ladybirds. *Population Ecology*, **49**, 15–27.

Martin, S. (1998). A population model for the ectoparasitic mite *Varroa jacobsoni* in honey bee (*Apis mellifera*) colonies. *Ecological Modelling*, **109**, 267–281.

Martin, S.J., Jenner, E.A., and Drijfhout, F.P. (2007). Chemical deterrent enables a socially parasitic ant to invade multiple hosts. *Proceedings of the Royal Society B: Biological Sciences*, **274**, 2717–2721.

Michelsen, A., Kirchner, W.H., Andersen, B.B., and Lindauer, M. (1986a). The tooting and quacking vibration signals of honeybee queens: a quantitative analysis. *Journal of Comparative Physiology A: Sensory, Neural, and Behavioral Physiology*, **158**, 605–611.

Michelsen, A., Kirchner, W.H., and Lindauer, M. (1986b). Sound and vibrational signals in the dance language of the honeybee, *Apis mellifera*. *Behavioral Ecology and Sociobiology*, **18**, 207–212.

Monnin, T., Malosse, C., and Peeters, C. (1998). Solid-phase microextraction and cuticular hydrocarbon differences related to reproductive activity in queenless ant *Dinoponera quadriceps*. *Journal of Chemical Ecology*, **24**, 473–490.

Morel, L. and Vander Meer, R.K. (1988). Do ant brood pheromones exist? *Annals of the Entomological Society of America*, **81**, 705–710.

Mori, A., Grasso, D.A., Visicchio, R., and Le Moli, F. (2000a). Colony founding in *Polyergus rufescens*: the role of the Dufour's gland. *Insectes Sociaux*, **47**, 7–10.

Mori, A., Visicchio, R., Sledge, M.F. *et al.* (2000b). Behavioural assays testing the appeasement allomone of *Polyergus rufescens* queens during host-colony usurpation. *Ethology, Ecology and Evolution*, **12**, 315–322.

Moritz, R.F.A., Kirchner, W.H., and Crewe, R.M. (1991). Chemical camouflage of the death's head hawkmoth (*Acherontia atropos* L.) in honeybee colonies. *Naturwissenschaften*, **78**, 179–182.

Moritz, R.F.A., Lattorff, H.M.G., Neumann, P., Kraus, F.B., Radloff, S.E., and Hepburn, H.R. (2005). Rare royal families in honeybees, *Apis mellifera*. *Naturwissenschaften*, **92**, 488–491.

Nash, D.R., Als, T.D., Maile, R., Jones, G.R. and Boomsma, J.J. (2008). A mosaic of chemical coevolution in a large blue butterfly. *Science*, **319**, 88–90.

Naug, D. and Arathi, H.S. (2007). Receiver bias for exaggerated signals in honeybees and its implications for the evolution of floral displays. *Biology Letters*, **3**, 635–637.

Neumann, P., Radloff, S.E., Pirk, C.W.W., and Hepburn, R. (2003). The behaviour of drifted Cape honeybee workers (*Apis mellifera capensis*): predisposition for social parasitism? *Apidologie*, **34**, 585–590.

Nielsen, J., Boomsma, J.J., Oldham, N.J., Petersen, H.C., and Morgan, E.D. (1999). Colony-level and season-specific variation in cuticular hydrocarbon profiles of individual workers in the ant *Formica truncorum*. *Insectes Sociaux*, **46**, 58–65.

Noble, D.G., Davies, N.B., Hartley, I.R., and Mcrae, S.B. (1999). The red gape of the nestling cuckoo (*Cuculus canorus*) is not a supernormal stimulus for three common hosts. *Behaviour*, **136**, 759–777.

North, R.D., Jackson, C.W., and Howse, P.E. (1997). Evolutionary aspects of ant-fungus interactions in leaf-cutting ants. *Trends in Ecology and Evolution*, **12**, 386–389.

Oldroyd, B.P. (1999). Development and behaviour of anarchistic honeybees. *Proceedings of the Royal Society B: Biological Sciences*, **266**, 1875–1875.

Oster, G.F. and Wilson, E.O. (1978). *Caste and Ecology in the Social Insects*, Princeton University Press, Princeton, NJ.

Palmer, K.A. and Oldroyd, B.P. (2003). Evidence for intracolonial genetic variance in resistance to American foulbrood of honey bees (*Apis mellifera*): further support for the parasite/pathogen hypothesis for the evolution of polyandry. *Naturwissenschaften*, **90**, 265–268.

Passera, L., Gilbert, M., and Aron, S. (2001). Social parasitism in ants: effects of the inquiline parasite *Plagiolepis xene* St. on queen distribution and worker production of its host *Plagiolepis pygmaea* Latr. *Insectes Sociaux*, **48**, 74–79.

Pasteels, J.M. (2007). Chemical defence, offence and alliance in ants–aphids–ladybirds relationships. *Population Ecology*, **49**, 5–14.

Pech, P., Fric, Z., and Konvicka, M. (2007). Species-specificity of the *Phengaris* (*Maculinea*)–*Myrmica* host system: fact or myth? (Lepidoptera: Lycaenidae; Hymenoptera: Formicidae). *Sociobiology*, **50**, 1–21.

Pernal, S.F., Baird, D.S., Birmingham, A.L., Higo, H.A., Slessor, K.N., and Winston, M.L. (2005). Semiochemicals influencing the host-finding behaviour of *Varroa destructor*. *Experimental and Applied Acarology*, **37**, 1–26.

Pierce, N.E. (1995). Predatory and parasitic Lepidoptera: carnivores living on plants. *Journal of the Lepidopterists Society*, **49**, 412 453.

Pierce, N.E. and Nash, D.R. (1999). The imperial blue, *Jalmenus evagoras* (Lycaenidae). In: R.L. Kitching, E. Scheermeyer, R.E. Jones and N.E. Pierce (eds), *Biology of Australian Butterflies—Monographs on Australian Lepidoptera*. CSIRO Publishing, Canberra.

Pierce, N.E., Kitching, R.L., Buckley, R.C., Taylor, M.F.J., and Benbow, K.F. (1987). The costs and benefits of cooperation between the Australian lycaenid butterfly, *Jalmenus evagoras*, and its attendant ants. *Behavioral Ecology and Sociobiology*, **21**, 237–248.

Pierce, N.E., Braby, M.F., Heath, A. *et al.* (2002). The ecology and evolution of ant association in the Lycaenidae (Lepidoptera). *Annual Review of Entomology*, **47**, 733–771.

Provost, E., Cerdan, P., Bagneres, A.G., Morgan, E.D., and Riviere, G. (1992). Role of the queen in *Messor barbarus* colony signature. *Biology and Evolution of Social Insects*, pp. 195–202. Leuven University Press, Leuven.

Pyle, R., Bentzien, M., and Opler, P. (1981). Insect conservation. *Annual Review of Entomology*, **26**, 233–258.

Queller, D.C. and Goodnight, K.F. (1989). Estimating relatedness using genetic markers. *Evolution*, **43**, 258–275.

Queller, D.C. and Strassmann, J.E. (2002). The many selves of social insects. *Science*, **296**, 311–313.

Radchenko, A. and Elmes, G.W. (2003). A taxonomic revision of the socially parasitic *Myrmica* ants (Hymenoptera : Formicidae) of the Palaearctic region. *Annals of Zoology*, **53**, 217–243.

Ratnieks, F.L.W. (1991). The evolution of genetic odor-cue diversity in social hymenoptera. *The American Naturalist*, **137**, 202–226.

Read, A.F. (1994). The evolution of virulence. *Trends in Microbiology*, **2**, 73–76.

Reeve, H.K., Ratnieks, F.L.W., and Keller, L. (1993). *Queen Number and Sociality in Insects*, Oxford University Press, Oxford.

Regnier, F.E. and Wilson, E.O. (1971). Chemical communication and 'propaganda' in slave-maker ants. *Science*, **172**, 267.

Rennie, J. (1834). Amazon ant. *Field Naturalist*, **2**, 203–208.

Richard, F.-J., Poulsen, M., Hefetz, A., Errard, C., Nash, D., and Boomsma, J. (2007). The origin of the chemical profiles of fungal symbionts and their significance for nestmate recognition in *Acromyrmex* leaf-cutting ants. *Behavioral Ecology and Sociobiology*, **61**, 1637–1649.

Savolainen, R. and Vepsalainen, K. (2003). Sympatric speciation through intraspecific social parasitism. *Proceedings of the National Academy of Sciences of the USA*, **100**, 7169–7174.

Schlick-Steiner, B.C., Steiner, F.M., Hottinger, H. *et al.* (2004). A butterfly's chemical key to various ant forts: intersection-odour or aggregate-odour multi-host mimicry? *Naturwissenschaften*, **91**, 209–214.

Schmid-Hempel, P. (1994). Infection and colony variability in social insects. *Philosophical Transactions of the Royal Society B: Biological Sciences*, **346**, 313–321.

Schmid-Hempel, P. and Ebert, D. (2003). On the evolutionary ecology of specific immune defence. *Trends in Ecology and Evolution*, **18**, 27–32.

Schönrogge, K., Barr, B., Wardlaw, J.C. *et al.* (2002). When rare species become endangered: cryptic speciation in myrmecophilous hoverflies. *Biological Journal of the Linnean Society*, **75**, 291–300.

Schönrogge, K., Wardlaw, J.C., Peters, A.J., Everett, S., Thomas, J.A., and Elmes, G.W. (2004). Changes in chemical signature and host specificity from larval retrieval to full social integration in the myrmecophilous butterfly *Maculinea rebeli*. *Journal of Chemical Ecology*, **30**, 91–107.

Schönrogge, K., Topham, E., Wardlaw, J.C. *et al.* (2005). Acoustical interactions between *Maculinea alcon* and its host ant. In: J. Settele, E. Kuehn and J.A. Thomas (eds), *Studies on the Ecology and Conservation of Butterflies in Europe. Vol. 2: Species Ecology Along a European Gradient:* Maculinea *Butterflies as a Model*. pp. 84–87. Pensoft, Sofia.

Schönrogge, K., Gardner, M.G., Elmes, G.W. *et al.* (2006). Host propagation permits extreme local adaptation in a social parasite of ants. *Ecology Letters*, **9**, 1032–1040.

Schultz, T.R., Bekkevold, D., and Boomsma, J.J. (1998). *Acromyrmex insinuator* new species: an incipient social parasite of fungus-growing ants. *Insectes Sociaux*, **45**, 457–471.

Seifert, B. (1993). Taxonomic description of *Myrmica microrubra* n. sp.—a social parasitic ant so far known as the microgyne of *Myrmica rubra* (L.). *Abhandlungen und Berichte des Naturkundemuseums Görlitz*, **67**, 9–12.

Settele, J., Kühn, E., and Thomas, J. (eds) (2005). *Studies on the Ecology and Conservation of Butterflies in Europe. Volume 2: Species Ecology along a European Gradient:* Maculinea *Butterflies as a Model*. Pensoft, Sofia.

Shannon, C.E. (1948). A mathematical theory of communication. *Bell System Technical Journal*, **27**, 379–423.

Sherman, P.W., Seeley, T.D., and Reeve, H.K. (1988). Parasites, pathogens, and polyandry in social Hymenoptera. *The American Naturalist*, **131**, 602–610.

Shykoff, J.A. and Schmid-Hempel, P. (1991). Parasites and the advantage of genetic variability within social insect colonies. *Proceedings of the Royal Society B: Biological Sciences*, **243**, 55–58.

Stadler, B. and Dixon, A.F.G. (2005). Ecology and evolution of aphid-ant interactions. *Annual Review of Ecology, Evolution and Systematics*, **36**, 345–372.

Stahls, G., Hippa, H., Rotheray, G., Muona, J., and Gilbert, F. (2003). Phylogeny of Syrphidae(Diptera) inferred from combined analysis of molecular and morphological characters. *Systematic Entomology*, **28**, 433–450.

Stankiewicz, A. and Sielezniew, M. (2002). Host specificity of *Maculinea teleius* Bgstr. and *M. nausithous* Bgstr. (Lepidoptera : Lycaenidae) the new insight. *Annales Zoologici*, **52**, 403–408.

Steiner, F.M., Schlick-Steiner, B.C., Konrad, H. *et al.* (2006). No sympatric speciation here: multiple data sources show that the ant *Myrmica microrubra* is not a separate species but an alternate reproductive morph of *Myrmica rubra*. *Journal of Evolutionary Biology*, **19**, 777–787.

Stuart, R.J. (1987). Transient nestmate recognition cues contribute to a multicolonial population structure in the ant, *Leptothorax curvispinosus*. *Behavioral Ecology and Sociobiology*, **21**, 229–235.

Summers, K., McKeon, S., Sellars, J., *et al.* (2003). Parasitic exploitation as an engine of diversity. *Biological Reviews*, **78**, 639–675.

Sumner, S., Hughes, W.O.H., and Boomsma, J.J. (2003a). Evidence for differential selection and potential adaptive evolution in the worker caste of an inquiline social parasite. *Behavioral Ecology and Sociobiology*, **54**, 256–263.

Sumner, S., Nash, D.R., and Boomsma, J.J. (2003b). The adaptive significance of inquiline parasite workers. *Proceedings of the Royal Society B: Biological Sciences*, **270**, 1315–1322.

Sumner, S., Aanen, D.K., Delabie, J., and Boomsma, J.J. (2004a). The evolution of social parasitism in *Acromyrmex* leaf-cutting ants: a test of Emery's rule. *Insectes Sociaux*, **51**, 37–42.

Sumner, S., Hughes, W.O.H., Pedersen, J.S., and Boomsma, J.J. (2004b). Ant parasite queens revert to mating singly. *Nature*, **428**, 35–36.

Sumner, S., Lucas, E., Barker, J., and Isaac, N. (2007). Radio-tagging technology reveals extreme nest-drifting behavior in a eusocial insect. *Current Biology*, **17**, 140–145.

Thomas, J.A. and Clarke, R.T. (2004). Extinction rates and butterflies: reply. *Science*, **305**, 1563–1564.

Thomas, J.A. and Elmes, G.W. (1998). Higher productivity at the cost of increased host-specificity when *Maculinea* butterfly larvae exploit ant colonies through trophallaxis rather than by predation. *Ecological Entomology*, **23**, 457–464.

Thomas, J.A. and Wardlaw, J.C. (1990). The effect of queen ants on the survival of *Maculinea arion* larvae in *Myrmica* ant nests. *Oecologia*, **85**, 87–91.

Thomas, J.A., Elmes, G.W., Wardlaw, J.C., and Woyciechowski, M. (1989). Host specificity among *Maculinea* butterflies in *Myrmica* ant nests. *Oecologia*, **79**, 452–457.

Thomas, J.A., Elmes, G.W., and Wardlaw, J.C. (1998). Polymorphic growth in larvae of the butterfly *Maculinea rebeli*, a social parasite of *Myrmica* ant colonies. *Proceedings of the Royal Society B: Biological Sciences*, **265**, 1895–1903.

Thomas, J.A., Schönrogge, K., and Elmes, G.W. (2005). Specializations and host associations of social parasites of ants. In: M. Fellowes, G. Holloway, and J. Rolff (eds), *Insect Evolutionary Ecology*, pp. 479–518. CABI Publishing, Wallingford.

Thompson, J.N. (2005). *The Geographic Mosaic of Coevolution*, University of Chicago Press, Chicago.

Tibbetts, E.A. (2002). Visual signals of individual identity in the wasp *Polistes fuscatus*. *Proceedings of the Royal Society B: Biological Sciences*, **269**, 1423–1428.

Tibbetts, E.A. and Dale, J. (2004). A socially enforced signal of quality in a paper wasp. *Nature*, **432**, 218–22.

Topoff, H. and Zimmerli, E. (1993). Colony takeover by a socially parasitic ant, *Polyergus breviceps*: the role of chemicals obtained during host-queen killing. *Animal Behaviour*, **46**, 479–486.

Tsutsui, N.D. (2004). Scents of self: the expression component of self/nonself recognition systems. *Annales Zoologici Fennici*, **41**, 713–727.

Vander Meer, R.K. and Morel, L. (1988). Brood pheromones in ants. In: J.C. Trager (ed.), *Advances in Myrmecology*, pp. 491–513. E. J. Brill, New York.

Vander Meer, R.K. and Morel, L. (1998). Nestmate recognition in ants. In: R.K. Vander Meer, M. Breed, M. Winston, and K.E. Espelie (eds), *Pheromone Communication in Social Insects: Ants, Wasps, Bees, and Termites*, pp. 79–103. Westview Press, Boulder, CO.

Vargo, E.L. (1998). Primer pheromones in ants. In: R.K. Vander Meer, M. Breed, M. Winston, and K.E. Espelie (eds), *Pheromone Communication in Social Insects: Ants, Wasps, Bees, and Termites*, pp. 293–313. Westview Press, Boulder, CO.

Vienne, C., Soroker, V., and Hefetz, A. (1995). Congruency of hydrocarbon patterns in heterospecific groups of ants: transfer and/or biosynthesis? *Insectes Sociaux*, **42**, 267–277.

Wada, A., Isobe, Y., Yamaguchi, S., Yamaoka, R., and Ozaki, M. (2001). Taste-enhancing effects of glycine on the sweetness of glucose: a gustatory aspect of symbiosis between the ant, *Camponotus japonicus*, and the larvae of the lycaenid butterfly, *Niphanda fusca*. *Chemical Senses*, **26**, 983–992.

Wallis, D.I. (1963). A comparison of the response to aggressive behavior in two species of ants, *Formica fusca* and *Formica sanguinea*. *Animal Behaviour*, **11**, 164–171.

Wasmann, E. (1886). Über die Lebensweise einiger Ameisengäste, I. *Deutsche Entomologische Zeitschrift*, **30**, 49–66.

Wasmann, E. (1888). Die Beziehungen zwischen fremden Ameisenarten in fremden Nestern. *Natur und Offenbarung*, **34**, 193–210.

Wasmann, E. (1889a). Die sklavenhaltenden Ameisen. *Natur und Offenbarung*, **35**, 1–11.

Wasmann, E. (1889b). Die sklavenhaltenden Ameisen. Fortsetzung und Schluss. *Natur und Offenbarung*, **35**, 471–486.

Wasmann, E. (1905). Ursprung und Entwickelung der Sklaverei bei den Ameisen. *Biologisches Centralblatt (Berlin)*, **25**, 117–127.

Way, M.J. (1963). Mutualism between ants and honeydew-producing Homoptera. *Annual Review of Entomology*, **8**, 307–344.

Werber, B. (1998) *Empire of the Ants*, Bantam Books, New York.

Wheeler, W.M. (1904a). A new type of social parasitism among ants. *Bulletin of the American Museum of Natural History*, **20**, 347–375.

Wheeler, W.M. (1904b). Social parasitism among ants. *American Museum Journal*, **4**, 74–75.

Wilson, E.O. (1965). Chemical communication in social insects—insect societies are organized principally by complex systems of chemical signals. *Science*, **149**, 1064–1071.

Wilson, E.O. (1968). Ergonomics of caste in social insects. *The American Naturalist*, **102**, 41–66.

Wilson, E.O. (1971). *The Insect Societies*. Harvard University Press, Cambridge, MA.

Winston, M.L. and Slessor, K.N. (1992). The essence of royalty: honey bee queen pheromone. *American Scientist*, **80**, 374–385.

Wojcik, D.P. (1989). Behavioral interactions between ants and their parasites. *The Florida Entomologist*, **72**, 43–51.

CHAPTER 5

Chemical communication and the coordination of social interactions in insects

Patrizia d'Ettorre and Allen J. Moore

5.1 Introduction: insects are a good model system

In the beginning there were chemicals, and they were the only possible vectors of messages. But chemicals as messengers worked, and now chemical communication is ubiquitous. Unicellular organisms continue to communicate mainly through chemicals as do cells in multicellular organisms. When chemicals are used in communication between cells 'within an individual body' they are called **hormones**; when they are used by separate individuals 'within a species' we refer to them as **pheromones**, originally defined as 'substances secreted to the outside by an individual and received by a second individual of the same species in which they release a specific action' (Karlson and Lüscher 1959). Chemicals can also act between individuals of different species and the picture may become so complicated that our need to be able to distinguish among the numerous and multifaceted biological functions of chemicals has given rise to a probably useful but somehow tedious terminology. However, the most commonly used terms need to be defined and they are summarized in Box 7.1 in Chapter 7.

In this chapter, we focus on pheromones and we adopt the broad definition of Wyatt (2003), which includes olfactory and contact chemoreception, substances that are transferred directly from signaller to receiver and chemical cues used in social recognition (e.g. cuticular hydrocarbon patterns in insects). It is particularly important to note that we will refer to pheromones to both describe **cues** and **signals** (for a definition of cues and signals see Chapter 1).

For those interested in chemical communication, insects are the most reliable—although unintentional—informants since they essentially live in a world of pheromones. Insects do use other sensory **modalities** to communicate, but often in conjunction with pheromones and integrating multiple modes of communication. A typical example is the courtship behaviour of *Drosophila*, composed of a sequence of actions during which males and females exchange auditory, visual, and chemical signals (Hall 1994). Insects are also suitable model organisms because of their typically short generation time, relatively easy rearing conditions, and the possibility of artificial selection. Despite their ease of use as an experimental system, we hope to convince the reader that insects are fascinating, behaviourally complex, and worthy of our respect. At a minimum, the next time you are walking around a forest we hope you consider the complex communication and social interactions all around you, and that this develops a desire to conserve rather than exterminate insects.

Insects are ecologically successful, tremendously diverse, and show the complete range of social structures, from 'simple' parental care to the **eusocial** condition. Here again—similar to the above mentioned case of pheromone classification—a laborious array of terms has been coined to describe the degrees of sociality in insects: pre-sociality, sub-sociality, semi-sociality, para-sociality, quasi-sociality, and so on. We completely agree with those authors who ask to abandon this over-complicated lexicon in favour of the more neutral term 'social' (Costa and Fitzgerald 1996; Wcislo 1997). However, we share with Costa (2006) the feeling that the term eusocial is so well established that it should probably be maintained for the most advanced societies of ants, bees, wasps, and termites, but we would better describe the 'other' social insects as a range of social strategies where each is characterized by its specific degree of adaptive complexity.

5.2 Coordination and communication

A successful way of understanding a particular social organization is to study its underlying communication strategies: without effective communication social interactions are simply impossible. Because communication is among the traits that require interactions to be expressed, it is subjected to **social selection**. Social selection occurs whenever the fitness of an individual depends, in part, on the phenotype of its social partners (West Eberhard 1979, 1983, 1984; Wolf *et al.* 1999). Social selection acts on *interacting phenotypes*; those phenotypes that have reduced or no meaning outside of a social context (such as social dominance or dominance status, courtship, and, of course, traits that function in communication). The form of social selection that is most commonly studied is **kin selection**, in which the interactants in the social group are related. There is a considerable body of theory involving the role of kin groups in evolution (see Frank 1998). However, interacting individuals need not be related, and there are different forms of social selection including kin selection, **sexual selection**, and various types of **natural selection** (Wolf *et al.* 1999).

Another consequence of the dependence of interacting phenotypes on social context is that the genetic influences of interacting phenotypes must be treated differently. In addition to *direct genetic effects*—those additive genetic effects that influence the phenotype independent of other genetic effects at the same or different loci—there is the potential influence of *indirect genetic effects* (A.J. Moore *et al.* 1997). Indirect genetic effects arise because the phenotype expressed in the interaction depends on the phenotype of the other individuals in the social group. Thus, the direct genetic effects on the phenotypes of the social partner contribute, indirectly, to the genetic make-up of the focal individual. The most familiar indirect genetic effect is the *maternal genetic effect* (Moore *et al.* 1998). However, many traits can be influenced by indirect genetic effects if they are expressed in a social setting (e.g. Linksvayer 2006). This is again easy to see with communication signals or cues. If a pheromone released by a male during courtship causes a response in the female with whom he is interacting, then the phenotype of the pheromone ('attractiveness') depends, in part, on the female. Likewise, the response of the female ('approach') depends in part on the pheromone of the male. Thus, genetic variation underlying the pheromone depends on genetic variation underlying female responsiveness, and female responsiveness depends, in part, on genetic variation underlying the pheromone. The most familiar traits of this sort are those influenced by sexual selection (traits influencing male attractiveness) but many other traits, and in particular all traits involved in communication, can be influenced by indirect genetic effects.

The consequence of social selection is potentially stronger response to selection and faster evolution (Wolf *et al.* 1999; see also Chapter 8). The consequence of indirect genetic effects is unpredictable evolutionary trajectories and again more rapid evolution (A.J. Moore *et al.* 1997). Further, traits influenced by indirect genetic effects may evolve even when there is no genetic variation in direct genetic effects as a result of being expressed during social interactions. Thus, traits such as those involved in communication can evolve in ways and directions unexpected for other traits, and increased complexity is easier and perhaps more likely to evolve. Simply put, social interactions matter in evolution.

Because fine-tuned coordination is required between the sender and receiver of a message, animal communication is often highly sophisticated and multifaceted. Here, we intend to challenge three widespread notions: that chemical communication is simple, that pheromone production is cheap, and that insects always use an easy to decipher 'one substance–one message' code. We instead propose that insects are ideal model systems for achieving an integrated understanding of the complexity of communication in general, both at the proximate and the ultimate level.

5.3 The complexity of chemical communication

The first pheromone to be isolated, bombykol (a sex attractant produced by the female silkworm moth *Bombyx mori*), was identified and characterized in 1959 (Butenandt *et al.* 1959); in the same year the term 'pheromone' was introduced (Karlson and Butenandt 1959; Karlson and Lüscher, 1959). At that time, common wisdom was that each insect species would produce and respond to a single pheromone component. It took almost two decades to find out that *B. mori* females produce a second pheromone component, bombykal, and that the ratio of the two components is crucial for its mode of action and specificity (Kasang *et al.* 1978). We are increasingly becoming aware that most insects use multicomponent pheromones and that single-component systems are extremely rare. Even when communication has been shown to rely on a single component, intrinsic characteristics of the molecules such as 'chirality' may play a role in determining pheromone specificity, with two enantiomers of the same molecule having antagonistic effects (see Box 5.1 for definitions). An instructive example is given by the sex pheromone of some scarab beetles. Females of two Japanese species that share a common habitat, *Anomala osakana* and *Popillia japonica*, each produce a different enantiomer of the same lactone. Males are highly attracted by the

Box 5.1 Chirality

Chirality (handedness, from the Greek 'kheir' meaning 'hand') is non-superimposability, an important asymmetry property. A very familiar example of chiral object is our right and left hands: they are mirror images that cannot superimpose, they do not match if put on the top of each other.

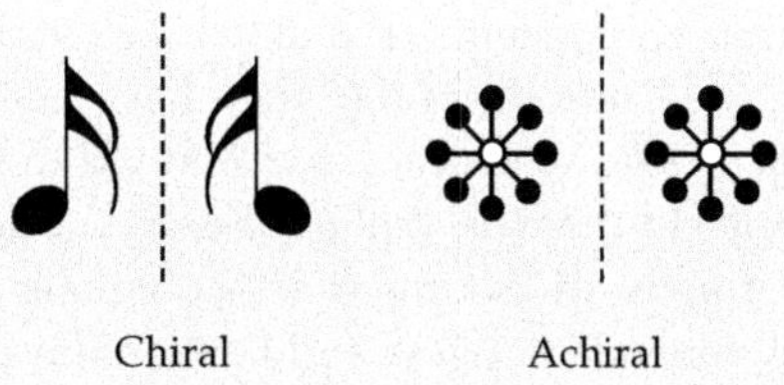

A chiral object, a non-chiral object, and their mirror images.

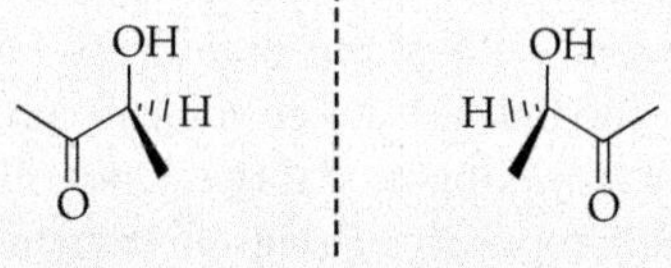

The two enantiomers (S and R) of 3-hydroxy-2-butanone (see also Fig. 5.3).

Two molecules that are mirror images and non-superimposable are called **enantiomers** (optical **antipodes**).

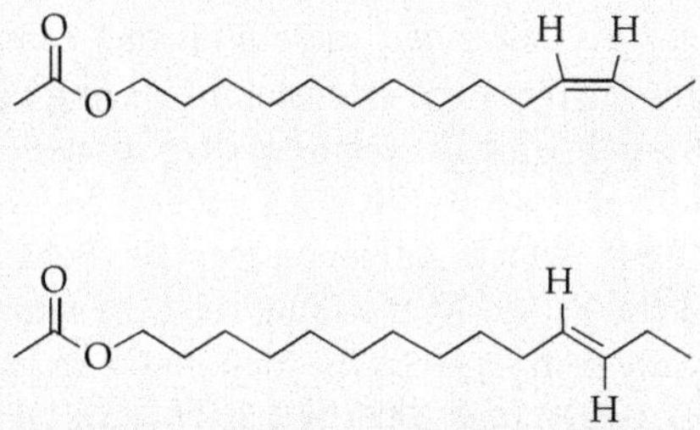

Two diastereomers: (*Z*)-tetradec-11-enyl acetate (top) and (*E*)-tetradec-11-enyl acetate (bottom) (see also Section 5.3.2).

Enantiomers are a type of **stereoisomer**: molecules that have the same molecular formula, the same atom-to-atom connections, but different spatial arrangements. Another group of stereoisomers are called **diastereomers**; these are not mirror images of each other. Usually, the term diastereomer is applied to molecules with more than one stereogenic centre (chiral centre). However, *cis–trans* isomers are also diastereomers.

conspecific female pheromone but are inhibited even by small quantities of its antipode. The (*S*)- and (*R*)-enantiomers have reverse roles in the two species and males have receptors tuned to both enantiomers. The author of this fascinating study refers to this and similar phenomena as 'the ultimate refinement in chemical communication' (Leal 1996). We share his view.

5.3.1 Multiple chemical messages

The *B. mori* female pheromone nicely exemplifies *multicomponent signals*: those unimodal signals using a single sensory modality (e.g. chemical) but more than one component (e.g. bombykol and bombykal). These kinds of signals are also called 'unimodal composite signals'. Some representative examples of pheromones as multicomponent signals are listed in Table 5.1.

A functional distinction shall be made between multicomponent and *multimodal signals*, which are characterized by the use of more than one sensory modality (e.g. visual plus chemical) and one or more component per modality. Again, the courtship behaviour of *Drosophila melanogaster*, during which both males and females use multimodal communication involving visual, acoustic, olfactory, gustatory, and tactile signals, provides an example. In *Drosophila*, successful and complete courtship involves the extension of a wing by the male (visual) which is then rapidly vibrated. This vibration creates a sound, the so-called 'love song' (Hall 1994), which influences female behaviour around the males. At the same time the female is chemically stimulated by the male cuticular hydrocarbons (Grillet *et al.* 2006) perhaps in part because the vibrating wing wafts odours of the male to the female. Courtship behaviour often involves different modalities. In *Bombus terrestris*, visual cues attract males from a distance, but at close range sex pheromones, and thus olfactory cues, are responsible for eliciting copulation behaviour (Krieger *et al.* 2006).

Multimodal signals are also called 'multisensory' signals and have been classified in different ways. Simplification is unavoidable when dissecting a complex biological phenomenon to restrict its elements in fixed categories, but we believe that the classification of multimodal signals proposed by Partan and Marler (1999, 2005) can be useful.

In recent decades there has been increasing attention given to the complexity of animal communication as expressed in multiple signals, especially multimodal signals (multisensory communication, cf. Bradbury and Vehrencamp 1998) as a way of ensuring accurate and honest signals. At first glance it is not obvious how and why increased complexity leads to greater honesty in signals. As a result, a number of theoretical papers have addressed the question of why animals use multiple displays, multiple ornaments for instance. Although our scope here is not an extensive review of the literature, we believe it is useful to briefly mention some of the leading hypotheses for the evolution and stability of multiple signals. Hebets and Papaj (2005) provide a review listing various hypotheses and suggested tests for the evolution of signal complexity.

Hasson (1989, 2000) and Møller and Pomiankowski (1993) were among the first to consider why there might be multiple displays functioning as a single signal. Hasson (1989) suggests that because receivers are not perfect, multiple signals would function to amplify the signal and make perception more reliable. Møller and Pomiankowski (1993) using a comparative approach and bird data, and Johnstone (1995, 1996), using a game theory approach, show that multiple signals may evolve because while one ornament may not provide accurate information, multiple ornaments are reliable (honest). When different components of a signal are 'redundant' (have the same meaning), they have been called *backup signals*, when they are 'non-redundant', we can refer to them as *multiple messages*. Underlying all of this is an assumption that signals are condition dependent. Iwasa and Pomiankowski (1994) further showed that these results are most stable when the costs of assessment are not too great.

Signal diversity can be gained within a single modality as well. Many organisms have limited sensory capabilities, or ecological constraints on different signalling modalities. Under these conditions, the complexity of a signal—which is required for signal accuracy and honesty—can be gained by adding components within the same modality and developing multicomponent signals. While some

Table 5.1 Representative examples of multicomponent pheromones. In each of these examples, there are multiple chemicals that influence the effectiveness of the pheromone, but the communication occurs within a single sensory modality (chemoreception). Examples are selected to illustrate the breadth of compounds that can act as pheromones, the breadth of species where multicomponent pheromones play a role, and the breadth of behavioural contexts in which pheromones influence behaviour. However, this is not a comprehensive or inclusive list. See http://www.pherobase.com/ for a sample of the potential complexity of pheromones in diverse taxa.

Taxa	Pheromone producer	Pheromone	Behavioural Effect or Role	Reference
Drosophila species	Males	Cuticular hydrocarbons	Species specificity, male attractiveness, condition	1
Cabbage looper moths (*Trichoplusia ni*)	Females	Acetates	Species specific mate attraction	2, 3
American cockroach (*Periplaneta americana*)	Both	Cuticular hydrocarbons	Attraction (distance) and aggregation (contact)	4
Lobster cockroach (*Nauphoeta cinerea*)	Males	Cuticular hydrocarbons, complex volatiles	Male dominance & attractiveness	5, 6
Burrower bug (*Sehirus cinctus*)	nymphs	Monoterpenes	Signal hunger, attract mother (begging signals)	7
Burying Beetle (*Nicrophorus vespilloides*)	Males	Cuticular hydrocarbons	Male breeding & parental care status	8
Honeybee (*Apis mellifera*)	Females	Multiple volatile compounds	Social organisation	9
Pharaoh ant (*Monomorium pharaonis*)	Females	Monomorines	Ant foraging trails	10
Red-legged salamanders (*Plethodon shermani*)	Males	Protein mixture	Increase female receptivity	11
Garter Snake (*Thamnophis sirtalis parietalis*)	Females	Saturated and unsaturated methyl ketones	Signal female size; influence male mate choice	12
House mice (*Mus musulus*)	Males	Major urinary proteins	Individual and social recognition	13, Chapter 6

1) Ferveur, J-F. (2005). Cuticular hydrocarbons: their evolution and roles in *Drosophila* pheromonal communication. *Behavior Genetics*, **35**, 279–295.

2) Bjostad, L.B., and Roelofs, W.L. (1983). Sex pheromone biosynthesis in *Trichoplusia ni*: Key steps involve delta-11 desaturation and chain-shortening. *Science*, **220**, 1387–1389.

3) Linn, C.E. Jr, Bjostad, L.B., Du, J. and Roelofs, W.L. (1984). Redundancy in a chemical signal: Behavioral responses of male *Trichoplusia ni* to a 6-component sex pheromone blend. *Journal of Chemical Ecology*, **10**, 1635–1658.

4) Said, I., Gaertner, C., Renou, M. and Rivault, C. (2007). Perception of cuticular hydrocarbons by the olfactory organs in *Periplanata americana* (L.) (Insecta: Dictyoptera). *Journal of Insect Physiology*, **51**, 1384–1389.

5) Sirugue, D., Bonnard, O., Quere, J.L., Farine, J.-P. and Brossut, R. (1997). 2-Methylthiazolidine and 4-ethyl-2-methoxyphenol, male sex pheromone components of the cockroach *Nauphoeta cinerea* (Dictyoptera, Blaberidae): a reinvestigation. *Journal of Chemical Ecology*, **18**, 2261–2276.

6) Everaerts, C., Fenaux-Benderitter, F., Farine, J.-P. and Brossut, R. (1997). Male dominant/subordinate relationships, cuticular profiles and sex pheromone in *Nauphoeta cinerea* (Dictyoptera, Blaberidae). *Insectes Sociaux*, 44: 277–287.

7) Kölliker, M.J., Chuckalovcak, P., Haynes, K.F. and Brodie E.D.III (2006). Maternal food provisioning in relation to condition-dependent offspring odours in burrower bugs (*Sehirus cinctus*). *Proceedings of the Royal Society of London B*, **273**, 1523–1528.

8) Steiger, S., Peschke, K., Francke, W. and Müller, J.K. (2007). The smell of parents: breeding status influences cuticular hydrocarbon pattern in the burying beetle *Nicrophorus vespilloides*. *Proceedings of the Royal Society of London B*, **274**, 2211–2220.

9) Slessor, K.N., Winston, M.W. and Le Conte, Y. (2005). Pheromone communication in the honeybee (*Apis mellifera* L.). *Journal of Chemical Ecology*, **31**, 2731–2745.

10) Jackson, D.E., Martin, S.J., Ratnieks, F.L.W. and Holcombe, M. (2007). Spatial and temporal variation in pheromone composition of ant foraging trails. *Behavioral Ecology*, **18**, 444–450.

11) Houck, L.D., Palmer, C.A., Watts, R.A., Arnold, S.J., Feldhoff, P.W. and Feldhoff, R.C. (2007). A new vertebrate pheromone, PMF, affects female receptivity in a terrestrial salamander. *Animal Behaviour*, **73**, 315–320.

12) Lemaster, M.P. and Mason, R.T. (2002). Variation in a female sexual attractiveness pheromone controls male mate choice. *Journal of Chemical Ecology*, **28**, 1269–1285.

13) Hurst J.L., Payne, C.E., Nevison, C.M., Marie, A.D., Humphries, R.E., Robertson, D.H., Cavaggioni, A. and Beynon, R.J. (2001). Individual recognition in mice mediated by major urinary proteins. *Nature*, **414**, 631–4.

of the hypotheses for multicomponent signals may differ (Hebets and Papaj 2005), the principles overlap.

5.3.2 Multicomponent pheromones

Although insects do use multimodal communication, here we focus on multicomponent chemical signals and their ample opportunities for informative and reliable communication strategies. How do insects achieve complex signalling by using chemicals? A simple answer is by using more than one chemical substance at the same time, as we have seen in the case of silkworm moths. The first multicomponent pheromone was described in a species of bark beetles, where a blend of three terpenes is used by males as sex attractant (Silverstein *et al.* 1966). Ipsenol, ipsdienol, and *cis*-verbenol were totally inactive when tested in laboratory bioassays as single compounds, even at high concentrations, but they gave a clear response when tested in mixtures. This is a remarkable example of the synergistic effect of multicomponent pheromones: individual constituents may elicit little response, but the blend is highly attractive. It also emphasizes the difficulties faced when assaying pheromones. As acknowledged by the authors of this pioneering study: "An apparent inactive fraction cannot be discarded until it has been tested in combination with other fractions" (Silverstein *et al.* 1966). Thus, synergy is an essential characteristic of chemical blends that should be borne in mind when attempting to analyse pheromone function (cf. Wyatt 2003).

In multicomponent chemical signals, relationships among the components within blends are often of crucial importance. Our understanding of the relationship among different pheromone components is dominated by studies of ratios. Several species may produce different ratios of the same chemicals, the quintessential examples being moth sex pheromones, which are usually a blend of two, three, or more components used in long-distance communication. Males are particularly sensitive to the specific component ratios emitted by conspecific females but are usually not attracted by a different ratio of the same chemicals, typical of a different species. An interesting case is that of six species of ermine moths, genus *Yponomeuta*, where females produce a blend of (*E*) and (*Z*) isomers of the same acetate (see Box 5.1). Three of these species produce approximately the same ratios of the two isomers, but live on different host plants and thus the males do not have the risk of being attracted by the 'wrong' female. However, the three species that share the same host plant use very different ratios of the two compounds (Löfstedt 1993). Common wisdom had been that pheromone blends in moths are under stabilizing selection that would not favour any possible shift from the norm characteristic of a given species. However, Roelofs *et al.* (2002) proposed that a saltational speciation event occurred in an ancestral *Ostrinia* population due to a major switch in the sex pheromone blend. The activation of a previously present but unexpressed gene controlling desaturation of pheromone molecules produced a new pheromone and a new species. Male butterflies may also produce multicomponent sex pheromones, which act during short-range courtship. For example, in *Bicyclus anynana*, the sex pheromone contains three components (two fatty acid derivatives and one terpenoid; C. Nieberding, personal communication).

There are limitations in investigating ratios, however, particularly in statistical analyses (see Packard and Boardman (1999) for a discussion of the problems of ratios as a general issue in biology). In addition, ratios are sufficient for understanding relationships among pairs of compounds, but can become awkward when there are more than two compounds in a blend. Which ratios are of interest? Which is the denominator and which the numerator? If there is a major component and minor components, as in ermine moths (Löfstedt *et al.* 1991) or in perhaps the most extensively studied species of insects in pheromone research, the cabbage looper *Trichoplusia ni* (Bjostad *et al.* 1984), then it may be possible to form a series of ratios with a common denominator that allows comparisons. However, it is not clear that 'major' and 'minor' components are universal or even common outside of moths. Nevertheless, it is still common and useful to use the term 'ratio' colloquially to mean 'relationship among pheromone components'. This need not imply or dictate a specific statistical approach.

An alternative approach is to use multivariate statistical techniques that provide a linear combination of characters, such as principal component analysis (Neems and Butlin 1995). There are a number of advantages to such an approach (see Blows and Allan 1998). First, analyses are facilitated because the new traits are statistically independent. Second, the linear combinations often produce interpretable vectors, and allow the testing of a priori hypotheses among traits such as morphological and functional integration among components (Cheverud *et al.* 1989; Cheverud 1996). Moore (1997) provides an example of this approach and explores questions of morphological and functional integration applied to pheromones.

The approach of multivariate statistics, although useful, remains mostly correlative without confirmatory manipulative experiments. A multivariate approach has been extensively used in trying to identify nestmate recognition cues in social insects. Here, the actual signal or cue is not a particular compound or a group of compounds, but possibly *the difference between ratios of compounds* in a blend. There is substantial evidence that social insects discriminate nestmates and non-nestmates by perceiving patterns of cuticular hydrocarbons, which are usually colony-specific (review in Lenoir *et al.* 1999, 2001; Vander Meer and Morel, 1998; Singer 1998; Hefetz 2007; see also Chapter 4). Supplementation experiments, in which synthetic hydrocarbons added to an insect cuticle specifically modify the reaction elicited in nestmates, are then used to directly demonstrate the role of different classes of hydrocarbons (e.g. Dani *et al.* 2001, 2005). Alternatively, researchers using an inert support, such as glass beads covered with a single compound or hydrocarbon mixtures, have elucidated the specific role of hydrocarbons as nestmate recognition cues, at least in a few species (Akino *et al.* 2004; Greene and Gordon 2007). Deciphering the recognition code of social insects has been proven to be particularly challenging, in part because cuticular hydrocarbon patterns may simultaneously contain information on three different levels of recognition (Fig. 5.1). Between-colony recognition should allow rapid recognition of friends (nestmates) and enemies (non-nestmates), but its effectiveness can be constrained by the colony kin structure, for example how many times the queen mated (Boomsma *et al.* 2003). Within the colony, cuticular hydrocarbon patterns distinctive of different classes of individuals can, for instance, enhance foraging efficiency (cf. Greene and Gordon 2003), signal fertility (Hefetz 2007; Peeters and Liebig 2008), or enable the suppression of **selfish** workers by policing (see Section 5.3.3). Finally, in some particular circumstances, cuticular hydrocarbons may contain enough information to permit individual identity, as in the case of co-founding ant queens that establish dominance hierarchies with division of labour (d'Ettorre and Heinze 2005; Dreier *et al.* 2007). Much experimental work remains to be done to unravel the perception mechanisms and the information processing underlying recognition at each of these three levels. Though we are beginning to understand the neurobiology of olfaction in the honey bee, many questions remain open (see Chapter 7).

Multicomponent communication is common in social insects, especially ants, which have been appropriately defined 'walking chemical factories' full of exocrine glands producing a variety of chemicals (Hölldobler and Wilson 1990). Interesting examples of composite alarm and recruitment pheromone are reviewed by Hölldobler (1995), who also highlighted the modulatory functions of

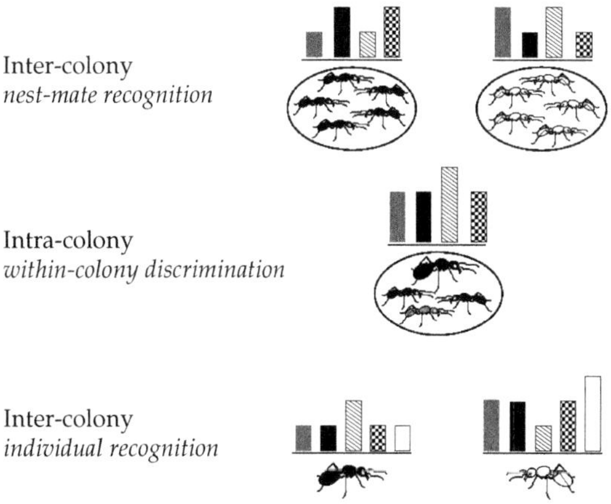

Figure 5.1 Levels of recognition in social insects. The recognition cues are represented by the chemical signature depicted by a schematic gas-chromatogram where each bar represents a chemical compound (e.g. a cuticular hydrocarbon).

multicomponent signals: "communication in ant societies is often based on multi-component signals, on nested levels of variation in chemical and other cues, which feature both anonymous and specific characteristics" (Hölldobler 1995). A pheromone may be composed by one substance which is common to several species or to all colonies of the same species, but this substance may be coupled with other substances which are species specific or with blend of substances, the ratio of which can be colony specific. There are also surprising cases in which chemical trail markers have been shown to be specific to individuals (e.g. *Pachycondyla tesserinoda*; Jessen and Maschwitz 1985, 1986).

A remarkable variation in the blend of a trail pheromone between individuals belonging to two behavioural subcastes has been recently reported in the pharaoh ant, *Monomorium pharaonis*. These ants make extensive use of pheromone trails between the nest and foraging sites. Trails contain chemicals secreted both by the Dufour's (farnal) and poison gland (called monomorines, which are highly stable, low-volatile compounds). Jackson *et al.* (2007) showed that the relative abundance of two monomorines, M1 and M3, on foraging trails varies over time during trail build-up but it is also accompanied by a spatial variation in the M3:M1 ratio along the trail (low close to food and high near the nest). These variations are due to the production of different ratios of pheromone blend by different kinds of foragers. The so-called 'pathfinders', foragers specializing in relocating long-lived trails, have a low M3:M1 ratio and use it to mark sections of pheromone trail networks for future explorations. Conversely, 'non-pathfinders' mark trails during early stages of foraging on a new food source. Interestingly, monomorines have multiple roles in pharaoh ants, since they also act as venoms and repellents against insect competitors. This leads us to the introduction of *multipurpose* (multifunctional) signals.

5.3.3 Multipurpose chemical signals

The same blend of chemicals can be used in different ways to convey different messages, as illustrated by the case of the trail pheromone of eastern tent caterpillars. This is a mixture of two steroids used electively: when searching for food, the caterpillars lay light or intermittent 'exploratory trails', but when a good food source is found, they lay a persistent 'recruitment trail' while going back to their tent (cf. Costa 2006). Thus, multicomponent pheromones can serve more than one purpose, as predicted by the *pheromonal parsimony* assumption (Blum 1996), the obvious example being the honey bee queen pheromone, which has both primer and releaser effects for the workers and is also a sex attractant for drones (Slessor *et al.* 1988, 2005; see also Chapter 7). In the ant *Pachycondyla inversa*, there is evidence that one of the hydrocarbons present on the queens' cuticle and detected by the workers' antennae, the branched alkane 3,11-diMeC_{27}, acts as a fertility signal advertising the presence of the queen and thus inhibiting ovary development in workers (Heinze *et al.* 2002; d'Ettorre *et al.* 2004a). The same substance is more abundant on queen-laid eggs than on worker-laid eggs (d'Ettorre *et al.* 2004b) and it is not transferable by simple contact between the two kinds of eggs (d'Ettorre *et al.* 2006). In this species, worker policing (cf. Ratnieks 1988) does occur: workers discriminate between queen- and worker-laid eggs and kill the latter (d'Ettorre *et al.* 2004b; van Zweden *et al.* 2007). Thus, 3,11-diMeC_{27} appears to be a robust and honest chemical strategy with the double function of signalling a queen's fertility and protecting queen-laid eggs from policing (Fig. 5.2). Unfortunately, unequivocal direct proof of one (or more) chemical compound acting as queen signal in any ant species is still lacking. In the case of *P. inversa*, the ideal experiment would show that worker-laid eggs treated with the synthetic putative queen signal (3,11-diMeC_{27}) become more acceptable by adult workers than control eggs do, but first experiments of this kind have so far given contradictory results (van Zweden *et al.* in preparation).

Another example of a pheromone with multiple purposes occurs in the cockroach *Nauphoeta cinerea* (Fig. 5.3). The male-produced pheromone of this species, which is composed primarily of 3-hydroxy-2-butanone, 2-methylthiazolidine, and 4-ethyl-2-methoxyphenol, serves as a social signal and has evolved as a result of sexual selection. However, the role of this pheromone in

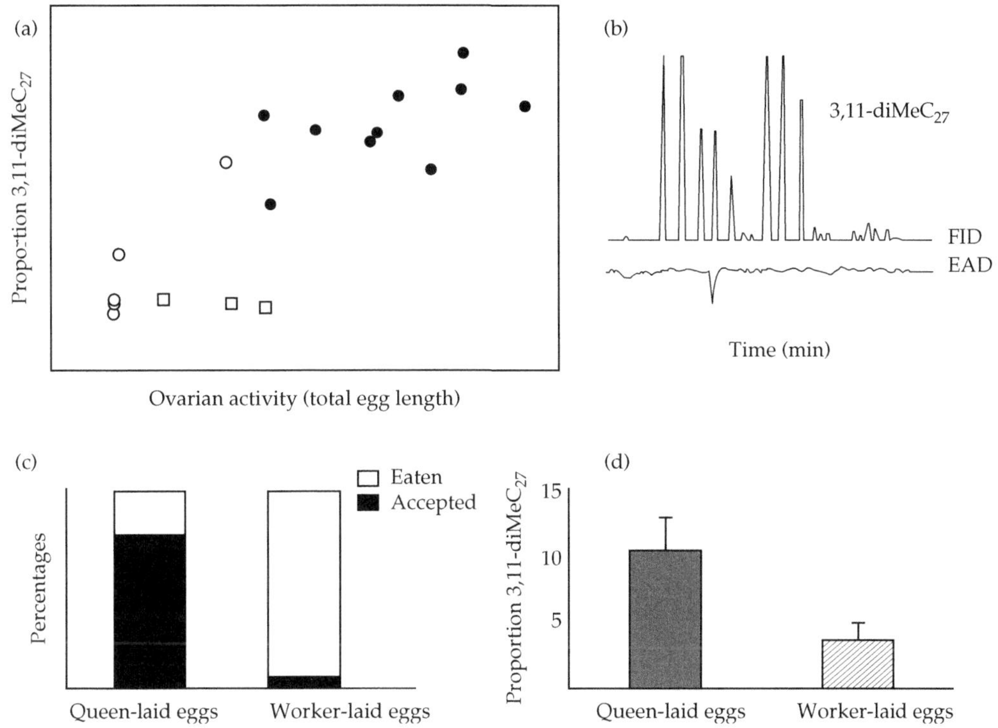

Figure 5.2 A possible multi-purpose signal, the hydrocarbon 3,11-diMeC_{27} produced by queens of the ant *Pachycondyla inversa*. The compound is present on the ant cuticle and it is correlated with ovarian development (a) mature queens (filled circles), founding queens (open squares), virgin queens (open circles). 3,11-diMeC_{27} is detected by the workers' antennae (b) (FID = Flame Ionization Detector, the signal of the gas-chromatograph; EAD = Electro-antenno-detector, the response of the antenna). Worker-laid eggs are policed by other workers in queenright discriminator colonies (c) and the compound is more abundant on the surface of queen-laid eggs than on worker-laid eggs (d). Figures adapted from d'Ettorre et al. (2004a, b).

attractiveness to females does not overlap completely with male status (Moore and Moore 1999). Females prefer males with high levels of 3-hydroxy-2-butanone regardless of the levels of the other two components. A pheromonal **badge of status** determines dominance (P.J. Moore *et al.* 1997; Moore *et al.* 2002), where males with higher amounts of 2-methylthiazolidine and 4-ethyl-2-methoxyphenol relative to 3-hydroxy-2-butanone are dominant. The amount of pheromone components produced by males is genetically influenced (Moore 1997; Moore *et al.* 2002). It is known that 2-methylthiazolidine and 4-ethyl-2-methoxyphenol are developmentally and genetically correlated, while 3-hydroxy-2-butanone is developmentally and genetically independent (Moore *et al.* 1995; Moore 1997). Thus, although the pheromone is the main trait that influences sexual selection, there are trade-offs in the social signal that confers attractiveness and high social status (Moore and Moore 1999).

5.4 Chemical components as composite traits

Pheromones are at least as complex as other signals using other modalities such as sound or colour. Like sound and colour, pheromones are composed of multiple components with varying functions. Sound, for example, has different properties such as pressure, frequency, temporal structure, and medium (substrate-borne or airborne). Colour patches are three-dimensional structures with many layers and different pigment types,

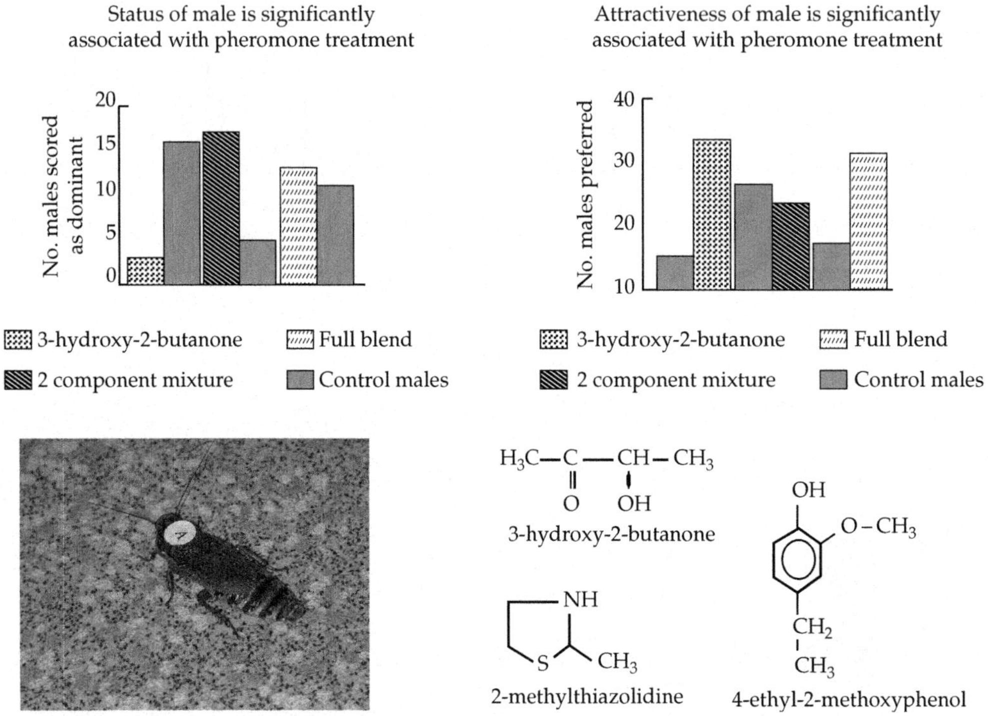

Figure 5.3 The multicomponent pheromone blend of *Nauphoeta cinerea* has different effects under two different mechanisms of sexual selection. Using phenotypic manipulations of the pheromone, A. J. and P. J. Moore show that males with higher levels of 2-methylthiazolidine and 4-ethyl-2-methoxyphenol, which are genetically correlated, and lower levels of 3-hydroxy-2-butanone, which is genetically independent, are more likely to dominate other males. In contrast, females prefer males with higher levels of 3-hydroxy-2-butanone. The pheromone blend is therefore under balancing sexual selection. Figures adapted from Moore *et al.* (1997) and Moore and Moore (1999) by P. J. Moore.

with patches varying in brightness, size, shape, and location (Endler and Mielke 2005). In addition, colour varies in chroma (colour saturation) and hue (spectral shape). Thus, multiple trait evolution theory can be applied to each component of a single colour patch (Grether *et al.* 2004; Endler and Mielke 2005). Moreover, the simultaneous production of two different pigments (pigment symmetry) could be a signal component itself because it is related to the metabolic capacity to coordinate the production of two different pigments (signal of quality). When considering the variation of each component, colour patches would result in being ornaments nested within ornaments where each component may evolve independently or as a functionally integrated suite of traits (Endler and Mielke 2005). Pheromones share this complexity. Pheromones can thus vary in concentration, composition, and structural make-up. Or, to make a direct comparison with colour, a single chemical component can be a composite trait itself, characterized by its concentration (absolute amount, analogous to chroma) and stereochemistry (comparable to hue). Multiple components constituting a pheromone blend can lead to synchronization of chemical production of the pheromone constituents, providing an additional variable similar to pigment symmetry. Finally, like the variation in number and location of colour patches, pheromones can vary in the site of release, with deposition on a substrate, release from a specialized gland, or emission from the entire body as in cuticular hydrocarbons.

Therefore, we propose to extend Grether *et al.*'s (2004) recursive approach to pheromones and to start widening our view by considering individual chemical substances as variables nested inside variables. We believe that appreciating and disentangling the complexity of pheromones is the key to revealing the underestimated potential for reliability that appears to be hidden in chemical signals and its evolutionary stability.

5.4.1 Costs of chemical signals

According to the handicap principle, in order to carry reliable messages, pheromones have to meet three key characteristics of signals: (1) they have to be costly to the individual sending them; (2) the cost has to be more of a burden to a dishonest communicator than to a honest one; (3) there must be a logical relationship between the specific cost of the signal and the message conveyed by the signal (see Zahavi and Zahavi 1997 and Chapter 1). Pheromones come with costs and fit this description.

Pheromones are not merely cheap by-products of metabolism, but they can indeed be very expensive. One of the costs of pheromones derives from the fact that the production of a specific chemical—or mixture of chemicals—requires a particular starting material that is not always easily accessible. One of the first examples of a plant chemical transformed into an insect pheromone was shown in the same species of bark beetles cited above (Section 5.3.2), *Ips paraconfusus*, in which a constituent of the oleoresin system of the host tree *Pinus ponderosa* is converted by the male beetle into ipsenol and ipsdienol (Hendry *et al.* 1980).

There is an extensive literature on the evolution of unpalatability by phytophagous insects (reviewed in e.g. Bowers 1992). Many insects feed on toxic plants but only some insects have developed the ability to sequestrate plant secondary metabolites and use them as defensive compounds. This may be limited by the ability to locate a suitable host plant or plant tissue, given the variation in quality and amount of secondary metabolites among plant populations (e.g. see Brower *et al.* (1982, 1984) for a study of emetic potency of monarch butterflies). Moreover, recycling plant metabolites requires specific physiological adaptations involving the loss of detoxification mechanisms allowing the circulation of the defensive compounds inside the insect body, while at the same time protecting the insect tissue and organs from these toxic chemicals. Other unpalatable insects synthesize defensive compounds *de novo*, but here again there are costs associated with the production of specific enzymes and the prevention of autotoxicity.

In general, 'production' and 'maintenance' costs are linked to the biological effects of some constituents of pheromones, like the above-mentioned terpenes and phenols, which are toxic chemicals and thus ensure signal honesty ('handicaps' in Zahavi and Zahavi (1997); 'strategic' costs in Maynard Smith and Harper (2003)). Another kind of maintenance cost may be seen in the cockroach *N. cinerea*. Dominant males signal their status, but this has to be supported by aggression. Dominant males lunge at, chase, and bite subordinate males (Schal and Bell 1983). While the pheromone may reduce these costs when males vary greatly in status, males of similar signal status often engage in vigorous and highly aggressive and damaging fights (Kou *et al.* 2006). There is a linear relationship between *N. cinerea* pheromone and strength of interactions (Kou et *al.* 2006). However, the direct costs of producing pheromones have been shown in cotton aphids (*Aphis gossypii*), where small individuals produce more pheromone at the cost of growth (Byers 2005). In the Hawaiian fruit fly, *Drosophila gryimshawi*, increased pheromone calling decreases male lifespan (Johansson *et al.* 2005).

5.5 Concluding remarks

Pheromones involved in communication during social interactions are likely to be complex. This complexity has at least three sources. First, like many other signals, the complexity of pheromonal signals arises from the need to convey honest information. Complex honest signals often involve multimodal signalling, but can also involve multicomponent signals. Second, evolution of traits involved in interactions result in interacting phenotypes, which are influenced by both direct and indirect genetic effects, and are typically influenced by complex forms of social selection. The net result of this is unpredictable but rapid evolutionary trajectories toward an optimum, as interacting phenotypes fulfil multiple roles and functions in structuring social interactions. Finally, because pheromonal social signals can influence multiple behaviours, there is a potential for balancing selection like we see in the pheromone of the cockroach *N. cinerea*. Similarly, recognition in social insects is subjected to contrasting selection pressures. At one end, efficient nestmate recognition requires highly heritable recognition cues (heritable specific blend of the cuticular hydrocarbon mixture). At the other end, avoidance of disruptive nepotism selects against individual kin recognition within a colony (recognition of full-sisters and half-sisters in colonies with multiple mated queens or with more than one queen, cf. Boomsma *et al.* 2003), thus counter-balancing the selection for more heritable recognition cues. Wherever signals are playing multiple roles, and selection occurs in opposing directions, the result will be apparent stabilizing selection on continuously distributed phenotypes. This will result in increased variation being maintained in the signal. Therefore, despite a strong relationship with fitness, the complexity of the signal will persist.

For the reasons expressed at the beginning of this chapter, we believe that insects are ideal model organisms for studying the complexity of chemical communication. The 'chemical language' of insects is based on 'words' made of atoms instead of letters that can be combined in molecules to build 'sentences' with different meaning. However, the analogy with human language is approximate at best, unless we were to discover something similar to a 'grammar' of chemicals and arbitrary symbols (cf. Chapters 14 and 16) when insects use molecules. We hope to stimulate further research aimed at understanding proximate and ultimate aspects of insect pheromones, and their complexity, in the framework of social selection. The general principles of chemical communication are similar in different model systems. Chapter 6, 'Chemical communication in societies of rodents', gives a beautiful example of a multicomponent and multipurpose chemical signal in the house mouse. Here the major urinary proteins (a highly variable set of small lipocalin proteins) communicate information about invariable genetic identity and variable status of the animal.

By being innovative in our application of the comparative approach and trying to override the traditional barriers across taxa, we will gain an integrated and interdisciplinary understanding of chemical communication in general. This will eventually lead us to a better understanding of the neglected importance of chemical communication in humans. For instance, some constituents of human axillary secretions have been proposed as candidate human pheromones, but their influence on our behaviour is still controversial. A recent elegant study, performed with an ecologically valid experimental paradigm, described a clear effect of an androgen steroid (4,16-androstadien-3-one) on female perception of male attractiveness; thus, this compound could act as a modulatory pheromone in humans (Saxton *et al.* 2007). The complexity of axillary secretions, however, begs for further integrated studies aiming at deciphering the multicomponent and possibly multipurpose nature of human pheromones.

Summary

Effective communication, often involving pheromones, is a fundamental component of social life. Communication requires interactions to be expressed and it is convenient to consider communication within the context of the theory of interacting phenotypes—those phenotypes that have reduced or no meaning outside of a social context. Pheromonal communication will therefore be subject to social selection and indirect genetic effects and is often highly sophisticated and multifaceted, allowing fine-tuned coordination of messages from senders and receivers. Pheromones can be characterized by nested levels of variation: a multicomponent structure in which individual components contain additional source of variation. An integrated understanding of communication by multicomponent chemical signals provides insight into the evolution of social signals in general. Insects are ideal model systems to investigate and disentangle the complexity of pheromones and reveal the underestimated potential for reliability that appears to be hidden in chemical signals and their evolutionary stability.

Acknowledgements

We thank Koos Boomsma, David Nash, Christer Löfstedt, Trish Moore, and Craig Roberts for their valuable comments. John Nielsen drew the molecules and helped with the text of Box 5.1. Mauro Patricelli drew the ants in Fig. 5.1 and the example in Box 5.1. PdE is grateful to Amotz Zahavi for having taken the burden of prolonged discussions on chemical signals during his visit to Copenhagen.

References

Akino, T., Yamamura, K., Wakamura, S., and Yamaoka, R. (2004). Direct behavioural evidence for hydrocarbons as nestmate recognition cues in *Formica japonica*. *Applied Entomology and Zoology*, **39**, 381–387.

Bjostad, L.B., Linn, C.E., Du, J.-W., and Roelofs, W.L. (1984). Identification of new sex pheromone components in *Trichoplusia ni*, predicted from biosynthetic precursors. *Journal of Chemical Ecology*, **10**, 1309–1323.

Blows, M.W. and Allan, R.A. (1998). Levels of mate recognition within and between two *Drosophila* species and their hybrids. *The American Naturalist*, **12**, 826–837.

Blum, M.S. (1996). Semiochemical parsimony in the Arthropoda. *Annual Review of Entomology*, **41**, 353–374.

Boomsma, J.J., Nielsen, J., Sundström, L., Oldham, N.J., Tentschert, J., Petersen, H.C., and Morgan, E.D. (2003). Informational constraints on optimal sex allocation in ants. *Proceedings of the National Academy of Sciences of the USA*, **100**, 8799–8804.

Bowers, M.D. (1992). The evolution of unpalatability and the cost of chemical defense in insects. In: B.D. Roitberg and M.B. Isman (eds), *Insect Chemical Ecology: an Evolutionary Approach*, pp. 216–244. Chapman and Hall, New York.

Bradbury, J.W. and Vehrencamp, S.L. (1998). *Principles of Animal Communication*. Sinauer Associates, Sunderland, MA.

Brower, L.P., Seiber, J.N., Nelson, C.J., Lynch, S.P., and Tuskes, P.M. (1982). Plant-determined variation in the cardenolide content, thin-layer chromatography profiles, and emetic potency of monarch butterflies, *Danaus plexippus* reared on the milkweed, *Asclepias eriocarpa* in California. *Journal of Chemical Ecology*, **8**, 579–633.

Brower, L.P., Seiber, J.N., Nelson, C.J., Lynch, S.P., and Holland, M.M. (1984). Plant determined variation in the cardenolide content, thin-layer chromatography profiles, and emetic potency of monarch butterflies, *Danaus plexippus,* reared on the milkweed plants in California: two *Asclepias speciosa*. *Journal of Chemical Ecology*, **10**, 601–639.

Butenandt, A., Beckmann, R., Stamm, D., and Hecker, E. (1959). Über den Sexuallockstoff den Seidenspinners *Bombyx mori. Reindarstellung und Konstitution. Zeitschrift für Naturforschung*,**14b**, 283–284.

Byers, J.A. (2005). A cost of alarm pheromone production in cotton aphids, *Aphis gossypii*. *Naturwissenschaften*, **92**, 69–72.

Cheverud, J.M. (1996). Developmental integration and the evolution of pleiotropy. *American Zoologist*, **36**, 44–50.

Cheverud, J.M., Wagner, G.P., and Dow, M.M. (1989). Methods for the comparative analysis of variation patterns. *Systematic Zoology*, **38**, 201–213.

Costa, J.T. (2006). *The Other Insect Societies*. Harvard University Press, Cambridge, MA.

Costa, J.T. and Fitzgerald, T.D. (1996). Developments in social terminology: semantic battles in a conceptual war. *Trends in Ecology and Evolution*, **11**, 285–289.

Dani, F.R., Jones, G.R., Destri, S., Spencer S.H., and Turillazzi S. (2001). Deciphering the recognition signature within the cuticular chemical profile of paper wasps. *Animal Behaviour*, **62**, 165–171.

Dani F.R., Jones, G.R., Corsi, S., Beard, R., Pradella, D., and Turillazzi, S. (2005). Nestmate recognition cues in the honey bee: differential importance of cuticular alkanes and alkenes. *Chemical Senses*, **30**, 477–489.

Dreier, S., van Zweden, J.S., and d'Ettorre, P. (2007). Long term memory of individual identity in ant queens. *Biology Letters*, **3**, 459–462.

Endler, J.A. and Mielke, P.W. Jr (2005). Comparing entire colour patterns as birds see them. *Biological Journal of the Linnean Society*, **86**, 405–431.

d'Ettorre, P. and Heinze, J. (2005). Individual recognition in ant queens. *Current Biology*, **15**, 2170–2174.

d'Ettorre, P., Heinze, J., Schulz, C., Francke, W., and Ayasse, M. (2004a). Does she smell like a queen? Chemoreception of a cuticular hydrocarbon signal in the ant *Pachycondyla inversa*. *Journal Experimental Biology*, **207**, 1085–1091.

d'Ettorre, P., Heinze, J., and Ratnieks, F.L.W (2004b). Worker policing by egg eating in the ponerine ant *Pachycondyla inversa*. *Proceedings of the Royal Society B: Biological Sciences*, **271**, 1427–1434.

d'Ettorre, P., Tofilski, A., Heinze, J., and Ratnieks, F.L.W. (2006). Non-transferable signals on ant queen eggs. *Naturwissenschaften*, **93**, 136–140.

Frank, S.A. (1998). *Foundations of Social Evolution*. Princeton University Press, Princeton, NJ.

Greene, M.J. and Gordon, D.M. (2003). Cuticular hydrocarbons inform task decisions. *Nature*, **423**, 32.

Greene, M.J. and Gordon, D.M. (2007). Structural complexity of chemical recognition cues affects the perception of group membership in the ants *Linephithema humile* and *Aphaenogaster cockerelli*. *Journal of Experimental Biology*, **210**, 897–905.

Grether, G.F., Kolluru, G.R., and Nersissian, K. (2004). Individual colour patches as multicomponent signals. *Biological Reviews*, **79**, 583–610.

Grillet, M., Dartevelle, L., and Ferveur, J.F. (2006). A *Drosophila* male pheromone affects female sexual receptivity. *Proceedings of the Royal Society Series B: Biological Sciences*, **27**, 315–323.

Hall, J.C. (1994). The mating of a fly. *Science*, **264**, 1702–1714.

Hasson, O. (1989). Amplifiers and the handicap principle in sexual selection: a different emphasis. *Proceedings of the Royal Society of London Series B, Biological Sciences*, **235**, 383–406.

Hasson, O. (2000). Knowledge, information, biases and signal assemblages. In: Y. Espmark, T. Amundsen, and G. Rosenqvist (eds), *Animal Signals. Signalling and Signal Design in Animal Communication*, pp. 445–463. The Royal Norwegian Society of Sciences and Letters. The Foundation of Tapir Publishers, Trondheim, Norway.

Hebets, E.A. and Papaj, D.R. (2005). Complex signal function: developing a framework of testable hypotheses. *Behavioural Ecology and Sociobiology*, **57**, 197–214.

Hefetz, A. (2007). The evolution of hydrocarbon pheromone parsimony in ants: (Hymenoptera: Formiciade)—interplay of colony odor uniformity and odor idiosyncracy. A review. *Myrmecological News*, **10**, 59–68.

Heinze, J., Stengl, B., and Sledge, M.F. (2002). Worker rank, reproductive status and cuticular hydrocarbon signature in the ant *Pachycondyla* cf. *inversa*. *Behavioral Ecology and Sociobiology*, **52**, 59–65.

Hendry, L.B., Piatek, B., Browne, L.E., Wood, D.L., Byers, J.A., Fish, R.H., and Hicks, R.A. (1980). *In vivo* conversion of a labelled host plant chemical to pheromones of the bark beetle, *Ips paraconfusus*. *Nature*, **284**, 485.

Hölldobler, B. (1995). The chemistry of social regulation: multicomponent signals in ant societies. *Proceedings of the National Academy of Sciences of the USA*, **92**, 19–22.

Hölldobler, B. and Wilson, E.O. (1990). *The Ants*. Harvard University Press, Cambridge, MA.

Iwasa Y. and Pomiankowski A. (1994). The evolution of mate preferences for multiple sexual ornaments. *Evolution*, **48**, 853–867.

Jackson, D.E., Martin, S.J., Francis L.W., Ratnieks F.L.W., and Holcombe, M. (2007). Spatial and temporal variation in pheromone composition of ant foraging trails. *Behavioral Ecology*, **18**, 444–450.

Jessen, K. and Maschwitz, U. (1985). Individual specific trails in the ant *Pachycondyla tesserinoda* (Formicidae, Ponerinae). *Naturwissenschaften*, **72**, 549–550.

Jessen, K., and Maschwitz, U. (1986). Orientation and recruitment behavior in the ponerine ant *Pachycondyla tesserinoda* (Emery) laying of individual-specific trails during tandem running. *Behavioral Ecology and Sociobiology*, **19**, 151–155.

Johansson, B.G., Jones, T.M., and Widemo, F. (2005). Cost of pheromone production in a lekking *Drosophila*. *Animal Behaviour*, **69**, 851–858.

Johnstone, R.A. (1995). Honest advertisement of multiple qualities using multiple signals. *Journal of Theoretical Biology*, **177**, 87–94.

Johnstone, R.A. (1996). Multiple displays in animal communication: 'backup signals' and 'multiple messages'. *Philosophical Transactions of the Royal Society B: Biological Sciences*, **351**, 329–338.

Karlson, P. and Butenandt, A. (1959). Pheromones (ectohoromones) in insects. *Annual Review of Entomology*, **4**, 39–58.

Karlson, P. and Lüscher, M. (1959). 'Pheromones': a new term for a class of biologically active substances. *Nature*, **183**, 55–56.

Kasang, G., Kaissling, K-E., Vostrowsky, O., and Bestmann, H.J. (1978). Bombykal, a second pheromone component of the silkworm moth *Bombyx mori*. *Angewandte Chemie International Edition*, **17**, 60.

Kou, R., Chen, S-C., Chen, Y-R., and Ho, H-Y. (2006). 3-hydroxy-2-butanone and the first encounter fight in the male lobster cockroach, *Nauphoeta cinerea*. *Naturwissenschaften*, **93**, 286–291.

Krieger, G.M., Duchateau, M.J., van Doorn, A., Ibarra, F., Francke, W., and Ayasse, M. (2006). Identification of queen sex pheromone components of the bumblebee *Bombus terrestris*. *Journal of Chemical Ecology*, **32**, 453–471.

Leal, W.S. (1996). Chemical communication in scarab beetles: Reciprocal behavioral agonist-antagonist activities of chiral pheromones. *Proceedings of the National Academy of Sciences of the USA*, **93**, 12112–12115.

Lenoir, A., Fresneau, D., Errard, C., and Hefetz, A. (1999). Individuality and colonial identity in ants: the emergence of the social representation concept. In: C. Detrain, J.L. Denebourg, and J. Pasteels (eds), *Information Processing in Social Insects*, pp. 219–237. Birkhauser, Basel, Switzerland.

Lenoir, A., d'Ettorre, P., Errard, C., and Hefetz A. (2001). Chemical ecology and social parasitism in ants. *Annual Review of Entomology*, **46**, 573–599.

Linksvayer, T.A. (2006). Direct, maternal, and sibsocial genetic effects on individual and colony traits in an ant. *Evolution*, **60**, 2552–2561.

Löfstedt, C. (1993). Moth pheromone genetics and evolution. *Philosophical Transactions of the Royal Society B: Biological Sciences*, **340**, 167–177.

Löfstedt, C., Herrebout, W.H., and Menken, S.B.J. (1991). Sex pheromones and their potential role in the evolution of reproductive isolation in small ermine moths (Yponomeutidae). *Chemoecology*, **2**, 20–28.

Maynard Smith, J. and Harper, D. (2003). *Animal Signals*. Oxford University Press, New York.

Moore, A.J. (1997). The evolution of social signals: morphological, functional and genetic integration of the sex pheromone in *Nauphoeta cinerea*. *Evolution*, **51**, 1920–1928.

Moore, A.J. and Moore, P.J. (1999). Balancing sexual selection through opposing mate choice and male competition. *Proceedings of the Royal Society B: Biological Sciences*, **266**, 711–716.

Moore, A.J., Reagan, N.L., and Haynes, K.F. (1995). Conditional signalling strategies: effects of ontogeny, social experience and social status on the pheromonal signal of male *Nauphoeta cinerea*. *Animal Behaviour*, **50**, 191–202.

Moore, A.J., Brodie, E.D. III, and Wolf, J.B. (1997). Interacting phenotypes and the evolutionary process. I. Direct and indirect genetic effects of social interactions. *Evolution*, **51**, 1352–1362.

Moore, A.J., Wolf, J.B., and Brodie, E.D. III (1998). The influence of direct and indirect genetic effects on the evolution of behavior: sexual selection meets maternal effects. In: T.A. Mousseau and C.W. Fox (eds), *Maternal Effects as Adaptations* , pp. 22–41. Oxford University Press, Oxford.

Moore, A.J., Haynes, K.F., Preziosi, R.F., and Moore, P.J. (2002). The evolution of interacting phenotypes: genetics and evolution of social dominance. *The American Naturalist*, **160**, S186–S197.

Moore, P.J., Reagan-Wallin, N.L., Haynes K.F., and Moore, A.J. (1997). Odour conveys status on cockroaches. *Nature*, **389**, 25.

Møller, A.P. and Pomiankowski, A. (1993). Why have birds got multiple sexual ornaments? *Behavioral Ecology and Sociobiology*, **32**,167–176.

Neems, R.M. and Butlin, R.K. (1995). Divergence in cuticular hydrocarbons between parapatric subspecies of the meadow grasshopper, *Chorthippus parallelus* (Orthoptera, Acrididae). *Biological Journal of the Linnean Society*, **54**, 139–149.

Packard, G.G. and Boardman, T.J. (1999). The use of percentages and size-specific indices to nomalize physiological data for variation in body size: wasted time, wasted effort? *Comparative Biochemistry and Physiology A*, **122**, 37–44.

Partan, S. and Marler, P. (1999). Communication goes multimodal. *Science*, **283**, 1272–1273.

Partan, S. and Marler, P. (2005). Issues in the classification of multimodal communication signals. *The American Naturalist*, **166**, 231–245.

Peeters, C. and Liebig, J. (2008, in press). Fertility signaling as a general mechanism of regulating reproductive division of labor in ants. In: J. Gadau and J. Fewell (eds), *Organization of Insect Societies—Integrated Research Projects*. Harvard University Press, Cambridge MA.

Ratnieks, F.L.W. (1988) Reproductive harmony via mutual policing by workers in eusocial Hymenoptera. *The American Naturalist*, **132**, 217–236.

Roelofs, W.L., Liu, W., Hao, G., Jiao, H., Rooney, A.P., and Linn, C.E. (2002). Evolution of moth sex pheromones via ancestral genes. *Proceedings of the National Academy of Sciences of the USA*, **99**, 13621–13626.

Saxton, T.K., Little, A.C., and Roberts, S.C. (2007). Ecological validity in the study of human pheromones. In: J.L. Hurst, R.J. Beynon, S.C. Roberts, and T. Wyatt (eds), *Chemical Signals in Vertebrates 11*, pp. 111–120. Springer, New York.

Schal, C. and Bell, W.J. (1983). Determinants of dominant-subordinate interactions in males of the cockroach *Nauphoeta cinerea*. *Biology of Behaviour*, **8**, 117–139.

Silverstein, R.M., Rodin, J.O., and Wood, D.L. (1966). Sex attractants in frass produced by male *Ips confusus* in ponderosa pine. *Science*, **154**, 509–510.

Singer, T.L. (1998). Role of hydrocarbons in the recognition system of insects. *American Zoologist*, **38**, 394–405.

Slessor, K.N., Kaminski, L.A., King, G.G.S., Borden, J.H., and Winston, M.L. (1988). Semiochemical basis of the retinue response to queen honey bees. *Nature*, **332**, 354–356.

Slessor, K.N., Winston, M.W., and Le Conte, Y. (2005). Pheromone communication in the honeybee (*Apis mellifera* L.). *Journal of Chemical Ecology*, **31**, 2731–2745.

Vander Meer, R.K. and Morel, L. (1998). Nestmate recognition in ants. In: R.K. Vander Meer, M.D. Breed, K. Espelie, and M.L. Winston (eds), *Pheromone Communication in Social Insects. Ants, Wasps, Bees and Termites*, pp. 79–103. Westview Press, Boulder, CO.

Wcislo, W.T. (1997). Social terminology: what are words worth? *Trends in Ecology and Evolution*, **12**, 161.

West-Eberhard, M.J. (1979). Sexual selection, social competition and evolution. *Proceedings of the American Philosophical Society*, **123**, 222–234.

West-Eberhard, M.J. (1983). Sexual selection, social competition, and speciation. *Quarterly Review of Biology*, **58**, 155–183.

West-Eberhard, M.J. (1984). Sexual selection, competitive communication and species-specific signals. In: T. Lewis (ed.), *Insect Communication*, pp. 283–324. Academic Press, New York.

Wolf, J.B., Brodie, E.D. III, and Moore, A.J. (1999). Interacting phenotypes and the evolutionary process. II. Selection resulting from social interactions. *The American Naturalist*, **153**, 254–266.

Wyatt, T.D. (2003). *Pheromones and Animal Behaviour: Communication by Smell and Taste.* Cambridge University Press, Cambridge.

Zahavi, A. and Zahavi, A. (1997). *The Handicap Principle: a Missing Piece of Darwin's Puzzle.* Oxford University Press, Oxford.

van Zweden, J.S., Fürst, M.A., Heinze, J., and d'Ettorre, P. (2007). Specialization in policing behaviour among workers of the ant *Pachycondyla inversa*. *Proceedings of the Royal Society B: Biological Sciences*, **274**, 1421–1428.

CHAPTER 6

Chemical communication in societies of rodents

Jane L. Hurst and Robert J. Beynon

6.1 Introduction

Scent is the main form of communication between rodents. In part, this may be due to the need for small animals to remain hidden from larger predators, as extensive acoustic communication may be disadvantageous and vision will be limited when animals are active largely within covered sites or in darkness. However, scents have another major advantage for communication that sets them apart from visual or acoustic **signals**. While body scents are important for intimate and immediate communication when two or more individuals interact, scents can also be deposited in the environment and can persist in the absence of the signaller, often over extended periods. Such scent signals are not directed towards specific recipients but are broadcast to any other animals in the locality. This type of communication can thus provide a much broader advertisement of an animal's location, status, and identity to others that is likely to underpin both competitive and sexual interactions between animals in the local population.

There are three facets to the information content in scent marks. First, they provide information through the chemical components of the scent. Secondly, the spatial and temporal pattern of scent deposition provides further complexity. Finally, since scent signals are often deposited in response to competition, with one animal countermarking the scents of another, there is also information in the pattern of scents deposited by different individuals. All of these qualities must be captured in a cocktail of chemicals of remarkable diversity, which is to be expected given the requirement for species-specific communication. In particular, to provide information about a particular owner when the owner is likely to be elsewhere, scents must provide information on the genetic identity of the owner (species, sex, individual, and perhaps kinship). Such information needs to be invariant and unaffected by phenotypic modulations such as disease or reproductive status, or by changes in scent as it ages after deposition. Molecules providing information about genetic identity are thus likely to be 'hard-coded' products of the genome of each animal. Additional information concerning the animal's status (such as health and sexual status) changes throughout life and needs much more plastic encoding. This will thus be communicated by molecules that are variably expressed according to metabolism and hormonal state.

One of the challenges in chemical communication studies lies in linking the molecules that encode information with the use of that information. This demands a multidisciplinary approach that relates chemical characterization with behavioural responses. So far, the necessary extensive molecular characterization of scent marks combined with detailed behavioural analyses has been achieved to a significant extent only in the house mouse (*Mus musculus domesticus*). The availability of genetically defined inbred laboratory strains together with a fully sequenced mouse genome and the ability to test hypotheses under semi-natural conditions using captive wild animals has provided considerable advantages. Further, while many rodent species use multiple specialized scent glands as well as urine and faeces in communication, the focus on urine scents by mice provides a simpler system for analysis. We therefore focus here on the biology of scent communication in the house mouse, bringing in relevant evidence from other species where appropriate, before finally exploring the implications of these findings for other species.

6.2 Strategies for molecular approaches in chemical communication

An important aspect of understanding signalling mechanisms requires a parallel and comprehensive evaluation of the molecular species that are involved in the process, and the behavioural consequences that they elicit. Moreover, individual components can play different roles, either acting *directly* (the immediate conveyers of information, such as **pheromones**) or *indirectly*, such as the molecules that bind and elicit a slow release of pheromones (Armstrong *et al.* 2005) or enzymes that might catalyse the formation (Miyazaki 2008) or indeed, removal of semiochemicals. Whilst it is possible to assess chemical signalling without molecular identification (and indeed, that is more often the case than not), a clear understanding of the molecular events that are involved not only enhances understanding of the process, but reconciles molecular events with behavioural observations to provide new insight into function.

Notwithstanding the phenomenal achievements in functional genomics in recent years, there is very little opportunity to capitalise upon this knowledge in mammalian species other than the mouse, or possibly the rat. Most species that have been the subject of semiochemical studies are 'orphan species' and there is no immediate plan to sequence the genome of such organisms (http://www.ensembl.org). However, with the advent of novel, high-throughput sequencing technology (Ryan *et al.* 2007) we can anticipate that in future we will gain new genomes at rates hitherto unimagined. In the interim, a compelling case can be made for molecular characterization of scent secretions themselves, driven by analytical chemistry and biochemistry, particularly in the characterization of peptides, proteins, and low-molecular-weight metabolites. The methodologies are driven by high-resolution separation techniques (chromatography or electrophoresis) succeeded by detailed structural analysis of the separated molecules (usually mass spectrometry). Space precludes any detailed discussion of these methodologies, but recent volumes in the *Chemical Signals in Vertebrates* series (Mason *et al.* 2005, Hurst *et al.* 2008) provide ample pointers to the relevant technologies.

Low-molecular-weight semiochemicals can be volatile or involatile. For volatile molecules the separation is achieved by gas chromatography, a method in which different volatiles are separated by their tendency to stick to the coating of a fine capillary as they are swept along on a stream of inert gas (helium). Involatile molecules are most effectively resolved by liquid chromatography, where properties such as polarity and charge are used to effect separation. In both separation methods, maximum advantage is gained when the outflow of the separation can be directly coupled to the mass spectrometer, optimizing sensitivity resolution (the ability to analyse multiple components in the separation stream). In protein characterization, an additional separation approach is widely used, namely gel electrophoresis, in which proteins are separated on the basis of charge or molecular weight (one-dimensional electrophoresis) or both (two-dimensional electrophoresis). This separation method cannot be coupled directly to mass spectrometry and proteins must be recovered from the gel and treated before analysis by mass spectrometry. Balanced against that is the opportunity that gel electrophoresis provides to generate a rapid visual snapshot of the complexity of the protein content in a sample (Beynon *et al.* 2008). For protein characterization, the techniques of proteomics are used. Although the methods are often thought to apply to complex protein mixtures, they can be applied with considerable effect to simpler mixtures, or even single proteins (Karr 2008). Proteins can be analysed in their intact state or fragmented chemically or enzymically (digested) to produce discrete non-overlapping fragments (peptides) that can themselves be analysed further, after separation. Typically, strategies for protein characterization use progressively more complex techniques to gain information of increasing depth (Robertson *et al.* 2005). How does this process change if the genome sequence is known? For the low-molecular-weight semiochemicals surprisingly little, since these are often species-specific molecules and identification of the genes that encode enzymes for their synthesis would be challenging. For the protein components of the semiochemicals, the potential is greater. For example, identification of a protein is much easier if the genome sequence encoding that protein is known. For the mouse, knowledge of the genome allows us to reconcile the observed data on urinary proteins that are predicted from genome analysis. However, with a highly polymorphic family such as the major urinary proteins (MUPs) this is less of advantage, because of the disparity between MUPs in inbred mouse and wild-caught mice. Thus, for the C57BL/6J inbred mouse (the first rodent used for genome sequencing) we have been able to demonstrate which of the MUP genes are expressed in liver. Indeed, this can largely be achieved by simply measuring the mass of the proteins in urine by mass spectrometry—there is an excellent correlation between the masses of proteins that are predicted from the genome data and those that are observed, once adjustments are made for known protein processing events (Robertson *et al.* 1996, 2007; Armstrong *et al.* 2005).

6.3 Competitive signalling and assessment of social status

6.3.1 Territorial advertisement

The most obvious use of scent in rodents and other terrestrial vertebrates is for territory marking (for related discussions of olfactory communication in invertebrates see Chapters 5 and 7). Rodents typically deposit scent marks all round their territories, often using specific behaviours such as anal dragging or flank rubbing to deposit scents from specialized glands, although most species also exploit urine and faeces as sources of scent that can be widely distributed (Brown and McDonald 1985). Scent marks are particularly suited to the role of advertising territory ownership because the spatial pattern and density of scent marks provide physical proof of ownership: only animals that successfully dominate an area can ensure that their scent marks predominate there (Gosling 1982; Hurst 1993). This requires scent marks to carry information about the individual owner's identity, allowing others to identify the territory owner simply by matching scents in the surrounding area with those of animals encountered within the area (Gosling and McKay 1990). Thus scent marks provide a reliable signal of territory ownership as long as scents are individually distinctive (see below), regardless of any further information in the quality of an animal's scent. The presence of scent marks from competitors provides additional information about how effectively an owner defends its territory. Fresh scent marks from competitors indicate that an owner is not defending its territory effectively, which can lead to increased challenges by other competitors (Hurst 1993) and reduce the attractiveness of a territory owner as a potential mate (Johnston *et al.* 1997, Rich and Hurst 1998, 1999). Thus, owners typically spend much time investigating and refreshing the scents around their territory and will rapidly countermark any such scent challenges. This temporal pattern of scent deposition (i.e. the relative freshness of scents), as well as the spatial distribution of marks, provides other animals with a summary record of competitive challenges between animals and, most importantly, the outcome of those challenges. Competitor scents may be over-marked to provide physical evidence of the scent most recently deposited, such as hamster vaginal scent marking (Wilcox and Johnston 1995; Cohen *et al.* 2001). This tactic may be most appropriate for discrete scent marks that are deposited in a small number of sites. More typically, fresh scents are placed near to the ageing scents of a competitor, requiring animals to discern the relative freshness of scent marks from different individuals (Johnston and Schmidt 1979; Humphries *et al.* 1999; Rich and Hurst 1999). Scent mark refreshment rates are particularly high at the borders between neighbouring territories. This probably reflects the need for animals to ensure that their scent marks remain fresh in the immediate vicinity of competitors' scent marks, as others use these scent marks to assess the relative competitive ability of scent owners (Rich and Hurst 1999).

6.3.2 Scent mark age

Although scent marks clearly have the capacity to provide useful information about the age of scents, and thus how long ago the scent was deposited, very little attention has been paid to the chemical changes in scent marks as they age and how much information animals can gain from this. Our studies

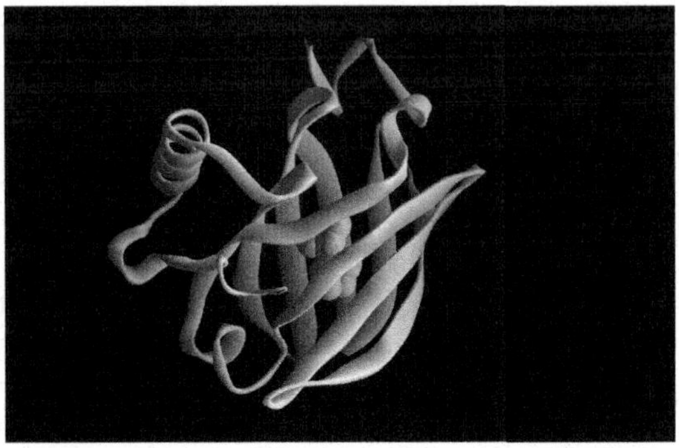

Figure 6.1 Cartoon structure of a major urinary protein. The MUP are members of the lipocalin family of proteins, which comprise a barrel-like structure bounding a central cavity or calyx, in which can be bound a pheromone molecule. The entrapment of the pheromone within the protein allows the signal to persist in the scent mark for longer than would be possible if it was freely volatile.

of urine scents and scent marking in house mice suggest that males may have evolved a number of tactics to maximize the apparent freshness of their scent marks, particularly in relation to competitor scents. Long-lasting signals that require only infrequent refreshment might seem most appropriate for advertising territory ownership, but labile signals provide much more reliable proof of current occupancy and the freshness of scents relative to those of any competitors (Hurst *et al.* 2001a). Volatile components are gradually lost from scents as they age. However, as the amount of any volatile component in a scent mark also depends on the amount of scent deposited in the first place, receivers cannot assess the age of a mark from the concentration of volatile scents alone. However, scents also contain involatile components that play an important role in the dynamics of scent ageing. The urine of house

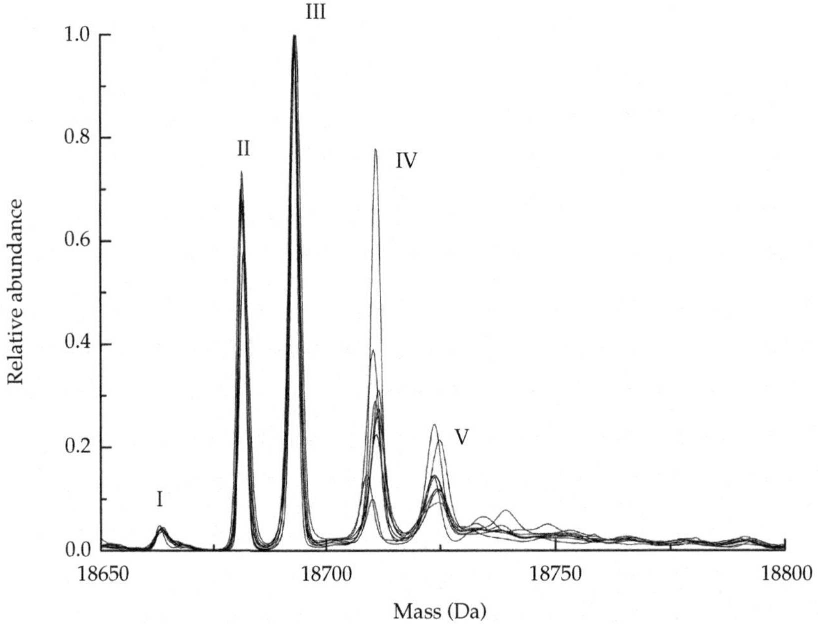

Figure 6.2 Intact mass profiling of MUPs in mouse urine. Mass spectrometry can be used with as little as one microlitre of urine to create a profile of the masses of MUPs. The profiles of nine wild male mice of the same MUP genotype (defined by tracking haplotypes from parents to offspring using microsatellite markers, Sherborne *et al.* 2007) living in large semi-natural enclosures are superimposed, normalized relative to mass peak III. Whilst the relative intensity of mass peaks I and II show little variation between individuals, the intensities of peaks IV and V are considerably more variable. We conjecture that these latter proteins are regulated by development or status. Moreover, these peaks are not centered around the same mass, consistent with developmental expression of additional proteins.

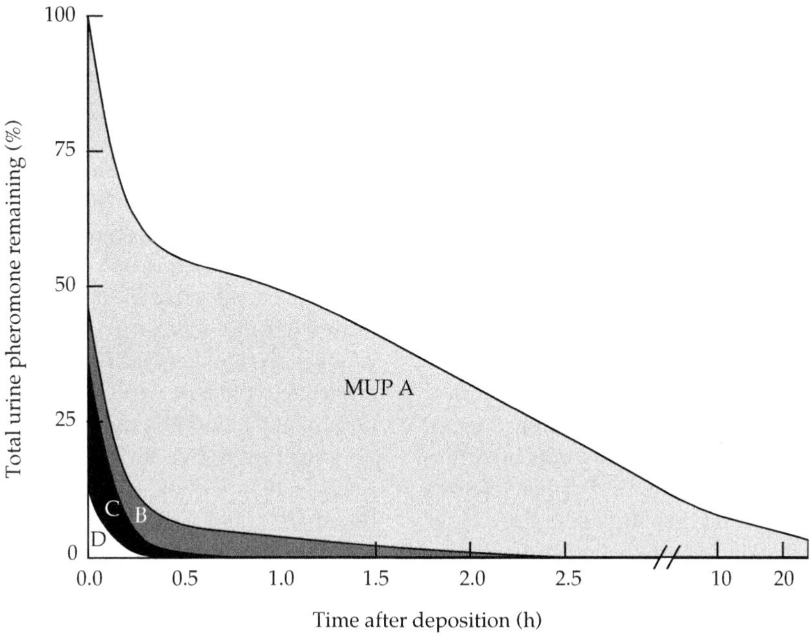

Figure 6.3 Protein binding as a slow release device to delay release of pheromones from scent marks. The male specific pheromone 2-sec-butyl 4,5 dihydrothiazole is bound to different extents to four groups (A to D) of MUPs in mouse urine. In the absence of protein binding, the volatile pheromone is lost within a few minutes. Most of the pheromone is bound to a male specific protein (A) that releases the ligand slowly. The other three groups (B–D) bind almost as much pheromone as A, but are not able to retain it for as long. This mixture of slow and fast releasing proteins results in a complex release profile that ensures that the signal is displayed by the scent mark for an extended period.

mice contains a high concentration of small lipocalin proteins called MUPs (Finlayson and Baumann 1958; Beynon and Hurst 2004). These proteins are produced in the liver and efficiently filtered into urine where their only known function is for scent communication. Mice of both sexes excrete a substantial quantity of MUPs, but adult males produce three to four times as much as females, resulting in sticky deposits as the urine dries. MUPs are barrel-shaped proteins (Fig. 6.1) with a central cavity that binds small lipophilic molecules including a number of volatile male pheromones (Bacchini *et al.* 1992; Robertson *et al.* 1993; Novotny *et al.* 1999).

When not bound to MUPs, free volatiles are lost from scent marks in a few minutes (Robertson *et al.* 2001), but those bound by MUPs continue to be released gradually and can be detected by animals over at least 24 hours (Hurst *et al.* 1998; Humphries *et al.* 1999). Thus, by excreting a high concentration of MUPs, males extend the longevity of scent signals that are still highly volatile and can be detected as an airborne scent. Further, as the amount of involatile protein in a scent mark does not change through time, and each protein molecule can bind only one volatile ligand molecule, the proportion of MUPs that contain ligands might be used as an indicator of scent age regardless of how much scent was actually deposited (Hurst *et al.* 2001a).

MUPs are products of a highly polymorphic cluster of genes on mouse chromosome 4, with individual mice each expressing approximately 8–14 different MUP isoforms (e.g. Fig. 6.2). As inbred laboratory mice have homozygous MUP types, they express a smaller number of 5–7 urinary MUPs (Robertson *et al.* 1996).

Although most variation between MUPs is located on the surface of the molecule (Beynon *et al.* 2002), a few MUPs exhibit variation within the central cavity, leading to different binding affinities for volatile ligands (Marie *et al.* 2001; Armstrong *et al.* 2005). This appears to be particularly the case for some MUPs that are expressed only by adult males, under androgen control (Knopf *et al.* 1983). One MUP in particular is male-specific and is responsible for binding the most abundant male pheromone in urine, 2-*sec*-butyl-4,5-dihydrothiazole, slowing its release from scent marks over a 24-hour period (Fig. 6.3) (Armstrong *et al.* 2005). Two other MUPs (uMUP-X and uMUP-VII) sequenced from laboratory mice have an amino acid substitution in the central cavity that alters ligand binding (Marie *et al.* 2001). We now know that both of these are also expressed only by males (unpublished data), although no specific ligand for these MUPs has yet been identified. Males may thus alter the release rates of pheromones and thus the 'ageing' of their scents by differential expression of these MUPs. Further research is needed to understand how this alters the relative attractiveness of scents to females, and whether males adjust their tactics of scent refreshment according to the volatile pheromone release characteristics associated with specific MUP patterns.

Scent marking patterns reflect the importance of maintaining fresh scent marks rather than the absolute amount of scent in a mark. Many rodents deposit their urine scent in numerous small spots and streaks which not only helps to distribute scent throughout the territory, but also maximizes the refreshment rate in each location. The presence of fresh intruder scents stimulates male house mice to increase their own rate of scent marking to countermark. When faced with a competitor's scent mark, they do not attempt to deposit a larger scent mark than the competitor's. Instead they deposit small scent marks that are repeatedly refreshed over the next few hours, so that their own scent remains fresh while the competitor's scent ages (Hurst and Beynon 2004). Further, although territory owners countermark any fresh scent marks from an intruder, mice deposit even more scent marks near to ageing intruder scent marks. By doing this, their own fresh scent will draw attention to a site where this is clearly fresher than the intruder's (Humphries *et al.* 1999).

6.3.3 Advertisement of subordinate status

Not all males are successful territory owners and some low-quality or young males choose to live as subordinates in another male's territory. However, dominant territory owners appear to force subordinates to advertise their subordinate status to be accepted within the territory (Hurst *et al.* 2001a). Defeated males show a dramatic and immediate change in their scent marking behaviour (Desjardins *et al.* 1973; Rozenfeld *et al.* 1987), reducing their rate of scent marking in the presence of a dominant male's scent while continuing to deposit some scent marks. The quality of a subordinate male's scent also changes, due to reduced production of male-specific volatile scents that are attractive to females and induce aggression from dominant males. Thus, subordinate male scents are no longer attractive to females (e.g. Bronson and Caroom 1971; Jemiolo *et al.* 1991). Further, resident dominant males will only tolerate individual subordinates that deposit scent marks in the territory that indicate their subordinate status (Hurst *et al.* 1993). As scent marks act as broadcast signals available to all other animals using the area, dominant territory owners prevent subordinates from hiding their low status from females, thus tolerating only those males that are not likely to be competitors for mating opportunities.

6.4 Individual identity recognition

In some situations, it may be sufficient to identify whether animals belong to a particular class, such as 'own species' versus 'not own species' for conspecific identification, 'male' or 'female' to identify potential mates or competitors, 'familiar' versus 'unfamiliar' to recognize unknown intruders, and so on. This type of simple categorization may be sufficient for many non-social species, if the same response is appropriate to all individuals of a particular category (e.g. all same-sex conspecifics are intruders that need to be excluded from the territory; all opposite-sex conspecifics are potential mates, etc.). However, the ability to recognize individuals, and to be readily recognized by others, is important in many contexts, especially among social species that form distinct relationships. Individual recognition allows animals to modify their social response according to information gained and remembered about a particular individual (Dale *et al.* 2001; Lai and Johnston 2002;

Thom and Hurst 2004). Animals might use general characteristics to distinguish any discriminable difference between individuals for short-term identity recognition. Scents are affected by a large number of genetic and environmental factors such that, at any one time point, there are always likely to be incidental differences between scents from different individuals. However, reliable recognition requires the use of **cues** that are *fixed* characteristics of an individual that do not change, for example, with the vagaries of physiological or metabolic status or with environmental changes such as diet. Animals that themselves benefit from being recognized are likely to evolve effective signals for communicating their identity to others (Johnstone 1997; Thom and Hurst 2004). This is particularly important for animals that communicate through scent marks, which are physically separated from the owner and will only provide useful information about the owner if others can identify who the scent owner is. To be reliable fixed signatures of individual identity, signals are thus likely to be genetically determined and to have sufficient polymorphism to distinguish between individuals (Boyse *et al.* 1987), while also being relatively insensitive to environmental influences and variation in individual condition (Thom and Hurst 2004).

6.4.1 Evidence from inbred laboratory rodents

Over the past 30 years, attention has focused almost exclusively on the **polygenic** and highly polymorphic major histocompatibility complex (MHC) as a source of genetically determined scents that differ between individuals (often termed 'individuality odours' or 'individual odourtypes'). Genes within the MHC are highly polymorphic and have a well-established function in self/non-self recognition in the immune system. MHC genotype also appears to influence an individual's scent in a wide range of species including humans (Brown and Eklund 1994; Wedekind *et al.* 1995; Olsen *et al.* 1998; Reusch *et al.* 2001; Jacob *et al.* 2002; Penn 2002; Olsson *et al.* 2003) (see also Chapter 9, on human mate choice). Direct evidence that MHC contributes to discriminable differences in rodent urinary scents comes from highly inbred strains of laboratory rodents, where all other genetic differences have been eliminated (Yamaguchi *et al.* 1981; Singh 2001; Carroll *et al.* 2002). Other 'background' genes contribute to discriminable scent differences (Yamazaki *et al.* 1986; Carroll *et al.* 2002), but the extreme polymorphism of MHCs between individuals in wild populations has led to a widespread assumption that MHC odours provide the main basis for individual recognition through scents in rodents and possibly other vertebrates (e.g. Boyse *et al.* 1987, Brown 1995; Yamazaki *et al.* 1999; Singh 2001; Brennan and Zufall 2006). Molecular analysis reveals that MHC type influences numerous volatile urinary metabolites (Schaefer *et al.* 2001; Willse *et al.* 2006), but metabolites reflect the combination of MHC and genetic background rather than providing a distinct MHC signature (Willse *et al.* 2006). Further, as urinary volatile metabolites are influenced by many non-genetic as well as genetic factors, they are not likely to provide an invariant individual identity signature (Hurst *et al.* 2005a). As an alternative mechanism, peptide receptors in both the vomeronasal organ (Leinders-Zufall *et al.* 2004) and main olfactory epithelium (Spehr *et al.* 2006) may detect the peptides that bind to MHC class I proteins. As the set of peptide ligands bound by MHC proteins will be defined by the specific MHC type of the animal, this might allow specific recognition of MHC type independently of other factors that influence scents (Brennan and Zufall 2006). However, the presence of such MHC peptide ligands in rodent urine (distinct from any other peptides) still remains to be demonstrated.

Evidence from laboratory rodents thus suggests that MHC genotype has a major influence on individual urinary scents. Nevertheless, the ability to discriminate differences between scents is not sufficient evidence to indicate that these differences are used for individual or other genetic identity recognition. Tests must demonstrate that an individual's identity has been recognized (a perceptual process), reflected by a predictable functional difference in behaviour towards equivalent individuals that will only be shown if individuals are recognized (Thom and Hurst 2004; Hurst *et al.* 2005a). Furthermore, animals must be able to identify individuals against the normally variable genetic and environmental backgrounds typical of natural populations, not when this is the only genetic difference between animals held under constant laboratory conditions (Thom and Hurst 2004).

6.4.2 MUP polymorphism among wild mice

Our recent studies using functional tests of recognition among wild house mice (*Mus musculus domesticus*) strongly challenge the assumption that variation in MHC scents makes a significant contribution to individual recognition, at least in mice. Instead, mice use MUPs, a specialized set of scent communication proteins in their urine, as a polymorphic marker to recognize individuals. The pattern of MUPs expressed by each individual is a direct reflection of their MUP genotype and is remarkably resistant to variation in genetic background (Fig. 6.2). The polymorphism in MUP genotype among wild mice is such that only some very closely related animals share the same MUP phenotype. This provides each individual with a distinctive MUP signature that is highly resistant to degradation in urine scent marks (Beynon and Hurst 2004). This is in dramatic contrast to the lack of variation among strains of laboratory mice, as we have found only two different patterns among many classical strains that have been derived from separate lineages (S. A. Cheetham, J. L. Hurst, and R. J. Beynon, data in preparation). This is consistent with findings from mitochondrial DNA analysis which indicate that all of the classical strains of laboratory mice have been derived from a single female (Ferris *et al.* 1982) and thus represent only a tiny fraction of the normal genetic variability among individual wild mice.

6.4.3 Recognition of own versus other individual scents

To investigate whether MUP signatures are used for individual recognition, we tested whether male wild mice use urinary MUP patterns to recognize their own scent marks and those of other males. As only very close relatives share the same MUP pattern among wild mice, we used sib males that had inherited the same or different MUP signatures on a random genetic background. When adult males encounter scent marks from another male in their territory they rapidly countermark, repeatedly depositing fresh countermarks in the vicinity of the competitor's scent (see Section 6.3). However, when scent marks are from another male with the same MUP pattern as their own, they do not increase their rate of scent marking and spend no more time in the vicinity of these marks than when their own urine is introduced (Hurst *et al.* 2001b). To prove that males respond to MUP patterns rather than some other genetic difference linked to MUP type, we altered the pattern of MUPs in a male's own urine using a recombinant MUP (artificially synthesized in yeast but known to be expressed by mice) while maintaining a normal urinary MUP concentration. As predicted, this caused males to countermark their own urine scent. Mice thus use MUP pattern to recognize individual ownership of urine scent marks (own or other male), and are sensitive to even a single MUP difference, while ignoring all other non-MUP differences that contribute to urinary scent profiles.

At first sight this is surprising in view of the abundant literature showing that mice are sensitive to even very small genetic differences between urine donors in scent discrimination studies. However, discrimination studies focus only on whether animals can detect a difference between scents, not what (if anything) that difference means. Animals may be trained to discriminate between two scents using rewards, regardless of whether they have any natural motivation to respond. Alternatively, tests assess the duration of spontaneous sniffing investigation when animals encounter a novel scent, which indicates that animals recognize that a scent is novel and are motivated to gather information from the unfamiliar scent (reviewed by Thom and Hurst 2004). To assess *recognition* of the individual owner, we focused only on the subsequent response to scents once animals had gained information through sniffing the scents, by measuring their countermarking response and the total time spent in the vicinity of the stimulus (inappropriately termed 'investigation' in Hurst *et al.* 2001b). We have subsequently repeated this study, again introducing urine streaks into the individual territories of wild male mice, to confirm that mice detect differences in scents regardless of MUP pattern. The results, shown in Fig. 6.4, confirm that any other male urine stimulates prolonged

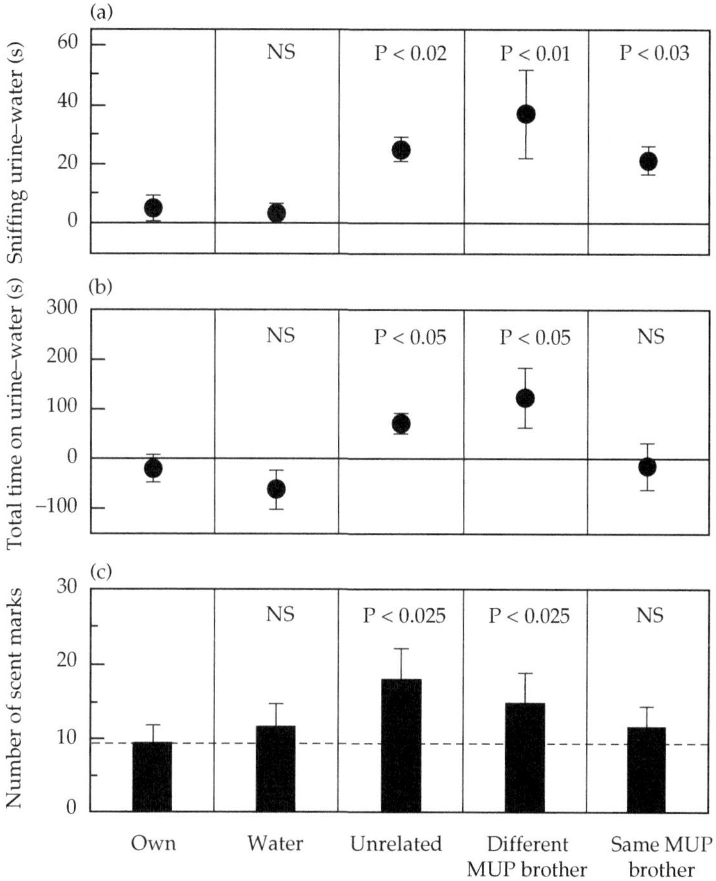

Figure 6.4 Male mouse response when urine from different individual males is introduced into their territory. Urine (10μl) from specific male donors was introduced into enclosures housing single wild male house mice (n = 12), together with an equivalent water control placed on the opposite side of the enclosure for 30min. The urine and water stimuli were streaked onto 15 × 15cm tiles covered in absorbent paper in turn covered by an open mesh grid (3.5 × 3.5 cm with 3 × 2mm apertures) that mice could push their noses through—this allowed measurement of close contact sniffing (nose in contact with the mesh). We measured (a) time spent sniffing the urine minus the water control presented simultaneously; (b) total time spent on the urine minus water tile; (c) the number of urine marks deposited on the urine stimulus tile, scanned under ultraviolet light using a FluorS imager. Urine scents were from the resident male (own), or from other males housed equivalently in enclosures that were unrelated, or were sibs that expressed either a different or same MUP pattern as the subject. In the water control test, both tiles were streaked with water. Stimuli were presented in a different order to each male, with at least 24h between tests. Probability values are given for Wilcoxon matched pair tests comparing the response to each stimulus with response to own urine.

contact sniffing regardless of MUP pattern, while own highly familiar urine does not (Fig. 6.4a). This sniffing occurs during the first few visits to a novel scent and is required to pump involatile scents to the vomeronasal organ (Luo *et al.* 2003). Confirming our previous study, however, having gained information from the scent, mice only spend significantly more time in the vicinity and countermark those containing a different MUP pattern to their own (Fig. 6.4b,c). Notably, mice must be able to contact the scent during investigation to recognize the owner. This is consistent with detection of the involatile MUP–ligand complexes through the vomeronasal system (Nevison *et al.* 2003).

To investigate the role of MHC in this response more closely, we used inbred strains of laboratory mice that differ (1) only in MHC, (2) in genetic background including MUP type but not MHC, or (3) both in MHC and genetic background (Hurst *et al.* 2005a). This confirmed that MHC differences are not sufficient or required for recognition of own and other male scent marks. Interestingly, urine of a different MHC type stimulated more prolonged investigation only when this matched both the MHC and genetic background of a familiar neighbour competitor. Thus, MHC scents contribute to the recognition of familiar scent profiles that may have important social significance to animals (i.e. recognition of familiarity in scents that induces further investigation, rather than recognition of owner identity). The fact that MHC scents were discriminated only in combination with background genes corresponds to molecular analyses showing that MHC type and genetic background interact in determining urinary volatile profiles (Willse *et al.* 2006). Nonetheless, discrimination of scent familiarity stimulates only further close investigation of scents. It is the information gained from the MUP pattern during such investigation that determines whether males recognize their own and other male scents.

6.4.4 Individual recognition

The recognition of own versus other male scents is a special case of individual recognition, where own scent is clearly much more familiar to the subject than the scents of other individuals. To examine the more general case of how animals recognize equivalent individuals (not including themselves) we switched our focus to a different functional response. Females need to recognize different individual male scent owners to assess their quality as potential mates from their territorial scent marks. Studies using laboratory mice suggest that females are sensitive to the MHC type of male urine donors (e.g. Egid and Brown 1989; Eklund 1997; Jordan and Bruford 1998; Penn and Potts 1998b) and this may be important in determining mate choice (see Section 6.6 below). To investigate whether MHC and/or MUP type contributes to the recognition of individual males by females, we used the well-established preference of females for males that have countermarked a competitor's scent (Johnston *et al.* 1997; Rich and Hurst 1999; Fisher *et al.* 2003). In an 'identity learning' phase, females were presented with urine streaks from two unrelated and unfamiliar wild male mice, deposited 24 hours apart as if one male had countermarked the other. This creates the illusion of a status difference between two equivalent males that should cause females to be more attracted to the owner of the countermark if they can recognize the individual identities of the males (note that females can identify males from fresh or 24-hour aged urine scents; Cheetham 2006). Under natural conditions, this attraction may provide females with the direct benefits of a well-defended male territory and/or indirect benefits of good genes for their offspring if they mate with a successful territory owner. We tested individual recognition by giving females a choice between the two male scent owners presented behind separate mesh barriers. As an unambiguous measure of attraction, we assessed the time that females spent gnawing and pulling at the barriers attempting to get through to each male, as well as time spent in close proximity and time sniffing through the barriers to gain further scent information from each male. When the two males differed from each other across the genome (including MHC and MUP types), females easily recognized and were attracted to the countermarking male. However, differences in MHC type were not used for individual recognition. This again depended on a difference in MUP type between the two males, regardless of all the other genetic differences between wild mice (Cheetham *et al.* 2007). Recognition also required contact with the scent during initial investigation, suggesting detection of involatile MUPs through the vomeronasal system. Non-MUP differences in scents stimulated further investigation of the males, but this did not result in individual recognition if males did not differ in MUP type (Cheetham *et al.* 2007).

Thus, mice use the distinct polymorphic patterns of MUPs in urine to recognize individuals in these two very different social contexts, regardless of other genetic differences that influence an individual's scent including MHC. It is not yet known how mice detect MUP patterns in scent. MUP receptors have yet to be identified, but as a recombinant MUP elicits electrical activity in the vomeronasal organ

(Kimoto *et al.* 2007), and both synthetic MUP peptides and MUPs stripped of their natural ligands trigger ovulation through the vomeronasal system (More 2006), it seems likely that differences in MUPs are detected directly rather than through the pattern of ligands differentially bound to MUPs. Most recently, a class of receptors that respond to MUPs in the absence of bound ligands has been identified (Chamero *et al.* 2007).

6.5 A model of social recognition through scents

Mice use the involatile pattern of MUPs to recognize individuals, but this does not mean that volatile scents play no role in the recognition process. Indeed, airborne volatiles are essential for the detection of scents, alerting animals at a distance to the presence and location of a scent source. In addition, the pattern of airborne volatiles, detected through the main olfactory system, appears to play a key role in assessing whether scents are familiar or unfamiliar. When airborne scents are novel, they stimulate approach and close investigation of the scent source to gain further information (Humphries *et al.* 1999). This detection of novel stimuli activates a pumping mechanism to deliver scents to the vomeronasal organ (Meredith 1994) when animals make direct nasal contact with a scent source (Luo *et al.* 2003). This allows detection of the involatile MUP identity signature for owner identification. However, detection of scents through the vomeronasal system is slow. Animals first have to make contact with the scent source, followed by 2–3 seconds to deliver water-soluble stimuli to the vomeronasal organ (Luo *et al.* 2003), while ligand-mediated activation of vomeronasal organ receptor neurons takes a further second (Holy *et al.* 2000). There is thus a latency of 3–4 seconds from initial contact until activity in the accessory olfactory bulb is increased and peak responses to stimuli can take around 18 seconds (Luo *et al.* 2003). This is in stark contrast to the much more rapid processing times of other sensory systems, including the detection of volatile odorants through the main olfactory system. Although the accessory olfactory system allows specific detection of complex social stimuli, it is not a very efficient processing system. This is likely to be costly in many social situations where recognition needs to be rapid and close physical contact may need to be avoided, for example when encountering a potentially dangerous competitor. However, prolonged contact investigation is needed only when a scent is unfamiliar—animals do not need to sniff closely at every familiar individual or scent mark in their environment to recognize it. Instead, animals learn to recognize familiar volatile scents associated with familiar individuals. (See detailed discussion of communication **networks** in Chapter 3.)

Close contact investigation of fresh unfamiliar scents provides an opportunity to learn an association between the involatile ownership signal detected through the vomeronasal system, and volatile scents detected simultaneously through the main olfactory system (Guo *et al.* 1997; Hurst *et al.* 2005b). If the same pattern of airborne volatiles is encountered subsequently, a receiver that has learnt the associated involatile fixed identity signature will not need to approach and detect the signature again. Thus, when animals are presented repeatedly with the same scent, the initially prolonged close sniffing declines dramatically after only one or two presentations (e.g. Singh *et al.* 1987). However, the precise pattern of volatiles in scent sources like urine is very complex and influenced by many factors. When an animal undergoes a metabolic shift that changes its volatile odour profile, such as a change in diet or bacterial flora (Brown 1995), the airborne scent is again perceived as novel. This activates the drive to associate the new volatile scent with an individual identity signature through close contact investigation. Similarly, a forgotten association between volatile profile and identity signature stimulates reinvestigation and the association is relearnt. Fresh scent marks in the environment provide an easy opportunity to learn this association without having to make direct contact with the scent donor itself. Thus, animals can learn to recognize the volatile scents of a resident territory owner through investigation of its scent marks, and will be able to recognize and avoid this highly aggressive individual from airborne scents detected at a distance (Hurst 1993).

This learnt association between involatile and volatile components appears to be a general feature involved in identity recognition, not only of individuals but also of the opposite sex. Naïve female mice show an innate attraction to male scents when they can contact involatile components of the scent, but they are only attracted to the airborne volatiles from male scents once they have learnt an association through prior contact with the involatile components of male scent (Moncho-Bogani *et al.* 2002). Male mice that have encountered females artificially odorized with perfume subsequently emit ultrasonic courtship vocalizations to the volatile perfume itself, apparently associating the perfume with recognition of a female mouse (Nyby *et al.* 1978).

6.6 Kin recognition

The ability to recognize kin is a critical process for the avoidance of inbreeding if animals are likely to meet kin of the opposite sex in adulthood (Pusey and Wolf 1996) (see also discussion of kin recognition for bacteria in Chapter 2). Kin recognition could also allow adults to establish cooperative relationships that promote the reproductive success of their relatives, increasing an individual's **inclusive fitness** (Hamilton 1964). One way to recognise kin is to learn phenotypic cues of individuals during development (familial imprinting) because in most social systems these animals are likely to be relatives. Familiarity is dynamic and learned, and because of this plasticity, may be of limited use for recognizing kin in later adulthood or kin that were not encountered during rearing. However, the use of polymorphic genetic markers would allow animals to recognize close kin based on phenotypic similarity to themselves and perhaps also to known familiar relatives (such as the animal's familiar mother). Thus, kin recognition could involve self-referent matching at one or more highly polymorphic genetic markers, and could also involve familial imprinting on the genetic markers of close relatives learnt during rearing. There is considerable debate as to whether such genetic markers could evolve specifically to promote cooperative and altruistic behaviour towards kin (Gardner and West 2007). However, the need to avoid inbreeding, or to promote offspring variation in host defence mechanisms against parasites, could drive the evolution of polymorphic genetic markers for kin recognition. Animals might then take advantage of these genetic markers to establish cooperative relationships between kin (Rousset and Roze 2007).

6.6.1 Evidence from laboratory rodents and hybrids

Despite the obvious benefits of kin recognition, there is surprisingly little empirical evidence for such genetic recognition mechanisms in animals (Gardner and West 2007). The one example widely cited is the use of scents associated with MHC in mice. Yamazaki and colleagues first noticed an MHC-based mate preference among congenic strains of laboratory mice that differed only at MHC (Yamazaki *et al.* 1976). Some strains prefer to mate disassortatively, i.e. with mice that have a different MHC type. As those sharing the same genotype at such a highly polymorphic marker in natural populations are very likely to be closely related, matching of MHC-based scents could be used to recognize close kin and reduce the probability of inbreeding (Brown and Eklund 1994; Potts *et al.* 1994). Offspring might gain additional fitness benefits since heterozygosity at the MHC itself may confer increased resistance to pathogens (Penn 2002; but see Ilmonen *et al.* 2007). Indeed, this has led to the hypothesis that self/non-self recognition mechanisms in the adaptive immune system (such as MHC) might have evolved from an ancient sensory mechanism to evaluate relatedness in the context of **sexual selection** (Boehm 2006). Studies in which pups are cross-fostered between strains suggest that mice remember the familiar MHC scents of their mother through familial imprinting to avoid inbreeding rather than matching their own scents to potential mates (self-referent matching) (Yamazaki *et al.* 1988; Penn and Potts 1998a). However, although MHC-based mate choice is so widely cited, supporting evidence is surprisingly limited.

The strength and direction of MHC preferences among laboratory mice varies considerably between strains, with some strains showing disassortative preferences, some preferring to mate assortatively, and others showing no preference

(Jordan and Bruford 1998). As we discussed above, inbred strains of laboratory rodents present considerable problems for addressing the mechanisms that underlie genetic identity recognition. This is particularly the case when trying to understand mechanisms of kin recognition and inbreeding avoidance. Not only do these animals have highly abnormal genetic homogeneity and derive from an extremely small pool of hybrid founders, they have also been subject to strong artificial selection for ease of breeding (including inbreeding) in the laboratory over many generations, protected from the pressures of **natural selection**. The relevance of such studies to normal behaviour is thus highly questionable (Manning *et al.* 1992a). Genetic recognition needs to be demonstrated against the variable genetic and environmental backgrounds typical of natural populations, using animals with normal behaviour. To address some of these issues, Potts and colleagues (Potts *et al.* 1991, 1994; Penn and Potts 1998a) crossed inbred laboratory strains with wild mice to derive subjects with MHC **haplotypes** from laboratory strains but with 50% of the rest of their genome from wild mice. These hybrid animals are still relatively inbred and express a reduced set of homozygous and heterozygous laboratory-derived MHC types, but behave more like wild mice. When they were allowed to breed freely in semi-natural enclosures, there was a deficit in the frequency of MHC homozygous offspring compared with random mating expectations, consistent with MHC disassortative mating (Potts *et al.* 1991). This classic study provides the main evidence that MHC is used as a marker for disassortative mating by mice and possibly other animals. A bias in communal nesting partnerships among MHC-similar females in these populations (Manning *et al.* 1992b) is also consistent with the use of MHC as a genetic marker of kinship. However, interpretation of this study remains controversial because of key limitations in experimental design and analysis (Pomiankowski and Pagel 1992; Brown and Eklund 1994; Hughes and Hughes 1995; Jordan and Bruford 1998; Penn 2002; though see Potts *et al.* 1992). Offspring were typed only for MHC, so parentage could not be assigned and, crucially, parental differences in overall relatedness and other genes that might contribute to inbreeding avoidance could not be assessed. As the inbred strains used to derive MHC and half of the genome of these hybrid mice also have two different MUP types, the deficit in MHC homozygous offspring could have arisen from a correlation between MHC and MUP types in the derivation of the founder lines (Sherborne *et al.* 2007). However, the potential role of MUPs or other polymorphic genetic markers that might reflect kinship has not been examined among laboratory mice.

6.6.2 Kin recognition among wild mice

Following our discovery that MUPs provide the genetic identity signature that underlies individual recognition in wild mice, we tested whether this could also be used for kin recognition in the context of inbreeding avoidance. We also wanted to confirm that MHC scents are used for inbreeding avoidance in animals with normal genetic variation, which had not previously been tested. MUPs are encoded by a tightly linked cluster of at least 30 genes and pseudogenes and are inherited as a haplotype (a linked set of genes) similar to the inheritance of MHC haplotypes. These two gene clusters are so polymorphic that only very close relatives are likely to share the same MUP or MHC genotypes. While this means that they could both provide useful genetic markers of kinship detectable through scent, the correlation between similarity at one of these markers and similarity across the genome means that it is essential to control for differences in relatedness to test which particular markers are used. Further, as only closely related animals are likely to share MHC and/or MUP types in normal outbred populations and inbreeding avoidance is only expected to evolve between very close relatives where there is a high risk of inbreeding depression (Kokko and Ots 2006), inbreeding avoidance needs to be examined among related animals. We therefore created captive populations of wild mice that each consisted of a mixture of full sibs and paternal half sibs, allowed to breed freely in large (250 m^2) semi-natural enclosures. After 15 weeks, we captured 483 offspring from 81 original founders (split across four separate populations), which were genotyped to establish parentage and, independently, their MHC and MUP haplotypes.

This study reveals that wild house mice use self-referent matching of MUP patterns to avoid inbreeding but found no evidence that MHC sharing influences mate selection (Sherborne *et al.* 2007). Analysis of the number of successful matings by each female with each available male, and the number of offspring per mating, revealed that wild mice show no avoidance of mates with the same MHC genotype (whether one or both haplotypes are shared). Instead, inbreeding avoidance is fully explained by a very strong deficit in successful matings between mice sharing both MUP haplotypes ($P = 0.002$). Sharing both haplotypes is a good indicator that mice are very closely related, but sharing only a single haplotype is not a good guide for identifying very close kin like full sibs (Sherborne *et al.* 2007). Correspondingly, there was no deficit in mating between those sharing just one MUP haplotype. Although it has previously been proposed that animals might increase the range of relatives avoided through behavioural imprinting on the separate haplotypes carried by their own mother (Penn and Potts 1998a), mate choice was not influenced by behavioural imprinting on maternal MHC or MUP haplotypes. Such a strategy would include many animals that are not close relatives, considerably reducing the choice of mates in the local population even when inbreeding depression is not likely. The deficit in successful matings between those sharing the same MUP genotype could have been caused by pre-copulatory mate choice or post-copulatory mechanisms as we were unable to observe mating behaviour under such naturalistic conditions. However, as MUP type is clearly detectable through urinary scents, pre-copulatory mate selection is the most likely mechanism underlying this inbreeding avoidance.

This study has some important implications. Direct evidence for the use of MHC scents for kin recognition and mate selection derives largely from studies of laboratory or hybrid mice with abnormal genetic backgrounds and social experience. However, wild mice use a specialized set of species-specific urinary proteins to avoid inbreeding and did not use other genetic markers to improve their level of inbreeding avoidance. The strong consistency between MUP genotype and phenotypic expression (see Fig. 6.2), together with evidence from individual recognition studies, indicates that MUP patterns provide stable signatures of individual genetic identity that are easily recognized. However, this system is imperfect for recognizing all close kin. Those that have the same MUP type are very likely to be close kin and are avoided. Nonetheless, in fully outbred populations most full sibs will not share both MUP haplotypes and thus will not be recognized. The same would apply to any other highly polymorphic marker, including MHC. Mice could improve the number of animals correctly identified as close kin if they used matching at either MUP or MHC, but they do not do this. They failed to avoid mates of the same MHC type, even though 50% of the offspring from such matings will be MHC homozygous and might have poorer parasite defence mechanisms (though see Ilmonen *et al.* 2007). This suggests that MHC scents may not be easily recognized under naturalistic conditions. However, the use of a single genetic marker (MUP) to avoid inbreeding under natural conditions may be more effective than it seems initially. First, even if female discrimination against related males is relatively weak (i.e. each male is avoided by only some of his close relatives), theoretical modelling predicts that this discrimination against kin drives male-biased dispersal from natal areas in **polygynous** mating systems (Lehmann and Perrin 2003). This will separate close kin of the opposite sex, reducing inbreeding through spatial separation even between pairs that do not share the same MUP type. Correspondingly, young males have the highest levels of dispersal in house mouse populations (Pocock *et al.* 2005). In our experiment, we deliberately prevented dispersal, removing this additional mechanism that would further prevent inbreeding between close kin more generally, driven by MUP sharing. Secondly, when dispersal does not separate all close kin of the opposite sex (e.g. where there are barriers to dispersal or where locally abundant resources reduce the tendency to disperse), outbred animals may mate with close kin that they cannot recognize. As animals become more inbred, the proportion of close relatives that share the same MUP type will increase, in line with the sharing of many other genes between close relatives. Thus MUP sharing will have an increasing impact on inbreeding avoidance as animals become more inbred and the inbreeding load increases. Notably, animals should only avoid mating with kin when the inbreeding load exceeds a substantial threshold (Lehmann and Perrin 2003; Kokko and Ots 2006).

The use of MUP as a genetic marker for inbreeding avoidance will promote genome-wide heterozygosity including MHC. Our results suggest that observed deficits in MHC homozygous offspring in studies of hybrid laboratory × wild mouse crosses are most likely due to correlations between MHC and MUP types in the founder lines. This might explain why biases have not been very consistent in these MHC experiments. For example, to investigate familial imprinting on MHC scents, Penn and Potts (1998a) produced two lines with different MHC types by crossing wild mice with two different laboratory strains. However, theses strains differed not only at MHC but also across the genome including MUP type. Further research is required to examine whether wild mice not only avoid mates with the same MUP type as themselves (through self-referent matching) but also those matching both MUP haplotypes of their highly familiar mother learnt through familial imprinting.

6.7 Species-specific signalling

The appeal of MHC scents as a likely genetic marker of individual identity and kinship arises from the fact that MHC is highly polymorphic in a wide range of vertebrates and thus might provide a common mechanism of individual and kin recognition through scent that could apply to many different species. Instead, house mice use a specialized set of communication proteins to achieve this. How widespread is this mechanism likely to be among rodents? It is relatively straightforward to address this question with analyses of increasing complexity, starting with global analyses of the magnitude and complexity of the protein complement of the scent secretion, progressing to identification of the protein components and evaluation of the relationship between the protein components and the non-proteinaceous components of the scent mark. To illustrate this strategy, we have embarked upon a survey of urinary lipocalins in rodents, and although the data accrued thus far are limited, a picture of surprising complexity has already emerged (Table 6.1). Most of the murid species studied to date produce MUP-like lipocalins (Beynon *et al.* 2008), although in two species (Roborovski hamster and bank vole) the urinary lipocalin more closely resembles aphrodisin, a lipocalin in hamster vaginal fluid that stimulates copulatory behavior in male hamsters (Briand *et al.* 2004). Sexual dimorphism ranges from exclusive expression in males (Norway rat) to equal protein production in males and females (Roborovski hamster). Notably, we have not yet discovered evidence for the dramatic polymorphism that is observed in house mouse urinary MUPs in other species, although we have observed individual variation in the field vole. Thus, urinary lipocalins seem to play very different roles depending on the species, which may relate to both social structure and environment. Not all species live in complex social groups like house mice, and those that are much less social may have no requirement to identify individuals and kin in adulthood. For example, the closely related grassland mice *Mus macedonicus*, thought to be largely solitary, express a much simpler pattern of a single predominant MUP with no individual variation (Robertson *et al.* 2007). In these species, the simple urinary MUP pattern might be most critical for species and/or sex recognition. It is possible that other mechanisms may operate for individual recognition, or animals may rely on short-term familiarity with scents rather than true recognition of individuals.

It is important to recognize that not all communication proteins are urinary, including the MUPs. For example, the rat genome contains as many MUP genes as the mouse, but only one or two proteins are presented in the urine of male rats. However, other glands express MUPs, raising the opportunity for protein-mediated individuality coding in different scent secretions such as that from the preputial glands which expresses more MUPs than are found in urine (Beynon *et al.* 2008). Urine may not be an ideal medium for communication in many habitats such as grassland, where water-soluble proteins deposited in the environment would be readily lost in rain or dew. For such species, an apolar, water-insoluble signal might be anticipated, possibly derived from other scent glands. As yet, little is understood about the

Table 6.1 Urinary protein expression in rodents

Species	Habitat/ social structure	Urinary protein?	Molecular heterogeneity
Mus musculus domesticus (house mouse)	Largely commensal with humans, broad range of habitats. Large family-based social groups of both sexes.	Male: Yes Female: Yes Male > Female	Extensive, considerable variation between individuals
Mus macedonicus	Grassland. Large territories and highly intolerant of conspecifics.	Male: Yes Female: Yes	Each of five individuals express a single predominant MUP protein
Mus spretus	Grassland with thick undergrowth. Males and females share same area, males occupy non-overlapping territories, females in small groups	Male: Yes Female: Yes	Eleven males show same pattern of three MUPs. Expression more variable among females (n = 5).
Rattus norvegicus (Norway rat)	Largely commensal with humans, broad range of habitats. Large family-based social groups of both sexes.	Male: Yes Female: No Male >> Female	Nine wild caught individuals express the same two major MUPs.
Phodopus roborovskii (Roborovsky hamster)	Desert steppe. Thought to be highly social.	Male: Yes Female: Yes Male = Female	Both captive-bred males and females express the same lipocalin that is more like aphrodisin than MUP
Clethrionomys glareolus (bank vole)	Woodland and thick undergrowth. Males overlapping home ranges, females non overlapping	Male: Yes Female: Yes Male >> Female	Males express a lipocalin that is more like aphrodisin than MUP
Microtus agrestis (field vole)	Grassland, preference for meadow. Male home ranges overlap with other males, females exclusive home ranges.	Male: Yes Female: Yes Male = Female	Protein of correct size for lipocalin, no further characterisation. Evidence for variation between individuals.
Micromys minutus (harvest mouse)	Tall grass, reed beds. Small, overlapping home ranges, both sexes intolerant of conspecifics.	Male: Yes Female: Yes	Lipocalin like protein.
Meriones unguiculatus (Mongolian gerbil)	Semi-desert steppes. Live in family groups.	Male: Yes Female: Yes Male = Female	Protein of correct size for lipocalin, no further characterisation.
Sciurus carolinensis (grey squirrel)	Mature woodland. Overlapping home ranges, territorial.	Male: No Female: No	

MUP, major urinary protein.

different information provided by various scent sources in the same animal, nor their functional significance in rodent communication.

6.8 Conclusion

Our programme of research with house mice usefully defines a strategy for similar work to be applied to other species. It is particularly important to be sensitive to the risks of over-extrapolation of laboratory mouse studies to functional explanations in natural rodent populations. To illustrate, the evidence for MHC encoded social recognition, predominantly based on laboratory mouse studies, is challenged by studies of wild mice. This has revealed that MUPs provide a specialized set of communication proteins that integrate genetic identity information (species, sex, individual, and kinship) with more plastic information on the animal's status through the binding and time-dependent release of low-molecular-weight pheromones. The molecular characterization of scents is an important facet of these studies and it is possible to define a sequence of investigations of increasing complexity, requiring more sophisticated instrumentation, allowing for rapid survey analyses followed by detailed molecular characterization. Progression through this hierarchy of molecular characterizations should be driven by the outcomes of behavioural experiments that maintain a functional perspective and the link between biochemistry and behaviour.

Future investigation should focus on a series of questions: Are scent marks being used? What are the social contexts and function (for example, is communication competitive, sexual, between parents and offspring)? What are the scent secretions that are being used in each context? What is the molecular composition of the scent secretion? Do the same compounds, chemically synthesized, elicit the same response? This 'bottom-up' progression, driven foremost by behavioural investigation and backed up by high-resolution analytical approaches, offers the greatest opportunity to unravel the enormous complexity of signals, their sources, and their uses in social communication. Understanding the roles played by multiple sources of scents in many rodent species remains one of the main challenges for the future.

Summary

Scents play a central role in rodent societies, communicating information about identity (species, sex, individual, kinship) and status (social, reproductive, health, age). This requires the interaction between volatile and involatile molecular components of scents, the spatial deposition pattern of scent marks, and time of deposition. The major histocompatibility complex (MHC) and major urinary proteins (MUPs) are both highly polymorphic systems that contribute to scents. Most studies have focused on MHC in inbred laboratory rodents. However, studies of wild rodents are revealing that MUPs provide a species- and sex-specific genetic identity signature that also underlies individual and kin recognition in house mice. MUPs are mediators of both identity and current status information. Although the MHC contributes to the recognition of familiar scents, there is little evidence that it provides direct information about genetic identity.

Acknowledgements

We thank members of both the Mammalian Behaviour and Evolution Group and the Proteomics and Functional Genomics Group, whose hard work and tireless dedication have contributed to many of the discoveries and ideas presented in this chapter. We are also pleased to acknowledge productive collaborations with Dr Steve Paterson (University of Liverpool) and Professor Bill Ollier (University of Manchester). This work was supported by grant funding from the Biotechnology and Biological Sciences Research Council.

References

Armstrong, S.D., Robertson, D.H.L., Cheetham, S.A., Hurst, J.L., and Beynon, R.J. (2005). Structural and functional differences in isoforms of mouse major urinary proteins: a male-specific protein that preferentially binds a male pheromone. *Biochemical Journal*, **391**, 343–350.

Bacchini, A., Gaetani, E., and Cavaggioni, A. (1992). Pheromone binding proteins of the mouse, *Mus musculus*. *Experientia*, **48**, 419–421.

Beynon, R.J. and Hurst, J.L. (2004) Urinary proteins and the modulation of chemical scents in mice and rats. *Peptides*, **25**, 1553–1563.

Beynon, R.J., Veggerby, C., Payne, C.E., Robertson, D.H., Gaskell, S.J., Humphries, R.E., and Hurst, J.L. (2002). Polymorphism in major urinary proteins: molecular heterogeneity in a wild mouse population. *Journal of Chemical Ecology*, **28**, 1429–1446.

Beynon, R.J., Turton, M.J., Robertson, D.H.L., Armstrong, S.D., Macnicoll, A., Humphries, R.E., and Hurst, J.L. (2008). Urinary lipocalins in Rodenta: is there a generic model? In: J.L. Hurst, R.J. Beynon, S.C. Roberts, and T. Wyatt (eds), *Chemical Signals in Vertebrates 11*, pp. 37–50. Springer, New York.

Boehm, T. (2006). Co-evolution of a primordial peptide-presentation system and cellular immunity. *Nature Reviews Immunology*, **6**, 79–84.

Boyse, E.A., Beauchamp, G.K. and Yamazaki, K. (1987). The genetics of body scent. *Trends in Genetics*, **3**, 97–102.

Brennan, P.A. and Zufall, F. (2006). Pheromonal communication in vertebrates. *Nature*, **444**, 308–315.

Briand, L., Trotier, D., and Pernollet, J.C. (2004). Aphrodisin, an aphrodisiac lipocalin secreted in hamster vaginal secretions. *Peptides*, **25**, 1545–1552.

Bronson, F.H. and Caroom, D. (1971). Preputial gland of the male mouse; attractant function. *Journal of Reproductive Fertility*, **25**, 279–282.

Brown, J.L. and Eklund, A. (1994). Kin recognition and the major histocompatibility complex—an integrative review. *The American Naturalist*, **143**, 435–461.

Brown, R.E. (1995). What is the role of the immune system in determining individually distinct body odours? *International Journal of Immunopharmacology*, **17**, 655–661.

Brown, R.E. and McDonald, D.W. (1985) *Social Odours in Mammals*, Vols 1 and 2. Oxford, Clarendon Press.

Carroll, L.S., Penn, D.J., and Potts, W.K. (2002). Discrimination of MHC-derived odors by untrained mice is consistent with divergence in peptide-binding region residues. *Proceedings of the National Academy of Sciences of the USA*, **99**, 2187–2192.

Chamero, P., Marton, T.F., Logan, D.W., Flanagan, K., Cruz, J.R., Saghatelian, A., Cravatt, B.F., and Stowers, L. (2007). Identification of protein pheromones that promote aggressive behaviour. *Nature*, **450**, 899–902.

Cheetham, S.A., Thom, M.D., Jury, F., Ollier, W.E.R., Beynon, R.J., and Hurst, J.L. (2007). The genetic basis of individual recognition signals in the mouse. *Current Biology*, **17**, 1771–1777.

Cheetham, S.C. (2006). Chemical communication in the house mouse: linking biochemistry and behaviour. Unpublished PhD thesis, University of Liverpool.

Cohen, A.B., Johnston, R.E., and Kwon, A. (2001). How golden hamsters (*Mesocricetus auratus*) discriminate top from bottom flank scents in over-marks. *Journal of Comparative Psychology*, **115**, 241–247.

Dale, J., Lank, D.B., and Reeve, H.K. (2001). Signaling individual identity versus quality: a model and case studies with ruffs, queleas, and house finches. *The American Naturalist*, **158**, 75–86.

Desjardins, C., Maruniak, J.A., and Bronson, F.H. (1973). Social rank in the house mouse: differentiation revealed by ultraviolet visualisation of urinary marking patterns. *Science*, **182**, 939–941.

Egid, K. and Brown, J.L. (1989). The major histocompatibility complex and female mating preferences in mice. *Animal Behaviour*, **38**, 548–550.

Eklund, A. (1997). The major histocompatibility complex and mating preferences in wild house mice (*Mus domesticus*). *Behavioral Ecology*, **8**, 630–634.

Ferris, S.D., Sage, R.D., and Wilson, A.C. (1982). Evidence from mtDNA sequences that common laboratory strains of inbred mice are descended from a single female. *Nature*, **295**, 163–165.

Finlayson, J.S. and Baumann, C.A. (1958). Mouse proteinuria. *American Journal of Physiology*, **192**, 69–72.

Fisher, H.S., Swaisgood, R.R., and Fitch-Snyder, H. (2003). Countermarking by male pygmy lorises (*Nycticebus pygmaeus*): do females use odor cues to select mates with high competitive ability? *Behavioral Ecology and Sociobiology*, **53**, 123–130.

Gardner, A. and West, S.A. (2007). Social evolution: the decline and fall of genetic kin recognition. *Current Biology*, **17**, R810–R812.

Gosling, L.M. (1982). A reassessment of the function of scent marking in territories. *Zeitschrift Fur Tierpsychologie—Journal of Comparative Ethology*, **60**, 89–118.

Gosling, L.M. and Mckay, H.V. (1990). Competitor assessment by scent matching—an experimental test. *Behavioral Ecology and Sociobiology*, **26**, 415–420.

Guo, J., Zhou, A., and Moss, R.L. (1997). Urine and urine-derived compounds induce c-fos mRNA expression in accessory olfactory bulb. *Neuroreport*, **8**, 1679–1683.

Hamilton, W.D. (1964). The genetical evolution of social behaviour. *Journal of Theoretical Biology*, **7**, 1–16.

Holy, T.E., Dulac, C., and Meister, M. (2000). Responses of vomeronasal neurons to natural stimuli. *Science*, **289**, 1569–1572.

Hughes, A.L. and Hughes, M.K. (1995). Natural selection on the peptide-binding regions of major histocompatibility complex-molecules. *Immunogenetics*, **42**, 233–243.

Humphries, R.E., Robertson, D.H.L., Beynon, R.J., and Hurst, J.L. (1999). Unravelling the chemical basis of competitive scent marking in house mice. *Animal Behaviour*, **58**, 1177–1190.

Hurst, J.L. (1993). The priming effects of urine substrate marks on interactions between male house mice, *Mus musculus domesticus* Schwarz and Schwarz. *Animal Behaviour*, **45**, 55–81.

Hurst, J.L. and Beynon, R.J. (2004). Scent wars: the chemobiology of competitive signalling in mice. *BioEssays*, **26**, 1288–1298.

Hurst, J.L., Fang, J.M., and Barnard, C.J. (1993). The role of substrate odors in maintaining social tolerance between male house mice, *Mus musculus domesticus*. *Animal Behaviour*, **45**, 997–1006.

Hurst, J.L., Robertson, D.H.L., Tolladay, U., and Beynon, R.J. (1998). Proteins in urine scent marks of male house mice extend the longevity of olfactory signals. *Animal Behaviour*, **55**, 1289–1297.

Hurst, J.L., Beynon, R.J., Humphries, R.E., Malone, N., Nevison, C.M., Payne, C.E., Robertson, D.H.L., and Veggerby, C. (2001a). Information in scent signals of competitive social status: the interface between behaviour and chemistry. In: A. Marchelewska-Koj, D. Muller-Schwarze, and J. Lepri (eds), *Chemical Signals in Vertebrates 9*, pp. 43–52. Plenum Press, New York.

Hurst, J.L., Payne, C.E., Nevison, C.M., Marie, A.D., Humphries, R.E., Robertson, D.H., Cavaggioni, A., and Beynon, R.J. (2001b). Individual recognition in mice mediated by major urinary proteins. *Nature*, **414**, 631–634.

Hurst, J.L., Thom, M.D., Nevison, C.M., Humphries, R.E., and Beynon, R.J. (2005a). MHC odours are not required or sufficient for recognition of individual scent owners. *Proceedings of the Royal Society B: Biological Sciences*, **272**, 715–724.

Hurst, J.L., Thom, M.D., Nevison, C.M., Humphries, R.E., and Beynon, R.J. (2005b). The 'scents' of ownership. In: R.T. Mason, M.P. Lemaster, and D. Muller-Schwarze (eds), *Chemical Signals in Vertebrates 10*, pp. 199–208. Springer, New York.

Hurst, J.L., Beynon, R.J., Roberts, S.C., and Wyatt, T. (eds) (2008). *Chemical Signals in Vertebrates 11*, New York, Springer.

Ilmonen, P., Penn, D.J., Damjanovich, K., Morrison, L., Ghotbi, L., and Potts, W.K. (2007). Major histocompatibility complex heterozygosity reduces fitness in experimentally infected mice. *Genetics*, **176**, 2501–2508.

Jacob, S., McClintock, M.K., Zelano, B., and Ober, C. (2002). Paternally inherited HLA alleles are associated with women's choice of male odor. *Nature Genetics*, **30**, 175–179.

Jemiolo, B., Xie, T.M., and Novotny, M. (1991). Socio-sexual olfactory preference in female mice: attractiveness of synthetic chemosignals. *Physiology and Behavior*, **50**, 1119–1122.

Johnston, R.E. and Schmidt, T. (1979). Responses of hamsters to scent marks of different ages. *Behavioral and Neural Biology*, **26**, 64–75.

Johnston, R.E., Sorokin, E.S., and Ferkin, M.H. (1997). Female voles discriminate males' over-marks and prefer top-scent males. *Animal Behaviour*, **54**, 679–690.

Johnstone, R.A. (1997). Recognition and the evolution of distinctive signatures: when does it pay to reveal identity? *Proceedings of the Royal Society B: Biological Sciences*, **264**, 1547–1553.

Jordan, W.C. and Bruford, M.W. (1998). New perspectives on mate choice and the MHC. *Heredity*, **81**, 239–245.

Karr, T.L. (2008). Application of proteomics to ecology and population biology. *Heredity*, **100**, 200–206..

Kimoto, H., Sato, K., Nodari, F., Haga, S., Holy, T.E., and Touhara, K. (2007). Sex- and strain-specific expression and vomeronasal activity of mouse ESP family peptides. *Current Biology*, **17**, 1879–1884.

Knopf, J.L., Gallagher, J.F., and Held, W.A. (1983). Differential multihormonal regulation of the mouse major urinary protein gene family in the liver. *Molecular and Cellular Biology*, **3**, 2232–2240.

Kokko, H. and Ots, I. (2006). When not to avoid inbreeding. *Evolution*, **60**, 467–475.

Lai, W.S. and Johnston, R.E. (2002). Individual recognition after fighting by golden hamsters: a new method. *Physiology and Behavior*, **76**, 225–239.

Lehmann, L. and Perrin, N. (2003). Inbreeding avoidance through kin recognition: choosy females boost male dispersal. *The American Naturalist*, **162**, 638–652.

Leinders-Zufall, T., Brennan, P., Widmayer, P., Chandramani, P., Maul-Pavicic, A., Jager, M., Li, X.H., Breer, H., Zufall, F., and Boehm, T. (2004). MHC class I peptides as chemosensory signals in the vomeronasal organ. *Science*, **306**, 1033–1037.

Luo, M., Fee, M.S. and Katz, L.C. (2003). Encoding pheromonal signals in the accessory olfactory bulb of behaving mice. *Science*, **299**, 1196–1201.

Manning, C.J., Potts, W.K., Wakeland, E.K., and Dewsbury, D.A. (1992a). What's wrong with MHC mate choice experiments? In: R.L. Doty and D. Muller-Schwarze (eds), *Chemical Signals in Vertebrates 6*, pp. 229–235. Plenum, New York.

Manning, C.J., Wakeland, E.K., and Potts, W.K. (1992b). Communal nesting patterns in mice implicate MHC genes in kin recognition. *Nature*, **360**, 581–583.

Marie, A.D., Veggerby, C., Robertson, D.H.L., Gaskell, S.J., Hubbard, S.J., Martinsen, L., Hurst, J.L., and Beynon, R.J. (2001). Effect of polymorphisms on ligand binding by mouse major urinary proteins. *Protein Science*, **10**, 411–417.

Mason, R.T., Lemaster, M., and Muller-Schwarze, D. (eds) (2005). *Chemical Signals in Vertebrates 10*, New York, Springer.

Meredith, M. (1994). Chronic recording of vomeronasal pump activation in awake behaving hamsters. *Physiology and Behavior*, **56**, 345–354.

Miyazaki, M. (2008). The biological function of cauxin, a major urinary protein of the domestic cat (*Felis catus*). In: J.L. Hurst, R.J. Beynon, S.C. Roberts, and T. Wyatt (eds), *Chemical Signals in Vertebrates 11*, pp. 51–60. Springer, New York.

Moncho-Bogani, J., Lanuza, E., Hernandez, A., Novejarque, A., and Martinez-Garcia, F. (2002). Attractive properties of sexual pheromones in mice: innate or learned? *Physiology and Behavior*, **77**, 167–176.

More, L. (2006). Mouse major urinary proteins trigger ovulation via the vomeronasal organ. *Chemical Senses*, **31**, 393–401.

Nevison, C.M., Armstrong, S., Beynon, R.J., Humphries, R.E., and Hurst, J.L. (2003). The ownership signature in mouse scent marks is involatile. *Proceedings of the Royal Society B: Biological Sciences*, **270**, 1957–1963.

Novotny, M.V., Ma, W., Wiesler, D., and Zidek, L. (1999). Positive identification of the puberty-accelerating pheromone of the house mouse: the volatile ligands associating with the major urinary protein. *Proceedings of the Royal Society B: Biological Sciences*, **266**, 2017–2022.

Nyby, J., Whitney, G., Schmitz, S., and Dizinno, G. (1978). Postpubertal experience establishes signal value of mammalian sex odor. *Behavioral Biology*, **22**, 545–552.

Olsen, K.H., Grahn, M., Lohm, J., and Langefors, A. (1998). MHC and kin discrimination in juvenile Arctic charr, *Salvelinus alpinus* (L.). *Animal Behaviour*, **56**, 319–327.

Olsson, M., Madsen, T., Nordby, J., Wapstra, E., Ujvari, B., and Wittsell, H. (2003). Major histocompatibility complex and mate choice in sand lizards. *Proceedings of the Royal Society B: Biological Sciences*, **270**, S254–S256.

Penn, D.J. (2002). The scent of genetic compatibility: sexual selection and the major histocompatibility complex. *Ethology*, **108**, 1–21.

Penn, D. and Potts, W. (1998a). MHC-disassortative mating preferences reversed by cross-fostering. *Proceedings of the Royal Society B: Biological Sciences*, **265**, 1299–1306.

Penn, D. and Potts, W.K. (1998b). Untrained mice discriminate MHC-determined odors. *Physiology and Behavior*, **64**, 235–243.

Pocock, M.J.O., Hauffe, H.C., and Searle, J.B. (2005). Dispersal in house mice. *Biological Journal of the Linnean Society*, **84**, 565–583.

Pomiankowski, A. and Pagel, M. (1992). Sexual selection and MHC genes. *Nature*, **356**, 293–294.

Potts, W.K., Manning, C.J., and Wakeland, E.K. (1991). Mating patterns in seminatural populations of mice influenced by MHC genotype. *Nature*, **352**, 619–621.

Potts, W.K., Manning, C.J., and Wakeland, E.K. (1992). Sexual selection and MHC genes—reply. *Nature*, **356**, 294.

Potts, W.K., Manning, C.J., and Wakeland, E.K. (1994). The role of infectious disease, inbreeding and mating preferences in maintaining MHC genetic diversity: an experimental test. *Philosophical Transactions of the Royal Society B: Biological Sciences*, **346**, 369–378.

Pusey, A. and Wolf, M. (1996). Inbreeding avoidance in animals. *Trends in Ecology and Evolution*, **11**, 201–206.

Reusch, T.B., Haberli, M.A., Aeschlimann, P.B., and Milinski, M. (2001). Female sticklebacks count alleles in a strategy of sexual selection explaining MHC polymorphism. *Nature*, **414**, 300–302.

Rich, T.J. and Hurst, J.L. (1998). Scent marks as reliable signals of the competitive ability of mates. *Animal Behaviour*, **56**, 727–735.

Rich, T.J. and Hurst, J.L. (1999). The competing countermarks hypothesis: reliable assessment of competitive ability by potential mates. *Animal Behaviour*, **58**, 1027–1037.

Robertson, D.H.L., Beynon, R.J., and Evershed, R.P. (1993). Extraction, characterization and binding analysis of two pheromonally active ligands associated with major urinary protein of house mouse (*Mus musculus*). *Journal of Chemical Ecology*, **19**, 1405–1416.

Robertson, D.H., Cox, K.A., Gaskell, S.J., Evershed, R.P., and Beynon, R.J. (1996). Molecular heterogeneity in the major urinary proteins of the house mouse *Mus musculus*. *Biochemical Journal*, **316**, 265–272.

Robertson, D.H.L., Marie, A.D., Veggerby, C., Hurst, J.L., and Beynon, R.J. (2001). Characteristics of ligand binding and release by major urinary proteins. In: A. Marchlewska-Koj, D. Muller-Schwarze, and J. Lepri (eds), *Chemical Signals in Vertebrates 9*, pp. 169–176. Plenum Press, New York.

Robertson, D.H.L., Cheetham, S.A., Armstrong, S.D., Hurst, J.L., and Beynon, R.J. (2005). Characterization of proteins in scent marks: proteomics meets semiochemistry. In: R.T. Mason, M. Lemaster, and D. Muller-Schwarze (eds), *Chemical Signals in Vertebrates 10*, pp. 183–198. Springer, New York.

Robertson, D.H.L., Hurst, J.L., Searle, J.B., Gunduz, I., and Beynon, R.J. (2007). Characterization and comparison of major urinary proteins from the house mouse, *Mus musculus domesticus*, and the aboriginal mouse, *Mus macedonicus*. *Journal of Chemical Ecology*, **33**, 613–630.

Rousset, F. and Roze, D. (2007). Constraints on the origin and maintenance of genetic kin recognition. *Evolution*, **61**, 2320–2330.

Rozenfeld, F.M., Le Boulange, E., and Rasmont, R. (1987). Urine marking by male bank voles (*Clethrionomys glareolus* Schreber, 1780; Microtidae, Rodentia) in relation to their social rank. *Canadian Journal of Zoology*, **65**, 2594–2601.

Ryan, D., Rahimi, M., Lund, J., Mehta, R., and Parviz, B.A. (2007). Toward nanoscale genome sequencing. *Trends in Biotechnology*, **25**, 385–389.

Schaefer, M.L., Young, D.A., and Restrepo, D. (2001). Olfactory fingerprints for major histocompatibility complex-determined body odors. *Journal of Neuroscience*, **21**, 2481–2487.

Sherborne, A.L., Thom, M.D., Paterson, S., Jury, F., Ollier, W.E.R., Stockley, P., Beynon, R.J., and Hurst, J.L. (2007). The genetic basis of inbreeding avoidance in house mice. *Current Biology*, **17**, 2061–2066.

Singh, P.B. (2001). Chemosensation and genetic individuality. *Reproduction*, **121**, 529–539.

Singh, P.B., Brown, R.E., and Roser, B. (1987). MHC antigens in urine as olfactory recognition cues. *Nature*, **327**, 161–164.

Spehr, M., Kelliher, K.R., Li, X.H., Boehm, T., Leinders-Zufall, T., and Zufall, F. (2006). Essential role of the main olfactory system in social recognition of major histocompatibility complex peptide ligands. *Journal of Neuroscience*, **26**, 1961–1970.

Thom, M.D. and Hurst, J.L. (2004). Individual recognition by scent. *Annales Zoologici Fennici*, **41**, 765–787.

Wedekind, C., Seebeck, T., Bettens, F., and Paepke, A. J. (1995). MHC-dependent mate preferences in humans. *Proceedings of the Royal Society B: Biological Sciences*, **260**, 245–249.

Wilcox, R.M. and Johnston, R.E. (1995). Scent counter-marks: specialized mechanisms of perception and response to individual odors in golden hamsters (*Mesocricetus auratus*). *Journal of Comparative Psychology*, **109**, 349–356.

Willse, A., Kwak, J., Yamazaki, K., Preti, G., Wahl, J.H., and Beauchamp, G.K. (2006). Individual odortypes: interaction of MHC and background genes. *Immunogenetics*, **58**, 967–982.

Yamaguchi, M., Yamazaki, K., Beauchamp, G.K., Bard, J., Thomas, L., and Boyse, E.A. (1981). Distinctive urinary odors governed by the major histocompatibility locus of the mouse. *Proceedings of the National Academy of Sciences of the USA*, **78**, 5817–5820.

Yamazaki, K., Boyse, E.A., Mike, V., Thaler, H.T., Mathieson, B.J., Abbott, J., Boyse, J., Zayas, Z.A., and Thomas, L. (1976). Control of mating preferences in mice by genes in the major histocompatibility complex. *Journal of Experimental Medicine*, **144**, 1324–1335.

Yamazaki, K., Beauchamp, G.K., Matsuzaki, O., Bard, J., Thomas, L., and Boyse, E.A. (1986). Participation of the murine X and Y chromosomes in genetically determined chemosensory identity. *Proceedings of the National Academy of Sciences of the USA*, **83**, 4438–4440.

Yamazaki, K., Beauchamp, G.K., Kupniewski, D., Bard, J., Thomas, L., and Boyse, E.A. (1988). Familial imprinting determines H-2 selective mating preferences. *Science*, **240**, 1331–1332.

Yamazaki, K., Beauchamp, G.K., Singer, A., Bard, J., and Boyse, E.A. (1999). Odortypes: their origin and composition. *Proceedings of the National Academy of Sciences of the USA*, **96**, 1522–1525.

CHAPTER 7

Neurobiology of olfactory communication in the honeybee

Maria Gabriela de Brito-Sanchez*, Nina Deisig*, Jean-Christophe Sandoz, and Martin Giurfa

7.1 Introduction

The fundamental importance of chemical communication in biological systems is long-established. Indeed, olfactory **signals** play a vital role in most groups of animals both for survival and reproduction. To cite just a few examples, sexual partners are located via sex **pheromones** that can be detected on the basis of a reduced number of molecules, food sources are found via kairomones, which can be learned as a predictor of food, reliable food sources can be marked by means of attractive pheromones, conspecifics can be gathered or dispersed using aggregation or repellent pheromones, respectively, and potential enemies or noxious events can be signalled by means of alarm pheromones. In all of these interactions (Box. 7.1) and in others not covered by this short list of examples, the olfactory system is indispensable.

Pheromones are volatile chemicals used for communication between individuals of the same species. Karlson and Luscher (1959) first defined them as '...substances which are secreted to the outside by an individual and received by a second individual of the same species, in which they release a specific reaction, for example, a definite behaviour or a developmental process'. Later, the definition was modified in order to incorporate the beneficial aspect of intraspecific communication that pheromones mediate (Rutowski 1981). Although at the beginning pheromones were assumed to be unique substances, now it is widely accepted that they are mainly blends of compounds so that variation in the ratios of components may define signals characteristic of different species (Mustaparta 1996).

Although pheromones are known throughout the animal kingdom, most of our knowledge on these substances has been derived from research on insects. Besides the fundamental aspects of pheromone research, some of which are reviewed in this chapter for the honeybee, *Apis mellifera*, pheromones have attracted much attention due to their applied value. Pheromones are widely used as control tools to fight against insect pests. They do not damage other animals, nor do they pose health risks to humans. Pheromones can be used to lure the pests into traps and thereby reduce the amount of insecticide applied on cultures (Jutsum and Gordon 1989; El-Sayed *et al.* 2006). Additionally, pest control can also be achieved by applying compounds that interfere with pheromone communication (Plettner 2002).

Insect pheromones are secreted by exocrine glands and are transmitted to conspecifics in vapour form. In some species, like the silk moth, *Bombyx mori*, only a few molecules of sex pheromone are necessary to produce an orientation response of the male towards the emitting female (Kaissling 1987) which emphasizes the extraordinary sensitivity of the neural olfactory system of certain insects for detecting pheromones. This reduced amount of substance can be detected over long distances so that the active 'air space' of a female (i.e. the range in which its pheromone will

*These authors contributed equally to this work.

Box 7.1 Classification of compounds having an intra- or interspecific informational value

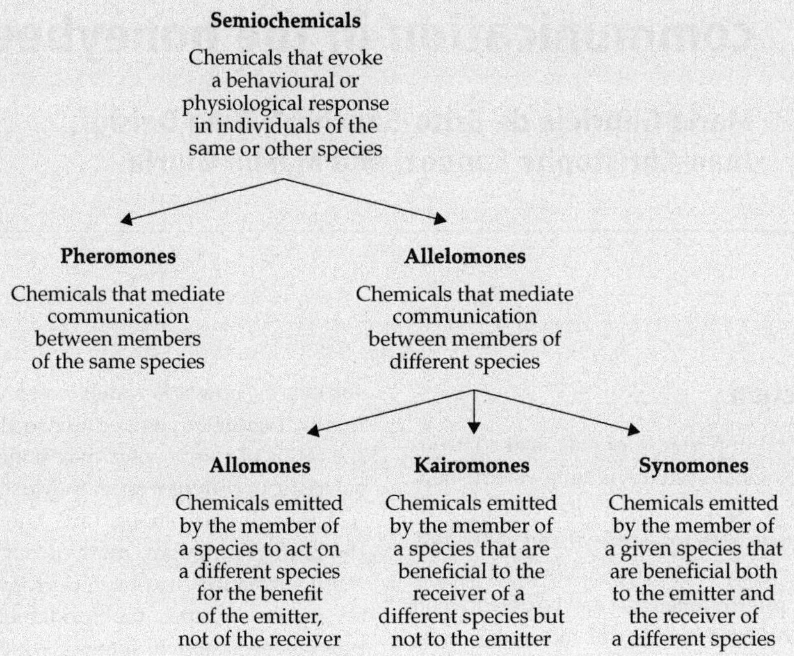

be an effective sex signal) can be several kilometres long and over hundred metres in diameter (Wilson 1970). Any male entering this space, which usually adopts the form of an odour plume at whose base is the female, will show upwind anemotaxis, i.e. it will fly towards the odorant source in a typical zig-zag flight that involves turning commands triggered when the insect loses the plume (Olberg 1983).

7.2 Pheromone communication in a model insect, the honeybee (*Apis mellifera*)

Honeybees, along with many other insects, employ a rich repertoire of pheromones to ensure intraspecific communication in several behavioural contexts (Free 1987). The social organization of a honeybee colony is widely determined by chemical signals that are actively produced and transmitted by the queen, the adult workers at various tasks and life stages, the brood, and possibly the drones. Other forms of communication are well known in the honeybee, such as the dance language that is a stereotyped behaviour used to recruit foragers to food, water, and nest cavities (von Frisch 1967). Such behaviour, even in its highest complexity, constitutes only a small fraction of the communication systems that operate in a functioning colony, the other **channels of communication** being mainly mediated by pheromones (for additional details on social insects see Chapter 4, 5, and 10).

The wealth of identified pheromones used by honeybees and their functions is unparalleled in the insect literature (Free 1987). This may be due to extensive investigations stimulated by the beneficial nature of this insect pollinator but also because of its inherent interest due to its social and cognitive complexity (Giurfa 2007). As it would be impossible to list all known honeybee pheromones here (see Free 1987), we will limit our survey to those pheromones that are best known and most studied in each caste because of their fundamental biological functions.

7.2.1 Queen pheromones

The complex social organization of a hive depends on how the queen controls its environment. The queen, the only fertile female in the colony, must indeed communicate her presence and manifest her influence by means of a mixture of substances released mainly by her mandibular glands. This queen mandibular pheromone (QMP) reinforces social cohesion and has a *releaser effect* because it has an immediate effect on the behaviour of the receiver bees, attracting and enticing them to lick and antennate the queen (the so-called 'retinue behaviour'). It also has a *priming effect*, because, in the long-term, it changes the physiology of the receiver worker bees, inhibiting the development of their ovaries.

QMP was originally considered to be a unique substance, 9-oxo-(*E*)-2-decenoic acid (9-ODA) (Barbier and Lederer 1960; Butler *et al.* 1961) but this compound is considerably less effective than a natural queen extract for inhibiting ovary development, or for inducing retinue behaviour. Indeed, later studies revealed the existence of at least four additional components (Slessor *et al.* 1988), including two enantiomers of 9-ODA's biosynthetic precursor, (*R*)- and (*S*)-9-hydroxy-(E)-2-decenoic acid (9-HDA), and two other compounds, methyl *p*-hydroxybenzoate (HOB) and 4-hydroxy-3-methoxyphenylethanol (HVA) (for further details on the nature of enantiomers see Chapter 5). The four compounds are only weakly active when presented alone, but when formulated with 9-ODA they provide a source nearly as attractive as an extract from the mandibular glands in which the five compounds are produced (Slessor *et al.* 1988).

More recently, novel components have been extracted from several glandular sources, which act in synergy with the QMP components to elicit full retinue behaviour: methyl oleate, coniferyl alcohol, hexadecane-1-ol, and linolenic acid (Keeling *et al.* 2003). From these compounds, only coniferyl alcohol is found in the mandibular glands. The combination of these four compounds and the five QMP compounds is called the queen retinue pheromone (QRP). These nine compounds are important for the retinue attraction of worker bees around their queen. The queen pheromone is, therefore, like other bee pheromones, a complex blend which is most effective when all components are present in appropriate ratios in the blend.

7.2.2 Worker pheromones

Workers have to perform different tasks depending on their age. Bees of intermediate age are usually allocated to guarding and defence tasks at the hive entrance. Not surprisingly, therefore, bees present several alarm pheromones which are released when confronting potential noxious stimuli or enemies (Breed *et al.* 2004). The main alarm pheromone is released by the Koschevnikov gland, which is situated near the sting shaft. This pheromone consists of more than 40 highly volatile chemical compounds, including isopentyl acetate, (*Z*)-11-eicosenol, butyl acetate, 1-hexanol, 1-butanol, 1-octanol, hexyl acetate, octyl acetate, and 2-nonanol (Boch *et al.*, 1962; Collins and Blum 1982, 1983; Picket *et al.* 1982). Release of this pheromone causes the other bees to sting or attack.

Another important alarm pheromone, 2-heptanone, is released by the mandibular glands (Shearer and Boch 1965) and exerts a repellent action on potential intruders and robbers from other hives. Additionally, it has been suggested that it can be used by foragers, which are usually older bees, to mark recently depleted flowers whose immediate revisit has to be avoided (Giurfa and Núñez 1992).

Other pheromones are used by workers to attract and recruit other workers to attractive places. The pheromone of the Nasanov gland (Free 1987; Winston 1987) is released from a the gland whose opening is situated on the dorsal surface of the seventh abdominal tergum. This pheromone is a complex blend in which geraniol, nerol, (*E*,*E*)-farnesol, (*E*)- and (*Z*)-citral, and geranic and nerolic acid are the principal components (Pickett *et al.* 1980). The Nasonov pheromone is used in a variety of circumstances in which it releases attraction and aggregation of receiver workers. It is also used to mark the entrance of the nest, to mark profitable food and water sources, and to elicit swarming behaviour (i.e. to recruit nestmates to a new nesting cavity).

7.2.3 Drone pheromones

Drones, i.e. male bees, seem to have pheromones that attract other flying drones to promote aggregations at sites suitable for mating with virgin queens. Such substances may help establish aerial congregation sites (Free 1987). Male aggregation pheromones have been identified from the mandibular gland of some hymenopteran species (Ayasse *et al.* 2001), and Lensky *et al.* (1985) suggested that a similar gland exists in drone honeybees. However, in honeybees, such a possibility has not yet been investigated.

7.2.4 Brood pheromones

The brood pheromone is a mixture of ethyl and methyl esters of the common fatty acids palmitic, linoleic, linolenic, stearic, and oleic acids, produced by larvae. Four of these methyl esters constitute a signal from the larvae to stimulate the workers to cap the cells prior to pupation (Le Conte *et al.* 1990). The methyl esters of palmitic and oleic acid were most effective in recruiting workers to this activity. In addition, differences in the relative proportions of esters represent a chemical signature of larval age. A component of this blend, methyl linolenate, induces enhanced provisioning of new queen cells with royal jelly, facilitating the development of new queen larvae (Le Conte *et al.* 1995). Two of the brood esters with demonstrated releaser effects, ethyl palmitate and methyl linolenate, also have primer effects, partially inhibiting the ovarian development of worker bees, at least in an experimental situation in which the workers are isolated from a queen and brood (Arnold *et al.* 1994; Mohammedi *et al.* 1998).

Having mentioned some examples of honeybee pheromones that underline the richness and complexity of the chemical repertoire used by this insect to mediate intraspecific communication, we will focus on the question of neural pheromonal processing. We will analyse current knowledge of how pheromonal signals, both sexual and social, are processed in the olfactory nervous system of the honeybee, from the periphery to the more central levels. In this context, it is worth mentioning the two main hypotheses proposed for olfactory coding, the *labelled-line* and the *across-fibre pattern* hypotheses (Fig. 7.1). Both hypotheses attempt to explain olfactory coding in terms of activity in populations of neurons. According to the labelled-line theory, each receptor is highly specific, responds to one olfactory stimulus or a very limited range of stimuli, and sends a direct 'line' to the central nervous system to communicate information about this particular odorant. According to the across-fibre pattern theory, each receptor is less specific and responds to a wider range of olfactory stimuli; therefore, the entire population of odorant-responsive neurons is coding the olfactory information.

Labelled-line processing has the advantage of providing very precise knowledge about a limited number of odorants because each separate channel is dedicated to one (or a few) odorant(s). On the other hand, given the natural constraints of neural systems, it cannot code all possible odorants in the environment. Labelled-line processing is therefore a good system for detecting and recognizing a few stimuli with a crucial biological value for the animal, but not for general odour coding. Conversely, the combinatorial across-fibre processing can code a much greater number of odorants with the same number of neurons, but may be much less specific in its ability to detect a particular odorant, especially at low concentration. We will see that in the honeybee, which presents a remarkable number of pheromones, each with various components, such a segregation is not evident. In presenting the different physiological steps of pheromone processing in the honeybee, we will discuss whether or not pheromone processing follows labelled lines or across-fibre patterns.

7.3 The basic organization of the honeybee olfactory system

Considerable previous work on the honeybee has provided a thorough description of its olfactory circuits, making this insect a model system for studying the neural basis of pheromone detection and processing (e.g. Mobbs 1982; Abel *et al.* 2001; Müller *et al.* 2002; Gronenberg 2001; Kirschner *et al.* 2006). Here we describe the basic organization of the olfactory system of the honeybee (Fig. 7.2). We

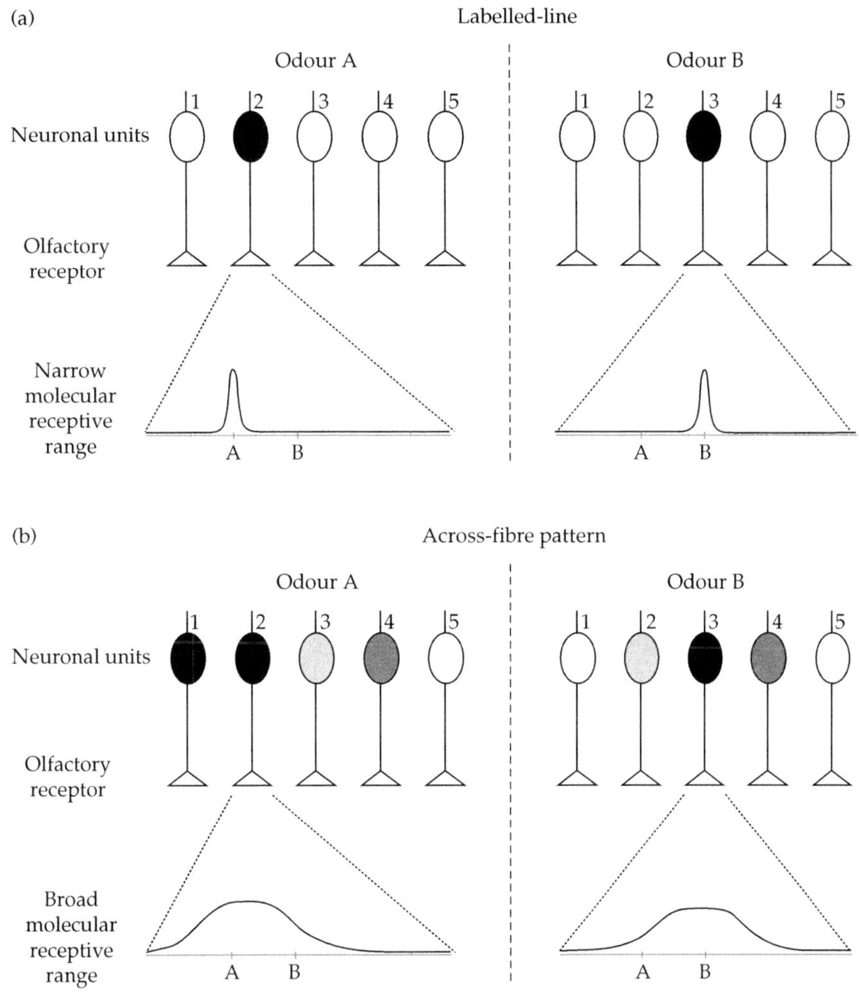

Figure 7.1 Schematic representation of two theories of odour processing. A simplified olfactory system (without lateral connections) is presented, with five different olfactory receptors. For each theory (labeled-line and across-fiber pattern), neural activation for two different odours A and B is represented, with darker shadings of grey indicating increasing neural activity (white: inactive, black: most activated). Below each graph, the molecular receptive range of chosen receptor neurons is shown, according to a putative chemical dimension along which chemically similar odorants would be placed near each other on the scale (such a dimension could be the number of carbon atoms in an aliphatic chain). **a) Labelled-line**: each receptor has a *narrow molecular receptive range*, i.e. it is activated by a single (or very few) odorant(s). The two different odours, A and B, are each detected by only one receptor, which activates only one neuronal unit for each odour. Differentiation between A and B does not need further processing, but only five different odorants can be thus encoded. **b) Across-fibre pattern**: each receptor has a *broad molecular receptive range*, i.e. it can be activated by a wide range of different odours. The five different receptors have different—but broad—receptive ranges. In our example, odour A will activate several neuronal units, although with different intensities depending on the receptor. Receptor 2 will be highly activated by odour A, but only slightly by odour B. Conversely, receptor 3 shows the opposite response profile. Among the other receptors, some will be equally activated by the two odours (receptor 4), others will show a contrasted response (i.e. responding to A but not to B; receptor 1), while others will not be activated at all by either odour (receptor 5). This system allows the fine coding of many odours, but differentiation among odours needs additional downstream processing as the representation of each odour is contained in the combination of activations of the different neuronal units.

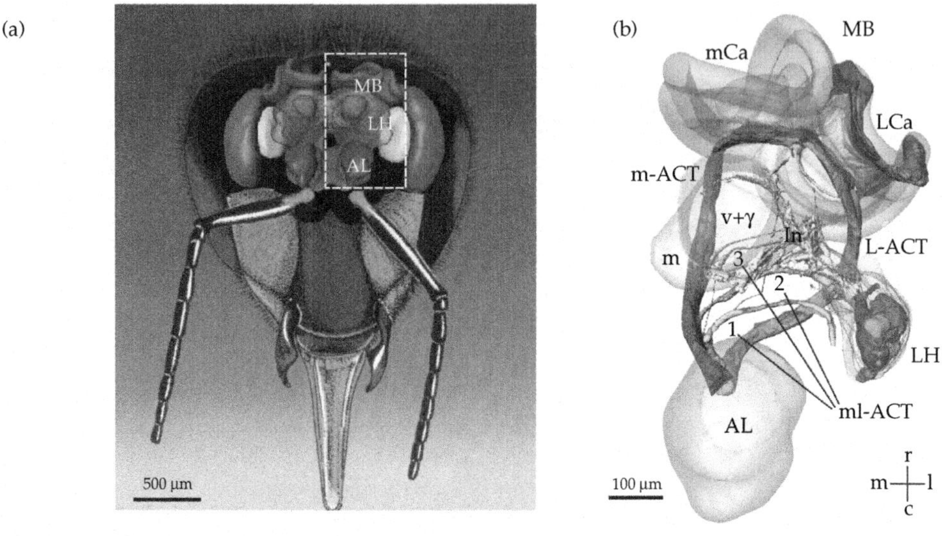

Figure 7.2 The basic organization of the honeybee olfactory system. a) Frontal view of the brain with the main olfactory centers: AL: antennal lobe, LH: lateral horn, MB: mushroom body; b) Three-dimensional reconstruction of the olfactory circuit, corresponding to the broken line square in A: information leaving the AL projects to the calyces of the MB (mCa: medial calyx; lCa: lateral calyx) via two main tracts (m-ACT: medial antenno-cerebral tract and l-ACT: lateral antenno-cerebral tract); The three ml-ACTs (mediolateral antenno-cerebral tracts) 1–3 branch off the m-ACT sequentially and innervate the lateral protocerebral lobe to form the lateral network (ln) that spans from the vertical lobe (v+γ) to the LH (Fig. 3B adapted from Kirschner *et al.* 2006 by kind courtesy of Wolfgang Rössler).

Box 7.2 Abbreviations

ORN, olfactory receptor neuron
Or, olfactory receptor
AL, antennal lobe
MB, mushroom body
KC, Kenyon cell
LH, lateral horn
PN, projection neuron
uPN, uniglomerular projection neuron
mPN, multiglomerular projection neuron
ACT, antenno-cerebral tract
QMP, queen mandibular pheromone
QRP, queen retinue pheromone
9-ODA, 9-oxo-(*E*)-2-decenoic acid
9-HDA, 9-hydroxy-(*E*)-2-decenoic acid
HOB, methyl *p*-hydroxybenzoate
HVA, 4-hydroxy-3-methoxyphenylethanol

focus on honeybee workers as the basic blueprint or 'bauplan', and from there discuss pheromone processing separately for each caste and/or sex. Box 7.2 summarizes the abbreviations used.

Peripheral odour detection starts at the level of the olfactory receptor neurons (ORNs) located within sensillae (cavities or evaginations) on the antennae. Odorant molecules reach the neurons either by diffusing through a receptor haemolymph located in the sensillum cavity and surrounding the ORNs or carried by odorant-binding proteins that bring them through the receptor haemolymph to the surface of the ORN membrane. The odorant molecule will then bind to a molecular olfactory receptor (Or) in the membrane providing that structural coincidence exists between odorant molecule and Or structure.

When ORNs are stimulated by the appropriate ligand molecule, the olfactory message is further conveyed via the antennal nerve, composed of the axons of ORNs, to the antennal lobe (AL), the primary olfactory centre in the insect brain. The antennal nerve (AN) splits into six sensory tracts upon entrance to the AL. Four of these tracts (T1–T4) innervate distinct areas in the AL. The two remaining tracts (T5, T6) bypass the AL and project to the antennal mechanosensory centre in the deutocerebrum (called the dorsal lobe, DL), the suboesophageal ganglion (SOG), and the caudal protocerebrum. The AL is compartmentalized in spheroidal neuropil units called *glomeruli*. Glomeruli are the anatomical and functional units of the AL and constitute sites of synaptic interaction between axons of ORNs expressing the same odorant receptor or with similar odour specificities, local inhibitory interneurons connecting glomeruli, and projection neurons that relay the olfactory message processed at the level of the AL to higher-order centres such as the lateral horn (LH) and the mushroom bodies (MBs). Five antenno-cerebral tracts (ACTs) of projection neurons leave the honeybee AL toward the MBs and the LH (Mobbs 1982; Abel *et al.* 2001). The medial and lateral output tracts (m- and l-ACT) are made of axons of uniglomerular projection neurons (uPNs) that convey information both to the MBs and LH (Bicker *et al.* 1993; Abel *et al.* 2001; Müller *et al.* 2002; Brandt *et al.* 2005). Interestingly, glomeruli transmit their information to higher centres via either one of these main tracts, but not both (Abel *et al.* 2001; Kirschner *et al.* 2006). Apart from these two tracts, three smaller mediolateral ACTs (ml-ACTs) project only to the LH and surrounding protocerebral areas and contain mainly axons of multiglomerular projection neurons (mPNs) (Fonta *et al.* 1993).

Axonal terminals of uPNs are relayed to densely packed MB-intrinsic neurons, the Kenyon cells (KCs; 170,000 per MB). The MBs present cup-shaped regions termed calyces, which receive input from olfactory and visual pathways, and probably also from mechanosensory and gustatory pathways (Strausfeld 2002). MB calyces are anatomically and functionally subdivided into the basal ring, collar, and lip (Mobbs 1982; Gronenberg 2001; Strausfeld 2002). The lip region and the inner half of the basal ring receive olfactory input, whereas the collar and the outer half of the basal ring receive visual input (Gronenberg 2001). Interestingly, further segregation according to the origin of the uPNs (m- or l-ACT neurons) has recently been shown using double staining: PNs of the m-ACT innervate the peripheral part of the olfactory basal-ring region, whereas PNs belonging to the l-ACT innervate the central part of the basal ring (Kirschner *et al.* 2006).

The second major target area of both the m- and l-ACT uPNs is the LH. In addition to the uPN innervation, the LH receives input from mPNs via the ml-ACTs, which were shown to project to the lateral protocerebrum only (Fonta *et al.* 1993). Similar to the olfactory input of the MB calyx, the LH shows a PN tract-specific compartmentalization, with at least four subcompartments: one receives exclusively projections of m-ACT uPNs, while others receive mixed input from m- and l-ACT PNs, from l-ACT and ml-ACT PNs, or from the latter type alone (Kirschner *et al.* 2006). While the function of the LH is still unclear (see Section 7.5.2), it is known that the MBs are involved in further processing of olfactory signals and in olfactory learning and memory, as well as the combination of olfactory information with other sensory **modalities** (Giurfa 2003, 2007).

Although the anatomical description of central projections within olfactory circuits is very good, functional knowledge of these pathways, in particular concerning the possible segregation between pheromonal and non-pheromonal processing, is still in its infancy. The clearly structured and segregated organization of the bee olfactory system, with several parallel olfactory pathways from the AL to MBs and LH, could reflect a labelled-line organization, with pheromonal processing following specific pathways, different from those used to code non-pheromonal odorants. In the next sections we will focus on functional analyses which support or challenge this hypothesis and on olfactory specializations arising from differences in caste or sex.

7.4 Pheromonal processing at the peripheral level: sensillae and olfactory receptors on the antenna

7.4.1 The sensillae

The sensillae containing ORNs are located on the antennae of the honeybee. Different types of sensillae have been traditionally distinguished based on their morphological features (Esslen and Kaissling 1976): sensillae placodea, ampulacea, coeloconica, basiconica, campaniforme, and trichodea. Of all these types, sensillae placodea, which appear in the form of poreplates, are the main olfactory antennal structures in the honeybee (Esslen and Kaissling 1976). Each sensillum placodeum consists of an oval cuticular plate (6 μm × 9 μm) with numerous minute pores. Each sensillum is innervated by 5 to 35 neurons (Esslen and Kaissling 1976; Kelber *et al.*, 2006). There is a remarkable dimorphism between workers and drones with respect to the relative number of sensillae of each type and to the total number of sensory cells. Workers have far more sensillae trichodea while drones lack sensillae basiconica and have far more sensillae placodea than the worker. The flagellum surface of the drone is twice as large as that of the worker and has seven times as many sensilla placodea (18,000 compared with 2600) (Esslen and Kaissling 1976; Brockmann and Brückner 2001). The worker flagellum has a poreplate-free zone on the side facing the head which is densely packed with non-innervated hairs. In the corresponding zone, the drone has a lower density of poreplates than elsewhere on its antennal flagellum (Fig. 7.3).

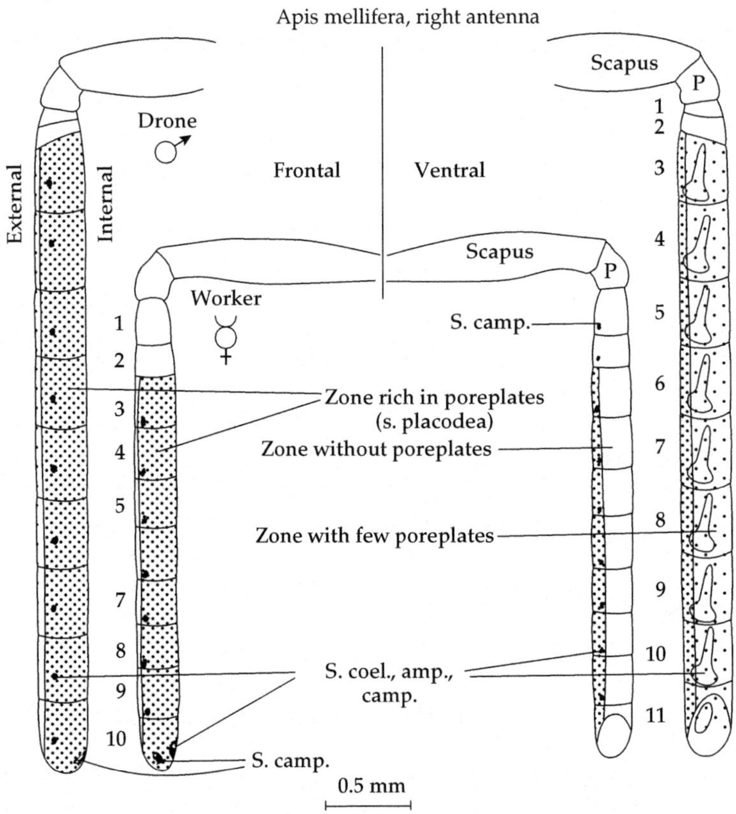

Figure 7.3 Scheme of the worker (♀) and drone (♂) right antenna in frontal (left) and ventral (right) view. The highly dotted, less dotted and empty segments correspond to segments rich, moderately rich and deprived of sensillae placodea (poreplates), respectively (adapted from Esslen and Kaissling, 1976).

Electrophysiological recordings have revealed that the receptor neurons of the sensillae placodea respond to a variety of plant and flower odorants as well as to the components of the honeybee pheromones (Lacher and Schneider 1963; Esslen and Kaissling 1976; Vareschi 1971). In particular, olfactory neurons within sensillae placodea of drones respond to 9-ODA, the main QMP component.

7.4.2 The receptors

As mentioned above (Section 7.3), an ORN presents molecular Ors in its membrane, which allow binding of odorant molecules. Ors are G protein-coupled receptors presenting seven transmembrane regions, with a ligand-binding domain in the plane of the membrane. In insects, one kind of Or is generally present per ORN (Dahanukar *et al.* 2005), which confers specific odour responses to the neuron. Taking advantage of the recent availability of the honeybee genome (The Honeybee Genome Sequencing Consortium 2006), Robertson and Wanner (2006) showed that honeybees present a remarkable expansion of the insect Or family relative to the repertoires of the fruit fly *Drosophila melanogaster* and mosquito *Anopheles gambiae*, which have 62 and 79 Ors, respectively. Indeed, a total of 170 Or genes were annotated in the bee, of which seven are pseudogenes. This number coincides with the number of glomeruli present in the antennal lobe (~165; Galizia *et al.* 1999a), thus supporting the one-receptor/one-neuron/one-glomerulus relationship (Dahanukar *et al.* 2005).

Which molecular specificity do these Or genes present? Although neurogenetic studies performed in *D. melanogaster* have shown that an Or confers the odour response spectrum to its olfactory neuron and other response properties such as the spontaneous firing rate, the temporal dynamic of the response, and whether the response is excitatory or inhibitory (Hallem *et al.* 2004), less is known about Or specificity in honeybees. An exception is the case of the Or for the QMP component 9-ODA, which has recently been identified (Wanner *et al.* 2007). Wanner *et al.* (2007) identified four candidate sex pheromone Ors from the honeybee genome based on their biased expression in drone antennae with respect to that in worker antennae. This number coincides with the number of macroglomeruli in the drone antennal lobe (Arnold *et al.* 1985; Brockmann and Brückner 2001; Sandoz 2006; and see below). In other insect models, macroglomeruli respond specifically to the female-produced sex pheromones and the number of macroglomeruli generally corresponds to the number of sex pheromone components. The pheromone responsiveness of these four Ors preferentially expressed in the drone antennae was studied by expressing them in *Xenopus* oocytes, which allowed electrophysiological characterization of their molecular specificity. One of the Ors (called AmOr11) specifically responded to 9-ODA and not to any of the other QRP (see Section 7.2.1) components, worker pheromones, or floral odours. Interestingly, the other three Ors preferentially expressed in the drone antennae could not be linked to other queen pheromone components.

The existence of an Or specifically tuned to 9-ODA is coincident with previous suggestions on Or olfactory tuning based on the electroantennogram (EAG) technique, in which a global neuronal response to odorants is recorded at the level of the whole antenna. By testing queen pheromone components, Brockmann *et al.* (1998) found that EAG responses of worker and drone antennae differ so that workers exhibit a generalized response to numerous odour compounds, both pheromonal and non-pheromonal, while drones have an EAG profile more specifically tuned to 9-ODA. This difference indicates that already at the peripheral level, dramatic differences exist between drones and workers in terms of their processing of pheromonal and non-pheromonal odorants. While males exhibit a clear olfactory specialization, consistent with their exclusive reproductive role in the hive, worker bees process pheromonal and non-pheromonal odorants in a similar way, consistent with their use of these different signals in different behavioural contexts.

So far no other receptor genes have been found for other queen pheromone components. This result is surprising because despite the EAG results mentioned above, behavioural experiments have shown the effect of other QMP components on male attractiveness. For instance, Brockmann *et al.* (2006) tested the attractiveness of 9-ODA compared with mixtures of 9-ODA and the three other most abundant components in virgin queen mandibular gland secretions: 9-HDA, HOB, and (2*E*)-10-hydroxydecenoic acid (10-HDA). No differences in the number of drones attracted over a distance to a rotating dummy baited with 9-ODA or the respective mixtures were found. However, adding 9-HDA and 10-HDA, or 9-HDA, 10-HDA, and HOB to 9-ODA increased the number of drones actually making contact with the baited dummy. It was therefore suggested that 9-HDA and 10-HDA may be additional components of the sex pheromone blend of *A. mellifera*, at least for short-range attraction. An exhaustive functional screening of drone Ors should therefore yield evidence for the existence of other Ors tuned to these additional queen pheromone components. A clear dimorphism between workers and drones might exist with respect to the presence of these receptors. Alternatively, new queen pheromone components may exist which have not yet been identified (see Section 7.5.1 below).

7.5 Pheromonal processing at the central level

7.5.1 Processing at the antennal lobe

A clear sexual dimorphism is also found at the central level, especially at the level of the first olfactory centre, the antennal lobe (AL). Most anatomical descriptions of the AL have been performed on workers (e.g. Pareto 1972; Flanagan and Mercer 1989; Galizia *et al.* 1999a), whilst data on drones and queens are rather scarce (Arnold *et al.* 1985, 1988).

7.5.1.1 Workers

In workers, the AL consists of ~165 glomeruli (Fig. 7.4), which—as mentioned above—are the functional units of the AL, where intense synaptic connections between different neuronal populations are found. Glomeruli are intensively interconnected by about 4000 local inhibitory interneurons that mediate global and lateral inhibition (Fonta *et al.* 1993; Sachse and Galizia 2002). Processed olfactory information leaves the AL through 800 projection neurons, toward higher-order brain centres like the mushroom bodies and the lateral protocerebral lobe (Mobbs 1982; Abel *et al.* 2001).

Calcium imaging was first successfully applied to record neural activity induced by odour stimulations in the AL of worker bees (Joerges *et al.* 1997). The basic principle of this recording technique resides in visualizing with fluorescent dyes the increase of intracellular calcium (coming from the extracellular medium and/or released from intracellular stores) following neuronal excitation. Such dyes bind to free calcium, thereby changing their fluorescence excitation or emission properties.

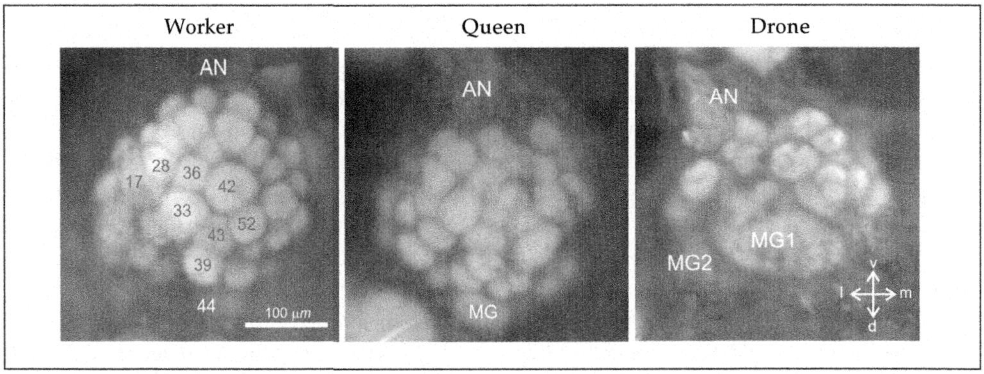

Figure 7.4 Anatomy of the antennal lobe in the three honeybee castes. Anatomical staining (4% neutral red) of the left antennal lobes of a worker (left), a queen (middle) and a drone (right). The lobes are shown in frontal view, in the position in which they can be accessed during calcium imaging recordings. About ¼ of the antennal lobe is thus accessible (30 to 40 glomeruli in workers). Note the presence of an enlarged glomerulus in workers and queens (labeled '44' and 'MG' respectively, see text) and two conspicuous macroglomeruli in the drone (labeled 'MG1' and 'MG2'). AN: antennal nerve; MG: macroglomerulus; v: ventral; l: lateral; m: medial; d: dorsal.

Using a dedicated setup with a highly sensitive CCD camera and an epifluorescence microscope, it is possible to monitor calcium concentration changes from brain structures throughout odour stimulations (Galizia *et al.* 1997; Galizia and Vetter 2004). To record neural activity in this way, bees are fixed in a recording chamber, and the head capsule is carefully opened. Membranes and trachea covering the brain are removed, and a calcium-sensitive fluorescent dye (for instance Calcium Green 2 AM) is bath-applied onto the brain. After about an hour's staining, the brain is rinsed with saline solution and the bee is placed under an upright fluorescence microscope in front of an odour stimulation device. In the AL, odours were found to elicit combinatorial activity patterns across glomeruli (Joerges *et al.* 1997; see Fig. 7.5a) and odour quality is represented by a specific distributed code, conserved between individuals (Galizia *et al.* 1999b; Sachse *et al.* 1999). In that sense, odorant coding in the worker AL corresponds to an across-fibre pattern rather than to a labelled line processing, which is particularly important for the processing of olfactory mixtures (Deisig et al. 2006), especially in honeybees, which learn and recognize complex floral odour blends during foraging.

Activity patterns in the AL clearly correspond to a perceptual representation of odorants, since physiological similarity between activity patterns correlates well with perceptual similarity measured from bees' generalization performances after conditioning to a wide spectrum of selected odours

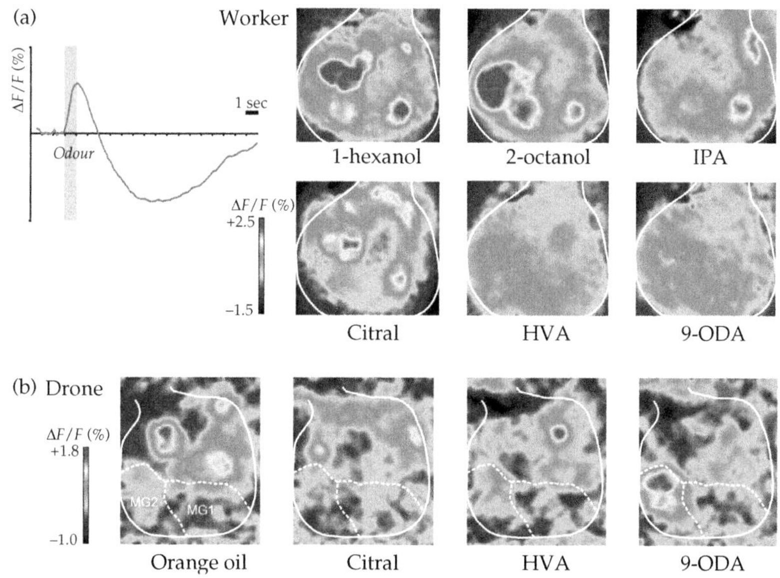

Figure 7.5 Physiological responses of the antennal lobe in workers and drones. a) Calcium imaging recording (using bath-applied Calcium Green 2AM) in a worker bee. **Upper left**: upon odour delivery (grey bar), a biphasic fluorescence signal is observed in active glomeruli, with a first fast positive component (max after ~1 sec), followed by a slow—highly spatially correlated—negative component (minimum after 8–10 sec). **Right**: Odour activity maps, showing for each pixel in a false-color code the amplitude of the biphasic signal. General odours (1-hexanol and 2-octanol) and social pheromone (isopentyl acetate (IPA) and citral) elicit combinatorial activity in the imaged glomeruli. Note that the glomeruli activated by the pheromones can be active in response to general odours and vice-versa. By contrast, no clear signals appeared with components of the queen mandibular pheromone (here HVA and 9-ODA). We believe that the glomeruli responsible for processing of these signals are in other—yet non-imaged—parts of the antennal lobe. b) Calcium imaging recordings (using bath-applied Calcium Green 2AM) of antennal lobe activity in a drone bee. The odour activity maps are calculated as in a). The position of the two accessible macroglomeruli is overlaid on the maps (white). General odours (here a complex blend, orange essential oil) and social pheromones (here citral) induce activity in ordinary-sized glomeruli, i.e. on the medio-ventral side of the antennal lobe. Interestingly, the major component of the queen mandibular pheromone, 9-ODA, which is involved in the attraction of males towards queens during nuptial flight, is specifically detected by the most voluminous macroglomerulus of the drone antennal lobe, MG2. By contrast, the component HVA, the pheromonal role of which is only proven in workers, induces activity in an ordinary-size glomerulus (for details, see Sandoz, 2006). (see Plate 3)

(Guerrieri *et al.* 2005). In imaging studies (Galizia *et al.* 1999b; Sachse *et al.* 1999; N. Deisig *et al.* unpublished data), along with floral odours, several pheromonal components have been presented to worker bees, such as worker-emitted (social) pheromones like geraniol and citral (aggregation; see above), isopentyl acetate and 2-heptanone (alarm; see above), or components of the queen mandibular pheromone. The common picture emerging from these studies for the honeybee worker is that both floral odours and social pheromones induce clear responses in the AL, each compound eliciting activity in a combination of glomeruli, irrespective of their social or floral nature (see Fig. 7.5a). Moreover, QMP presented either as a blend (Galizia *et al.* 1999b) or as its separate components (our recordings, see Fig. 7.5a) induces very little activity in these glomeruli.

When comparing the signals corresponding to social pheromones and floral odours, visual observation does not allow the isolation of any glomerulus that could be specifically involved in pheromone processing. In particular, it was observed that the signal induced by 2-heptanone, a ketone with an aliphatic chain of seven carbons that acts as an alarm pheromone (see above), elicits an activity pattern that appears halfway between those of 2-hexanone (six carbons) and 2-octanone (eight carbons) (Sachse *et al.* 1999), two non-pheromonal molecules. Thus, at first glance, signals to 2-heptanone seem to belong to a continuum of increasing chain length within ketones. However, careful similarity measures performed by Sachse *et al.* (1999), taking into account the entire odour pattern, clearly showed that the 2-heptanone signal has a particularity: whereas all the molecules tested in this study (belonging to functional groups like alcohols, aldehydes, and ketones, with chain lengths between five and ten carbons) showed a clear similarity relationship depending on chain length (longer chains supporting more similarity between odours from different functional groups), 2-heptanone showed a dramatic drop in similarity relative to odours of the same chain length (see Fig. 6C in Sachse *et al.* 1999). In other words, the pattern of 2-heptanone was more distinct from those of other odours than what would be predicted from the chain length/similarity rule indicated above. What can be concluded from these observations? Social pheromones in the honeybee (but also in other Hymenoptera) are rather common molecules, which bees certainly encounter in their natural environment, in particular as components of floral aromas (see Knudsen *et al.* 1993). It is therefore not astonishing that these molecules are detected by the general olfactory system and induce glomerular activity patterns in the same regions of the AL as floral odours. Pheromones could thus be detected by this part of the olfactory system of workers as a *general odour* because of their common chemical structure. However, the fact that glomerular activity to the social pheromones is observed in one AL region thought to be responsible for the coding of general odours does not preclude that other regions of the AL (belonging to ORN tracts T2, T3, or T4) could be dedicated to the detection of these social odours as *pheromones*. Clues as to which AL region could be involved in social or queen pheromone processing are still too scarce to allow us to make predictions. New physiological experiments should be performed, measuring responses from the other glomerular populations using optical imaging. Because of their position in the brain, glomeruli from the tracts T2–T4 are not easily accessible to conventional microscopy, and have mostly been ignored in imaging studies (Galizia and Menzel 2001). The recent development of multiphotonic microscopy, which allows deeper imaging within brain structures with good spatial and temporal resolution, could be one solution to this problem. Additionally, the electrophysiological study of individual neurons coupled to their precise labelling and glomerulus identification (Abel *et al.* 2001; Müller *et al.* 2002; Galizia and Kimmerle 2004) should be systematically pursued, using a standardized odour list containing all the different classes of odorants.

7.5.1.2 Drones

In drones, the tracts of olfactory receptor neurons are thicker but project into a smaller number of glomeruli than in workers (Arnold *et al.* 1985). Most of them (~103) correspond to glomeruli of a similar size to those of workers ('ordinary' glomeruli). However, the most dramatic difference between the drone and the worker AL is the presence in the drone of four hypertrophied glomeruli, the macroglomeruli (Arnold *et al.* 1985; see Fig. 7.4). Their increased volume and their anatomical similarity to the macroglomerular complexes found in males of several moth species (e.g. Kaissling 1987), where they are involved in the detection and processing of female pheromone components, suggests that macroglomeruli in honeybee drones could play a similar role and serve in the detection and processing of queen pheromonal components (Arnold *et al.* 1985; Masson and Mustaparta 1990). Recently, one of us has used *in vivo* calcium imaging to study responses to pheromonal and general odours by the drone antennal lobe (Sandoz 2006; Fig. 7.5b). Two out of four macroglomeruli and about 20 ordinary glomeruli on the frontal surface of the antennal lobe, all belonging to the T1 tract of ORNs, were accessible to our recordings. We found that the macroglomerulus MG2, which is the most voluminous of the drone ALs, responds specifically to the main queen pheromone component 9-ODA but not to other social and floral odours tested. This result therefore confirmed the hypothesis formulated by Arnold *et al.* (1985) and fits well with previous electrophysiological studies showing that an important part of the drone peripheral olfactory system is dedicated to the detection of 9-ODA (Kaissling and Renner 1968; Vareschi 1971; Skirkevičiene and Skirkevičius 1994; Brockmann *et al.* 1998; Vetter and Visscher 1997; and see above). This result also coincides with molecular studies on antennal Ors in drones (Wanner *et al.* 2007; and see above). In those studies, four Ors were found that were preferentially expressed in the antennae of drones but only one could be assigned to 9-ODA. ORNs expressing the 9-ODA-specific Or probably send their axon terminals directly to MG2. This would correspond to a labelling-line processing strategy rather than to an across-fibre pattern strategy.

Our calcium imaging studies (Sandoz, 2006) did not detect any activity in the other accessible macroglomerulus, MG1, when drones were stimulated with queen mandibular components, social pheromones, or floral odours. We believe that MG1 may respond to other queen components that were not present in our samples. Note also that the molecular studies mentioned above (Wanner *et al.* 2007) could also not assign the other three Ors to other QMP components. In fact, until now, apart from 9-ODA, the search for queen pheromonal components has mainly focused on creating blends able to accurately reproduce workers'—but not drones'—behaviour (Slessor *et al.* 1988; Keeling *et al.* 2003). Therefore, even if 9-ODA is clearly the main attractant for drones, the question of possible co-attractants is still mostly unsolved. Since the initial description of a queen mandibular extract able to reproduce the retinue behavior of worker bees (9-ODA, 9-HDA, HVA, and HOB; Slessor *et al.* 1988), novel components have been found which act in synergy with the former ones: methyl oleate, coniferyl alcohol, hexadecane-1-ol, and linolenic acid (Keeling *et al.* 2003). Such components should be tested in imaging conditions, as should complete queen extracts. In our work, 9-HDA failed to induce consistent signals. Moreover, responses to HVA and HOB occurred mainly in two ordinary glomeruli, which clearly responded to general odorants (Sandoz 2006), suggesting that the responses obtained to HVA and HOB are due to their detection by the general olfactory system and not by a pheromonal subsystem. As explained above, 9-HDA, HOB, and HVA were isolated for their role on worker behaviour, and may not have a pheromone value for drones in nature. So far, only one study has found drone attraction to 9-HDA (Butler and Fairey, 1964) but two subsequent studies failed to reproduce this result (Blum *et al.* 1971; Boch *et al.* 1975). As also mentioned above, though, 9-HDA could play a role as co-attractant with 9-ODA at a short range (Brockmann *et al.* 2006). The value of HVA and HOB for drones is also questionable: virgin queens do not produce any HVA and very little HOB in comparison to mated queens (Plettner *et al.* 1997). Therefore, these compounds could be necessary only for the induction of workers' retinue behavior by mated queens, and not for drone attraction to virgin queens. This emphasizes again the particular case of the queen pheromone in honeybees, in which the same components can have different roles depending on the receiver being a drone or a worker. Caution is, however, necessary because two macroglomeruli of the drone AL (MG3 and MG4) were not accessible to our imaging study and may respond to these queen signals despite the fact that the corresponding Ors have not been identified at the level of the drone antennae (Wanner *et al.* 2007). As discussed above for workers, the use of multiphotonic microscopy and electrophysiological measurements of labelled neurons could help us to understand more about the respective roles of the four macroglomeruli in the drone AL.

7.5.1.3 Queens

In queens, females genetically identical to workers but which were fed differently, both qualitatively and quantitatively, during development, AL organization is similar to that of workers (Arnold *et al.* 1988; our observations). Despite clear anatomical differences (queens have a much longer abdomen, different mouthparts, a lack of pollen-collecting structures on the legs, fewer antennal pore plates, and a different development of the glandular system compared with workers; Winston 1987) the AL of queens has a similar number of glomeruli to that of workers (~155 in queens vs. ~165 in workers; Fig. 7.4). One particularly conspicuous glomerulus of the dorsal region belonging to the tract T1 shows a volume about three to four times larger than those of other glomeruli. Its volume, relative to the rest of the antennal lobe, is higher than in workers (in which it is already a voluminous glomerulus; Flanagan and Mercer 1989; Galizia *et al.* 1999a), especially in the case of mated queens. For these reasons, it has been postulated that this glomerulus could represent a female macroglomerulus, which could be dedicated to the recognition of species-specific signals (Arnold *et al.* 1988). It is interesting to note that this putative macroglomerulus is placed in a similar position in the antennal lobe of queens (and workers) as three out of the four drone macroglomeruli, i.e. on the more dorsal part of the T1 region. Future behavioural work associated with physiological work using calcium imaging should study the possible involvement of this female putative macroglomerulus in drone or brood pheromone detection and processing.

7.5.2 Processing in higher-order olfactory centres: the mushroom body (MB) and lateral horn (LH)

In both the fruit fly and the honeybee, higher-brain centres like the MBs are known to be important sites for olfactory-based decision-making and house olfactory memory traces (Hammer and Menzel 1998; Zars 2000; Davis 2005). Very little is known, however, about how pheromonal information is represented and processed therein. We have seen that we could only demonstrate the existence of a labelled-line in the case of the drone, for 9-ODA from ORNs to the AL. In workers, such a labelled-line may exist in AL regions that are not yet accessible to optical imaging, but proof is still lacking. On the other hand, we have ample demonstration of across-pattern processing. Thus, general odorants are processed in terms of an across-fibre pattern both in workers and in drones. More surprisingly, our recordings until now suggest that it may also be the case for social pheromones, for which no dedicated glomeruli have been yet found. One interesting hypothesis is that specific recognition of pheromones, in particular the social ones, would take place at higher processing levels, i.e. after the AL networks processed odour information.

As found in the drone, there appears to be a clear segregation between 9-ODA-responding glomeruli and floral/social odour-responding glomeruli. Based on the clear anatomical segregation of projections from the AL to higher centres described above (see Section 7.3), a possible hypothesis would be that 9-ODA and non-pheromonal information would follow different pathways, and that specific projection areas within the drone MBs and LH are devoted to the processing of this QMP component. Likewise, in the case of the worker, pheromone information could follow a different route and project to specific areas within higher-order centres. Although evidence is missing in the case of drones, electrophysiological recordings performed on worker PNs have shown that no clear separation exists between floral odours and worker-emitted pheromones (Sun *et al.* 1993; Abel *et al.* 2001; Müller *et al.* 2002; Galizia and Kimmerle 2004). PNs from both the l-ACT and the m-ACT (see Section 7.2) respond to floral odours and also to odours like geraniol (aggregation), citral (aggregation), isoamyl acetate (alarm), or 2-heptanone (alarm), making it unlikely that one of these two tracts is specifically involved in pheromonal processing. Rather, it is currently believed that these two main tracts of PNs are involved in general odour coding, providing higher areas with differential information: m-ACT neurons would code odours by latency differences or patterns of inhibitory and excitatory phases while l-ACT neurons would code odours by spike-rate differences (Müller *et al.* 2002). Moreover, it must be noted that most projection neurons of the m- and l-ACT are uniglomerular, so that if we do not find specific glomeruli for pheromones in workers, it is unlikely that some of these PN would carry a specific pheromonal signal to higher centres.

In the worker MBs, calcium imaging recordings have shown that odours also evoke combinatorial activity patterns, as in the AL (Szyszka *et al.* 2005). However, the MB spatial patterns are consistently sparser than those found at the AL. Such a sparsity of odorant representation occurs in the transmission from PNs to Kenyon cells (KCs), the constitutive neurons of MBs. Many PNs feed onto each Kenyon cell, but imaging has shown that activated KCs are highly odour specific and exhibit sharpened temporal responses, probably due to the presence of a broad loop of inhibitory neurons acting on KCs (Szyszka *et al.* 2005). In the locust, such an inhibitory input on KCs, which also results in sparsity of odour representation in the KCs, is provided by gabaergic neurons from the lateral horn (Pérez Orive *et al.* 2002). How this sparsity and KC temporal sharpening affects pheromone vs. floral odour representation in the MBs remains unknown. It is conceivable that particular Kenyon cells would 'recognize' specific combinations of activated projection neurons, which would indicate that the detected stimulus is a pheromone. In the only study performed on the worker MBs (Szyszka *et al.* 2005), only non-pheromonal odours were tested (1-hexanol, 2-octanol, limonene, linalool, and peppermint oil), so that future work is necessary to look for the existence of such pheromone-specific Kenyon cells. It must be emphasized that MB extrinsic neurons (i.e. neurons that act as output of the MBs), like the Pe-1 neuron, typically respond to pheromones (2 heptanone, citral; Rybak and Menzel 1998) as do MB feedback neurons—i.e. neurons that having beeen output from the MBs feed again onto these structures (geraniol, citral; Grünewald 1999). However, they do not respond specifically to these compounds, therefore it is not clear whether such responses are related to the general olfactory system, or to pheromone processing *per se*.

Less information is available from ml-ACT neurons which project to the LH and lateral protocerebrum. These neurons could be especially interesting in the context of pheromone processing because they are mostly multiglomerular, and could thus detect a specific pattern of activity and code the pheromonal nature of stimuli. The available data suggest that they can respond to pheromones like geraniol or citral, but these stimuli mostly provoke an inhibition of spontaneous activity (Abel *et al.* 2001). Thus, although the bee olfactory system clearly presents several parallel olfactory processing lines (m-, l-, and ml-ACT and their corresponding projection areas), functional data do not point to the dedication of one of these pathways to pheromonal processing. This, however, does not preclude that within a particular pathway, pheromone-dedicated neurons could project to a specific area within the MB or LH.

The role of the honeybee LH in olfactory processing is currently unknown. Due to its unstructured arrangement, which renders functional recordings of neural activity difficult, less attention has been paid to this structure. In the fruit fly, however, recent neuroanatomical work could reconstruct putative maps of olfactory input to the MBs and to the LH (Jefferis *et al.* 2007). In this species, the response spectra of individual ORNs to odours are known (Hallem *et al.* 2004; Hallem and Carlson 2006). Moreover, glomeruli receiving input from ORNs carrying each receptor have been carefully mapped (Couto *et al.* 2005) and the exact projection of individual uniglomerular PNs from identified glomeruli has been retraced so that it is now possible to predict where information gained by each olfactory receptor is projected to. Interestingly, reconstructed olfactory maps at the level of the LH predict a clear segregation between candidate pheromone-responsive PNs and fruit odour-responsive PNs (Jefferis *et al.* 2007). Such functional segregation was not apparent in the MBs, although PNs from different glomeruli also project there in at least 17 different areas (Jefferis *et al.* 2007). These data, which still await physiological confirmation, suggest that, at least in the *Drosophila* LH, particular subregions may code the biological nature of olfactory stimuli. If a similar organization of the olfactory circuit exists in fruit flies and honeybees (and there are indeed several remarkable parallels), one could expect the honeybee LH to exhibit pheromone-processing regionalization. Here again, novel anatomical and physiological experiments are required to address this question.

7.6 Conclusion

The honeybee possesses an extremely rich and complex social communication system, which relies heavily upon olfactory communication and uses a wide repertoire of different pheromones, produced mainly by the workers and queen. The behavioural and physiological effects of these pheromones on workers and drones are well established, and some of them can be measured in the laboratory. The intensive previous work that has described the different elements of the olfactory circuits, as well as the accessibility of this animal model to physiological recording methods such as electrophysiology and optical imaging, makes it an ideal model for study of the neurobiological basis of pheromonal processing. The emerging picture is that of a system in which both dedicated labelled-line processing and across-fibre pattern coding coexist in the same species. Labelled-line processing applies to the case of 9-ODA in drones, starting with a dedicated antennal Or (Wanner *et al.* 2007) and continuing with a highly 9-ODA-specific macroglomerulus in the AL (Sandoz 2006). Such a dedicated pathway remains to be followed at higher levels of the drone olfactory circuit, in particular in the MBs and LH. Moreover, it is probable that more such labelled lines exist in the drone for other components of the queen sex pheromone, as there are three other drone-specific Ors and three other macroglomeruli, for which we do not know the key odorants. Across-fibre patterning certainly applies to the processing of non-pheromonal odours (flower odours) and may also apply to social pheromone components both in drones and workers, as shown by imaging recordings. In other traditional models for the study of pheromonal vs. non-pheromonal processing, like moths, there is growing evidence that pheromonal processing also contains a certain combinatorial quality, in particular within the male macroglomerular complex (Christensen and Hildebrand 2002). These authors even proposed that the pheromonal system, due to its crucial role in reproduction, represents an exaggeration of the structure and functioning of the general olfactory system but is based on a similar logic, including combinatorial coding. The case of the honeybee, at least regarding social pheromones, seems to confirm this view. It must be emphasized here that bees' behavioural responses to social pheromones are quite specific (Free 1987). Thus, it is probable that specific social-pheromone processing units do exist somewhere in the bee brain, which give these pheromones their specific meaning for the animal. Several possibilities exist. First, because until now optical recordings have been limited to about 40 glomeruli at the frontal surface of the AL, there may still be social pheromone-specific glomeruli in this structure. Conversely, another system may exist, in which pheromone-specific units would only be found in higher-brain centres like MBs or LH, i.e. a system combining peripheral across-fibre patterning with more central labelled-line units. Because studies on odorant processing in the bee central nervous system have mainly focused on the AL, areas like the MBs and LH have been left mostly unexplored with respect to the question of a possible differential treatment of pheromonal vs. non pheromonal odorants therein. We expect that the development of novel microscopic techniques (like multiphoton microscopy allowing accessing neural structures in depth) and molecular genetics will help improving our knowledge of the neural basis of social olfactory communication in honeybees.

Summary

Pheromones are chemical substances mediating intraspecific communication in a variety of behavioural contexts. Honeybees constitute a historical model for the study of pheromonal communication in insects so that much is known about the chemical structure of various pheromones, the context in which they are released, and the physiological effects they can exert on receiver bees of different castes. Here we discuss the neurobiology of pheromone processing in the honeybee brain, from peripheral antennal detection to central-level processing.

We focus on modern electro- and optophysiological recording techniques at different stages of the honeybee olfactory circuit and ask whether or not the traditional distinction between labelled-line and across-fibre pattern processing applies to pheromone processing as compared to non-pheromonal odours. We propose new avenues of research for stimulating future work in this area.

References

Abel, R., Rybak, J., and Menzel, R. (2001). Structure and response patterns of olfactory interneurons in the honeybee, *Apis mellifera. Journal of Comparative Neurology*, **437**, 363–383.

Arnold, G., Budharugsa, S., and Masson,C. (1988). Organization of the antennal lobe in the queen honey bee *Apis mellifera* L. (Hymenoptera: Apidae). *International Journal of Insect Morphology and Embryology*, **17**, 185–195.

Arnold, G., Le Conte, Y., Trouiller, J., Hervet, H., Chappe, B., and Masson, C. (1994). Inhibition of worker honeybee ovaries development by a mixture of fatty acid esters from larvae. *Comptes Rendus de l'Académie des Sciences, Paris, Série III*, **317**, 511–515.

Arnold, G., Masson, C., and Budharugsa, S. (1985). Comparative study of the antennal lobes and their afferent pathways in the worker bee and the drone (*Apis mellifera*). *Cell Tissue Research*, **242**, 593–605.

Ayasse, M., Paxton, R.J., and Tengo, J. (2001). Mating behavior and chemical communication in the order Hymenoptera. *Annual Review of Entomology*, **46**, 31–78.

Barbier, M. and Lederer, E. (1960). Structure chimique de la substance royale de la reine d'abeille *Apis mellifera* L. *Comptes Rendus de l'Académie des Sciences, Paris, Série III*, **251**, 1131–1135.

Bicker, G., Kreissl, S., and Hofbauer, A. (1993). Monoclonal-antibody labels olfactory and visual pathways in Drosophila and Apis brains. *Journal of Comparative Neurology*, **335**, 413–424.

Blum, M.S., Boch, R., Doolittle, E., Tribble, M.T., and Traynham, J.G. (1971). Honey bee sex attractant: conformational analysis, structural specificity, and lack of masking activity of congeners. *Journal of Insect Physiology*, **17**, 349–364.

Boch, R., Shearer, D.A., and Stone, B.C. (1962). Identification of iso-amyl acetate as an active component in the sting pheromone of the honey bee. *Nature*, **195**, 1018–1020.

Boch, R., Shearer, D.A., and Young, J.C. (1975). Honeybee pheromones: field tests of natural and artificial queen substance. *Journal of Chemical Ecology*, **1**, 133–148.

Brandt, R., Rohlfing, T., Rybak, J. *et al.* (2005). Three-dimensional average-shape atlas of the honeybee brain and its applications. *Journal of Comparative Neurology*, **492**, 1–19.

Breed, M.D., Guzman Novoa, E., and Hunt, G.J. (2004). Defensive behavior of honey bees: organization, genetics, and comparisons with other bees. *Annual Review of Entomology*, **49**, 271–298.

Brockmann, A. and Brückner, D. (2001). Structural differences in the drone olfactory system of two phylogenetically distant *Apis* species, *A. florea* and *A. mellifera*. *Naturwissenschaften*, **88**, 78–81.

Brockmann, A., Brückner, D., and Crewe, R. (1998). EAG responses of worker and drone honeybees *Apis mellifera* to pheromones of the mandibular gland of the queen. *Naturwissenschaften*, **85**, 283–285.

Brockmann, A., Dietz, D., Spaethe, J., and Tautz, J. (2006). Beyond 9-ODA: sex pheromone communication in the European honey bee *Apis mellifera* L. *Journal of Chemical Ecology*, **32**, 657–667.

Butler, C.G. and Fairey, E.M. (1964). Pheromones of the honeybee: biological studies of the mandibular gland secretion of the queen. *Journal of Apicultural Research*, **3**, 65–76.

Butler, C.G., Callow, R.K., and Johnson, N.C. (1961). The isolation and synthesis of queen substance, 9-oxo-dec-trans-2-enoic acid, a honeybee pheromone. *Proceedings of Royal Entomological Society*, **155**, 417–432.

Christensen, T.A. and Hildebrand, J.G. (2002). Pheromonal and host-odor processing in the insect antennal lobe: how different? *Current Opinion of Neurobiology*, **12**, 393–399.

Collins, A.M. and Blum, M.S. (1982). Bioassay of compounds derived from the honeybee sting. *Journal of Chemical Ecology*, **8**, 463–470.

Collins, A.M. and Blum, M.S. (1983). Alarm responses caused by newly identified compounds derived from the honeybee sting. *Journal of Chemical Ecology*, **9**, 57–65.

Couto, A., Alenius, M., and Dickson, B.J. (2005). Molecular, anatomical, and functional organization of the *Drosophila* olfactory system. *Current Biology*, **15**, 1535–1547.

Dahanukar, A., Hallem, E.A., and Carlson, J.R. (2005). Insect chemoreception. *Current Opinion in Neurobiology*, **15**, 423–430.

Davis, R.L. (2005). Olfactory memory formation in *Drosophila*: from molecular to systems neuroscience. *Annual Review of Neuroscience*, **28**, 275–302.

Deisig, N., Giurfa, M., Lachnit, H., and Sandoz, J.C. (2006). Neural representation of olfactory mixtures in the honeybee antennal lobe. *European Journal of Neuroscience*, **24**, 1161–1174.

El-Sayed, A.M., Suckling, D.M., Wearing, C.H., and Byers, J.A. (2006). Potential of mass trapping for long-term pest management and eradication of invasive species. *Journal of Economical Entomology*, **99**, 1550–1564.

Esslen, J. and Kaissling, K.E. (1976). Zahl und Verteilung antennaler Sensillen bei der Honigbiene (*Apis mellifera* L.). *Zoomorphology*, **83**, 227–251.

Flanagan, D. and Mercer, A.R. (1989). An atlas and 3-D reconstruction of the antennal lobes in the worker honey bee, *Apis mellifera* (Hymenoptera, Apidae). *International Journal of Insect Morphology and Embryology*, **18**, 145–159.

Fonta, C., Sun, X.J., and Masson, C. (1993). Morphology and spatial-distribution of bee antennal lobe interneurons responsive to odors. *Chemical Senses*, **18**, 101–119.

Free, J.B. (1987). *Pheromones of Social Bees*. Comstock, Ithaca, NY.

von Frisch, K. (1967). *The Dance Language and Orientation of Bees*. Harvard University Press, Cambridge, MA.

Galizia, C.G. and Kimmerle, B. (2004). Physiological and morphological characterization of honeybee olfactory neurons combining electrophysiology, calcium imaging and confocal microscopy. *Journal of Comparative Physiology A*, **190**, 21–38.

Galizia, C.G. and Menzel, R. (2001). The role of glomeruli in the neural representation of odours: results from optical recording studies. *Journal of Insect Physiology*, **47**, 115–130.

Galizia, C.G., and Vetter, R.S. (2004). Optical methods for analyzing odor-evoked activity in the insect brain. In: T.A. Christensen (ed.), *Methods in Insect Sensory Neuroscience*, pp. 345–388. CRC Press, Boca Raton, FL.

Galizia, C.G., Joerges, J., Küttner, A., Faber, T., and Menzel, R. (1997). A semi-in-vivo preparation for optical recording of the insect brain. *Journal of Neuroscience Methods*, **76**, 61–69.

Galizia, C.G., McIlwrath, S.L., and Menzel, R. (1999a). A digital three-dimensional atlas of the honeybee antennal lobe based on optical sections acquired by confocal microscopy. *Cell Tissue Research*, **295**, 383–394.

Galizia, C.G., Sachse, S., Rappert, A., and Menzel, R. (1999b). The glomerular code for odor representation is species specific in the honeybee *Apis mellifera*. *Nature Neuroscience*, **2**, 473–478.

Giurfa, M. (2003). Cognitive neuroethology: dissecting non-elemental learning in a honeybee brain. *Current Opinion of Neurobiology*, **13**, 726–735.

Giurfa, M. (2007). Behavioral and neural analysis of associative learning in the honeybee: a taste from the magic well. *Journal of Comparative Physiology A*, **193**, 801–824.

Giurfa, M. and Núñez, J.A. (1992). Honeybees mark with scent and reject recently visited flowers. *Oecologia*, **89**, 113–117.

Gronenberg, W. (2001). Subdivisions of hymenopteran mushroom body calyces by their afferent supply. *Journal of Comparative Neurology*, **435**, 474–489.

Grünewald, B. (1999). Morphology of feedback neurons in the mushroom body of the honeybee, *Apis mellifera*. *Journal of Comparative Neurology*, **404**, 114–126.

Guerrieri, F., Schubert, M., Sandoz, J.C., and Giurfa, M. (2005). Perceptual and neural olfactory similarity in honeybees. *PLoS Biology*, **3**, e60.

Hallem, E.A. and Carlson, J.R. (2006). Coding of odors by a receptor repertoire. *Cell*, **125**, 143–160.

Hallem, E.A., Ho, M.G., and Carlson, J.R. (2004). The molecular basis of odor coding in the *Drosophila* antenna. *Cell*, **117**, 965–979.

Hammer, M. and Menzel, R. (1998). Multiple sites of associative odor learning as revealed by local brain microinjections of octopamine in honeybees. *Learning and Memory*, **5**, 146–156.

Jefferis, G.S., Potter, C.J., Chan, A.M. *et al.* (2007). Comprehensive maps of *Drosophila* higher olfactory centers: spatially segregated fruit and pheromone representation. *Cell*, **128**, 1187–1203.

Joerges, J., Küttner, A., Galizia, G.C., and Menzel, R. (1997). Representation of odours and odour mixtures visualized in the honeybee brain. *Nature*, **387**, 285–288.

Jutsum, A.R. and Gordon, R.F.S. (1989). *Insect Pheromones in Plant Protection*. Wiley & Sons, New York.

Kaissling, K.E. (1987). In: K. Colbow (ed.), *R. H. Wright Lectures on Insect Olfaction*. Simon Fraser University, Burnaby, British Columbia.

Kaissling, K.E. and Renner, M. (1968). Antennale Rezeptoren für Queen Substance und Sterzelduft bei der Honigbiene. *Zeitschrift für vergleichende Physiologie*, **59**, 357–361.

Karlson, P. and Luscher, M. (1959). Pheromones: a new term for a class of biologically active substances. *Nature*, **183**, 55–56.

Keeling, C.I., Slessor, K.N., Higo, H.A., and Winston, M.L. (2003). Isolation and identification of new components of the honey bee (*Apis mellifera* L.) queen retinue pheromone. *Proceedings of the National Academy of Sciences of the USA*, **100**, 4486–4491.

Kelber, C., Rössler, W., and Kleineidam, C.J. (2006). Multiple olfactory receptor neurons and their axonal projections in the antennal lobe of the honeybee *Apis mellifera*. *Journal of Comparative Neurology*, **496**, 395–405.

Kirschner, S., Kleineidam, C.J., Zube, C., Rybak, J., Grünewald, B., and Rössler, W. (2006). Dual olfactory pathway in the honeybee, *Apis mellifera*. *Journal of Comparative Neurology*, **499**, 933–952.

Knudsen, J.T., Tollsten, L., and Bergström, L.G. (1993). Floral scents—a checklist of volatile compounds isolated by head-space techniques. *Phytochemistry*, **33**, 253–280.

Lacher, V. and Schneider, D. (1963). Elektrophysiologischer Nachweis der Riechfunktion von Porenplatten auf den Antennen der Drohne und der Arbeitsbiene (*Apis mellifera* L.). *Zeitschrift für vergleichende Physiologie* **47**, 274–278.

Le Conte, Y., Arnold, G., Trouiller, J., Masson, C., and Chappe, B. (1990). Identification of a brood pheromone in the honeybee. *Naturwissenschaften*, **77**, 334–336.

Le Conte, Y., Sreng, L., and Poitout, S.H. (1995). Brood pheromone can modulate the feeding behavior of *Apis mellifera* workers (Hymenoptera: Apidae). *Journal of Economical Entomology*, **88**, 798–804.

Lensky, Y., Cassier, P., Notion, M., Delorme-Joulie, C., and Levinsohn, M. (1985). Pheromonal activity and fine structure of the mandibular glands of honeybee drones (*Apis mellifera* L.) (Insecta, Hymenoptera, Apidae). *Journal of Insect Physiology*, **31**, 265–276.

Masson, C. and Mustaparta, H. (1990). Chemical information processing in the olfactory system of insects. *Physiological Reviews*, **70**, 199–245.

Mobbs, P.G. (1982). The brain of the honeybee *Apis mellifera*. 1. The connections and spatial-organization of the mushroom bodies. *Philosophical Transactions of the Royal Society of London B: Biological Sciences*, **298**, 309–354.

Mohammedi, A., Paris, A., Crauser, D., and Le Conte, Y. (1998). Effect of aliphatic esters on ovary development of queenless bees (*Apis mellifera* L.). *Naturwissenschaften*, **85**, 455–458.

Müller, D., Abel, R., Brandt, R., Zockler, M., and Menzel, R. (2002). Differential parallel processing of olfactory information in the honeybee, *Apis mellifera* L. *Journal of Comparative Physiology A*, **188**, 359–370.

Mustaparta, H. (1996). Central mechanisms of pheromone information processing. *Chemical Senses*, **21**, 269–275.

Olberg, R.M. (1983). Pheromone-triggered flip-flopping interneurons in the ventral nerve cord of the silkworm moth, *Bombyx mori*. *Journal of Comparative Physiology A*, **152**, 297–307.

Pareto, A. (1972). Spatial-distribution of sensory antennal fibers in central nervous-system of worker bees. *Zeitschrift für Zellforschung und mikroskopische Anatomie*, **131**, 109–140.

Pérez Orive, J., Mazor, O., Turner, G.C., Cassenaer, S., Wilson, R.I., and Laurent, G. (2002). Oscillations and sparsening of odor representations in the mushroom body. *Science*, **297**, 359–365.

Pickett, J.A., Williams, I.H., Martin, A.P., and Smith, M.C. (1980). Nasonov pheromone of the honeybee, *Apis mellifera* L. (Hymenoptera: Apidae). Part 1. Chemical characterization. *Journal of Chemical Ecology*, **6**, 425–434.

Pickett, J.A., Williams, I.H. and Martin, A.P. (1982). (Z)-11-eicosen-1-ol, an important new pheromonal component from the sting of the honey bee, *Apis mellifera* L. (Hymenoptera, Apidae). *Journal of Chemical Ecology*, **8**, 163–175.

Plettner, E. (2002). Insect pheromone olfaction: new targets for the design of species-selective pest control agents. *Current medicinal chemistry*, **9**, 1075–1085.

Plettner, E., Otis, G.W., Wimalaratne, P.D.C *et al.* (1997). Species- and caste determined mandibular gland signals in honeybees (*Apis*). *Journal of Chemical Ecology*, **23**, 363–377.

Robertson, H.M. and Wanner, K.W. (2006). The chemoreceptor superfamily in the honey bee, *Apis mellifera*: expansion of the odorant, but not gustatory, receptor family. *Genome Research*, **16**, 1395–1403.

Rutowski, R.L. (1981). The function of pheromones. *Journal of Chemical Ecology*, **7**, 481–483.

Rybak, J. and Menzel, R. (1998). Integrative properties of the Pe1-neuron, a unique mushroom body output neuron. *Learning and Memory*, **5**, 133–145.

Sachse, S. and Galizia, C.G. (2002). Role of inhibition for temporal and spatial odor representation in olfactory output neurons: a calcium imaging study. *Journal of Neurophysiology*, **87**, 1106–1117.

Sachse, S., Rappert, A., and Galizia, C.G. (1999). The spatial representation of chemical structures in the antennal lobe of honeybees: steps towards the olfactory code. *European Journal of Neuroscience*, **11**, 3970–3982.

Sandoz, J.C. (2006). Odour-evoked responses to queen pheromone components and to plant odours using optical imaging in the antennal lobe of the honey bee drone *Apis mellifera* L. *Journal of Experimental Biology*, **209**, 3587–98.

Shearer, D.A. and Boch, R. (1965). 2-heptanone in the mandibular gland secretion of the honey bee. *Nature*, **206**, 530.

Skirkevičiene, Z. and Skirkevičius, A. (1994). Worker bee and drone (*Apis mellifera* L.) behavior and functional reorganization of their olfactory receptors. *Pheromones*, **4**, 83–92.

Slessor, K.N., Kaminski, L.-A., King, G.G.S., Borden, J.H., and Winston, M.L. (1988). Semiochemical basis of the retinue response to queen honey bees. *Nature*, **332**, 354–356.

Strausfeld, N.J. (2002). Organization of the honey bee mushroom body: representation of the calyx within the vertical and gamma lobes. *Journal of Comparative Neurology*, **450**, 4–33.

Sun, X.J., Fonta, C., and Masson, C. (1993). Odour quality processing by bee antennal lobe interneurones. *Chemical Senses*, **18**, 355–377.

Szyszka, P., Ditzen, M., Galkin, A., Galizia, C.G., and Menzel, R. (2005). Sparsening and temporal sharpening of olfactory representations in the honeybee mushroom bodies. *Journal of Neurophysiology*, **94**, 3303–3313.

The Honeybee Genome Sequencing Consortium (2006). Insights into social insects from the genome of the honeybee *Apis mellifera*. *Nature*, **443**, 931–949.

Vareschi, E. (1971). Duftunterscheidung bei der Honigbiene—Einzelzell-Ableitungen und Verhaltensreaktionen. *Zeitschrift für vergleichende Physiologie*, **75**, 143–173.

Vetter, R.S. and Visscher, P.K. (1997). Influence of age on antennal response of male honey bees, *Apis mellifera*, to queen mandibular pheromone and alarm pheromone component. *Journal of Chemical Ecology*, **23**, 1867–1880.

Wanner, K.W., Nichols, A.S., Walden, K.K., Brockmann, A., Luetje, C.W., and Robertson, H.M. (2007). A honey bee odorant receptor for the queen substance 9-oxo-2-decenoic acid. *Proceedings of the National Academy of Sciences of the USA*, **104**, 14383–14388.

Wilson, E.O. (1970). *Chemical communication within animal species*. In: B. Sandheimer and J.B. Simeone (eds), *Chemical Ecology*, Academic Press, New York.

Winston, M.L. (1987). *The Biology of the Honey Bee*. Harvard University Press, Cambridge, MA.

Zars, T. (2000). Behavioral functions of the insect mushroom bodies. *Current Opinion in Neurobiology*, **10**, 790–795.

CHAPTER 8

Rapid evolution and sexual signals

Marlene Zuk and Robin M. Tinghitella

8.1 Introduction

For most of the last century and a half evolution was assumed to happen over a large timescale, requiring thousands upon thousands of generations and generally at least hundreds of thousands, if not millions, of years for changes in gene frequency to become established in populations. Over the last decade, however, an increasing number of studies have demonstrated much more rapid adaptive evolution in natural populations, sometimes called 'contemporary evolution' or 'evolution in ecological timescales' (Carroll 2007). Many of these examples are of changes in response to **selection** by human-induced disturbance, such as insecticide resistance in insects (Labbe *et al.* 2005) or industrial melanism (Cook 2003). Reznick and Ghalambor (2001) characterize rapid evolution in five types of traits: morphology, physiology, life history, phenology, and behaviour. For example, the seed-eating soapberry bug (*Jadera hamatoloma*) has shown repeated genetic changes in beak length and numerous other **fitness** characters over just a few dozen generations as it colonized new plant hosts in North America and Australia (Carroll 2007 and references therein). Indeed, many if not most examples of rapid evolution are associated with the colonization of new habitats or circumstances with novel, and particularly novel and heterogeneous, environments likely to supply the strong **directional selection** seen in most cases of contemporary evolution (Reznick and Ghalambor 2001).

This topic has received a great deal of recent interest, with a special issue of the journal *Functional Ecology* devoted to exploring contemporary adaptation in nature (see *Functional Ecology* 2007, **21**(3)). Rapid evolution is of interest not just because it seems to contradict conventional wisdom, but because study of the rate at which change is incorporated in the genes of a population sheds light on the way in which selection acts. For example, complex traits have generally been thought to be subject to stabilizing selection, which adds to their inertia (Moore 1997). Directional selection is usually thought to be most significant in rapid evolution (Ghalambor *et al* 2007). Yet some complex traits, particularly those associated with **phenotypic plasticity**, have evolved quite rapidly (Ghalambor *et al.* 2007). The growing interest in the effect of climate change on evolution has also renewed interest in how contemporary response to environmental change can mediate threats to existing populations and communities (Zimmer 2003; Kinnison and Hairston 2007).

What about behaviour, and more specifically communication and signalling, in the context of rapid evolution? Although a few behavioural examples of rapid evolution have been documented, these are mainly found in migratory or anti-predator behaviour (Reznick and Ghalambor 2001). Surprisingly, changes in communication networks or in **signals** are rarely noted in the literature on contemporary evolution. Likewise, work on behaviour and evolution does not tend to consider the rate of evolutionary change. Yet there are many reasons to expect behavioural traits in general, and communication elements in particular, to be especially prone to change over a short time frame. Signals are obviously essential in mate recognition, and such recognition in turn plays a vital role in the divergence of populations and eventual reproductive isolation; rapid species formation is one of the key areas of study in contemporary evolution. As traits that are often accompanied by a set of behaviours, signals are frequently plastic or facultative in their expression, which can further pave the way for modifications over a short time span (Ghalambor *et al.* 2007). These modifications may become incorporated into the genome through a variety of mechanisms (West-Eberhard 2003; Moczek 2007). Indeed, Gleason and Ritchie (1998) suggested that sexual signals might be expected to evolve more quickly than sexual isolation itself, which in turn evolves more quickly than post-mating isolation, and found support for this idea in a laboratory study of *Drosophila willistoni*. Many changes in social behaviour, such as increased predator avoidance by groups of animals, will require not only a shift in the anti-predator behaviour itself but also in the signals enabling it to occur.

Here we explore the role of signals in rapid evolution, and place this phenomenon in the broader context of the way in which behaviour itself influences evolution. For the purposes of this chapter we consider contemporary or rapid evolution to have occurred within 100 generations, a widely accepted criterion (Ghalambor *et al.* 2007). We focus on sexual signals, those used in courtship and mating. A recent provocative review by Svensson and Gosden (2007) found that surprisingly few secondary sexual characteristics have evolved rapidly, despite theoretical grounds for expecting the opposite, and we consider the potential reasons for this paucity of examples. We also reconsider the role of such secondary sexual characteristics as signals, and ask whether the behaviour that necessarily accompanies the function of most sexual signals facilitates or hinders evolutionary change. Because behavioural plasticity is frequently discussed in the context of brain size, and large brains are often implicated in rapid evolution of humans and other primates, we review the literature on the interaction between learned signals, behavioural flexibility, and brain size, and suggest that an overly 'vertebrate-centric' viewpoint may have unnecessarily constrained the development of theory in this area. Finally, we discuss the implications of rapid signal evolution for conservation and invasion biology as well as for speciation. We believe that behaviour, particularly with regard to signalling, has been overlooked in studies of contemporary adaptation, and a renewed focus on this area can provide insight into which traits are likely to exhibit rapid evolution and why.

8.2 Secondary sexual characters and signalling

Part of the reason why scientists studying rapid evolution have not explicitly considered communication is the disconnect that often exists between the study of morphological traits and their function. Workers in animal communication often focus on signal design, and hence rarely consider signals in multiple modalities (see also Chapter 3), whereas evolutionary biologists may examine how selection has acted on a signal but do not take the mechanisms involved in producing that signal into account (Zuk 1991; Moore 1997). Thus researchers in sexual selection would think nothing of studying both song and plumage in the context of, say, female choice. In the context of animal communication, however, visual signals are often studied in a different framework, and by scientists asking different types of questions, from auditory or chemical signals. In addition, sexually selected traits such as plumage colour are not always considered to be signals, although they fit the standard definition of changing the environment in a way that allows them to convey information (Endler 1993). A more synthetic approach, however, recognizes that information about how signals are generated and received, as well as how the environment influences their perception, will help in understanding the evolution of communication (Endler 1993). We consider virtually all secondary sexual characters to be signals.

Furthermore, signals can exhibit morphological integration, in which the various components depend upon each other in the production of the whole (Moore 1997). This relationship among the structures and processes necessary for producing a signal can mean that stabilizing selection is more prominent in the evolution of social signals (Moore 1997), a force that can impede rapid evolution, which often relies on directional selection.

Finally, sexual signals need to be recognized in order to be effective, which means that any change in their structure or production has to be accompanied by a concomitant change in the receiver, or at least the ability to recognize the new form of the signal. Although at first it might seem as if this requirement for evolution in two parties rather than just one would hinder rapid evolution, this need not necessarily be the case. Rapid evolution of smell and taste receptor genes was documented during host specialization in *Drosophila sechellia* (McBride 2007), and there is no reason to suppose that such parallel adaptation is unusual. Similarly, *Drosophila serratia* in a novel environment evolved changes in both the cuticular hydrocarbons used as signals during mate choice and in the female preference for the chemicals (Rundle *et al.* 2005). Alternatively, as we detail below, pre-existing behavioural plasticity can facilitate the spread of a mutation in genes associated with signalling.

8.3 When do traits evolve rapidly?

Reznick and Ghalambor (2001) surveyed empirical examples of rapid evolution in natural populations to identify the conditions likely to promote rapid adaptive change. Intriguingly, they found that nearly all examples of contemporary adaptation were associated with 'colonization' events that fell into one of two broad population circumstances: (1) introduction to new environments and subsequent isolation, or (2) local adaptation within a heterogeneous environment. Colonizing populations are thought to be subject to rapid evolution because they often consist of very few individuals and are subject to genetic drift, they encounter novel selection pressures in their new environments, and they can experience rapid population growth after introduction (MacArthur and Wilson 1967; Carson and Templeton 1984). Among the ecological factors influencing rapid evolution in Reznick and Ghalambor's (2001) study were new hosts and food resources, changes in the physical environment, and altered predation and competition regimes. Additionally, both types of colonization (introduction followed by isolation and local adaptation in heterogeneous environments) were often associated with anthropogenic change. If not due to intentional or unintentional movement of organisms from place to place, many cases of rapid adaptive evolution were the result of human-imposed environmental heterogeneity. For instance, herbicide and insecticide resistance frequently arise in the patchy environments to which they are prescribed (Reznick and Ghalambor 2001). Human-imposed variation in the environment is such a strong force in rapid adaptive change that multiple reviews have been dedicated to singular phenomena such as industrial melanism (Kettlewell 1973; Majerus 1998). This plethora of rapid evolutionary change in naturally selected traits, particularly in response to anthropogenic influences, leads us to wonder why there seems to be a scarcity of rapid evolution in signals, particularly those subject to **sexual selection** (Svensson and Gosden 2007). Do the factors influencing the evolution of naturally selected traits also influence communication, particularly between the sexes? If so, this seems a potent source of variation that might lead to reproductive isolation, potentially driving speciation.

8.4 Do these general principles apply to sexual signals?

We argue that, generally, the circumstances favouring rapid evolutionary change in all traits (morphology, life history, behaviour, phenology, etc.) apply equally well to signals. Predators and parasites can impose costs on signalling animals if signals are conspicuous to natural enemies, the physical environment can affect transmission efficiency of signals, and competing signallers can favour character displacement. If it is true that signals are subject to the same influences as other types of traits, we should also find rapid evolution of sexual signals in response to divergent physical environments, predator communities, and competition. Below we consider an example in which each of these ecological factors has driven the evolution of sexual signalling to illustrate the generality of their influence. Indeed, human intervention is important in many of these examples as well.

8.4.1 The physical environment

One important anthropogenic change of great recent interest is the effect of urbanization. Clearly, urban environments have different physical characteristics from the native environments of most organisms. These novel habitat characteristics are likely to supply new selection pressures for the plants and animals that make attempts to live in them. In the early 1980s, a population of dark-eyed juncos (*Junco hyemalis*) was established on the University of California, San Diego campus, and within only eight generations they experienced a 22% reduction in the white of the tail (a signal used in territorial fights and in courtship) (Yeh 2004). This change in signalling was extremely rapid, on a par with some of the fastest changes in naturally selected traits. Although the exact selection pressure responsible for the change is unknown, several environmental characteristics differ between their native montane and introduced coastal habitats that are suspect. These include, but are not limited to, components of the physical environment such as climate, vegetation type, and light levels (Yeh 2004).

8.4.2 Predators and parasites (mortality)

The influence of predators and parasites on the sexual signals of their hosts and food sources is well understood (see also Chapters 5 and 10 for the role of parasites in non-sexual signals). Generally, in the presence of predators and parasitoids, signallers engage in less risky behaviour and signals are modified to be less conspicuous (Zuk and Kolluru 1998). We know, for instance, that the songs of field crickets attract the attention of deadly parasitoid flies (Cade 1975; Zuk *et al* 1993), the 'whine-chuck' songs of Túngara frogs attract predatory fringe-lipped bats (Ryan *et al.* 1982), and the vibrant colour patterns of guppies are attractive to predatory fish (Endler 1980). In high-predation streams male guppies are less colourful than their counterparts in low-predation streams and they engage in fewer courtship interactions, instead opting for less conspicuous gonopodial thrusts (Magurran and Seghers 1990). In some parasitized populations of the field cricket, *Teleogryllus oceanicus*, males modify their calling behaviour relative to that of unparasitized populations, calling more frequently when parasitoid flies are not active, and delivering songs that are shorter overall and less attractive to the deadly parasitoid (Zuk *et al.* 1993). If a colonization event introduces organisms to environments with altered predator regimes relative to the source population, changes in signals and associated behaviour are likely to follow. This is in fact the case with the cricket, which was introduced to Hawaii (from Australia or the Pacific) where it co-occurs with the parasitoid fly, which was similarly introduced but from North America (Zuk *et al* 1993; Lehmann 2003).

8.4.3 Competition

Finally, competitive interactions are known to select for character displacement (Schluter 1996). One can imagine competition between organisms signalling simultaneously and at the same frequency or wavelength or with a similar suite of chemicals favouring a change in timing or modality of signal. These types of competition should select for divergent mating signals in sympatry and divergent receiver preferences to avoid inappropriate matings (Gerhardt 1994). One recent example in green tree frogs (*Hyla cinerea*) found stronger female preferences and differentiation of male advertisement signals in areas of sympatry with a closely related congener than in areas of allopatry (Höbel and Gerhardt 2003). In this example, behaviour associated with signalling showed displacement as well, with males of the two species also signalling from perches of different heights in sympatric portions of their range.

8.5 Sexual selection and rapid evolution

If the selection pressures most likely to influence rapid evolutionary change apply equally to sexual signals, as we suggest they do, we expect to find empirical evidence for contemporary evolution of secondary sexual characteristics. A recent review by Svensson and Gosden (2007) found only 11 such examples of rapid evolutionary change in the wild in only 10 species, and nearly all of these are from studies published since 2000 (Table 8.1). The selection pressures responsible for rapid evolution of secondary sexual traits included such things as

Table 8.1 Rapid evolution of sexual signals in natural populations. This table contains examples of sexual signals that have exhibited rapid evolutionary change, including traits that have a known genetic component as well as cases in which only phenotypic change in response to selection has been demonstrated

Organism	Trait	Selective agent	Trait type	Source
Barn swallow (*Hirundo rustica*)	Tail length	Climate change	Q	Moller and Szep (2005)
Collared flycatcher (*Ficedula albicollis*)	Forehead patch size	Climate change	Q	Garant *et al.* (2004)
Collared flycatcher (*Ficedula albicollis*)	Forehead patch size	Climate change	Q	Hegyi *et al.* (2006)
Dark-eyed junco (*Junco hyemalis*)	White on tail feathers	Climate change, signaling environment	Q	Yeh (2004)
Dark-eyed junco (*Junco hyemalis*)	Song frequency range	Acoustic environment	Q	Slabbekoorn *et al.* (2007)
Great tit (*Parus major*)	Song frequency range	Acoustic environment	Q	Slabbekoorn and Peet (2003)
House finch (*Carpodacus mexicanus*)	Song frequency range	Acoustic environment	Q	Fernandez-Juricic *et al.* (2005)
Song sparrow (*Melospiza melodia*)	Song frequency range	Acoustic environment	Q	Wood and Yezerinac (2006)
Field cricket (*Teleogryllus oceanicus*)	Male song (wing morphology)	Parasitoid fly	P	Zuk *et al.* (2006)
Blue-tailed damselfly (*Ischnura elegans*)	Female color morph frequency	Male mating harassment	P	Svensson *et al.* (2005)
Guppy (*Poecilia reticulata*)	Male coloration	Predatory fish	Q	Endler (1980)
Chinook salmon (*Oncorhynchus tshawytscha*)	Jaw length, hump size	Migration costs	Q	Kinnison *et al.* (2003)
Sockeye salmon (*Oncorhynchus nerka*)	Male body depth	Male-male competition	Q	Hendry *et al.* (2000) Hendry (2001)
Side-blotched lizard (*Uta stansburiana*)	Color morph frequencies	Intrasexual interactions	P	Sinervo and Lively (1996) Sinervo *et al.* (2000)
Bighorn sheep (*Ovis canadensis*)	Male weight, Male horn size	Trophy hunting	Q Q	Coltman *et al.* (2003)

Q refers to traits that are quantitative in nature; P refers to polymorphic traits.

climate change (Garant *et al.* 2004; Yeh and Price 2004; Möller and Szep 2005; Hegyi *et al.* 2006), migration costs (Kinnison *et al.* 2003), predators and parasitoids (Endler 1980; Zuk *et al.* 2006), and direct effects of human interactions (trophy hunting; Coltman *et al.* 2003). All of these fall into the categories of ecological influence identified by Reznick and Ghalambor (2001). So it seems the same pressures do apply equally well to sexual signals.

Why then are there so few examples of rapid evolution in sexual signals? In their review, Svensson and Gosden (2007) point out that the examples they did find were all published very recently, so perhaps this is just a new direction for study. But they also suggest several biological explanations worth considering. For instance, gene flow between populations subject to different selective regimes might constrain divergence. Genetic explanations are possible as well. Sexually selected traits could have low levels of additive genetic variance for selection to work on, or genetic covariance between traits could slow the rate of evolutionary change. Additionally, if sexually selected traits are subject to directional selection, and mutation reintroduces variation each generation, sexual selection could be purifying (Svensson and Gosden 2007).

We contend that in addition to these constraints, the apparent lack of rapid change in signalling is due to the necessary interaction of signals and behaviour—behaviour of both the signaller and the receiver. Signalling requires the organization of a suite of traits (the signal, the morphology necessary to produce and recognize the signal, and the behaviours associated with delivering and receiving the signal) and these traits might be expected to evolve in concert in much the same way we expect morphology and life history to evolve in concert. These characters form what Palmer *et al.* (2007) refer to as a 'functional complex', a set of characters that interact in a coordinated fashion to perform an intricate task. We might expect the various traits contributing to such a complex to evolve in similar ways (Watts *et al.* 2004). But if in order for one trait to change, another must change first, we instead end up with constraint.

A constraining role for behaviour is not a hard and fast rule, however. Behaviour is generally thought to be a particularly plastic trait, and has even been implicated in enhancing invasion ability because without plasticity population growth is likely to be very low or even negative in the early stages of colonization when adaptive evolution has not yet had time to act (Baldwin 1896; Holway and Suarez 1999; Yeh and Price 2004). Yeh and Price (2004) provided the first quantitative support for Baldwin's (1896) contention that plasticity was essential for population persistence when they demonstrated that in the recently founded coastal population of dark-eyed juncos (see above), females with longer breeding seasons (a highly plastic behavioural trait) had higher fitness (without a cost in future survival or reproductive effort) than females with shorter breeding seasons. They concluded that females begin breeding earlier in the coastal population because of the milder climate, and that without the increase in offspring production due to elongated breeding seasons the population would decline because the coastal juncos experience very high mortality during their first year.

In the 1930s, invasive red fire ants (*Solenopsis invicta*) were introduced to the southern United States from South America and have been an extremely successful invader. Changes in their mating system, apparently the result of genetic alteration during the bottleneck, have contributed to their success and illustrate how behavioural variation might contribute to invasion of a novel environment (Holway and Suarez 1999). When compared with native populations, introduced populations of the red imported fire ant have larger colonies with higher densities. In Argentina, colonies have few queens that are highly related to one another, but in the United States colonies have many queens (polygyne) that are relatively unrelated (Ross *et al.* 1996). Nestmate discrimination in the polygyne colonies is vastly reduced relative to that in Argentina, and this behavioural breakdown in territoriality allows an increase in population density that is key to their competitive ability during invasion (for further details of social insects see Chapters 4 and 5).

The facilitation of rapid evolutionary changes by behaviour is of particular interest to us. Over the last few years, we have documented the appearance and rapid spread of a wing mutation in a population of the Pacific field cricket, *Teleogryllus oceanicus*, on Kauai in the Hawaiian Islands that is subject to intense pressure from the acoustically orienting parasitoid fly *Ormia ochracea*. The cricket is native to Australia and the Pacific Islands, but was introduced to Hawaii; the fly, also introduced, is native to North America (Zuk *et al.* 1993). Gravid female flies locate calling males and deposit larvae on and around them. The larvae burrow inside the cricket and develop for 7–10 days, after which they emerge, killing the host (Cade 1975; Zuk *et al.* 1993). The mutation renders males unable to call, which protects them from the fly but has obvious repercussions for mating behaviour, since the crickets cannot produce their customary sexual signal (Zuk *et al.* 2006). The cricket and fly co-occur on three Hawaiian islands, but the mutation is only present in substantial numbers on Kauai, where it swept through the population in less than 20 generations (Zuk *et al.* 2006). The mutated males, which we call flatwings (see Fig. 8.1), are more attracted to playbacks of cricket song in the field than are normal-winged males, and we suggest that this satellite behaviour facilitates mate location for the silent males (Zuk *et al.* 2006). Breeding experiments show that the flatwing trait is inherited as a single, sex-linked gene (Tinghitella 2007).

Why do the flatwings behave differently from normal-winged males? One possibility is that another mutation in addition to the one that alters morphology arose concomitantly, with this second mutation altering the response of the males to the sounds of conspecifics. But we suggest a more plausible and parsimonious alternative, that pre-existing behavioural plasticity has allowed the mutation to spread. If male crickets, regardless of their wing morphology, respond to a lack of

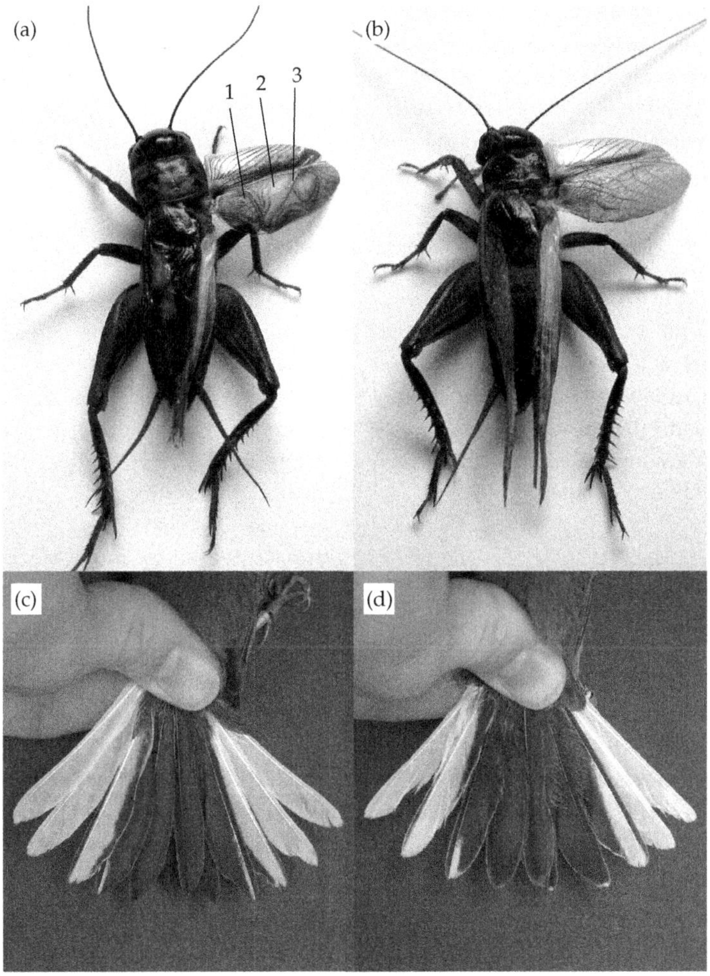

Figure 8.1 Contemporary evolution of sexual signals. Pictured are: male field crickets (*Teleogryllus oceanicus*) before (a) and after (b) the mutation to flatwing morphology; and male Dark-eyed Juncos (*Junco hyemalis*) from mountain (c) and coastal (d) populations. Note that the structures on the cricket wing that are used to produce song (file = 1, harp = 2, mirror = 3) are missing in the flatwing male field cricket. The coastal population of Dark-eyed Juncos experienced a 22% reduction in the amount of white in the tail relative to the ancestral mountain birds. *T. oceanicus* photos originally appeared in Tinghitella (2007) and junco photos are courtesy of J. Atwell and E. Snajdr.

mating success by moving closer to a nearby caller, they may enhance their own chances of encountering a female. Such movement is more likely to be triggered in the flatwings, simply because they are extremely unlikely to find a female since they cannot produce a calling song, but may well occur under other circumstances, such as a male finding himself in a poor microhabitat.

A similar scenario may account for the Kauai females' acceptance of males that cannot produce a courtship song, despite evidence from *T. oceanicus* and other closely related species that the courtship song is required for females to mount (Crankshaw 1979; Adamo and Hoy 1994; Libersat *et al.* 1994; Balakrishnan and Pollack 1996). Flatwing males perform normal courtship behaviours, but do not produce sound when vibrating their wings. Yet the thriving *T. oceanicus* population on Kauai suggests this problem, too, has been solved. Again, imagine that females have plastic behaviour depending on their mating history. If females that have not mated often or at all are less stringent in their requirement for a courtship song, they are more likely to be inseminated under a variety of circumstances that make mate availability low, including low population density as well as the flatwing mutation. Such a facultative response would allow the establishment of the mutation without the need for concomitant changes in the genes controlling female response to courtship. Were there not plasticity, we can imagine that the mutation to flatwings would simply not have spread—males and females would only rarely find one another (simply by encountering each other at random) and once in close proximity, flatwing males might have been altogether unacceptable to females.

8.6 Behaviour facilitates colonization of urban habitats—bird song as a potential example

We have already mentioned the importance of human impact for rapid evolutionary change, particularly regarding changed physical characteristics of the environment. We expect signal propagation in particular to be influenced by the transmission properties of the environment (Endler 1992), and recent work on bird song in urban environments may provide an example in which plasticity in behaviour has facilitated entry into a novel habitat type. Differences in the new environment and the population biology associated with colonization may further influence evolution of other components of an animal's biology once it is established.

Urban areas represent a new type of habitat that many animals, perhaps particularly birds, are successfully colonizing, but it follows that sound propagation in cities differs from that in their native environments (Patricelli and Blickley 2006). In urban environments, vegetation is replaced by concrete structures with reflective walls that produce echoes, as well as loud anthropogenic noise. Background noise affects which receivers can hear vocalizations, and thus many bird species have evolved signals that maximize the signal-to-noise ratio resulting in more effective transmission of the signal to the intended receiver (reviewed in Brumm and Slabbekoorn 2005). Loud background noise greatly affects the transmission of low-frequency sounds. Four species (great tits, house finches, song sparrows, and dark-eyed juncos) have shifted their songs to include only higher-frequency sounds by increasing the minimum frequency without changing the maximum frequency (Slabbekoorn and Peet 2003; Fernandez-Juricic *et al.* 2005; Wood and Yezerinac 2006; Slabbekoorn *et al.* 2007). It may be the case that these phenotypic differences are the result of genotypic variation, but it may also be that birds learn to sing successful songs by copying the songs of neighbours and dropping songs from the repertoire that do not result in a social response from neighbours (Slabbekoorn and den Boer-Visser 2006). If the latter, this behavioural flexibility (learning) may constrain rapid evolution because it shields the song from selection pressures imposed by city noise. However, it may also accelerate genotypic divergence if morphology, physiology, or neurology of the song 'functional complex' must change to accommodate production of the most successful song types. It is also conceivable that at some later date song templates will change via genetic assimilation. To this point, genetic changes have not been identified, but if we relax our criteria to include phenotypic change in response to selection, as Reznick and Ghalambor (2001) did our list of rapid evolutionary changes in sexual signalling expands to include the bird song examples discussed here (Table 8.1).

8.7 Brain size, plasticity, and rapid evolution

Communication, because it generally involves a receiver to respond to a signaller and vice versa, can be thought of as the quintessential example of behavioural flexibility and perhaps even phenotypic plasticity itself. Arguably, no other behaviour so routinely depends on a facultative response to stimuli (see also Chapter 6 regarding scent marking). If behavioural flexibility can provide fodder for rapid evolution, what in turn leads to behavioural flexibility? This topic has been of great interest to biologists, in no small part because behavioural innovation and the ability to respond to novel environments and colonize new habitats have been thought to be a hallmark of human evolution (Castro and Toro 2004; Sterelny 2007). Similarly, learning ability, the ability to change behaviour based on experience, is touted as a feature most highly developed in humans and the great apes, and the other animals that exhibit similar capacities, such as African grey parrots and New Caledonian crows, have been the subject of a great deal of research. A natural extension of this interest is the comparative study of brain size across taxa, with numerous authors attempting to explain variation in behavioural flexibility, behavioural innovation, the performance of complex behaviours, or related traits by comparing cranial volume, brain mass, or other measurements (Wyles *et al.* 1983; Healy and Rowe 2007; Sol *et al.* 2007).

For example, as part of a re-evaluation of the conventional wisdom that birds vary less anatomically than other vertebrates, Wyles *et al.* (1983) reported that birds as a whole have diversified relatively quickly. They posited 'behavioural drive' or 'behavioural selection' to account for this disparity among the vertebrates, suggesting that both innovations in behaviour without underlying genetic change and the social transmission of new skills allow a population to survive in novel environments or to move into new niches (Wyles *et al.* 1983). These non-genetic changes could then spread through a group and facilitate later morphological and genetic change. Such behavioural innovation and social transmission seemed to be particularly common among birds, especially songbirds, and Wyles *et al.* (1983) pointed out that relative brain size is larger in birds and mammals, and largest of all among songbirds and primates, both of which also exhibit high rates of anatomical evolution. They urged further study of the connection between brain size and rapid evolution in other groups of animals.

Indeed, such examinations have proliferated in recent years. Sol *et al.* (2007) found a significant negative correlation between the annual mortality rates for 236 bird species in the wild and brain size. They concluded that brain size allowed the birds to survive better, rather than the converse. Mortality was higher in species or populations living in agricultural or urban environments, suggesting that such disturbed habitats posed particular problems and that behavioural flexibility might help in circumventing them (Sol *et al.* 2007). Sol *et al.* (2007) also proposed that a 'cognitive buffer' in the form of a large brain could help animals survive environmental challenges by allowing behavioural flexibility to respond to new situations. A somewhat more targeted study found that birds with larger HVCs (higher vocal centers) in the brain also had larger repertoires (DeVoogd *et al.* 1993).

Healy and Rowe (2007) reviewed and highlighted some of the potential pitfalls of this approach. They point out that similar studies in the 1970s and early 1980s invoked diet as an explanatory variable for brain size in non-human primates and bats, but not birds; in the latter, altricial species were found to have larger brains as adults than precocial species (Bennett and Harvey 1985). These studies were largely abandoned until the last decade, when numerous comparisons of brain size and behavioural complexity or behavioural flexibility, including those cited above, emerged. Healy and Rowe (2007) further caution that brain size itself is measured differently in different studies and that the methods for collecting data also vary widely. Brain size, or the relative size of parts of the brain, can change over the course of an individual's lifetime; indeed, experience itself can alter brain architecture. In addition, behavioural complexity itself is difficult to define and hence objectively measure. Finally, they point out that few if any investigators have explained how a larger brain is used for more complex or flexible behaviour to occur (Healy and Rowe 2007).

This last point is of particular relevance to our interest here. In the flatwing cricket example described above, we suggest that plasticity in both female behaviour (willingness to mate with silent flatwing males) and male behaviour (acting as satellites to the remaining calling males) pre-dated the mutation to flatwings facilitating the spread of the alternative morphology (Zuk *et al.* 2006). This simple rule of thumb (respond to a lack of mating success by moving closer to a nearby caller) is no less an example of behavioural flexibility than many of the actions attributed to birds and mammals, and circumvents the need for a large or complex brain.

Indeed, large brains do not seem to be necessary for complex tasks, whether behavioural flexibility is required or not. In a 2007 study of web construction in miniaturized orb-weaving spiders, Eberhard found that even the smallest members of the group, with a mass of less than 0.005 mg, were able to repeat a single behaviour pattern with remarkable accuracy, and the ability to perform sophisticated behaviours was unrelated to body size (Eberhard 2007). Although it is unarguable that birds and mammals respond to their environments in complicated ways seldom seen in other taxa, we caution against a 'vertebrate-centric' point of view with regard to the evolution of signalling and the prospect of rapid evolution of behaviour in general. This broader viewpoint might also mean that rapid increases in brain size among human ancestors need not have been coupled to other important developments in human evolution, such as language or social cooperation.

8.8 Learning, behaviour, and rapid signal evolution

Behavioural flexibility is relevant to rapid signal evolution because many signals are either learned, which of necessity involves a short-term plastic response to stimuli, or occur in a milieu of associated behaviours, like the phonotaxis in crickets described above, in order for the signal itself to be effective. In addition, signals need to be recognized, which means that evolutionary changes in a signal's production or in the structure of a signal necessitate changes in the receiver as well. Finally, because behaviour can itself change over a very short time frame, it has an obvious role in rapid evolution. Behaviour can either facilitate or hinder the rate of evolution of a signal; without the male *T. oceanicus* behaviour that allowed silent males to encounter mates, the flatwing mutation would likely have quickly died out.

Bateson (2004) outlined four ways in which behaviour can change evolution. First and most simply, choices in, for example, the context of sexual selection may shape the evolution of secondary sexual characteristics and mating behaviour. Second, animals can alter their own environments, and these alterations can feed back upon behaviour. Third, behavioural modification by an individual can allow an animal to survive under conditions that might otherwise have been lethal. Fourth, animals may behave in a way that places them in new circumstances and hence new selective regimes.

A long-standing difficulty in finding a role for learning in evolution is how the learned behaviour is incorporated into genetic change. Several authors have suggested variants of the Baldwin effect, in which a trait is subject to 'genetic assimilation'; alternative versions have been termed 'genetic accommodation', but the idea is similar: a change in behaviour leads to more permanent changes in genes (Weber and Depew 2003; Price 2006; Moczek 2007). Bateson (2004) makes a particularly relevant point with regard to this issue in a counterargument to Depew's (2003) contention that if learning a behaviour were useful, selection to transform it into a genetically programmed behaviour instead would not be favoured. But learning is not a monolith of actions and responses that is either learned or unlearned; instead, it is composed of numerous subprocesses (Bateson 2004). If a learned phenotype exists, it can be modified at any level by removing some of the plasticity, and if the result is favoured by selection, the overall phenotype will change. Learning can link multiple unlearned behaviours through rules of thumb such as the one described for the crickets. This suggests that behaviour can indeed evolve rapidly, because the plasticity itself can remain intact, either wholly or in part. Learning ability itself can evolve, of course (Mery and Kawecki 2002), and at least in the laboratory it can do so remarkably quickly. Learned signals are therefore not precluded from rapid evolution.

Bateson (2004) also suggests that signals used in one context may be co-opted for use in another, citing defensive calls of burrowing owls (*Athene cunicularia*), which are very similar to sounds produced by the rattlesnakes that often co-occur with the owls in their burrows. The begging call in chicks of the closely related little owl (*Athene noctua*) is a hissing noise, and Rowe *et al.* (1986) suggested that this signal was co-opted by the congener. If social learning allows a new signal to be learned, and that signal is beneficial, it can spread rapidly through a group (Bateson 2004). Again, we suggest that such heretofore unconsidered mechanisms might provide numerous examples of rapid evolution.

Using avian colour patterns, which are often used as visual signals, as an example, Price (2006) outlined the ways in which phenotypic plasticity might interact with sexual selection to cross from a valley to an adaptive peak in a fitness landscape. Too much plasticity means that even if a population moves to a new peak, little or no genetic change will accompany the shift in phenotype, while too little plasticity means that the population remains stuck in the valley (Price 2006). Intermediate levels can facilitate adaptation if genetic assimilation occurs. Both carotenoids and melanin, two of the primary sources of plumage colour, can be expressed in a facultative manner (Price 2006). More generally, Behera and Nanjundiah (2004) showed that, at least in theory, phenotypic plasticity can favour rapid evolutionary change as long as selection acts on regulatory genes, without the need for any special mechanisms for genetic assimilation.

8.9 Does signal modality matter?

If sexual signals are as likely to undergo rapid change as other types of traits, are there certain signal modalities that are more prone to rapid change than others? Sexual signals, like other signals, are expressed in a multitude of ways, and detected visually, acoustically, chemically, and in combinations. We might expect one modality to be more constrained than the others. For instance, is there a greater degree of constraint on visual signals because morphology associated with these signals is highly conserved?

Though little work has addressed this question, we know of one study that suggests that acoustic signals have differentiated more rapidly than morphology. Mouse lemurs (*Microcebus* spp.) are particularly difficult to identify to species level using morphological characteristics, but they do have species-specific advertisement calls. A comparison of acoustic, morphological, and genetic differentiation among three European colonies revealed that acoustic differences were correlated with genetic differentiation in random amplification of polymorphic DNA (RAPD), but not with morphology (Zimmermann and Hafen 2001). The authors conclude that this suggests acoustic traits change at a more rapid rate than morphological traits. But no systematic examination of rate of signal change in relation to modality has been made.

The evolution of olfactory signals is probably less obvious to humans as observers, so rapid evolutionary change in olfactory signals may be overlooked. For instance, a salamander courtship **pheromone** gene (plethodontid receptivity factor, PRF) was completely replaced in the evolutionary history of the group despite stasis of the morphology (a pheromone-producing gland on the male's chin) and behaviour (a set of behaviours for transferring the pheromone from males to females) associated with delivering the pheromone (Watts *et al.* 2004; Palmer *et al.* 2007). So in this example at least, at the molecular level, pheromone composition evolves quickly. The authors discuss this change in olfactory communication in the context of the evolution of a signalling complex—the genes for pheromones change rapidly, but 'higher' levels of organization (the morphology and behaviour) have been retained.

Neither of these examples, however, can be considered contemporary evolution (occurring in fewer than 100 generations). What about our list of rapid evolutionary changes in sexual signalling? Is one modality of signal most prevalent amongst these examples? In Table 8.1, it is clear that morphology does not appear to be strongly constrained. In fact, the only acoustic or olfactory trait on the list is the loss of song in *T. oceanicus*; however, the change in this case was also morphological—a wing mutation eliminated the crickets' ability to call, but the behaviours associated with calling (stridulation) are retained. Because there is little here to suggest that one signal modality is more or less likely to evolve rapidly, we suggest a systematic review of the relative rates of change in different kinds of signalling traits.

8.10 The genetic architecture of rapid signal changes

If we do not find a plethora of examples of rapid evolutionary change in sexual signals, but the selective influences that facilitate rapid change in other types of traits apply equally well, perhaps there is some other constraint—what about the genetic architecture of sexual signals? Of the

11 examples of rapid evolution of secondary sexual traits, three are discrete polymorphisms with single gene control (Table 8.1). These include the colour morphs of side-blotched lizards and blue-tailed damselflies, and the flatwing–normal wing polymorphism in field crickets (Sinervo and Lively 1996; Sinervo *et al.* 2000; Svensson *et al.* 2005; Zuk *et al.* 2006). Colour morphs in the lizards and in the damselflies are controlled by autosomal loci and maintained by **frequency-dependent selection**. Experiments are under way to determine if this is the case with the field cricket wing polymorphism as well, and breeding experiments show that the mutation to flatwings is due to a change in a single sex-linked locus (Tinghitella 2007).

Theory predicts that the evolution of sexually dimorphic traits (like those subject to sexual selection) might be facilitated by sex linkage (Charlesworth *et al.* 1987; Reinhold 1998, 1999; Hurst and Randerson 1999). X-linked traits are expressed more often in the hemizygous sex (usually males) and this makes the rate of spread of X-linked alleles rapid (Avery 1984). While this suggests that sexually selected traits should be disproportionately linked to the sex chromosomes, an examination of the literature yielded mixed results, particularly in quantitative trait locus (QTL) analyses (Fairbairn and Roff 2006). Large X-effects have been noted in studies of the inheritance of mating signals, courtship behaviour, and post-zygotic reproductive isolation that distinguish species (Coyne and Orr 1989; Hollocher and Wu 1996; Yeh *et al.* 2006), but do not seem to be overwhelming in the rapid evolution of sexual signals.

If we compare the conclusions of Svensson and Gosden (2007) and those of Reznick and Ghalambor (2001) concerning the genetic architecture of rapid evolutionary change (in sexually selected and naturally selected traits, respectively), we find similar distributions of quantitative and single-gene changes. Reznick and Ghalambor's (2001) review includes a number of single-gene traits that have evolved quickly, particularly in local adaptation to anthropogenic influences like insecticides. It should be noted, however, that they did not consider rapid changes in gene frequency due to frequency- or density dependent selection, and some polymorphisms subject to these types of selection are simple allelic substitutions. In both cases, the majority of examples are of continuous, **polygenic traits**. We expect quantitative traits to evolve more slowly than traits controlled by genes of major effect because of the potential for interactions between genes (Orr 2000).

An additional, though little investigated, point is that genotype by environment interactions (G × E) are likely to be important for behavioural traits like signalling and reproductive isolation. Signals used in courtship and the behaviours associated with their expression may be dependent on ecological variation during development and adult stages. Etges *et al.* (2007) considered this important component of genetic architecture in a study of courtship song in *Drosophila mojavensis*. Using two populations of *D. mojavensis* that feed and breed on different host cacti, they determined not only that male song and courtship components (such as time to copulation) were influenced by multiple QTLs, but also by multiple G × E interactions. G × E interactions are important for the maintenance of genetic variation, but in this case also influence sexual selection and reproductive isolation, so they may slow the evolution of signals and subsequent reproductive isolation.

An important consideration in the evolution of sexually selected traits is the genetic coupling (linkage disequilibrium) expected between the genes for male traits and female preferences for them. If female preference must change before male traits can, this could slow rates of evolution. In an extreme example, consider the complete loss of a sexual signal. Despite the assumption that sexual signals are being maintained by female choice, their loss may be more widespread than has been thought (Wiens 2001). In order for a sexual signal to be lost, some other force like drift or selection must first overcome female preference, or the preference must be lost, reduced, or reversed (Wiens 2001). These additional steps will of course take more time. If, however, genes for signals and signalling behaviour are genetically correlated with those for female preference, we might expect rapid divergence of mating preferences and reproductive isolation. Supporting these ideas are cryptic species pairs that seem to have undergone very little genetic divergence, and have no post-reproductive isolation, but differ in mating signals (Otte 1994; Gray and Cade 2000; Shaw 2000; Uy and Borgia 2000). Alternatively, if female preferences tend to be plastic in response to ecological factors, as is the case in guppies exposed to predatory fish for instance (Gong and Gibson 1996), this type of plasticity might allow the rapid evolution of the male's signal. This is similar to our argument for the spread of the seemingly maladaptive flatwing morph in Hawaiian *T. oceanicus*.

8.11 Sexual selection and speciation

The divergence of mating signals has the potential to lead to reproductive isolation and eventually speciation (West-Eberhard 1983; Panhuis *et al.* 2001). If mating signals diverge rapidly in response to ecological factors similar to those that affect naturally selected traits, we might expect to find evidence of very rapid speciation events owing to sexual selection rather than natural selection. To identify such radiations, we should look for evidence of female mating preferences that are both involved in sexual selection and are important in species discrimination (Boake 2002). For instance, in Lake Victoria cichlids, female preferences for male coloration exert sexual selection within species and are important in reproductive isolation (Galis and Metz 1998; Seehausen and Van Alphen 1999). Other potential examples include cryptic species of field crickets (Gray and Cade 2000), two closely related species of bowerbirds (Uy and Borgia 2000), Hawaiian *Drosophila* (reviewed in Boake 2005), and three-spine sticklebacks (Boughman 2001). Of course the rate of speciation in these examples is not nearly as rapid as our examples of contemporary evolution, but if we can identify populations with currently diverging mating signals, identify the course of selection, and investigate the degree to which they are reproductively isolated, we may be able to gauge the importance of rapid signal changes for speciation in incipient species.

8.12 A role for gene flow

Gene flow, the movement of genes, can act in a manner similar to behaviour by either promoting or constraining divergence (Garant *et al.* 2007). Evidence from multiple taxa points to the constraining role of gene flow, when investigators find an inverse relationship between gene flow and adaptive divergence. For instance, morphological differentiation of three-spine sticklebacks is lowest between lake–stream pairs that are parapatric and experience higher gene flow (Hendry and Taylor 2004) and divergence increases with increasing distance from the lake, suggesting a constraining role for gene flow (Moore and Hendry 2005). One question, however, is whether gene flow is constraining adaptive divergence or adaptive divergence is constraining gene flow. Ecological speciation theory suggests that **divergent selection** across environments should select for greater adaptive divergence and this may decrease gene flow because introduced variants will be maladapted to new environments (Rundle and Nosil 2005).

Gene flow, however, can have positive effects on adaptive divergence as well. One such effect is to counteract the effects of inbreeding depression in small populations by introducing novel or rare alleles (Garant *et al.* 2007). It is possible, then, for gene flow to facilitate change, aiding in the persistence of populations and their adaptation to changing conditions, for example in heterogeneous environments, and promoting the spread of beneficial mutations among populations (Garant *et al.* 2007).

8.13 Conclusions and future directions

Adding animal signals to the list of candidates to be considered in discussions of contemporary evolution has the potential to enrich our understanding of rates of evolution and provide more links between the fields of animal behaviour and evolutionary biology. Because of the inherent plasticity of communication, signals provide an ideal testing ground for hypotheses about the Baldwin effect and other ideas about the incorporation of phenotypic plasticity into genetic change.

The potential for rapid evolution in signals is also relevant to current considerations of urbanization and the effect of anthropogenic change on animal behaviour. The shifts in bird song in urban environments documented by Slabbekoorn and others are only one example of an adaptation to a human-induced change in behaviour. It is possible that global climate change, because of its effects on habitat structure, could also influence selection on signals.

Even well-intentioned efforts to conserve species in captivity can have unforeseen consequences on signalling. For example, efforts to conserve fish by restricting catches to a minimum size appear to have resulted in selection for earlier reproduction and hence an overall smaller body size in a number of species (Zimmer 2003). It is easy to imagine a similar effect on plumage colour, chemical composition, or other signals. Simply placing a population in captivity could influence signal efficacy, so that when animals are released back into the wild they are no longer able to locate suitable mates. Captive breeding programmes may also relax selection by allowing animals to reproduce even when they cannot produce appropriate signals. A lack of acknowledgement of the speed with which signals can evolve might lead managers to discount the possibility of detrimental effects on the reproduction of the species they are trying to conserve.

Summary

Surprisingly few examples of rapid evolutionary change in behavioural traits have been documented in nature, yet circumstances favouring rapid evolution in other traits apply equally well to behaviour, including animal signals. Here, we consider the role of signals in rapid evolution and the way in which behaviour influences evolutionary change. Because communication involves interactions between individuals, changes in signal structure or production must be accompanied by change in the receiver, which means that behaviour itself can constrain evolution. Alternatively, behaviour may facilitate contemporary evolution. We review the literature as well as our work with a field cricket in which pre-existing behavioural plasticity apparently facilitated the spread of a mutation that silences males, simultaneously eliminating their sexual signal and protecting them from a parasitoid.

References

Adamo, S.A. and Hoy, R.R. (1994). Mating behaviour of the field cricket *Gryllus bimaculatus* and its dependence on social and environmental cues. *Animal Behaviour*, **47**, 857–868.

Avery, P.J. (1984). The population genetics of haplo-diploids and X-linked genes. *Genetical Research*, **44**, 321–342.

Balakrishnan, R. and Pollack, G.S. (1997). The role of antennal sensory cues in female responses to courting males in the cricket *Teleogryllus oceanicus*. *Journal of Experimental Biology*, **200**, 511–522.

Baldwin, J.M. (1896). A new factor in evolution. *The American Naturalist*, **30**, 441–451.

Bateson, P. (2004). Evolution and learning: The Baldwin effect reconsidered. *Biology and Philosophy*, **19**, 283–298.

Behera, N. and Nanjundiah, V. (2004). Phenotypic plasticity can potentiate rapid evolutionary change. *Journal of Theoretical Biology*, **226**, 177–184.

Bennett, P.M. and Harvey, P.H. (1985). Brain size, development and metabolism in birds and mammals. *Journal of Zoology*, **207**, 491–509.

Boake, C.R.B. (2002). Sexual signaling and speciation, a microevolutionary perspective. *Genetica*, **116**, 205–214.

Boake, C.R.B. (2005). Sexual selection and speciation in Hawaiian *Drosophila*. *Behavior Genetics*, **35**, 297–303.

Boughman, J.W. (2001). Divergent sexual selection enhances reproductive isolation in sticklebacks. *Nature*, **411**, 944–948.

Brumm, H. and Slabbekoorn, H. (2005). Acoustic communication in noise. *Advances in the Study of Behavior*, **35**, 151–209.

Cade, W. (1975). Acoustically orienting parasitoids: fly phonotaxis to cricket song. *Science*, **190**, 1312–1313.

Carroll, S.P. (2007). Brave New World: the epistatic foundations of native adapting to invaders. *Genetica*, **129**, 193–204.

Carson, H.L. and Templeton, A.R. (1984). Genetic revolutions in relation to speciation phenomena: the founding of new populations. *Annual Review of Ecology and Systematics*, **15**, 97–131.

Castro, L. and Toro, M.A. (2004). The evolution of culture: from primate social learning to human culture. *Proceedings of the National Academy of Sciences of the USA*, **101**, 10235–10240.

Charlesworth, B., Coyne, J.A., and Barton, N.H. (1987). The relative rates of evolution of sex chromosomes and autosomes. *The American Naturalist*, **130**, 113–146.

Coltman, D.W., O'Donoghue, P., Jorgenson, J.T., Hogg, J.T., Strobeck, C., and Festa-Bianchet, M. (2003). Undesirable evolutionary consequences of trophy hunting. *Nature*, **426**, 655–658.

Cook, L.M. (2003). The rise and fall of the Carbonaria form of the peppered moth. *Quarterly Review of Biology*, **78**, 399–417.

Coyne, J.A. and Orr, H.A. (1989). Two rules of speciation. In: D. Otte and J.A. Endler (eds), *Speciation and its Consequences*, pp. 180–207. Sinauer Associates, Sunderland, MA.

Crankshaw, O.S. (1979). Female choice in relation to calling and courtship songs in *Acheta domesticus*. *Animal Behaviour*, **27**, 1274–1275.

Depew, D.J. (2003). Baldwin and his many effects. In: B.H. Weber and D.J. Depew (eds), *Evolution and Learning: the Baldwin Effect Reconsidered*, pp. 3–32. MIT Press, Cambridge, MA.

DeVoogd, T.J., Krebs, J.R., Healy, S.D., and Purvis, A. (1993). Relations between song repertoire size and the volume of brain nuclei related to song—comparative analyses amongst oscine birds. *Proceedings of the Royal Society B: Biological Sciences*, **254**, 75–82.

Eberhard, W.G. (2007). Miniaturized orb-weaving spiders: behavioural precision is not limited by small size. *Proceedings of the Royal Society B: Biological Sciences*, **274**, 2203–2209.

Endler, J.A. (1980). Natural selection on color patterns in *Poecilia reticulata*. *Evolution*, **34**, 76–91.

Endler, J.A. (1992). Signals, signal conditions, and the direction of evolution. *The American Naturalist*, **139**, S125–S153.

Endler, J.A. (1993). Some general comments on the evolution and design of animal communication systems. *Philosophical Transactions of the Royal Society B: Biological Sciences*, **340**, 215–225.

Etges, W.J., de Oliveira, C.C., Gragg, E., Ortiz-Barrientos, D., Noor, M.A.F., and Ritchie, M.G. (2007). Genetics of incipient speciation in *Drosophila mojavensis*. I. Male courtship song, mating success, and genotype x environment interactions. *Evolution*, **61**, 1106–1119.

Fairbairn, D.J. and Roff, D.A. (2006). The quantitative genetics of sexual dimorphism: assessing the importance of sex-linkage. *Heredity*, **97**, 319–328.

Fernandez-Juricic, E., Poston, R., De Collibus, K. *et al.* (2005). Microhabitat selection and singing behavior patterns of male house finches (*Carpodacus mexicanus*) in urban parks in a heavily urbanized landscape in the Western U.S. *Urban Habitats*, **3**, 49–69.

Galis, F. and Metz, A.J. (1998). Why are there so many cichlid species? *Trends in Ecology and Evolution*, **13**, 1–2.

Garant, D., Forde, S.E., and Hendry, A.P. (2007). The multifarious effects of dispersal and gene flow on contemporary adaptation. *Functional Ecology*, **21**, 434–443.

Garant, D., Sheldon, B.C., and Gustafsson, L. (2004). Climatic and temporal effects on the expression of secondary sexual characters: genetic and environmental components. *Evolution*, **58**, 634–644.

Gerhardt, H.C. (1994). Reproductive character displacement of female mate choice in the grey treefrog, *Hyla chrysoscelis*. *Animal Behaviour*, **47**, 959–969.

Ghalambor, C.K., McKay, J.K., Carroll S.P., and Reznick D.N. (2007). Adaptive versus non-adaptive phenotypic plasticity and the potential for contemporary adaptation in new environments. *Functional Ecology*, **21**, 394–407.

Gleason, J.M. and Ritchie, M.G. (1998). Evolution of courtship song and reproductive isolation in the *Drosophila willistoni* species complex: do sexual signals diverge the most quickly? *Evolution*, **52**, 1493–1500.

Gong, A. and Gibson, R.M. (1996). Reversal of a female preference after visual exposure to a predator in the guppy, *Poecilia reticulata*. *Animal Behaviour*, **52**, 1007–1015.

Gray, D.A. and Cade, W.H. (2000). Sexual selection and speciation in field crickets. *Proceedings of the National Academy of Sciences of the USA*, **97**, 14449–14454.

Healy, S.D. and Rowe, C. (2007). A critique of comparative studies of brain size. *Proceedings of the Royal Society B: Biological Sciences*, **274**, 453–464.

Hegyi, G., Torok, J., Toth, L., Garamszegi, L.Z., and Rosivall, B. (2006). Rapid temporal change in the expression and age-related information content of a sexually selected trait. *Journal of Evolutionary Biology*, **19**, 228–238.

Hendry, A.P. (2001). Adaptive divergence and the evolution of reproductive isolation in the wild: an empirical demonstration using introduced sockeye salmon. *Genetica*, **112**, 515–534.

Hendry, A.P. and Taylor, E.B. (2004). How much of the variation in adaptive divergence can be explained by gene flow? An evaluation using lake-stream stickleback pairs. *Evolution*, **58**, 2319–2331.

Hendry, A.P., Wenburg, J.K., Bentzen, P., Volk, E.C., and Quinn, T.P. (2000). Rapid evolution of reproductive isolation in the wild: evidence from introduced salmon. *Science*, **290**, 516–518.

Höbel, G. and Gerhardt, H.C. (2003). Reproductive character displacement in the acoustic communication system of Green tree frogs (*Hyla cinerea*). *Evolution*, **57**, 894–904.

Hollocher, H. and Wu, C-I. (1996). The genetics of reproductive isolation in the *Drosophila simulans* clade: X vs autosomal effects and male *vs* female effects. *Genetics*, **143**, 1243–1255.

Holway, D.A. and Suarez, A.V. (1999). Animal behaviour: an essential component of invasion biology. *Trends in Ecology and Evolution*, **14**, 328–330.

Hurst, L.D. and Randerson, J.P. (1999). An exceptional chromosome. *Trends in Genetics*, **15**, 383–385.

Kettlewell, B. (1973). *The Evolution of Melanism: the Study of a Recurring Necessity*. Clarendon Press, Oxford.

Kinnison, M.T. and Hairston, N.G. Jr (2007). Eco-evolutionary conservation biology: contemporary evolution and the dynamics of persistence. *Functional Ecology*, **21**, 444–454.

Kinnison, M.T., Unwin, M.J., and Quinn, T.P. (2003). Migratory costs and contemporary evolution of reproductive allocation in male Chinook salmon. *Journal of Evolutionary Biology*, **16**, 1257–1269.

Labbe, P., Lenormand, T., and Raymond, M. (2005). On the worldwide spread of an insecticide resistance gene: a role for local selection. *Journal of Evolutionary Biology*, **18**, 1471–1484.

Lehmann, G.U.C. (2003). Review of biogeography, host range and evolution of acoustic hunting in Ormiini (Insecta, Diptera, Tachinidae), parasitoids of night-calling bushcrickets and crickets (Insecta, Orthoptera, Ensifera). *Zoologischer Anzeiger*, **242**, 107–120.

Libersat, F., Murray, J.A., and Hoy, R.R. (1994). Frequency as a releaser in the courtship song of two crickets, *Gryllus bimaculatus* (de Geer) and *Teleogryllus oceanicus*: a neuroethological analysis. *Journal of Comparative Physiology A*, **174**, 485–494.

MacArthur, R.H. and Wilson, E.O. (1967). *The Theory of Island Biogeography*. Princeton University Press, Princeton, NJ.

McBride, C.S. (2007). Rapid evolution of smell and taste receptor genes during host specialization in *Drosophila sechellia*. *Proceedings of the National Academy of Sciences of the USA*, **104**, 4996–5001.

Magurran, A.E. and Seghers, B.H. (1990). Risk sensitive courtship in the guppy (*Poecilia reticulata*). *Behaviour*, **112**, 194–201.

Majerus, M.E.N. (1998). *Melanism: Evolution in Action*. Oxford University Press, Oxford.

Mery, F. and Kawecki, T.J. (2002). Experimental evolution of learning ability in fruit flies. *Proceedings of the National Academy of Sciences of the USA*, **99**, 14274–14279.

Moczek, A.P. (2007). Developmental capacitance, genetic accommodation, and adaptive evolution. *Evolution and Development*, **9**, 299–305.

Möller, A.P. and Szep, T. (2005). Rapid evolutionary change in a secondary sexual character linked to climatic change. *Journal of Evolutionary Biology*, **18**, 481–495.

Moore, A.J. (1997). The evolution of social signals: morphological, functional and genetic integration of the sex pheromone in *Nauphoeta cinerea*. *Evolution*, **51**, 1920–1928.

Moore, J.S. and Hendry, A.P. (2005). Both selection and gene flow are necessary to explain adaptive divergence: evidence from clinal variation in stream stickleback. *Evolutionary Ecology Research*, **7**, 871–886.

Orr, H. (2000). Adaptation and the cost of complexity. *Evolution*, **54**, 13–20.

Otte, D. (1994). *The Crickets of Hawaii: Origin, Systematics, and Evolution*. The Orthopterists' Society, Academy of Natural Sciences of Philadelphia, Philadelphia.

Palmer, C.A., Watts, R.A., Houck, L.D., Picard, A.L., and Arnold, S.J. (2007). Evolutionary replacement of components in a salamander pheromone signaling complex: more evidence for phenotypic-molecular decoupling. *Evolution*, **61**, 202–215.

Panhuis, T.M., Butlin, R., Zuk, M., and Tregenza, T. (2001). Sexual selection and speciation. *Trends in Ecology and Evolution*, **16**, 364–71.

Patricelli, G.L. and Blickley, J.L. (2006). Avian communication in urban noise: causes and consequences of vocal adjustment. *The Auk*, **123**, 639–649.

Price, T.D. (2006). Phenotypic plasticity, sexual selection and the evolution of colour patterns. *Journal of Experimental Biology*, **209**, 2368–2376.

Reinhold, K. (1998). Sex linkage among genes controlling sexually selected traits. *Behavioral Ecology and Sociobiology*, **44**, 1–7.

Reinhold, K. (1999). Evolutionary genetics of sex-limited traits under fluctuating selection. *Journal of Evolutionary Biology*, **12**, 897–902.

Reznick, D.N. and Ghalambor, C.K. (2001). The population ecology of contemporary adaptations: what empirical studies reveal about the conditions that promote adaptive evolution. *Genetica*, **112–113**, 183–198.

Ross, K.G., Vargo, E.L., and Keller, L. (1996). Social evolution in a new environment: The case of introduced fire ants. *Proceedings of the National Academy of Sciences of the USA*, **93**, 3021–3025.

Rowe, M.P., Coss, R.G., and Owings, D.H. (1986). Rattlesnake rattles and burrowing owl hisses – a case of acoustic Batesian mimicry. *Ethology*, **72**, 53–71.

Rundle, H.D. and Nosil, P. (2005). Ecological speciation. *Ecology Letters*, **8**, 336–352.

Rundle, H.D., Chenoweth, S.F., Doughty, P., and Blows, M.W. (2005). Divergent selection and the evolution of signal traits and mating preferences. *PLoS Biology*, **11**, 1988–1995.

Ryan, M.J., Tuttle, M.D., and Rand, A.S. (1982). Bat predation and sexual advertisement in a neotropical anuran. *The American Naturalist*, **119**, 136–139.

Saifi, G.M. and Chandra, H.S. (1999). An apparent excess of sex- and reproduction-related genes on the human X chromosome. *Proceedings of the Royal Society B: Biological Sciences*, **266**, 203–209.

Schluter, D. (1996). Ecological causes of adaptive radiation. *The American Naturalist*, **148**, S40–S64.

Seehausen, O. and van Alphen, J.M. (1999). Can sympatric speciation by disruptive sexual selection explain rapid evolution of cichlid diversity in Lake Victoria? *Ecology Letters*, **2**, 262–271.

Shaw, K.L. (2000). Further acoustic diversity in Hawaiian forests: two new species of Hawaiian cricket (Orthoptera: Gryllidae: Trigonidiinae: Laupala). *Zoological Journal of the Linnean Society*, **129**, 73–91.

Sinervo, B. and Lively, C.M. (1996). The rock-paper-scissors game and the evolution of alternative male strategies. *Nature*, **380**, 240–243.

Sinervo, B., Svensson, E., and Comendant, T. (2000). Density cycles and an offspring quantity and quality game driven by natural selection. *Nature*, **406**, 985–988.

Slabbekoorn, H. and den Boer-Visser, A. (2006). Cities change the songs of birds. *Current Biology*, **16**, 2326–2331.

Slabbekoorn, H. and Peet, M. (2003). Birds sing at a higher pitch in urban noise. *Nature*, **424**, 267.

Slabbekoorn, H., Yeh, P., and Hunt, K. (2007). Sound transmission and song divergence: a comparison of urban and forest acoustics. *The Condor*, **109**, 67–78.

Sol, D., Szekely, T., Liker, A., and Lefebvre, L. (2007). Big-brained birds survive better in nature. *Proceedings of the Royal Society B: Biological Sciences*, **274**, 763–769.

Sterelny, K. (2007). Social intelligence, human intelligence and niche construction. *Philosophical Transactions of the Royal Society B: Biological Sciences*, **362**, 719–730.

Svensson, E.I. and Gosden, T.P. (2007). Contemporary evolution of secondary sexual traits in the wild. *Functional Ecology*, **21**, 422–433.

Svensson, E.I., Abbott, J., and Hardling, R. (2005). Female polymorphism, frequency dependence, and rapid evolutionary dynamics in natural populations. *The American Naturalist*, **165**, 567–576.

Tinghitella, R.M. (2007). Rapid evolutionary change in a sexual signal: genetic control of the mutation "flatwing" which renders male field crickets (*Teleogryllus oceanicus*) mute. *Heredity*, doi:10.1038/sj.hdy.6801069.

Uy, J.A.C. and Borgia, G. (2000). Sexual selection drives rapid divergence in bowerbird display traits. *Evolution*, **54**, 273–278.

Watts, R.A., Palmer, C.A., Feldhoff, R.C. *et al.* (2004). Stabilizing selection on behavior and morphology masks positive selection on the signal in a salamander pheromone signaling complex. *Molecular Biology and Evolution*, **21**, 1032–1041.

Weber, B.H. and Depew, D.J. (eds) (2003). *Evolution and Learning: the Baldwin Effect Reconsidered*. MIT Press, Cambridge, MA.

West-Eberhard, M. (1983). Sexual selection, social competition, and speciation. *Quarterly Review of Biology*, **58**, 155–183.

Wiens, J.J. (2001). Widespread loss of sexually selected traits. *Trends in Ecology and Evolution*, **16**, 517–523.

Wood, W.E. and Yezerinac, S.M. (2006). Song sparrow (*Melospiza melodia*) song varies with urban noise. *The Auk*, **123**, 650–659.

Wyles, J.S., Kunkel, J.G., and Wilson, A.C. (1983). Birds, behaviour, and anatomical evolution. *Proceedings of the National Academy of Sciences of the USA*, **80**, 4394–4397.

Yeh, P.J. (2004). Rapid evolution of a sexually selected trait following population establishment in a novel habitat. *Evolution*, **58**, 166–174.

Yeh, P. and Price, T.D. (2004). Adaptive phenotypic plasticity and the successful colonization of a novel environment. *The American Naturalist*, **164**, 531–542.

Yeh, S-D., Liou, S-R., and True, J.R. (2006). Genetics of divergence in male wing pigmentation and courtship behavior between *Drosophila elegans* and *D. gunungcola*. *Heredity*, **96**, 383–395.

Zimmer, C. (2003). Rapid evolution can foil even the best-laid plans. *Science*, **300**, 895.

Zimmermann, E. and Hafen, T.G. (2001). Colony specificity in a social call of mouse lemurs (*Microcebus spp.*). *American Journal of Primatology*, **54**, 129–141.

Zuk, M. (1991). Sexual ornaments as animal signals. *Trends in Ecology and Evolution*, **6**, 228–231.

Zuk, M. and Kolluru, G.R. (1998). Exploitation of sexual signals by predators and parasitoids. *Quarterly Review of Biology*, **73**, 415–438.

Zuk, M., Simmons, L.W., and Cupp, L. (1993). Calling characteristics of parasitized and unparasitized populations of the field cricket *Teleogryllus oceanicus*. *Behavioral Ecology and Sociobiology*, **33**, 339–343.

Zuk, M., Rotenberry, J.T., and Tinghitella, R.M. (2006). Silent night: Adaptive disappearance of a sexual signal in a parasitized population of field crickets. *Biology Letters*, **2**, 521–524.

CHAPTER 9

Communication of mate quality in humans

S. Craig Roberts

9.1 Introduction

Sociality affords individuals with greater opportunity for improving reproductive success through mate choice than will generally be possible for those of habitually solitary or socially monogamous species. Group-living individuals can attract more potential mates from a relatively large number of opposite-sex group members, or choose better ones, following the same principle that appears to drive the evolution of leks in some species, where males form aggregations to attract large numbers of females and females visit the aggregations solely for mating (Bradbury and Gibson 1983). Furthermore, the costs of mate choice, including time or energetic costs of travelling between potential mates, and the risk of injury or predation while doing so, are likely to be substantially lower in larger and mixed-sex groups (Reynolds and Gross 1990). Finally, mate choice within groups allows individuals to compare directly between potential mates and make choices based on relative qualities, which may be a quicker and more efficient mechanism for choice than relying on absolute trait values (Bateson and Healy 2005).

Just as in animal groups, opportunities exist in most human societies for individuals to view, track, and compare an array of potential mates; females are thought to be the choosier sex due to the relatively large costs of reproduction. Shaping of adult-like mate preferences starts well before first sexual encounters (Saxton *et al.* 2006); indeed, early exposure to individuals of the opposite sex can profoundly influence individual variation in preferences, whether it originates in general inbreeding avoidance mechanisms (Weisfeld *et al.* 2003) or through more subtle imprinting-like effects in which men and women are attracted to potential mates who share specific traits with their opposite-sex parent (Jacob *et al.* 2002; Little *et al.* 2003). Towards adulthood, there is further opportunity to gain experience from a variety of potential partners. For example, in most societies there is usually a considerable interval between first sexual intercourse and first marriage or cohabitation, from a mean of 5–7 years in most African countries to 7.5 and 10.6 years for the UK and USA, respectively (Wellings *et al.* 2006).

But if sociality offers particular opportunities for choice, it also requires individuals to sift through and process an array of available information about the quality or suitability of many different potential partners. Furthermore, the individuals doing the choosing must be sensitive to the vagaries of different kinds of information and to the context in which decisions are made. For example, we might not expect different kinds of information to be equally reliable about the quality of a potential mate, since some could be faked and some may be irrelevant to discrimination of mate quality. How then do males in particular communicate their own quality as a partner to potential mates, and how do females select a suitable partner from so many different males? The attributes used in mate choice are many and diverse, but include intelligence (including correlates like education and sense of humour; Miller 2000), personality (e.g. Buston and Emlen 2003), resources and wealth (Daly and Wilson 1983), and physical traits. Each of these characteristics can be reliable to a considerable degree, certainly when tracked over time, but physical traits are particularly hard to fake, even in the short term. This quality, coupled with generally high heritability and comparative dissociation from cultural effects, has led to a sustained and intense research effort in the past decade or two to catalogue the ways in which physical attributes influence judgements of attractiveness.

In this chapter I outline some of the recent progress that has been made towards a biological understanding of the communication of mate quality in human interactions, which closely follows and matches the findings of behavioural ecologists working on other animals. Just as in animal studies (e.g. Andersson 1994; Petrie 1994), evolutionary psychologists interested in human mate choice have particularly focused on physical characteristics that potentially indicate underlying good genes. These researchers conventionally use volunteers to rate the attractiveness of stimuli (e.g. face photographs) of other individuals. More recently, the scope of such studies has widened to incorporate behavioural patterns and expressions that may reveal similar information about genetic quality; that is, researchers might present a video, rather than a single photo. Questions remain as to how wide this behavioural scope can stretch and I conclude by exploring the extent to which dynamic **cues** can reliably reveal mate quality.

9.2 Physical traits and mate quality

Variation in male attractiveness is underpinned by variation in physical traits, most of which can be discerned at a distance and almost instantaneously, and acted upon without intimate contact. It has been hypothesized that at least some of these traits reveal underlying genetic quality. In the following sections, some key evidence to support this idea is described, but the reader is also referred to several recent wide-ranging reviews (Grammer *et al.* 2003; Gangestad and Scheyd 2005; Rhodes 2006; Roberts and Little 2008).

9.2.1 The face

Unsurprisingly, the face is the centrepiece of human attractiveness research. Facial perception is key to individual recognition, kin recognition, and judgements of mood, intention, and personality in others (Posamentier and Abdi 2003; Zhao *et al.* 2003), in addition to judgements of attractiveness (Thornhill and Gangestad 1999; Rhodes and Zebrowitz 2002). Facial judgements of opposite-sex individuals are also better correlated with overall impressions of their attractiveness than are judgements of body attractiveness (Peters *et al.* 2007). This is despite the fact that independent attractiveness ratings of face and body correlate very well, indicating that both are underpinned by congruent **hormonal** and genetic influence (Thornhill and Grammer 1999).

What facial characteristics are attractive? As Rhodes and Zebrowitz (2002) point out, there is no 'gold standard' in what makes any face attractive, many different factors appear to contribute. One of the most intensively studied aspects is fluctuating asymmetry, building on work in animals such as barn swallows, where males with the most symmetrical tails are preferred by females (Møller 1992). In similar fashion, facial symmetry is positively related to judgements of facial attractiveness (Grammer and Thornhill 1994). Some of the most persuasive studies are able to use elegant experimental designs by employing computer-graphics techniques and composite images to control for potential confounds (see Fig. 9.1) and hence manipulate symmetry to increase attractiveness (e.g. Rhodes *et al.* 1998; Little *et al.* 2001). However, although symmetry is correlated with attractiveness, individuals are not generally adept in assessing symmetry *per se* in faces (Scheib *et al.* 1999) and the ability to detect symmetry is dissociated from preference for it (Little and Jones 2006). Indeed, the symmetry–attractiveness relationship is maintained when attractiveness ratings are based on only the left or right half of a face, such that symmetry cues are withheld (Scheib *et al.* 1999). It would seem, therefore, that symmetry is unlikely to be the proximate mechanism that mediates judgements of attractiveness, rather it may simply correlate with one or more other attractive traits (Scheib *et al.* 1999; Penton-Voak *et al.* 2001). Scheib *et al.* suggest this may be facial masculinity (see also Gangestad and Thornhill 2003), but Penton-Voak *et al.* (2001), suggest facial skin condition would correlate more

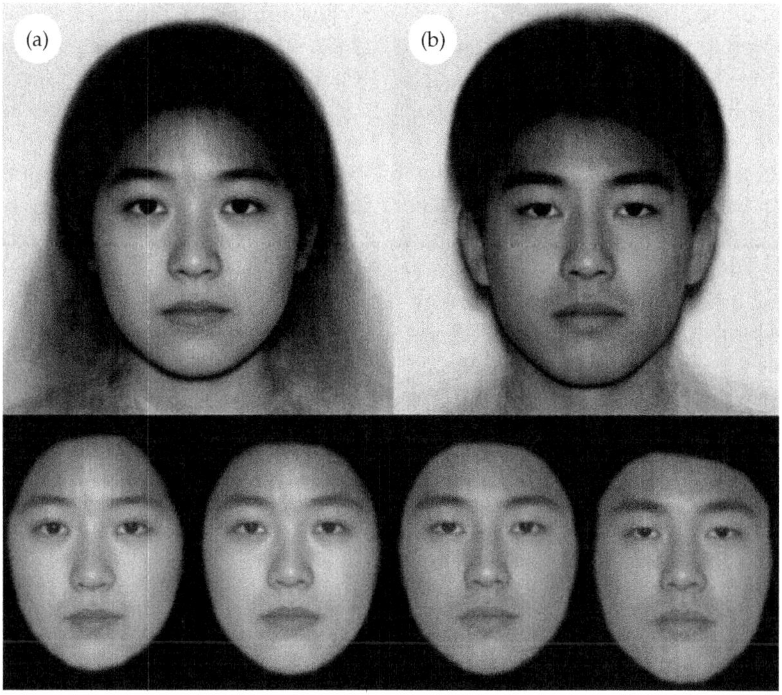

Figure 9.1 Use of facial composites in mate choice research. Progress in study of facial preferences has been facilitated by computer graphics techniques which enable the construction of 'average' or 'composite' faces by digitally blending photographs of individual faces. Here, 28 male and 28 female faces of young Japanese students (from Perrett *et al.* 1998) are used to construct a female composite (a) and male composite (b). In the lower panel, each composite has then been digitally morphed to render the face either feminised (the left-hand image of each pair) or masculinised (the right-hand image of each pair). These faces can then be presented as stimuli to assess preference for sexual dimorphism in faces. The same technique can be used to manipulate or control for other characteristics, such as symmetry. Reprinted by permission from Macmillan Publishers Ltd: Perrett DI *et al.*, Nature 394, 884–887, copyright 1998.

reliably with symmetry because both may be fundamentally linked to health (see also Jones *et al.* 2001). Indeed, in a subsequent study, Jones *et al.* (2004b) showed that symmetrical faces were judged to have healthier-looking skin than asymmetrical faces (whether real faces or composites).

A great deal of research has also focused on the extent to which attractiveness correlates with the degree of sexual dimorphism in the face. Perrett *et al.* (1998) produced composite images of men and women from either the UK or Japan, and morphed these in a masculine–feminine dimension (Fig. 9.1); subjects were then required to indicate the face they most preferred. More feminized versions of female faces were most attractive for men, while women preferred an intermediate level of facial masculinity, apparently because masculinization was associated with negative personality attributions associated with low paternal investment. Perceptions of facial masculinity are correlated with levels of testosterone (Penton-Voak and Chen 2004) and physical strength (Fink *et al.* 2007a), so it appears to be a reliable cue of physiological differences between men. Furthermore, facial masculinity also appears to be correlated with facial symmetry and thus to carry consistent information regarding genetic quality (Gangestad and Thornhill 2003). In support of this idea, highly dimorphic facial features co-vary with low susceptibility to infection (Thornhill and Gangestad 2006).

The extent to which individual women prefer facial masculinity varies with women's perception of their own attractiveness, in other words how well they perform in the mating marketplace (Little *et al.* 2001). In addition, there are systematic temporal changes in preference expression depending on the stage of the woman's menstrual cycle, such that preference shifts towards masculinity occur around ovulation, when women may benefit most in terms of indirect benefits from preferring and choosing males with good genes (Penton-Voak *et al.* 1999; Fig. 9.2).

More recently, the possibility that facial attractiveness could be underpinned by genetic heterozygosity has been investigated. To do this, Roberts *et al.* (2005b) asked women to rate the faces of men who had been genotyped at key loci in the major histocompatibility complex (MHC). Men who were heterozygous for all the genes under study were rated more attractive on average than men who were homozygous at one or more loci. This result is consistent with the general finding that perceived health of potential mates may be a reliable general influence on mate preferences in humans (Thornhill and Gangestad 1999; Grammer *et al.* 2003, 2005; Rhodes 2006). Indeed, further investigation showed that very similar results were obtained when women assessed the healthiness of small patches of skin excised from the facial photographs and that these healthiness ratings correlated with attractiveness ratings of the whole face. This finding was the first to directly link male genotype with perception of attractiveness, but shows consistent effects with other studies, notably those of Jones *et al.* (2004a,b), that use the healthiness of skin as an indicator of underlying good genes.

9.2.2 Body

Although bodily features have been less intensively studied than faces, similar approaches indicate that putative indicators of good genes are discriminable and considered attractive. Male body attractiveness correlates with higher physical **fitness** (Honekopp

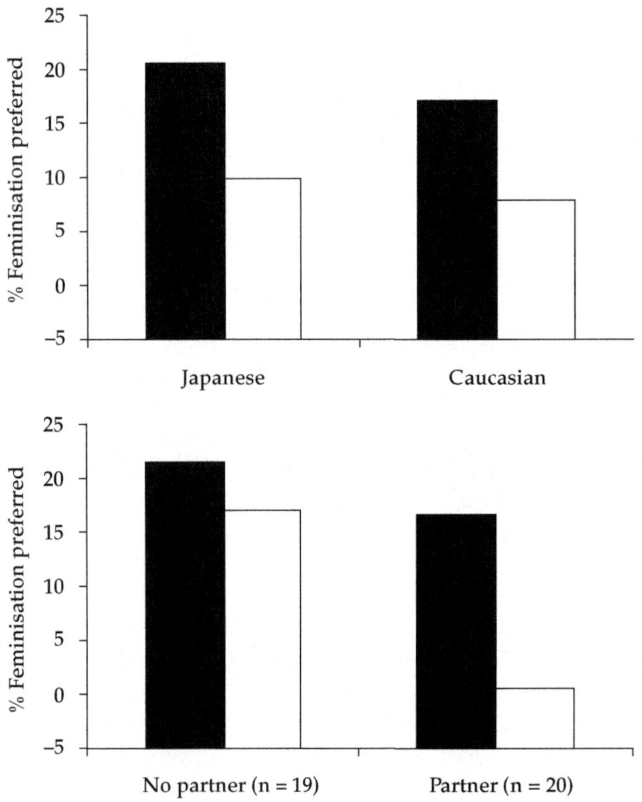

Figure 9.2 Conception risk and its effect on preference for femininity in male faces by Japanese females in Japanese and Caucasian faces (upper; open bars: non-fertile phase, shaded bars: fertile phase). Preferences during high and low conception risk phases for subjects with and without a partner (lower), data for Japanese and Caucasian faces combined. Redrawn from Penton-Voak *et al.*, 1999b.

et al. 2007) as well as with earlier onset of sexual activity and the number of short-term sexual partners (Rhodes *et al.* 2005). Fluctuating asymmetry in body traits correlates negatively with facial attractiveness (Gangestad *et al.* 1994), indicating an underlying connection between the two traits (see also Prokosch *et al.* (2005) for an extension of this principle to intelligence). Levels of fluctuating asymmetry also correlate negatively with weight (Manning, 1995), suggesting that only the most symmetrical males can achieve and maintain large body size. As in faces, low fluctuating body asymmetry correlates with low susceptibility to infection (Thornhill and Gangestad 2006).

For body height (Pawlowski *et al.* 2000), muscular body shapes (Barber 1995), and waist to hip ratio (Singh 1995), females tend to prefer intermediate levels and avoid extremes. Several studies demonstrate a general preference for a 'V-shape' torso with wider shoulders than waist (e.g. Hughes and Gallup 2003). Chest and body hair also appear to increase ratings of attractiveness (Dixson *et al.* 2003), perhaps because it **signals** high masculinity, dominance, and levels of testosterone. As in faces, the strength of preference expression for body markers such as height varies depending on the stage of the woman's cycle and the context of the judgements being made, namely her expectations of the length of the relationship (Pawlowski and Jasienska 2005).

A number of studies have examined another putative marker of testosterone, the ratio of the second to fourth digits (2D:4D), which is thought to reflect the level of the hormone experienced *in utero* (Manning 2002). Although this ratio is unlikely to be used as a proximate cue in judgements of mate attractiveness, we might predict that it will correlate with other phenotypic features if it indeed taps the same form of underlying quality. Consistent with this, 2D:4D ratio is correlated with facial asymmetry (Fink *et al.* 2004).

9.2.3 Voice

In one of the first studies that used vocal characteristics to study attractiveness, Collins (2000) found that men with voices characterized by having closely spaced, low-frequency harmonics were judged as being more attractive, older and heavier, more likely to have a hairy chest, and of a more muscular body type. Although estimates of male characteristics based on their voices were incorrect (with the exception of body weight), she concluded that the preference for deeper voices could be subject to **sexual selection** through female choice. Women agreed over which voices were attractive or unattractive. Subsequently, other studies have found significant correlations between body dimensions and vocal characteristics such as fundamental frequency and formant dispersion (see Evans *et al.* 2006), and these can be manipulated to increase attractiveness ratings (Feinberg *et al.* 2005). Male vocal attractiveness has been shown to be positively correlated with attractiveness of faces (Saxton *et al.* 2006), body shape (Hughes *et al.* 2004), and ratings of social dominance (Puts *et al.* 2006). Studies have also shown that individual women experience cyclical shifts in preference across the menstrual cycle, with preference for lower-pitched male voices around the time of ovulation (Feinberg *et al.* 2006). The attractiveness of women's voices also correlates with their facial attractiveness (Collins and Missing 2003).

9.2.4 Odour

Although humans have been described as microsmatic (having a poor sense of smell compared to other mammalian species), evidence is accumulating to show that odour can influence human behaviour to a greater extent than previously realized (Schaal and Porter 1991). Despite this, odour is perhaps the exception to the traits listed above in that it probably requires more intimate, close-range, and prolonged association in order to fully gauge its effects (for a complementary discussion of odour in rodents see Chapter 6 and for insects see Chapter 5). Nonetheless, in mate choice contexts, attractiveness ratings of body odour are correlated with body symmetry (Rikowski and Grammer 1999), facial attractiveness (Rikowski and Grammer 1999; Thornhill *et al.* 2003), heterozygosity at MHC genes (Thornhill *et al.* 2003), and measures of psychological dominance (Havlicek *et al.* 2005). Once again, these effects may be dependent upon the relationship context of the women judges and the phase of their menstrual cycle. Additionally, using synthetic steroid chemicals associated with male body odour, Cornwell *et al.* (2004) have demonstrated that the extent to which individual women prefer this masculine odour correlates with their expressed preference for masculinity in faces, indicating a concordant preference for potential good-gene indicators across two different sensory modalities.

9.3 Behavioural attributes

Research into non-physical traits or behaviour in human communication has a long history. Early work was mostly ethological in nature, focusing on general behavioural patterns and conducted by researchers who would describe themselves as anthropologists, psychologists, or ethologists. Notable examples can be found in Eibl-Eibesfeldt (1970), Hinde (1972), Mehrabian (1972), and Kendon *et al.* (1975). Similarly, work on human facial expression was led by psychologists such as Ekman (1982). The field has made enormous progress in the interceding years. Specialists use the terms paralanguage (non-verbal aspects of language, such as voice pitch and volume) and kinesics (facial expressions, postures, gestures, etc.) as the two non-verbal arms of the triune nature of human communication. These are now applied in an interdisciplinary fashion to a wide variety of communication contexts, including language interpretation, business communication, and the dramatic arts (e.g. Poyatos 2002).

Such progress has not been entirely matched by research into human courtship behaviour. Although there are many popular accounts of how to interpret body language, these are based on relatively few rigorous studies (cf. Grammer *et al.* 1999). Evolutionary psychologists who have thrown themselves headlong into the task of revealing the physical traits underlying human attractiveness have, with very few exceptions, appeared to ignore or even avoid these non-physical aspects of attraction and courtship, especially kinesics. This is probably largely due to a pervasive, but not completely accurate, perception that such communicative information can be easily faked and that it is therefore unreliable and unlikely to be used in decisions about mate choice. I will deal with this problem in more detail later in this chapter.

Whatever the reason for it, however, it is clear that this is an area which deserves more enquiry. The kinds of cues reviewed above, which are available almost at first sight, constitute only the foundation of an impression of a potential mate. This initial impression is then developed and refined, sometimes radically altered. Since almost all of the recent research addresses physical traits of unknown individuals—the zero-acquaintance paradigm—a large proportion of the complexity of mate choice may be being missed. Only in the past few years has the apparent reluctance of evolutionary psychologists to engage with kinesics been eroded, and it now receiving growing attention. Gangestad and Scheyd (2005), for example, include non-physical features as one of their five critical topics for future research in human mate choice. In the next section, I review some of the recent research which has begun to explore the contribution of kinesics to our understanding of perceptions of mate quality.

9.3.1 Communicative value of kinesic information

Humans possess the ability to infer critical socially relevant information from body posture, movement, and expression. A good example of this comes from Archer and Akert's (1977) study in which participants watched video clips of social interactions and were later asked questions about the clips, for each of which there was an unambiguously right or wrong answer. One such example was a clip of two women playing with a baby; at the end of the clip, participants were asked to judge which was the mother. A second group of participants were asked to make the same judgements but were shown only a verbal transcript of the encounters; they did not see the clips. While the second group actually performed worse than expected by chance, the group who both heard the conversations and observed the interactions did significantly better.

The potential for kinesic information to similarly provide useful information concerning mate value can be illustrated by a number of studies. A notable example is that of Brown *et al.* (2005), who set out to test whether dance, a dynamic display used in

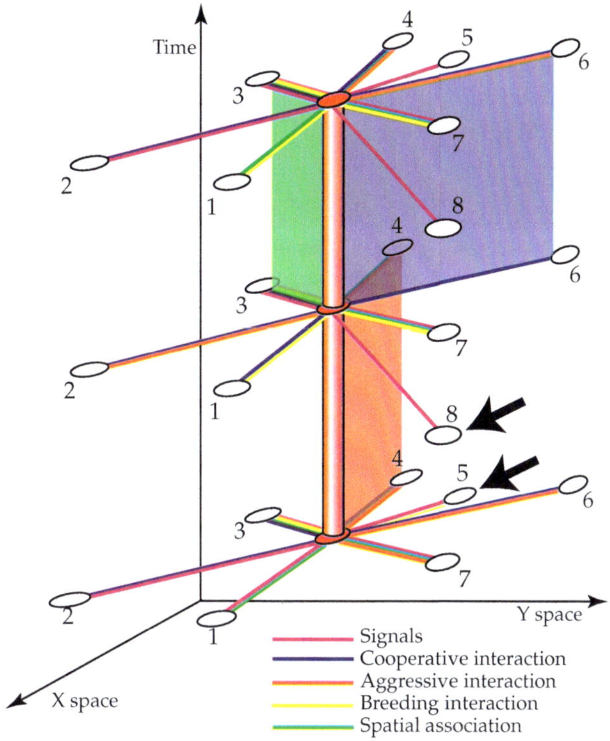

Plate 1 Graphic representation of the Rainbow Networks Model. At each moment in time the focal individual (red circle) is connected to a number of other individuals (open numbered circles) through different types of link. Some link types are more persistent (shaded colour areas) than others. Communication networks, which are based on signals (purple lines), change rapidly because individuals (n. 5 and 8, marked by arrows) join and leave the network according to their signalling and receiving status. Only bi-dimensional space (x and y) and time coordinate axes are represented for graphical simplicity. (see page 36)

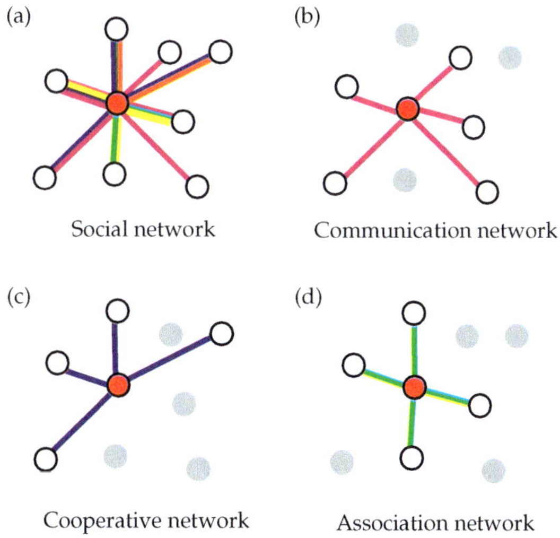

Plate 2 The social network (a) of the focal individual (red circle) can be filtered according to the different link types (b. signals; c. cooperation events; d. spatial association) to produce different types of network. Some individuals will be included (open circles) or excluded (shaded circles) from each network according to the type of interaction they have with the focal individual. The communication network (b) of an individual is particularly relevant for information flow since information exchange is the specific function of signals. (see page 37)

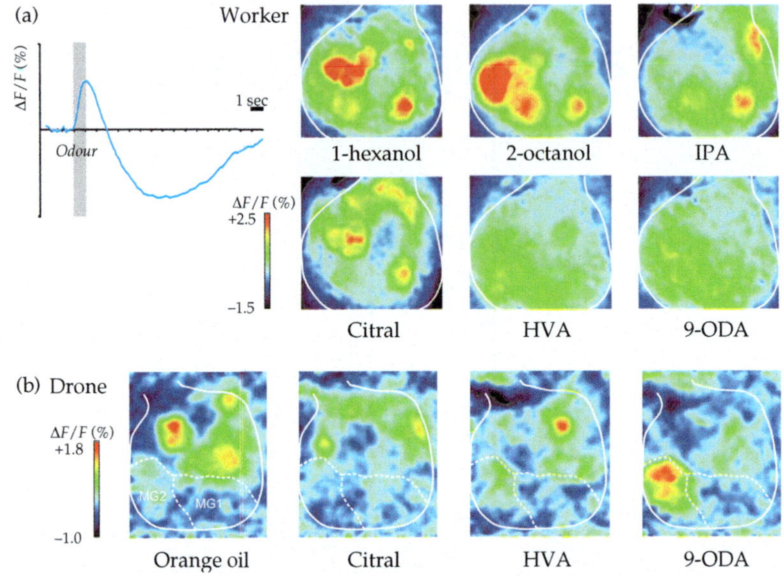

Plate 3 Physiological responses of the antennal lobe in workers and drones. a) Calcium imaging recording (using bath-applied Calcium Green 2AM) in a worker bee. **Upper left**: upon odour delivery (grey bar), a biphasic fluorescence signal is observed in active glomeruli, with a first fast positive component (max after ~1 sec), followed by a slow—highly spatially correlated—negative component (minimum after 8–10 sec). **Right**: Odour activity maps, showing for each pixel in a false-color code the amplitude of the biphasic signal. General odours (1-hexanol and 2-octanol) and social pheromone (isopentyl acetate (IPA) and citral) elicit combinatorial activity in the imaged glomeruli. Note that the glomeruli activated by the pheromones can be active in response to general odours and vice-versa. By contrast, no clear signals appeared with components of the queen mandibular pheromone (here HVA and 9-ODA). We believe that the glomeruli responsible for processing of these signals are in other—yet non-imaged—parts of the antennal lobe. b) Calcium imaging recordings (using bath-applied Calcium Green 2AM) of antennal lobe activity in a drone bee. The odour activity maps are calculated as in a). The position of the two accessible macroglomeruli is overlaid on the maps (white). General odours (here a complex blend, orange essential oil) and social pheromones (here citral) induce activity in ordinary-sized glomeruli, i.e. on the medio-ventral side of the antennal lobe. Interestingly, the major component of the queen mandibular pheromone, 9-ODA, which is involved in the attraction of males towards queens during nuptial flight, is specifically detected by the most voluminous macroglomerulus of the drone antennal lobe, MG2. By contrast, the component HVA, the pheromonal role of which is only proven in workers, induces activity in an ordinary-size glomerulus (for details, see Sandoz, 2006). (see page 129)

human courtship, might reveal information about mate quality. They specifically tested whether the dancing ability of men and women, as judged by participants from the same Jamaican population, correlated with body asymmetry. The results indicated a significant association between dancing ability and asymmetry, particularly in men, consistent with the idea that dance could be used as a reliable indicator of mate quality. Importantly, they employed in their study a form of motion-capture technology that enabled them to control for other visual cues such as attractiveness, symmetry, and height (Fig. 9.3), thus demonstrating a specific effect of motion and rhythm on the rater's perception of the video clips. A subsequent study which used simple video clips of dancers has also shown that men with low 2D:4D ratios (i.e. masculine ratios) were also judged more attractive than men with more feminized ratios (Fink *et al.* 2007b).

If non-physical cues such as movement are used in assessing mate quality in a way similar to that described for physical cues, then we might expect some degree of plasticity in judgements according to the context in which the cues are judged. Very little work has been done in this area, but Gangestad *et al.* (2004) have found just such an effect. Male participants in an experiment ostensibly about relationship formation, in which they were led to believe they were competing with another male for a date with an attractive female, were videoed responding to questions by the 'potential date' and then by the 'competitor' (both were confederates). Trained raters then coded responses in the two conditions for the number of certain non-verbal displays and behavioural tactics in order to obtain scores for each man on dimensions of 'social presence' (from interaction with the 'date') and 'direct intrasexual competitiveness' (interaction with the 'competitor'). A panel of independent female raters then observed segments of these video clips and made attractiveness judgements of the males as either a potential short-term or long-term partner. Gangestad *et al.* correlated the preference for men with high social presence and competitiveness with the menstrual cycle phase of the women raters, revealing that predilection for these male cues increased on high-fertility days, and only in the short-term relationship context. This result is reminiscent of those previously described for physical traits.

Very little research has examined the effect of male behavioural patterns on actual success in attracting female attention. However, Renninger *et al.* (2004) aimed to do just that in an observational study in a bar. They hypothesized that six specific behavioural patterns might be important in eliciting female attention: intrasexual touch, space-maximization movements, open body positioning, glancing, gesticulation, and automanipulation. Their study indicated that most of these behavioural patterns did indeed differ between men who were successful or not in establishing conversational contact with at least one woman, and that some of these were expressed at higher frequencies when women were present than when they were not.

9.3.2 Dynamic versus static stimuli

The previous section suggests that information available in dynamic, non-physical cues could potentially influence mate preferences in similar

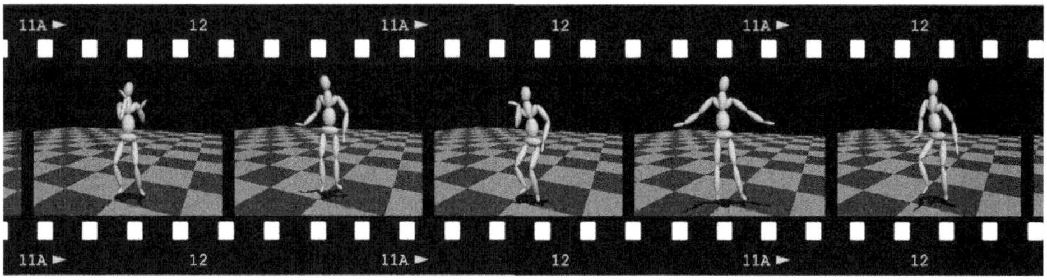

Figure 9.3 Example frames from video clips used by Brown *et al.* (2005) to show that dancing ability is correlated with body symmetry. The clips are manipulated from actual footage of dancers such that other physical cues (e.g. gender, identity, attractiveness) are removed. Image reproduced with permission of William M. Brown, Keith Grochow, Karen Liu, Zoran Popović and Robert Trivers.

fashion to purely physical traits. Is there then any advantage to using dynamic stimuli, which will usually be a more challenging logistical task? The answer would appear to be 'yes'. In a study which compared the contributions to judgements of overall attractiveness, Riggio *et al.* (1991) found that facial attractiveness (judged from static images) and expressive behaviour/style (gleaned from video clips) both independently explained overall judgements. Furthermore, Rubenstein (2005) reported a surprisingly low correlation between judgements of attractiveness based on video clips of participants reading from a card and judgements based on a single, standardized frame taken from the same clip. He concludes that the two types of presentation are evaluated in different ways and that dynamic displays in particular yield emotional content that is unavailable from static images. However, while these conclusions may well be true, the surprisingly low correlation between dynamic and static judgements may be an artefact of the between-subject design of this experiment: raters saw each face in only one format. In contrast, within-subject judgements are highly correlated (S. C. Roberts, unpublished data).

Although the dynamic stimulus studies potentially go further than those based on static stimuli, those of Riggio *et al.* and Rubenstein are based on the zero-acquaintance paradigm; in other words, they deal with the effects of kinesics over a very short period. This is entirely appropriate as an experimental design, but does little to tell us how these dynamic cues add or subtract to the initial judgement. This is certainly a question for the future, but an interesting study using actual acquaintances and attractiveness judgements neatly demonstrates the potential for describing longer-term development of initial impressions. Kniffin and Wilson (2004) explored the extent to which zero-acquaintance attractiveness ratings (e.g. from high-school yearbook photographs) compared with the attractiveness ratings of former classmates, who had obviously assimilated additional information about the people being judged. Although ratings by strangers accounted for a large proportion of the variance in the classmate ratings (60% for males rating females and 38% for females rating males), there was a proportion that remained unexplained, and this varied a great deal from person to person. In a separate experiment, they asked participants, who were unfamiliar with each other, to judge each other's attractiveness and other attributes at the beginning and end of a 6-week field course. Initial judgements of physical attractiveness accounted for a relatively small proportion of the variance in final judgements for females judging males, but a large proportion for males judging females.

Does dynamic information then provide better and more reliable estimates of underlying genetic quality than simple static cues that have been so widely used in recent research? This question may be difficult to answer, but one way might be to look for clearer relationships between attractiveness judgements and putative good-gene indicator traits when the judgements are made using dynamic stimuli. One such example is the relationship between 2D:4D ratio and attractiveness, which was positively correlated when attractiveness was judged from clips of short interactive conversations (Roney and Maestripieri 2004) but not in another study which used photographs (Neave *et al.* 2003). Alternatively, the answer may lie in the congruency of different categories of basic information, including that between statically and dynamically communicated cues (Johnson and Tassinary 2007).

9.3.3 The problem of honesty

It is clear that for most physical cues we have to make do with what we have. We cannot change the shape of our face or our body size, the sound of our voice, or the smell of our armpits. Modern cultural practices such as platform shoes, shoulder pads, facial cosmetics, perfumes, and cosmetic surgery allow us to tinker at the edges of our phenotypic inheritance, and the wide use of such practices certainly provides interesting insight into the design features of attractive bodies. However, these manipulations usually have relatively minor effect beyond very initial impressions and may often be rapidly 'discovered'. (Note that most studies of human indicator traits attempt to rigorously control for most of these cultural confounds.) The physical traits reviewed earlier in this chapter can thus be seen as providing reliable information about the underlying genetic quality of a potential partner, in the same way as the train of a peacock or odour of a mouse (Petrie 1994; Roberts and Gosling 2003).

But, as I have already touched upon, it is somewhat less clear that the information carried in dynamic displays can be considered reliable in this sense. This is probably the principal reason why approaches incorporating kinesic information have been so neglected in evolutionary psychology research (although to a lesser extent this may also be explained by differences in technological prerequisites). Although the relative imbalance in focus on static and dynamic cues is beginning to change, there remains engrained a sense that kinesics does not fully belong as a legitimate approach to the study of mate choice, or at best it can be relevant only in a narrowly defined subset of behavioural patterns. Neither a clear understanding of the limits to which kinesic cues can be used in reliable mate assessment, nor a conceptual framework which defines the conditions under which such cues can be reliable, currently exists.

Signalling theory proposes several mechanisms through which biological **signals** can be reliable (Maynard Smith and Harper 2003; see also Chapter 1). These include the ideas that signals are inherently costly to produce (the handicap principle), that a signal (costly or otherwise) which does not accurately reflect the quality of the signaller will invoke socially imposed costs, and that the very nature of some signals cannot possibly be faked (index signals: Maynard Smith and Harper 2003). Any example of kinesic or paralinguistic information being used as reliable information about mate quality would in all likelihood need to fulfil one of these three criteria. However, attributing an observed signal to a particular kind of cost is not straightforward: see, for example, the discussion concerning symmetry and ornaments in Maynard Smith and Harper (2003). If we take the example of the recently described link between fluctuating asymmetry and dancing ability of young men (Brown *et al.* 2005), it could be that dancing is a handicap, if the ability to dance well is energetically dependent and only the best males can afford to invest energy in sufficient quantity to produce a pleasing dance; or, perhaps more likely, it could be an index signal, if dancing ability is directly linked to symmetry or health. Thus, while costs are certainly important, it is evident that an understanding of signalling reliability at this level is not especially useful for defining behavioural patterns that can be used reliably in mate choice decisions.

I propose that an alternative, more productive and accessible, approach can be formulated by drawing on several emergent themes from the literature on the attractiveness of physical traits, which I reviewed earlier in this chapter (see also Roberts and Little 2008). If these themes were generalized to non-physical traits, we could make the following predictions:

1. Expression of reliable non-physical traits will correlate with attractive physical traits. The basis for this prediction is that if there is some variability between individuals in underlying genetic quality, and this quality is manifest in more than one phenotypic characteristic, then variability in the attractiveness dimension for the different traits should be intercorrelated. For example, ratings of male body odour (Rikowski and Grammer 1999; Thornhill *et al.* 2003) and voices (Saxton *et al.* 2006) co-vary with facial attractiveness ratings. Similar cross-correlations also occur between body (Thornhill and Grammer 1999) or voice attractiveness (Collins and Missing 2003) and women's faces. The same principle should apply to non-verbal behaviours if they indicate genetic quality: they should correlate with both attractive physical traits and also with other attractive non-physical traits.

2. Relative to same-sex individuals, trait expression should be temporally consistent. As we have discussed, physical cues such as masculine facial shape are, by their very nature, consistent over time. They may vary in the long-term, because of ageing or onset of disease for example, but on a day-to-day basis they are relatively constant. In contrast, non-verbal cues can alter much more fluidly, thus allowing the possibility of deception. If a non-verbal trait is to be potentially reliable, we would expect it to be consistently expressed over several independent occasions with similar social relevance. Patterns of expression of the trait may very likely vary with different social contexts (e.g. flirtatious behaviours will be more common in certain situations than others), but within these contexts the degree of expression relative to that of other males should be relatively consistent. This will be especially true where a degree of social policing or comparison is possible, as for example where potential competitors are present who are of both higher and lower mate value. Individual females do not necessarily need to observe specific males over this period of time for the trait to be reliable, since the premise of reliability in the sense of good genes is that it is relatively fixed. Furthermore it could be enforced by intrasexual selection.

3. Preference for the trait should be condition-dependent. Studies of physical traits repeatedly demonstrate that between-individual variation in the strength of preference for specific indicator traits correlates with measures of individual condition. For example, expression of preferences of facial traits such as masculinity, symmetry, and healthiness in male faces is predicted by the self-rated attractiveness or body attractiveness of female raters (Little *et al.* 2001; Penton-Voak *et al.* 2003; Jones *et al.* 2005). This makes sense if we visualize mate choice as a biological market, where differences in mate value dictate the range of choice open to individuals (e.g. Pawlowski and Dunbar 1999). In the same way, preferences for reliable non-verbal traits should also be expected to be condition-dependent.

4. Preference should be context-dependent. Within-individual shifts in preference for physical traits occur depending on the relationship context under which judgements are made, that is, whether raters are asked to make attractiveness judgements for short-term or long-term partners (e.g. Little *et al.* 2002; Roberts *et al.* 2005a). In addition, for females only, the raters' stage of menstrual cycle is also important (e.g. Penton-Voak *et al.* 1999; Havlicek *et al.* 2005). The theoretical basis for this is that preference shifts reflect the balance of indirect and direct benefits of choice (Gangestad and Simpson 2000). In the same way, if non-verbal cues reliably signal either genetic quality or some aspect of likely paternal investment, the same kinds of shifts in attention should be observed. Indeed, there is already some evidence for variation in preference shifts for non-verbal behaviour depending on the stage of the menstrual cycle (Gangestad *et al.* 2004).

5. Preferences should normally be culturally independent. There is more agreement regarding the attractiveness of physical traits across cultures and ethnicities than there is disagreement (Perrett *et al.* 1998; Langlois *et al.* 2000). While non-verbal behaviours may very well be more labile with respect to cultural influences, reliable components should be among the most robustly conserved across cultural and ethnic groups.

These five predictions provide a basis for extending the scope of investigation into phenotypic indicator traits from purely physical cues to include non-physical traits and behavioural patterns. Note that some of these predictions may also apply to other important aspects of mate choice which do not involve good-gene indicators, including preference for genetic dissimilarity or assortative preferences for traits such as personality. Thus, if a particular behavioural pattern does not fulfil most or all of these predictions, it remains possible that they could nonetheless reflect some important attribute other than good genes that may still be attractive and used in the formation of preferences. In this chapter I have not addressed assortative preferences because, although this is an important and emerging genetically based component of individual decisions (see for example Wedekind *et al.* 1995; Roberts and Gosling 2003; Mays and Hill 2004), the interpretation of mate quality is then specific to the individual. A review of this aspect of mate choice as it applies to humans has recently been published elsewhere (Roberts and Little 2008). The development of a set of predictions that pertains to such alternative considerations within mate choice is beyond the scope of this chapter but remains a challenging task for the future.

Finally, how should females behave in order to benefit from judicious choice based on non-physical traits? The reliability of any trait can be increased in three key ways. First, duration of observation of a particular male will be an important factor allowing ongoing adjustment of the estimate of mate quality. Second, the precision of this estimate is likely to be markedly increased if the target male(s) is unaware he is being observed (Locke 2005). Third, during the early stages of any initiated contact, females may probe the male to check whether their estimate appears reliable. Indeed this will be an ongoing process throughout courtship and beyond.

9.4 Conclusion

There is a huge diversity of approaches and volume of research that addresses the possibility in humans that underlying genetic quality can be perceived and acted upon. Physical characteristics, such as facial or bodily appearance, are generally accepted to be as useful proximate indicators of quality in humans as ornaments and weaponry are in animals. The influence of dynamic, non-physical traits is much more controversial, but the attention of researchers is gradually turning to this interesting area. Defining the limits of such enquiry, in terms of what kinds of behavioural traits can be useful and reliable cues of mate quality, is a key step. It is hoped that the predictions generated here will stimulate further progress.

Summary

In most human societies, individuals make comparisons between numerous potential mates. Recent research on the biological determinants of mate preferences explores the idea that attractive physical characteristics might be cues of underlying good genes. The first half of this chapter summarizes this work on physical cues of mate quality, including facial, bodily, vocal, and olfactory traits. In the second half, I speculate on how broad principles that arise out of this research might be directly transposed to helps us understand the potential effects of good genes on behaviour and 'body language'. The reliability of behavioural cues betraying mate quality is likely to determine how far we can apply biological interpretations on behaviour, and an outline for how researchers might tackle this issue is proposed.

Acknowledgements

I was stimulated to think about reliability of non-physical traits by staff and students at the Evolutionary Biology of Communication PhD course in June 2007 at the University of Copenhagen. I thank them all. I also thank John Lycett for helpful discussions.

References

Andersson, M. (1994). *Sexual Selection*. Princeton University Press, Princeton, NJ.

Archer, D. and Akert, R.M. (1977). Words and everything else: verbal and nonverbal cues in social interpretation. *Journal of Personality and Social Psychology*, **35**, 443–449.

Barber, N. (1995). The evolutionary psychology of physical attractiveness: asexual selection and human morphology. *Ethology and Sociobiology*, **16**, 395–424.

Bateson, M. and Healy, S.D. (2005). Comparative evaluation and its implications for mate choice. *Trends in Ecology and Evolution*, **20**, 659–664.

Bradbury, J. and Gibson, R. (1983). Leks and mate choice. In: P. Bateson (ed.), *Mate Choice*, pp. 109–138. Cambridge University Press, Cambridge.

Brown, W.M., Cronk, L., Grochow, K., Jacobson, A., Liu, C.K., Popovic, Z., and Trivers, R. (2005). Dance reveals symmetry especially in young men. *Nature*, **438**, 1148–1150.

Buston, P.M. and Emlen, S.T. (2003). Cognitive processes underlying human mate choice: the relationship between self-perception and mate preference in Western society. *Proceedings of the National Academy of Sciences of the USA*, **100**, 8805–8810.

Collins, S.A. (2000). Men's voices and women's choices. *Animal Behaviour*, **60**, 773–780.

Collins, S.A. and Missing, C. (2003). Vocal and visual attractiveness are related in women. *Animal Behaviour*, **65**, 997–1004.

Cornwell, R.E., Boothroyd, L., Burt, D.M. *et al.* (2004). Concordant preferences for opposite-sex signals? Human pheromones and facial characteristics. *Proceedings of the Royal Society B: Biological Sciences*, **271**, 635–640.

Daly, M. and Wilson, M. (1983). *Sex, Evolution and Behavior*. PWS Publishers, Boston, MA.

Dixson, A.F., Halliwell, G., East, R., Wignarajah, P., and Anderson, M.J. (2003). Masculine somatotype and hirsuteness as determinants of sexual attractiveness to women. *Archives of Sexual Behavior*, **32**, 29–39.

Eibl-Eibesfeldt, I. (1970). *Ethology, the Biology of Behaviour*. Holt, Rinehart, and Wilson, New York.

Ekman, P. (1982). *Emotion in the Human Face*. Cambridge University Press, Cambridge.

Evans, S., Neave, N., and Wakelin, D. (2006). Relationships between vocal characteristics and body size and shape in human males: an evolutionary explanation for a deep male voice. *Biological Psychology*, **72**, 160–163.

Feinberg, D.R., Jones, B.C., Little, A.C., Burt, D.M., and Perrett, D.I. (2005). Manipulations of fundamental and formant frequencies influence the attractiveness of human male voices *Animal Behaviour*, **69**, 561–568.

Feinberg, D.R., Jones, B.C., Law-Smith, M.J. *et al.* (2006). Menstrual cycle, trait estrogen level, and masculinity preferences in the human voice. *Hormones and Behavior*, **49**, 215–222.

Fink, B., Manning, J.T., Neave, N., and Grammer, K. (2004). Second to fourth digit ratio and facial asymmetry. *Evolution and Human Behavior*, **25**, 125–132.

Fink, B., Neave, N., and Seydel, H. (2007a). Male facial appearance signals physical strength to women. *American Journal of Human Biology*, **19**, 82–87.

Fink, B., Seydel, H., Manning, J.T., and Kappeler, P.M. (2007b). A preliminary investigation of the associations between digit ratio and women's perception of men's dance. *Personality and Individual Differences*, **42**, 381–390.

Gangestad, S.W. and Scheyd, G.J. (2005). The evolution of human physical attractiveness. *Annual Review of Anthropology*, **34**, 523–548.

Gangestad, S.W. and Simpson, J.A. (2000). The evolution of human mating: trade-offs and strategic pluralism. *Behavioral and Brain Sciences*, **23**, 573–587.

Gangestad, S.W. and Thornhill, R. (2003). Facial masculinity and fluctuating asymmetry. *Evolution and Human Behavior*, **24**, 231–241.

Gangestad, S.W., Thornhill, R., and Yeo, R.A. (1994). Facial attractiveness, developmental stability, and fluctuating asymmetry. *Ethology and Sociobiology*, **15**, 73–85.

Gangestad, S.W., Simpson, J.A., Cousins, A.J., Garver-Apgar, C.E., and Christensen, P.N. (2004). Women's preferences for male behavioral displays change across the menstrual cycle. *Psychological Science*, **15**, 203–207.

Grammer, K. and Thornhill, R. (1994). Human (*Homo sapiens*) facial attractiveness and sexual selection: the role of symmetry and averageness. *Journal of Comparative Psychology*, **108**, 233–242.

Grammer, K., Honda, M., Juette, A., and Schmitt, A. (1999). Fuzziness of nonverbal courtship communication unblurred by motion energy detection. *Journal of Personality and Social Psychology*, **77**, 487–508.

Grammer, K., Fink, B., Møller, A.P., and Thornhill, R. (2003). Darwinian aesthetics: sexual selection and the biology of beauty. *Biological Reviews*, **78**, 385–407.

Grammer, K., Fink, B., Møller, A.P., and Manning, J.T. (2005). Physical attractiveness and health: comment on Weeden and Sabini (2005). *Psychological Bulletin*, **131**, 658–661.

Havlicek, J., Roberts, S.C., and Flegr, J. (2005). Women's preference for dominant male odour: effects of menstrual cycle and relationship status. *Biology Letters*, **1**, 256–259.

Hinde, R.A. (1972). *Nonverbal Communication*. Cambridge University Press, Cambridge.

Honekopp, J., Rudolph, U., Beier, L., Liebert, A., and Muller, C. (2007). Physical attractiveness of face and body as indicators of physical fitness in men. *Evolution and Human Behavior*, **28**, 106–111.

Hughes, S.M. and Gallup, G.G. (2003). Sex differences in morphological predictors of sexual behavior—Shoulder to hip and waist to hip ratios. *Evolution and Human Behavior*, **24**, 173–178.

Hughes, S.M., Dispenza, F., and Gallup, G.G. (2004). Ratings of voice attractiveness predict sexual behavior and body configuration. *Evolution and Human Behavior*, **25**, 295–304.

Jacob, S., McClintock, M.K., Zelano, B., and Ober, C. (2002). Paternally inherited HLA alleles are associated with women's choice of male odor. *Nature Genetics*, **30**, 175–179.

Johnson, K.L. and Tassinary, L.G. (2007). Compatibility of basic social perceptions determines perceived attractiveness. *Proceedings of the National Academy of Sciences of the USA*, **104**, 5246–5251.

Jones, B.C., Little, A.C., Penton-Voak, I.S., Tiddeman, B.P., Burt, D.M., and Perrett, D.I. (2001). Facial symmetry and judgements of apparent health – support for a 'good genes' explanation of the attractiveness–symmetry relationship. *Evolution and Human Behavior*, **22**, 417–29.

Jones, B.C., Little, A.C., Burt, D.M., and Perrett, D.I. (2004a). When facial attractiveness is only skin deep. *Perception*, **33**, 569–576.

Jones, B.C., Little, A.C., Feinberg, D.R., Penton-Voak, I.S., Tiddeman, B.P., and Perrett, D.I. (2004b). The relationship between shape symmetry and perceived skin condition in male facial attractiveness. *Evolution and Human Behavior*, **25**, 24–30.

Jones, B.C., Little, A.C., Boothroyd, L. *et al.* (2005). Women's physical and psychological condition independently predict their preference for apparent health in faces. *Evolution and Human Behavior*, **26**, 451–457.

Kendon, A., Harris, R.M., and Key, M.R. (eds) (1975). *The Organization of Behavior in Face-to-face Interaction* Mouton, The Hague.

Kniffin, K.M. and Wilson, D.S. (2004). The effect of nonphysical traits on the perception of physical attractiveness – three naturalistic studies. *Evolution and Human Behavior*, **25**, 88–101.

Langlois, J.H., Kalakanis, L., Rubenstein, A.J., Larson, A., Hallamm, M., and Smoot, M. (2000). Maxims or myths of beauty? A meta-analytic and theoretical review. *Psychological Bulletin*, **126**, 390–423.

Little, A.C. and Jones, B.C. (2006). Attraction independent of detection suggests special mechanisms for symmetry preferences in human face perception. *Proceedings of the Royal Society B: Biological Sciences*, **273**, 3093–3099.

Little, A.C., Burt, D.M., Penton-Voak, I.S., and Perrett, D.I. (2001). Self-perceived attractiveness influences human female preferences for sexual dimorphism and symmetry in male faces. *Proceedings of the Royal Society B: Biological Sciences*, **268**, 39–44.

Little, A.C., Jones, B.C., Penton-Voak, I.S., Burt, D.M., and Perrett, D.I. (2002). Partnership status and the temporal context of relationships influence human female preferences for sexual dimorphism in male face shape. *Proceedings of the Royal Society B: Biological Sciences*, **269**, 1095–1100.

Little, A.C., Penton-Voak, I.S., Burt, D.M., and Perrett, D.I. (2003). Investigating an imprinting-like phenomenon in humans: partners and opposite-sex parents have similar hair and eye colour. *Evolution and Human Behavior*, **24**, 43–51.

Locke, J.L. (2005). Looking for, looking at: social control, honest signals and intimate experience in human evolution and history. In: P.K. McGregor (ed.), *Animal Communication Networks*, pp. 416–441. Cambridge University Press Cambridge.

Manning, J.T. (1995). Fluctuating asymmetry and body weight in men and women: implications for sexual selection. *Ethology and Sociobiology*, **16**, 145–153.

Manning, J.T. (2002). *Digit Ratio: a Pointer to Fertility, Behaviour and Health*. Rutgers University Press, New Brunswick, NJ.

Maynard Smith, J. and Harper, D. (2003). *Animal Signals*. Oxford University Press, Oxford.

Mays, H.L.J. and Hill, G.E. (2004). Choosing mates: good genes versus genes that are a good fit. *Trends in Ecology and Evolution.*, **19**, 555–559.

Mehrabian, A. (1972). *Nonverbal Communication*. Aldine, Chicago.

Miller, G. (2000). *The Mating Mind*. William Heinemann, London.

Møller, AP (1992). Female swallow preference for symmetrical male sexual ornaments. *Nature*, **357**, 238–240.

Neave, N., Laing, S., Fink, B., and Manning, J.T. (2003). Second to fourth digit ratio, testosterone and perceived male dominance. *Proceedings of the Royal Society B: Biological Sciences*, **270**, 2167–2172.

Pawlowski, B. and Dunbar, R.I.M. (1999). Impact of market value on human mate choice decisions. *Proceedings of the Royal Society B: Biological Sciences*, **266**, 281–285.

Pawlowski, B. and Jasienska, G. (2005). Women's preferences for sexual dimorphism in height depend on menstrual cycle phase and expected duration of relationship. *Biological Psychology*, **70**, 38–43.

Pawlowski, B., Dunbar, R.I.M., and Lipowicz, A. (2000). Evolutionary fitness—tall men have more reproductive success. *Nature*, **403**, 156.

Penton-Voak, I.S. and Chen, J.Y. (2004). High salivary testosterone is linked to masculine male facial appearance in humans. *Evolution And Human Behavior*, **25**, 229–241.

Penton-Voak, I.S., Perrett, D.I., Castles, D.L., Kobayashi, T., Burt, D.M., Murray, L.K., and Minamisawa, R. (1999). Menstrual cycle alters face preference. *Nature*, **399**, 741–742.

Penton-Voak, I.S., Jones, B.C., Little, A.C., Baker, S., Tiddeman, B.P., Burt, D.M., and Perrett, D.I. (2001). Symmetry, sexual dimorphism in facial proportions and male facial attractiveness. *Proceedings of the Royal Society B: Biological Sciences*, **268**, 1617–1623.

Penton-Voak, I.S., Little, A.C., Jones, B.C., Burt, D.M., Tiddeman, B.P., and Perrett, D.I. (2003). Female condition influences preferences for sexual dimorphism in faces of male humans (*Homo sapiens*). *Journal of Comparative Psychology*, **117**, 264–271.

Perrett, D.I., Lee, K.J., Penton-Voak, I.S. *et al.*(1998). Effects of sexual dimorphism on facial attractiveness. *Nature*, **394**, 884–887.

Peters, M., Rhodes, G., and Simmons, L.W. (2007). Contributions of the face and body to overall attractiveness. *Animal Behaviour*, **73**, 937–942.

Petrie, M. (1994). Improved growth and survival of offspring of peacocks with more elaborate trains. *Nature*, **371**, 598–599.

Posamentier, M.T. and Abdi, H. (2003). Processing faces and facial expressions. *Neuropsychology Review*, **13**, 113–143.

Poyatos, F. (2002). *Nonverbal Communication across Disciplines*. John Benjamins Publishing Company, Philadelphia, PA.

Prokosch, M.D., Yeo, R.A., and Miller, G.F. (2005). Intelligence tests with higher g-loadings show higher correlations with body symmetry: evidence for a general fitness factor mediated by developmental stability. *Intelligence*, **33**, 203–213.

Puts, D.A., Gaulin, S.J.C., and Verdolini, K. (2006). Dominance and the evolution of sexual dimorphism in human voice pitch. *Evolution and Human Behavior*, **27**, 283–296.

Renninger, L.A., Wade, T.J., and Grammer, K. (2004). Getting that female glance: patterns and consequences of male nonverbal behavior in courtship contexts. *Evolution and Human Behavior*, **25**, 416–431.

Reynolds, J.D. and Gross, M.R. (1990). Costs and benefits of female mate choice—is there a lek paradox? *The American Naturalist*, **136**, 230–243.

Rhodes, G. (2006). The evolutionary psychology of facial beauty. *Annual Review of Psychology*, **57**, 199–226.

Rhodes, G. and Zebrowitz, L.A. (2002). *Facial Attractiveness*. Ablex, Westport, CT.

Rhodes, G., Proffitt, F., Grady, J., and Sumich, A. (1998). Facial symmetry and the perception of beauty. *Psychonomic Bulletin and Review*, **5**, 659–669.

Rhodes, G., Simmons, L.W. and Peters, M. (2005). Attractiveness and sexual behavior: does attractiveness enhance mating success? *Evolution and Human Behavior*, **26**, 186–201.

Riggio, R.E., Widaman, K.F., Tucker, J.S., and Salinas, C. (1991). Beauty is more than skin deep—components of attractiveness. *Basic and Applied Social Psychology*, **12**, 423–439.

Rikowski, A. and Grammer, K. (1999). Human body odour, symmetry and attractiveness. *Proceedings of the Royal Society B: Biological Sciences*, **266**, 869–874.

Roberts, S.C. and Gosling, L.M. (2003). Genetic similarity and quality interact in mate choice decisions by female mice. *Nature Genetics*, **35**, 103–106.

Roberts, S.C. and Little, A.C. (2008). Good genes, complementary genes and human mate preference. *Genetica*, **132**, 309–321.

Roberts, S.C., Little, A.C., Gosling, L.M. *et al.* (2005a). MHC-assortative facial preferences in humans. *Biology Letters*, **1**, 400–403.

Roberts, S.C., Little, A.C., Gosling, L.M. *et al.* (2005b). MHC-heterozygosity and human facial attractiveness. *Evolution and Human Behavior*, **26**, 213–226.

Roney, J.R. and Maestripieri, D. (2004). Relative digit lengths predict men's behavior and attractiveness during social interactions with women. *Human Nature*, **15**, 271–282.

Rubenstein, A.J. (2005). Variation in perceived attractiveness—differences between dynamic and static faces. *Psychological Science*, **16**, 759–762.

Saxton, T.K., Caryl, P.G., and Roberts, S.C. (2006). Vocal and facial attractiveness judgments of children, adolescents and adults: the ontogeny of mate choice. *Ethology*, **112**, 1179–1185.

Schaal, B. and Porter, R.H. (1991). Microsmatic humans revisited—the generation and perception of chemical signals. *Advances in the Study of Behavior*, **20**, 135–199.

Scheib, J.E., Gangestad, S.W., and Thornhill, R. (1999). Facial attractiveness, symmetry, and cues to good genes. *Proceedings of the Royal Society B: Biological Sciences*, **266**, 1913–1917.

Singh, D. (1995). Female judgment of male attractiveness and desirability for relationships—role of waist-to-hip ratio and financial status. *Journal of Personality and Social Psychology*, **69**, 1089–1101.

Thornhill, R. and Gangestad, S.W. (1999). Facial attractiveness. *Trends in Cognitive Sciences.*, **3**, 452–460.

Thornhill, R. and Gangestad, S.W. (2006). Facial sexual dimorphism, developmental stability, and susceptibility to disease in men and women. *Evolution and Human Behavior*, **27**, 131–144.

Thornhill, R., Gangestad, S.W., Miller, R., Scheyd, G., McCullough, J.K., and Franklin, M. (2003). Major histocompatibility genes, symmetry and body scent attractiveness in men and women. *Behavioral Ecology*, **14**, 668–678.

Thornhill, R. and Grammer, K. (1999). The body and face of woman: one ornament that signals quality? *Evolution and Human Behavior*, **20**, 105–120.

Wedekind, C., Seebeck, T., Bettens, F., and Paepke, A.J. (1995). MHC-dependent mate preferences in humans. *Proceedings of the Royal Society B: Biological Sciences*, **260**, 245–249.

Weisfeld, G.E., Czilli, T., Phillips, K.A., Gall, J.A., and Lichtman, C.M. (2003). Possible olfaction-based mechanisms in human kin recognition and inbreeding avoidance. *Journal of Experimental Child Psychology*, **85**, 279–295.

Wellings, K., Collumbien, M., Slaymaker, E. *et al.* (2006). Sexual behaviour in context: a global perspective. *The Lancet*, **368**, 1706–1728.

Zhao, W., Chellappa, R., Phillips, P.J., and Rosenfeld, A. (2003). Face recognition: a literature survey. *ACM Computing Surveys*, **35**, 399–459.

CHAPTER 10

The extended phenotype within the colony and how it obscures social communication

David P. Hughes

10.1 Introduction

'This is a work of unabashed advocacy.' So Richard Dawkins began his seminal contribution to evolutionary biology entitled *The Extended Phenotype* (1982). So too, in a way, is this chapter. I do not wish to advocate the central theorem of the extended phenotype because that, in the intervening 25 years, has gained general, if not universal acceptance (Dawkins 2004). Rather, what I want to advocate is that the topic of this book, social communication, may be a misnomer if one 'individual' in a communication **network** is a chimera of two that have widely diverging interests. What this chapter is about is the intra-organismal **conflict** resulting from **parasitism** and how the phenotype, which is the originator of **signals** in all social communication, may in fact be an extended phenotype of the parasite. If so it will impact upon, and ultimately obscure, social communication.

I will advocate that if we reconsider the phenotypic features involved in social communication, be it bird song or the honeybee waggle dance, in the context of parasite extended phenotypes we will gain a greater understanding about how conflict within organisms shapes social communication among organisms. In order to do this we will need to view familiar phenomena from unfamiliar angles. Thus, I fall in step with Dawkins, whose book was not a laying down of facts to convince his audience of the generality of his theory; in fact it wasn't even a theory, but a *way to view facts* that he advocated.

I begin by defining the extended phenotype and provide examples both from biology generally and from the social insects specifically. Quite apart from the need to reacquaint readers with the extended phenotype we will need to precisely define our subject matter because of recent developments in evolutionary biology that centre on the extended phenotype. I will also briefly detail key information relating to the evolutionary biology of social insects which includes the ways in which they communicate. Once that is achieved I will encourage an alternative view of a colony of social insects as a single superorganism. This much maligned term has experienced a renaissance of late, and here I will discuss its usefulness for social communication.

10.2 The extended phenotype

The paradigm of the gene as the unit of **selection** emerged during a period of much debate between advocates of individual and group level selection and through the work of Hamilton (1963, 1964a,b). It was subsequently developed as a transparent concept by Dawkins in his **selfish gene** approach (Dawkins 1976) and became the foundation for sociobiological theory (Wilson 1975). As an historical aside it was recently emphasized that Wilson's sociobiological stance leant more towards group rather than individual selection and that it is Dawkins who deserves the major credit for the current association between sociobiology and gene level/individual selection (Segerstråle 2007). What the paradigm of the gene as the unit of selection states is that it is genes alone which are transferred between generations; the organisms in which genes reside and their phenotypes are the means by which transmission is secured. Organisms are vehicles and genes are replicators. **Natural selection** chooses among variation in phenotypes but the information encoding these phenotypes and, ultimately, the unit which is selected is the gene (see discussion by Mayr 1997).

The phenotype has principally been considered a trait of the individual organism. Examples are eye or flower colour, antler length, butterfly wing spots, behaviour, or chemical signals released into the air, to name just a few. But such foci only reflect the convenience with which we could study those easily visible attributes of organisms (Dawkins 1990). Increasing advances in cellular and chemical biology allow a fuller exploration of hitherto less obvious phenotypes of the organism such as the surface of cells, tissues, and organs (Chapter 12) or protein signatures in rodent urine (Chapter 6). Dawkins (1982) also advocated an additional level of the phenotype, but what was, and still remains, novel is that this additional level of phenotype is not physically attached to the organisms whose genes are encoding it, that is the *extended phenotype* (henceforth EP; see Fig. 10.1). The first of the three EPs to be considered was animal architecture which Von Frisch (1974) called 'frozen behaviour'. The work of Michael Hansell gives an excellent insight into this little studied, but fascinating, component of behaviour (Hansell 1996, 2004). The example *par excellence* is the beaver dam which is a physical representation of beaver behaviour that increases the **fitness** of the genes encoding the building behaviour. The second EP is parasite manipulation of host behaviour. This topic was elegantly reviewed by Janice Moore (2002). An exemplar of this field is the suicidal behaviour of crickets infected by hairworms, whereby they jump into water so the adult worm can impressively exit from the thrashing body of its drowning host (Thomas *et al.* 2002). This behaviour is controlled by parasite, and not host, genes (Biron *et al.* 2006). The third and final EP is action at a distance, and here a parasite example was used which is the manipulation of host behaviour by cuckoo chicks. In this case the chick is not physically associated with the host, as in the case of hairworms, but influences the expression of its behavioural phenotype nonetheless. Dawkins further discussed how action at a distance need not be confined to parasite–host relationships but can occur elsewhere, such as between conspecifics,

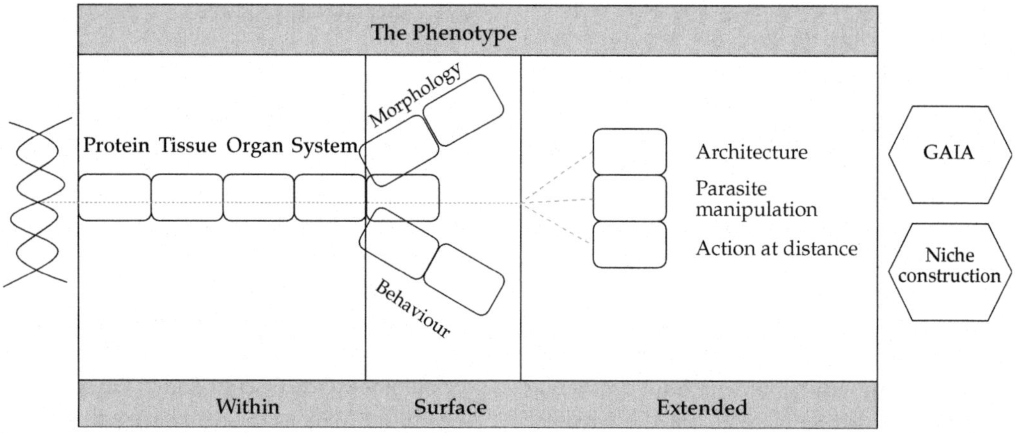

Figure 10.1 The phenotype, extended out from the gene, showing three levels: within the organism, on its surface and extended away from the organism. The dotted line links them all to the gene. Examples of morphology are colour and size and examples of behaviour are song and courtship dances. The box like representation communicates limits for the phenotype. Designations that have been illogically claimed as phenotypes are in hexagons outside the box.

as in **pheromone**-based social communication (Chapter 5).

The extension of the phenotype beyond the physical borders of the organism in which the encoding genes can be found is a logical one, just as examining other phenotypes that are inside the organism such as proteins, cells, tissues, and organs. There is a chain of phenotypes (Fig. 10.1) extending from the gene, and in this chapter I discuss the links that are beyond the organism's traditionally considered phenotype. Others in this volume (see Chapter 12) advocate moving backwards down through less obvious phenotypes within the organism such as proteins, cell products, and tissues. Yet despite its logic the language of *The Extended Phenotype* (Dawkins 1982) has not been universally adopted by biologists studying animal behaviour (Dawkins 2004). It has not, conversely, suffered from any sustained criticism of the fundamental positions. If anything it has recently gained increased general interest once more because of its suggested role in niche construction theory (Odling-Smee *et al.* 2003; Laland 2004; Laland and Sterelny 2006) (Fig. 10.1; and see also Chapter 15).

10.3 Social insects and their extended phenotypes

The ants, termites, wasps, and bees are the taxa we think of as the social insects. The technical term is **eusocial**, which is defined as having overlapping generations, cooperative care of the brood, and division of labour that typically means a reproductive division with the majority of individuals being sterile (Wilson 1971). There have been other definitions of eusociality (e.g. Crespi and Yanega 1995). There are many other taxa besides ants, termites, wasps, and bees in which we find eusociality; examples are mites, spiders, shrimp, thrips, aphids, beetles, and naked mole rats (Wilson 1971; Crespi and Cho 1997; Bennett and Faulkes 2000; Costa 2006). Even humans and pilot whales have been called eusocial (McAuliffe and Whitehead 2005; Foster and Ratnieks 2005). In this essay I will restrict myself to the traditionally defined social insects (ants, termites, wasps, and bees) since I know these best, but my arguments are applicable to the other taxa. Also, for convenience, and in line with most authors, I will use the term social insects, rather than eusocial insects (see also Chapter 5).

Social insects live in family-based groups where a minority of individuals reproduce (queens and kings) and the majority (the workers) are functionally sterile and collect resources to provision the offspring of the reproductives. Such **altruism** is considered adaptive for workers since the offspring are usually their full siblings and by helping they gain indirect fitness benefits (Hamilton 1963). The role of **kin selection**, as it is called, in the evolution and maintenance of such societies was recently challenged by E. O. Wilson, one of its early supporters (Wilson 2005; Wilson and Hölldobler 2005) but the prevailing view is still that kin selection is essential (Foster *et al.* 2006). Social insects live in colonies that vary in size from 10 individuals in hover wasp societies (Turillazzi 1991) to more that 10 million in army ant societies (Hölldobler and Wilson 1990). They can occupy living spaces ranging in size from an acorn (*Temnothorax*) to 3 m high mounds (termites). A great deal of communication goes on inside societies, and studies of social insects have been instrumental to the development of communication theory (Hölldobler and Wilson 1990; Ross and Matthews 1991; Seeley 1995; Abe *et al.* 2001). Obvious examples are communication of status in the linear dominance hierarchy first discovered in paper wasps (Pardi 1948; Turillazzi and West-Eberhard 1996), pheromone communication developed extensively in ants (Wilson 1959; Hölldobler 1995), language among insects in the honeybee waggle dance (Von Frisch 1968; Seeley 1995), and teaching (Franks and Richardson 2006; Richardson *et al.* 2007). Social insects communicate with other members of the society: signalling identity (which colony they belong to); soliciting of food by larvae and adults soliciting nutritious regurgitations from larvae; individuals signalling their reproductive status and their position in a hierarchy (submissive posture, **badge of status**) or describing the location and quality of food (waggle dance). Individuals also communicate with other societies: signalling identity (nest of origin), aggressive displays signalling fighting ability and resource ownership. Collective actions involving many individuals also have communicative roles and usually towards potential threats: Asian honey bees (*Apis dorsata*), which form a bee-curtain across their comb ripple *en masse* to confuse predatory birds (Kastberger and Sharma 2000), paper wasps (*Polistes*) dance *en masse* to threaten parasitoids (West-Eberhard 1969) and, most impressive of all to me, the production of sound up to 5 m away via cooperative wing beating (*Syanoeca surinama*, a wasp) against the inside of a corrugated carton nest to deter mammalian predators (Rau 1933). The latter report, which is anecdotal, has a recent parallel in African bees whose sound was shown to deter herds of elephants (King *et al.* 2007). So, social insects communicate in a wide range of modalities (**channels of communication**) among colony members, between colonies, and even towards predators and parasites.

This chapter is about parasites manipulating social insect behaviour and how that may affect the range of communication discussed above. But in a chapter about the EP (of parasites) I would be negligent if I didn't spend a few words on the EPs of the social insects themselves since they too are highly impressive. The first is animal architecture. The social insects are rivalled only by humans in their ability to construct artefacts. No bird's nest, spider's web, or caddis shell rivals the multifunctional cathedral mounds built by fungus-growing termites; these 3 m high, rock-like structures, standing in sun-baked desert brush, contain within them sophisticated natural air-conditioning units, crop-fungus-growing combs, brood nurseries, refuse piles, networks of passageways, and, at the centre, a rock-hard protective chamber in which the king and the 3000 eggs per day egg-laying machine that is the queen, reside (Abe *et al.* 2001). Even the multifunctional beaver dam with its homely lodge is merely a well-placed branch and hollowed out mound of dirt by comparison. A termite mound is all the more impressive when we recognize that the architectural feat exists as a greenhouse to grow a rainforest adapted fungus in such places as the dry savannah (Aanen and Eggleton 2005). The second of the EPs, that of parasite manipulation of host behaviour, does not have any examples among the social insects as none are internal parasites. But if we view the colony itself as a host, as it is correct to do because of **inclusive fitness** that merges the genetic interests of individuals with that the whole colony through kin selection (Sherman *et al.* 1988, p. 263; Schmid-Hempel 1998) then there are examples of internal parasites of the colony causing behavioural changes. These are '**social parasites**' that, through chemical, morphological, and behavioural means induce workers to care for the parasite's offspring in preference to their own (Chapter 4). Finally, genetic action at a distance is an EP with many interesting examples among social insects. My favourite is among queenless ants (*Dinoponera quadriceps*) in which there is a hierarchy such that the alpha 'worker' reproduces but beta, gamma, delta, and so on do not (Monnin *et al.* 2002). An overly ambitious beta worker who challenges a fit alpha risks being dobbed with a droplet that signals to the others to immobilize the insurgent, sometimes effective in excess of 24 hours. The genes of alpha produce a phenotypic behavioural response among gamma and delta workers, at a distance (see also discussion of chemicals in social parasitism in Chapter 4 and pheromones in Chapters 5 and 7). Another example is the aforementioned production of sound by wasps and bees to deter predators.

10.4 Superorganism and communication

Having provided a short background to social insects and their EPs I now want to discuss the valid use of the metaphor of the colony as a superorganism, since it is especially useful for considering ways in which communication can be obscured because of parasite EPs. The large sizes of social insect societies, the multiple examples of collective action, and the ways in which society members are often behaviourally or morphologically specialized for certain tasks, together with the localization of the colony in a bounded structure that is built by multiple individuals, has led to the view that the whole colony is a superorganism (Wheeler 1911). This view, though intuitively appealing, lost favour for two reasons. The first was the supposed conflict it had with individual or gene-level selection. However, no such conflict exists so long as the superorganism is viewed within the levels of selection framework (Bourke and Franks 1995, pp. 64–66; Reeve and Keller 1999). In addition we need to clarify the mechanistic/functional division in our approach (see Chapter 11). It is important to be clear and precise when adopting a metaphor, and it should be stressed that the superorganism, just like the organism itself, is not a replicator (Dawkins 1990). The second reason why the superorganism concept declined in popularity was the limitation of a primarily analogical approach (Hölldobler and Wilson 1990, p. 358). The concept was good, but not particularly useful when investigators proceeded to examine the fine details of colony life such as reproductive decision-making in the light of kin selection. That is because different individuals within a colony may have different goals. Colony members do not come into conflict over resource acquisition but can, and do, conflict over resource allocation (Boomsma and Franks 2006). A clear example is the conflict between workers and queens in hymenopteran societies over the sex ratio of the reproducing offspring; the former favour a 0.75 bias towards females and the latter an equal sex ratio (discussed extensively in Bourke and Franks 1995). There is also conflict between workers if one decides to reproduce, and here we see the evolution of policing behaviour (Ratnieks 1988) where workers 'police' the egg laying of other workers because it is in their genetic interests that only the queen reproduces. When examining such conflicts, the individual-level view is more useful than a superorganism view.

But in many activities individuals do cooperate and appear to be maximizing something that is usually colony survival or colony propagule production (Queller and Strassmann 2002). So, for example, in seasonally flooded Argentinian habitats, fire ant colonies make a raft of interlinked workers and float to safety; in choosing a new home, swarming bees migrate *en masse* as a single unit; and in rearing its crop fungus leaf-cutting ants have distinct morphological and behavioural castes that transport leaves from the forest to the food fungus in a 'Henry Ford factory-like' manner and then process the waste in an extraordinarily efficient division of labour (Anderson *et al.* 2002). In these cases multiple individuals cooperate because of shared interests and produce phenotypes that cannot be achieved individually. That is, the colony-level phenotype. Since the organism is neither the object of selection nor the replicator, but rather comprises cooperating genes that have resolved potential conflicts because of shared interests in gamete production (in which genes are packaged (Dawkins 1990); see also Chapters 12 and 13), then the apparent unity of the superorganism can be explained because it helps genes lever themselves into the next generation (see also Queller and Strassmann 2002).

Why do I want to use a term that is, for many, either a throwback to the bad old days of group selection or an extension too far of the phenotype (Dawkins 2004; Jablonka 2004; Laland 2004)? The fact is the colony has, at times, its own phenotype, such as self-assemblages, to perform tasks that are not possible for individuals (Oster and Wilson 1978, p. 10; see also Chapter 11). These include collective phenotypes such as rafting, choosing nest sites via **quorums**, thermoregulation through combined fanning, killing predators, and collecting large food items. In a review of this topic Anderson *et al.* (2002) identified 18 such self-assemblages. There is undoubtedly a genetic basis for this and no doubt natural selection acted upon variations in rafting ability, for example, to produce an optimal response to seasonally flooded habitats. This phenotype is not an extended one like the physical, abiotic nest walls but rather it is a cumulative effect of the coordinated actions of individuals. The colony-level behaviour we see is 'more than the sum of its parts' (Oster and Wilson 1978, p. 10). It has synergy (Chapter 11). In order to produce effective responses to collective goals (e.g. colony survival) individuals must cooperate irrespective of any gene-level conflicts they may have. They may be in conflict later on in the colony cycle (at the timing of reproduction) but when necessary for collective survival the cooperation is necessary and observed (see below the grey boxes in Fig. 10.2a,b). The desiderata, or interests, of the distinct members are aligned for a period of time (Dawkins 1990). Such cooperation requires communication among members of this kin-based network of individuals. But as emphasized already, members of a network may be chimeras of both social insect and parasite. Here conflicts are predicted because their goals are diametrically opposed. Some parasites have little interest in whether or not the colony to which its host belongs will survive or not; all it is

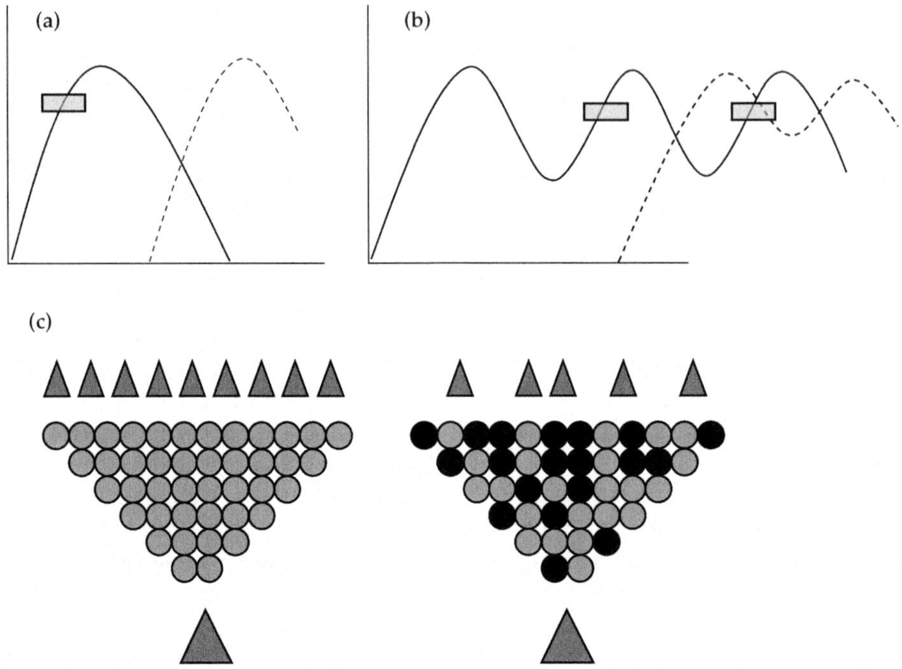

Figure 10.2 An idealised, bang-bang, mode of colony production for an annual a) and perennial b) social insect colony. Worker numbers (solid line) increase along y-axis until sufficient numbers are reached to achieve a switch to production of reproductives (dotted line). The difference between annual and perennial colonies is that worker numbers do not increase after production of reproductives (i.e. the colony dies) and that in the latter a few years may pass before reproductives are produced. Time (no units given) in on the x-axis. The grey box indicates the idealised switch point when worker numbers, sufficient for the production of sexuals, is made. Below this point the interests of colony members (queen and workers) are aligned and conflicts are not predicted except in special cases. In c) the number of workers (circles) builds up in an inverted pyramid fashion to produce new reproductives. The presence of chimeric individuals, workers infected by a manipulating parasite, is shown by black circles. These do not work but their presence in the colony will be detected their contribution to the production of reproductives (triangles) will not be realised, i.e. fewer triangles in the right hand panel of c).

trying to do is maximize is its own survival. Given that many of these colony-level activities (house-hunting, foraging, defending against predators) are risky pursuits (e.g. Schmid-Hempel and Schmid-Hempel 1984) then the conflict is more apparent. Whereas we generally do not see conflict in insect societies over resource acquisition (collecting food), but rather over resource allocation (to male vs. female larvae, to own vs. queen reproduction) (Boomsma and Franks 2006) the presence of parasites establishes a conflict scenario over resource acquisition since it entails an appreciable risk. The superorganism concept is therefore good because is forces us to remember the alignment of interests among non-infected colony members while at the same time erecting a category of aligned members into which the infected individuals may not always fit because of diverse desiderata of parasites within them. Parasitized individuals in the colony are the ultimate 'cheaters' of the cooperative hive but of course, unlike the more well-known selfish individuals that want to pursue their own interest (e.g. laying their own eggs), the infected individuals are vehicles for parasite genes. In the next section, I review what behaviourally modifying parasites these chimeric individuals contain.

10.5 Behaviourally modifying parasites of social insects

Among all the possible phenotypes expressed by the genes of social insects it is their behaviour that

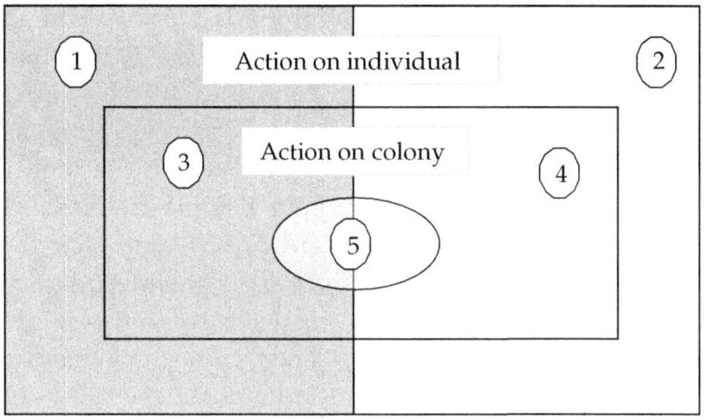

Figure 10.3 The dual host nature of social insects showing where the obvious effects of parasite manipulation can be on the individual (outer box) or on the colony (centre box). The shaded zones are interactions where the parasite benefits from manipulation and the un-shaded zones are where the host benefits. The innermost box represents the category of 'boring by-product' that are changes accompanying parasitism but which are not evolved manipulations by parasites or defence by hosts. Examples from each of the categories are 1) Nest desertion by individuals to promote parasite dispersal (e.g. Strepsiptera), 2) Nest desertion by individuals to retard parasite dispersal (e.g. Conopids), 3) Colony wide social confusion due to parasite presences and or manipulator compounds (ant warfare engineered by parasitoid) 4) Collective defence against parasite (wasp dancing against ovipositing parasites) 5) reduced individual activity and thus colony productivity if parasitism prevalence is high enough due to parasite induced lethargy (gregarines in ants).

is the most important when one considers sociality. Colonies do have fortified walls and individuals have an armoured cuticle and a battery of defensive compounds, but it was behaviour, and specifically altruistic behaviour, that seeded the growth of loosely banded individuals into colonies. Here I echo an important, though surprisingly neglected, assertion of behaviour as the evolutionary pacemaker (Baldwin 1896; Wilson 2000; West-Eberhard 2003) (see also Chapter 8). And behaviour remains the most important phenotype in the colony through such actions as division of labour, cooperative care of the brood, defence of the nest, and communication. Thus, in considering the EPs of parasites I will consider those parasites that alter the behaviour of their social insect hosts. I could have chosen parasites that cause colour changes (Trabalon *et al.* 2000) or size difference in infected individuals (Maeyama *et al.* 1994); but the effect on communication is less clear.

Parasites of social insects have provided prominent and compelling examples of parasite EPs where host behaviour is manipulated. The best known example is the 'brain-worm', which is a trematode that induces its intermediate ant host to leave the colony and climb blades of grass and bite hard (Carney 1969, references on pp. 55–57; Moore 2002). The final host is a grazing animal such as a sheep which is presumed to ingest ants along with the grass it is eating. So emblematic is this example that it 'made the cover' of Janice Moore's excellent review entitled *Parasites and the Behaviour of Animals* (Moore 2002). (Another manipulating parasite, the fungus *Cordyceps*, that also causes ants to bite onto vegetation similarly adorned the cover of Paul Schmid-Hempel's book, *Parasites in Social Insects* (Schmid-Hempel 1998).)

In reviewing here the range of parasites causing behavioural changes among the social insects it will be useful to introduce a schema (Fig. 10.3). There are five categories of behavioural modification in social insects:

1. The first is adaptive manipulation of individual host behaviour that favours parasite genes. The above-mentioned brain-worm is an example. For many horizontally transmitted or trophically transmitted parasites (i.e. where predation of the host is a necessary requirement for transmission) it is obligatory for the individual host to leave the colony, and in these cases nest desertion is the EP of the parasite: conopids, Strepsiptera, trematodes, cestodes, mermithid and rhabtid nematodes, Entomopthoralean and Clavicipitalean fungi (parasite associations with social insects was extensively reviewed in Schmid-Hempel (1998) so a full list of references is not presented here due to space constraints). In all cases the manipulation is a multistep process. Once outside the colony the host is often directed to a particular location where it performs a stereotypical activity: biting vegetation (fungi, trematodes), suicide in water (mermithid nematodes), digging to provide a diapause site for the parasite pupa (conopids), or inactivity in a prominent place to facilitate parasite mating (Strepsiptera) or predation by final host (cestodes, trematodes). In each of these cases the biology of the parasite and its mode of reproduction is such that nest desertion is interpreted as adaptive to the parasite: remaining in the nest would not lead to infection of other colony members because the parasite is not infective without that necessary departure outside the colony where it either mates or develops in a manner not possible in the colony (e.g. fungi growing through the cuticle or trematodes causing ants to bite).

2. The second category is adaptive manipulation of more than one individual (i.e. the colony) that favours parasite genes. The entry of social parasites into the colony can be accompanied by the release of chemicals that induce confusion among workers and prevent parasite exclusion (these propaganda signals are discussed in Chapter 4). Because the parasite is not internal to the individual host then we may view this as the action at a distance EP like the familiar example of cuckoo chicks manipulating their hosts to feed them. Indeed, the social parasites are often are called cuckoo wasps and ants. Recall the justification in considering the colony as a host, in addition to the individual (Sherman *et al.* 1988, p. 263; Schmid-Hempel 1998) which in this example means the EPs action at a distance and parasite manipulation of host behaviour can be used.

3. The third category switches the benefits of the parasite-associated behavioural change from the parasite to the host and into defensive behaviours. Individual social insects have a very large repertoire of defensive behavioural reactions against parasites. Most mundane, but evidently important for colony-level defence, is self-grooming, allogrooming, and, in the case of crop-rearing ants, grooming, or 'weeding', their **mutualistic** fungus (Cremer *et al.* 2007). More dramatic is cold-seeking behaviour by bees infected by conopid flies to retard parasite development; every night they move outside the high-temperature nest (Muller and Schmid-Hempel 1993). Conopid larvae live inside bees and cannot infect the siblings of the bee they are infecting, so this nightly self-exclusion is defensive in that it retards the parasite's growth and facilitates a longer working life for the infected individual.

4. The fourth category is also a defensive behaviour against parasites, but here it requires the coordinated action of multiple individuals to succeed. The 'dancing behaviour' of paper wasps in response to the presence of an ovipositing Ichneumonidae wasp is a good example (West-Eberhard 1969). Another example is construction behaviour where individuals cooperate to build satellite nests (Jeanne 1979) or walls to quarantine infected areas of the colony (Schultz *et al.* 2005) and even infected siblings (Epsky and Capinera 1988). The last one, that of construction, is an EP of ants and wasps as a defence against parasites and has nice parallels with avian construction such as oven and weaver bird nests as a defence against predators such as snakes (Hansell 2004).

5. The fifth category does not interpret the behavioural change as an adaptive trait of either the parasite or the host but rather as a 'boring by-product' of infection (coined by Dawkins 1990; see also Poulin 1994, 1998, 2000). This category, though important when considering the EPs of parasites, is often the least satisfactory: it is commonly the one advanced in objection to the adaptationist explanations (this point is touched upon by de Sousa in Chapter 16 when discussing the infamous Spandrels of San Marco by Gould and Lewontin 1979). We can think of lethargy or reduced flying ability when infected as possible examples (Kathirithamby and Hughes 2005). However unsatisfying to an adaptationist (Pigliucci and Kaplan 2000; Gardner *et al.* 2007), this category is very important because it can obscure communication.

So these are the five categories of behavioural changes among social insects due to the presence of parasites. I will now move onto the central thesis of this chapter, which is that such changes can significantly impact upon social communication. Recall that I said this is advocacy. We know that social insect nests are literally hives of complex communication and that this can be a phenotype at both the individual and the colony level. Now you are aware that parasites are imbedded in the bodies and chambers of both the nest occupants and the nest itself. I advocate viewing the nest from the perspective of the parasite's genes and asking what utility is there within this social communication network?

10.5.1 Scenarios where communication channels are potentially obscured

We have seen the ways in which behaviour can be changed at both the individual and the colony level; and to benefit either parasite or host (or neither in the case of boring by-products). There is no shortage of examples, and documenting behavioural change is particularly easy for social insects because as **central place foragers** with precisely defined roles within a kin-structured society any behavioural deviations from the norm are easily observed. For example, desertion; 'workers' not working or young workers performing tasks, or occupying areas, usually ascribed to older individuals and vice versa. The value of studying parasite EPs is not found in describing or cataloguing the behaviours but in understanding their consequences for both host and parasite fitness (Moore *et al.* 2005; Thomas *et al.* 2005). I have previously suggested that the complexity of insect societies offers rich rewards when studying the evolutionary consequences of manipulation (Hughes 2005). For example, recognizing that behavioural changes are surrogates of virulence (Read 1994) we could examine the cost of manipulated workers (or the cost of behavioural defence) on colony fitness (represented by the reduced output of reproductive individuals; see Fig. 10.2c). We could also ask what are the proximate-level mechanisms by which parasites exert such fine detailed control? And have the widely different taxa of parasites convergently evolved similar mechanisms to affect a common behaviour such as nest desertion? I mention these directions only in passing as profitable lines of research and now turn my attention to social communication. I will examine four scenarios where communication within the colony is potentially obscured and close with a fifth scenario concerning communication outside the colony.

10.5.1.1 Altruism

Despite the very obvious differences, to our eyes, between a cuckoo chick and its host chick, the cuckoo is provided with food. Clearly the host bird (a fairy wren for example) is losing out. Among social insect species the equivalent scenario would be interspecific social parasitism where one social insect, a cuckoo wasp, *Polistes sulcifer* for example, takes over the nest of another, *Polistes dominulus*, and the workers accept this alien queen (Cervo and Dani 1996). This type of parasitism, also called social parasitism, even occurs between orders with parasitic lycaenid caterpillars infecting ants nests and either being fed like cuckoo chicks or simply eating the ant brood (Pierce 1995; see also Chapter 4 and Pierce *et al.* 2002). In the case of cuckoos, cuckoo wasps, and parasitic caterpillars there is communication between host and parasites. Notwithstanding the Mafia hypothesis, where hosts 'cooperate' with parasites or face retaliatory behaviour (Zahavi 1979; Ponton *et al.* 2006; Hoover and Robinson 2007) it is not an adaptive strategy for hosts to feed parasites. The reason why hosts do feed alien chicks boils down to selection pressures on recognition mechanisms that are related to the prevalence of infection in the population at large (Winfree 1999). The communication system that exists is one of deception with the parasite either mimicking, camouflaging, appeasing, or over-stimulating the host (discussed at length in Chapter 4; see also Chapter 16 for a discussion on camouflage). For the social insects the external parasite (external that is to an individual host) may eventually come to 'look' like its host through camouflage or mimicry, which typically means either adopting or synthesizing the colony odour so as not to be discovered. But an external parasite is always different, and especially so from the start of the relationship when it just enters the colony. Thus the signal must be deceptive. What is important to understand is that the colony member which contains an internal parasite (category 1, Fig. 10.3) was not always infected. For the majority of its life it was a normal, uninfected, colony member, and kin selection favours cooperation, which for workers means feed a related individual until it is big enough to work and contribute, cooperatively, with similarly aged individuals to colony productivity. But once the individual is parasitized, i.e. becomes a chimera, then altruism directed towards it is the same as feeding a cuckoo chick. I will not discuss why colony members don't immediately recognize this changeling, but it is likely to be related to fitness costs, at the colony level, of lost investment, zero return because of no work and costs of discrimination. What is interesting to consider in the present context is communication.

Is an infected individual signalling? Just before it became infected the answer was yes. It signalled its status (larva, worker, male, queen) and, if an adult, its position in the hierarchy through chemical (e.g. Turillazzi and West-Eberhard 1996) and even visual modalities (Tibbetts 2002). Via pheromonal **cues** it can signal its reproductive status (Chapter 7). As a worker it signalled its task and in many cases information about colony and environmental resources (by orally exchanging liquid, called trophallaxis, colony members communicate if and where they have foraged, or if they are hungry). It also signalled to its nestmates and foreign colony members its colony of origin via the chemical odour present on its cuticle (see Chapter 5). All such signals were likely to be honest and to benefit all parties because the sender and the receivers had aligned interests (colony productivity). The parasitized individual, however, no longer has completely aligned interests with other colony members because many activities in the social insect life history are risky, such as working (Schmid-Hempel and Schmid-Hempel 1984), mating (Boomsma *et al.* 2005), or defending. If not risky then tasks are energetically costly. A parasite that is not transmitting itself, or reproducing, should not want its host to engage in risky tasks such as defending the nest, foraging, or mating; nor waste energy contributing to colony reproduction. A general effect of parasites is castration where hosts are prevented from reproducing while parasites channel resources into their own development (Poulin 2007). Because most individuals in societies are workers, and these are functionally sterile, then castration is not achieved by preventing workers from reproducing, which they wouldn't do anyway, but by preventing them from working since division of labour in the colony (superorganism) is the method by which reproduction is achieved using the reproductive castes. Parasitic castration of social insects is preventing the worker caste from working (but may involve associated physiological reduction of gametes which some workers can possess, but to my knowledge this has only once been investigated; Strambi *et al.* 1982). Castration, by behavioural modification, will benefit parasite fitness by (1) reducing the risk to the parasite vehicle and keep the parasite out of danger or (2) not wasting valuable energy. Whether this form of behavioural parasitic castration will reduce colony fitness (and the fitness of the individual that is parasitized through indirect effects) probably depends on how many other workers are likewise behaviourally castrated.

That parasites should not allow their vehicles to go outside the colony and engage in expensive tasks that divert resources away from parasite development is especially probable if we remember that the a mature colony is a protected fortress where predation is essentially zero (Keller and Genoud 1997). There are very few studies that have set out to test whether infected individuals do in fact work (Schmid-Hempel and Schmid-Hempel 1990; Schmid-Hempel and Muller 1991; Hughes *et al.* 2004a,b) and more data are required. Even if individuals do work and do repay the colony for its investment then it may transpire that they do so less efficiently, as in the case of parasites of bumblebees that affect flower choice and pollen load (Schmid-Hempel and Schmid-Hempel 1990; Schmid-Hempel and Muller 1991; Shykoff and Schmid-Hempel 1991; Schmid-Hempel and Stauffer 1998). The details are sure to vary with parasite taxa and ecological conditions but it is reasonable to suspect that parasitism prevents altruism towards other colony members and that the infected individual, by consuming colony resources for its growth and that of the parasite, is expressing the EP of the parasite at the expense of the society.

Let us realistically assume that infected individuals do not work, or that they are less efficient if they do work. We expect from evolutionary theory that non-cooperating individuals, i.e. cheats, are sanctioned (Chapter 2). One could speculate that in addition to expressing the extended phenotype of cheating behaviour (i.e. make your social insect host a lazy, non-working member of the society, because that is the safer and less energy-demanding course), the parasite has an additional agenda, which is to signal to colony members that its vehicle is not in fact a lazy individual but rather a productive member of the society. That is, do internal parasites deceive colony members at a distance as we know social parasites do (Chapter 4)? At the proximate level signals are being sent, and whether these are exact copies of signals normally transmitted between uninfected colony members or some *de novo* deceptive signal requires study. Besides the signals being sent (such as 'feed me even though I am actually a wolf in sheep's clothing') there potentially exists cues that could be used by non-infected colony members to decide the infection status of others. Such things are the aforementioned non-working behaviour, lethargy (Trabalon *et al.* 2000 and references therein), body distortions such as enlarged abdomens, and smaller wings in reproductives, following nematode infection (Maeyama *et al.* 1994), extruded pupae of Strepsiptera in ants (Hughes *et al.* 2003), or changes in hair growth pattern in wasps and bees making 'intersexes' (Salt 1927, 1931; Wcislo 1999). We know from other cooperative systems such as fish shoals that infected individuals can be easily recognized by the group via phenotypic assortment (Barber *et al.* 2000). If such cues have a realistic chance of evolving into a signal ('I am your kin but I am now infected and therefore non-working') then we would expect selection acting on the parasite to evolve deceptive signalling, such as camouflage. (With the obvious assumption that the prior cue resulted in sanctions against infected individuals, which, I suspect, would not always be the case. But that is beyond the present scope.) An interesting test of this would be determining whether the extruded pupa of internal strepsipterans (insects) infecting ants mimics the cuticular profile of the host because infected ants remain inside the nest even once the parasite has extruded through the cuticle, which means a large area of extruded parasite could be detected. This could be compared to the chemical profile of the extruded pupa of strepsipterans infecting wasps where the parasite extrudes the pupa only after the wasp has left the nest and thus the necessity of deceiving kin is gone (Hughes *et al.* 2003, 2004b). To conclude this section on altruism, is clear is that internal parasites disrupt the altruistic actions of the individuals they infect and this probably involves the evolution and transmission of signals that obscure social communication over who receives the benefits of altruism.

10.5.1.2 *Misdirected altruism*

When colonies of the Asian army ant, *Leptogenys distinguenda*, move home (a regular occurrence for such nomadic ants) they pick up a molluscan parasite of their colony in preference to their brood stage siblings (Witte *et al.* 2002). It appears that the mollusc produces an irresistible foam mass that the ants find very attractive (V. Witte, pers. comm.). This sort of super-normal signal (Dawkins and Krebs 1979) appears to be a common strategy among social parasites (see details in Chapter 4). It is also the situation to be found among brood parasites such as cuckoo chicks where 'cuckoos should be selfish because their greed is unconstrained by kinship' (Kilner and Davies 1999) A colony member responding to such signals by a social parasite is misdirecting altruism and deceptive communication is the explanation. But let us again consider the more subtle phenomenon of internal parasites of social insects (internal to individual social insects). A nursing worker that gives food items to brood members will be faced with situations of chimeric individuals. There are then two important points to consider. The first is whether increased begging is observed and the second is whether any non-parasitized individuals respond? We have no evidence at all for increased begging by infected brood and the only study to address this found no differential mass loss between infected and uninfected wasp brood, demonstrating that the parasite was not placing a high demand, but since the parasite was a strepsipteran that requires a morphologically intact adult wasp for future transmission then the absence of a high cost at the larval stage is reasonable (Hughes and Kathirithamby 2005). Begging responses of their hosts should be investigated in parasites that only infect larva or adults (but not both sequentially such as Strepsiptera).

The field of brood sibling rivalry has not been developed for social insects in the same way as for birds (Mock and Parker 1997). In general we might speculate that full siblings in hymenopterans societies with high relatedness (0.75, due to haplodiploidy) will not be selected to compete for food in quite the same way as 'normal' diploid siblings do. However, infected individuals no longer have aligned interests and should solicit more food without considering the kin of the vehicle they inhabit. Since the decision to feed a larva follows from communication of need by that larva, then parasitized larvae may send an honest signal but the receiver is duped because it is not feeding just a sibling, but also the parasite. The question, again, is not whether natural selection should allow nurses to recognize infected brood but rather whether communication has been obscured in cases where parasitism is obviously not detected? (Non-detection is evidenced by the fact that the infected individual has not been removed.) The answer would appear to be yes, and I will discuss the significance later.

Providing food to a begging larva is not the only form of altruism. Simply allowing an individual to reside within a cell is altruism and in cases of stress and low food availability nursing workers will remove and eat young individuals but allow older individuals to stay alive because of the higher investment in the latter (Hölldobler and Wilson 1990). In many cases where cells are used to rear brood (wasps and bees) the cell can be occupied by a parasite which often has completely consumed the former occupant (e.g. Ichneumonidae wasps and chalkbrood fungi are examples). Usually the parasite allows the host larvae to spin a silken cap and this seals off the cell (which precedes pupation in uninfected individuals). In some cases a behavioural defence is uncapping, where bees uncap the cell and remove the infected individual as we see in fungal infections (Schmid-Hempel 1998). However, no evidence exists that wasps uncap cells infected by Ichneumonidae wasps. Does this imply that the parasite within is signalling to the nurses that the cell is not parasitized?

10.5.1.3 Reproductive decision-making (because of cheaters)

There are three principal modes of colony development. The first is an annual cycle (Fig. 10.2a). For example, a bumblebee queen begins a nest in springtime after winter diapause. Her first eggs become workers and she feeds the subsequent larvae with metabolized body stores until they become adults and can take over the foraging, nursing, and building tasks of the colony, leaving her to the job of egg laying. When the workforce has built up to a sufficient level to ensure adequate resources the colony begins producing reproductives (males and females) that leave to mate with the new queens and overwinter before the cycle begins again. A second mode also involves this solitary founding but here the colony lasts many seasons; it is perennial (Fig. 10.2b, e.g. ants, termites, some wasps, and bees). The production of reproductives may not happen for several years as the colony builds up a sufficient number of workers. In both modes the colony can start from a single female and male (in hymenopterans the vehicle for male genes is sperm stored inside the female, while in termites the vehicle is a whole male), or with multiple females/males. The third method of colony development is to simply split a big colony in two just like a fissioning yeast cell (this happens in honeybees and army ants, for example). The benefit of colony fission is that during the early days of the new colony it already has a very large workforce. By contrast solitary founding is very risky and estimates of up to 99% failures are not unrealistic (Hölldobler and Wilson 1990).

Whatever the method of its beginning the goal of the colony is to produce reproductives, and at some point in the life of each colony the decision about when the time is correct will be made (Fig. 10.2). We know very little about the optimal decision-making process or what cues are used. The standard model is the 'bang–bang' mode where investment in reproductive workers begins within 'one half the lifetime of the last cohort of workers' (Oster 1976; Oster and Wilson 1978). That is, workers and reproductives are not produced synchronously but rather sequentially. The decision to begin sexual production can be viewed either as one by the queen who produces queen- or male-destined eggs; or by the workers who feed larvae either special food, or simply more food, thereby switching them from a worker-destined pathway to a queen-destined pathway. There are many taxon-specific details but the essential point is that the colony begins producing reproductives when the workforce is judged to be of sufficient size or when there are enough resources. To make this judgement must require some sort of internal communication and it is probably similar to **quorum sensing** in bacteria where a new action or phenotype occurs once bacterial numbers get over a set amount (Chapter 2).

What will be the role of chimeric individuals inside the colony for reproductive decision-making? They will not be collecting resources, and from the available evidence not tending the developing brood. Since the presence of the parasite divorces the infected colony member from its previously held common interests with its siblings, we are not surprised by this ensuing lack of altruism (Section 10.5.1.1 above). But it is unlikely that non-infected siblings know the infection status of each member of the colony due to the force of selection on evolving foolproof recognition systems (discussed above). More likely, infected individuals are recognized for what they are, another body in the colony that is expected to be collecting resources, tending brood, defending the nest, or acting as a reserve member for other tasks. Since the production of reproductives is a colony-level trait that is in the interest of all members, we would expect the signalling to be honest. I suggest that the presence of infected individuals changes the accuracy of information available preceding the decision to switch to producing reproductives. Let us realistically assume that the switch is a numbers game and relies upon a simple rule such as an encounter rate above a threshold value that will switch the colony into a reproductive phase (see Fig. 10.2c). Such a rule underlies quorum-based house-hunting in ants (Pratt *et al.* 2002) and bees (Seeley 1995; Seeley and Buhrman 1999; Seeley *et al.* 2006), and foraging decisions in ants (Greene and Gordon 2003) (see also Chapter 11). A colony above a certain size should switch to sexual production since the number of individuals obviously correlates with eventual resource acquisition and the transfer of those resources into reproductives. The presence of infected, non-working individuals, means that the information guiding the switch is not accurate and thus communication is obscured.

To sum up this section, there is increasing attention to colony-level decision-making in social insect evolutionary biology (Boomsma and Franks 2006), and hopefully further studies will begin to elucidate the cues, signals, and timing of reproduction by the colony. The presence of chimeric individuals and their effect on colony communication preceding reproductive switches should be taken into account when considering colony-level decision-making processes.

10.5.1.4 Colony-level decisions (defence, house-hunting, swarming)

In the previous section I briefly mentioned house-hunting by ants and honeybee colonies, which are excellent examples of a complex superorganism-level phenotype that is generally called collective decision-making or self-organization (see Chapter 11). It is similar to self-assemblages such as the rafting fire ants I also discussed above. The term self-organization refers to the observation that complex patterns of behaviour can be observed when multiple individuals cooperate, and that these behaviours are not expressed by single individuals that are alone. The 'whole is greater than the sum of its parts' (Oster and Wilson 1978, p. 10), meaning there is synergy (Chapter 11). There has been an increasing level of attention focusing on self-organisation, not just among social insects but other taxa such as locust marching behaviour, fish shoaling, and even the coordinated clapping of opera audiences (Camazine *et al.* 2001; Sumpter 2006; Garnier *et al.* 2007). Such phenomena can be adaptive at the individual level (selfish herd) and the group level (coordinated defence) or just an emergent phenomenon that is a by-product of coordinated actions (opera audiences). For social insect societies the self-organizing behaviour we see, and that includes self-assemblages, is a colony-level phenotype shaped by natural selection (Queller and Strassmann 2002). Quorum sensing-based house-hunting in ants and bees is obviously an optimal way to choose among nest sites of varying quality (Seeley *et al.* 2006; Visscher 2007). Other examples of collective behaviour, such as thermoregulation via coordinated movements, coordinated defence against invertebrate attackers, making trails, or coordinated efforts requiring multiple individuals physically linking together (building bridges, pulling chains, ladders, and bivouacs), would all appear to be colony-level phenotypes (Anderson *et al.* 2002).

To date studies have focused on a proximate-level, rather than ultimate, understanding of self-organization in biology generally and social insects specifically (Camazine *et al.* 2001; Boomsma and Franks 2006). This reflects the influence of mathematical and physical principles in developing algorithms that can explain the observed phenomenon (Sumpter 2006; see also Chapter 11). In my opinion, this lack of a functional-level approach has meant cheaters that benefit from the collective phenotype without investing in its production have not been considered deeply. Of course that is not a fault, and we need to consider many variants of non-cooperating individuals, whether because of parasitism or not, to better understand the fundamental question of how cooperation exists in the first place (see Chapter 2 and references therein). Obviously if cheating exists and it threatens a colony-level phenotype then one would expect punishment measures to evolve, as happens in the previously mentioned case of worker policing which is a colony-level defence to prevent selfish workers laying eggs against the collective interest (Ratnieks 1988; Wenseleers and Ratnieks 2006). But cheaters of course raise the important, though tautological, point that collective behaviour is the result of multiple individuals cooperating and these individuals are not all the same (Sumpter 2006). Recently, and perhaps for the first time, inter-individual variation was explicitly considered (Garnier *et al.* 2007) as a source of perturbation that could affect the outcome of self-organization. The authors imagine perturbations coming from outside or within the colony. Obviously, a within-colony source of variation among individuals that cooperate in collective behaviours is whether some of those individuals are chimeric and contain both parasite and social insect.

Two interesting examples can be considered here. In a swarm-founding nocturnal hornet from Southeast Asia (*Provespa anomala*) individuals infected by strepsipteran parasites (an internal parasitoid) were never found in swarms, though they were found in non-swarming stages of the colony (Matsuura 1999). Thus it appears that infected individuals don't participate in swarms. By contrast, in the collective defence of honeybee colonies against predatory hornets it is the virus-infected individuals that take the lead (Fujiyuki *et al.* 2004). Usually defence is a task performed by the oldest workers because they are the most expendable, so it is interesting that when honeybee colonies were presented with a predator (a hornet wasp) it was the middle-aged workers that would attack and these individuals had viral infections in their brain. The virus, called *Kakugo* meaning 'ready to attack' in Japanese, belongs to the same group as rabies so perhaps the high level of aggression in some way aids transmission (Fujiyuki *et al.* 2004). These two examples are tantalizing, but what is obviously required is many more data on the occurrence of infected individuals in collectively organized behaviours. Meanwhile, I feel the following two perspectives should be considered.

The first is whether or not infected individuals take part in collective activities. It will probably transpire that they do so long as it doesn't conflict with the desiderata, or desires, of the parasite within (Dawkins 1990). Clearly if the collective action is risky then we would not expect infected individuals to take part unless it facilitates transmission as in the probable case of the *Kakugo* virus above. The second and more interesting starting point is asking what communication occurs in collective organization and whether chimeric individuals can disrupt it? In most cases the answer will be probably be that chimeric individuals do not affect self-organization communication pathways because they do not assume the lead roles in self-organizing behaviour. For example, these lead roles are the scouts who search for nest sites, assess their suitability, and eventually communicate the location or quality of such sites (e.g. waggle dancing or laying pheromonal trails or physically leading individuals to the new site; see Chapter 11). But once a suitable location has been found through quorum sensing then the whole colony must move, and this requires communication between those who know the location of the new nest and those who communicate that knowledge (Seeley *et al.* 2006; Visscher 2007). Amazingly, ants actually teach others the location (Franks and Richardson 2006). In the latter situation an obvious question is how teaching varies when 'students' differ in their individual capability to learn? Can chimeric individuals learn the way to the new location? Do instructors give up if their students are lethargic zombies? Does this negatively feed back upon an individual's stimulus to teach since we typically assume that this self-organization behaviour is a set of simple rules coupled with negative and positive feedback (Chapter 15). All of this is speculation, but as we start to develop a better understanding of the optima of self-organization behaviour then the role of these chimeric individuals is worth considering because they may be key to understanding the rules of thumb present. Ultimately it will probably come down to a numbers game again (Fig. 10.2c). The effect of chimeric individuals is not observed in colony-level phenotypes such as house-hunting when their numbers are low. When infected individuals occur at high levels the collective breaks downs. A clear example of this it the parasitic Cape honeybee (*Apis mellifera capensis*) that can reproduce parthenogenetically inside African honeybee nests (*Apis mellifera scutellata*) and behave like a cancerous line that eventually destroys the colony (Martin *et al.* 2002). However, whether or not the effect on colony phenotype or colony fitness is felt the existence of chimeric individuals has the possibility to obscure social communication during collective behaviours.

10.5.1.5 Extra-nidal activities

In the last example of potential ways in which communication is obscured by the extended phenotype of parasites I consider activities outside the nest. Social insect workers leave the colony to forage for food, water, building materials, and when defending the fortress. Reproductives leave the colony to mate and establish new colonies (or in some cases re-enter their natal colony). 'Workers' possibly leave the colony to become reproductives (Reeve *et al.* 1998). The communicative behaviour of non-infected individuals away from the nest is little studied. For the most part it deals with the interactions among and between individuals at the colony boundaries (Hölldobler and Wilson 1990; Ross and Matthews 1991; Bourke and Franks 1995; Seeley 1995; Abe *et al.* 2001). Impressive examples are the 'border patrols' (Hölldobler 1979) and ritualized tournaments (Hölldobler 1981) where opposing colonies display their colonies' fighting abilities. Another example of colony boundary activity is the communication between individuals returning from foraging and those waiting to determine if they should forage following the transfer of chemical information (Greene and Gordon 2003). Definite extra-nidal activities are orientation either away from the nest or back to it as well as choice of food. When ants and bees navigate they pick up cues to allow them to return home or lay trails. Many EPs of parasites involve nest desertion (Fig. 10.3), and though this is pure speculation in the absence of any data it would appear likely that deserting individuals do not record landmark features or lay trails, since they do not intend to return. For food choice we do know that infected social insects can behave differently; for examples parasites of bumblebees can affect flower choice and pollen load (Schmid-Hempel and Schmid-Hempel 1990; Schmid-Hempel and Muller 1991; Shykoff and Schmid-Hempel 1991; Schmid-Hempel and Stauffer 1998).

In the context of communication what is interesting is if non-infected individuals interact at all with chimeric individuals when performing extra-nidal activities. Currently, due to a lack of data, we are not in a position to discuss how chimeric individuals may or may not impact upon competition between conspecific individuals away from the nest. But taking a cue from parasitological research among non-social taxa (such as trematodes in snails) the interactions between infected and non-infected social insects at feeding sites may be of great general interest. In the snail–trematode system uninfected individuals directly competed with parasitically castrated snails in intertidal mud flats leading the authors (Miura *et al.* 2006) to propose that we should view foraging ecology on mud flats as a battle between snails and trematodes (in snail bodies). This parasite's eye view is of course generally lacking in behavioural ecology (Poulin 2007), but foraging arenas, where competition is to be expected, are likewise areas of intense communication and the chimeric individuals (whether snail or social insect) can, once more, offer insights to general mechanisms in communication.

10.6 Conclusion

The social insects have been the subjects of pioneering work in communication and continue to be leading models systems in proximate and functional studies of communication, as a number of chapters in this volume attest (Chapters 4, 5, and 7). The rules that have been elucidated through their study have now been tested in diverse taxa from microbes to monkeys (Strassmann *et al.* 2000; Flack *et al.* 2006). But of course there are many details of communication in social insects that remain to be understood, and in particular a proximate-level understanding is lacking. One commonly applied approach when examining communication in diverse systems has been the use of naturally occurring or experimentally induced mutants, and a number of authors in this volume have discussed this: Haig when discussing inter- and intra-organismal communication (e.g. Prader–Willi and Angelmann syndrome, Chapter 12), Crespi when discussing language evolution (autism and affective psychotic spectrum, Chapter 13), Hurst and Benyon when discussing relevant contributions to scent marking (knockout mutants, Chapter 6), and Diggle *et al.* when discussing quorum sensing in bacteria (also knockout mutants, Chapter 2). In social insect research the use of 'knockouts', where part of the colony has been removed in what Wilson terms a sociectomy (Wilson 1985), has been a useful tool for examining effective colony-level phenotypes such as optimal caste ratios and foraging (Wilson 1983). The use of such a knockout approach has not, to my knowledge, been done specifically to examine inter-individual communication, but has recently been pioneered in social primates (Flack *et al.* 2006). It strikes me that the chimeric individuals I have discussed in this chapter, and the accompanying parasite EPs, are naturally occurring mutations in a highly complex communication network. Therefore, understanding the functional and proximate details of such obscured communication could offer multiple insights into general principles of communication; the goal of this volume.

Summary

Societies of social insects are paragons of communication. Multiple channels exist between different members and the transmitted information ranges from specifying the location of foraging areas to who controls reproduction. Whole colonies can also communicate with other colonies or even vertebrates. But what if the individuals within a society are not, in a word, themselves? Here I explore how adaptive manipulation of host behaviour by parasites, i.e. the extended phenotype of parasites, obscures social communication, and ask how it influences other members of the society. Since manipulated kin are at best cheaters and at worst potential infective agents can the society recognize them? Knowing how a highly complicated example of social communication is broken or subverted by parasites can provide considerable insight into the evolution of communication. I discuss conflict and communication in this system in the context of the debate over the nature of the organism.

Acknowledgements

I wish to thank Patrizia d'Ettorre, Koos Boomsma, and Anna M. Schmidt for comments and discussions. I am grateful to Frederic Thomas for constantly challenging me, and the field at large, to consider new directions in understanding parasite manipulation. I warmly thank my colleagues in the Centre for Social Evolution for such a stimulating work environment and the Marie Curie Actions for funding.

References

Aanen, D.K. and Eggleton, P. (2005). Fungus-growing termites originated in African rain forest. *Current Biology*, **15**, 851–855.

Abe, T., Bignell, D.E., and Higashi, M. (eds) (2001). *Termites: Evolution, Sociality, Symbiosis, Ecology*. Kluwer Academic Publishers, Dordrecht.

Anderson, C., Boomsma, J.J., and Bartholdi, J.J. (2002). Task partitioning in insect societies: bucket brigades. *Insectes Sociaux*, **49**, 171–180.

Baldwin, J.M. (1896). A new factor in evolution. *The American Naturalist*, **30**, 441–451, 536–553.

Barber, I., Hoare, D., and Krause, J. (2000). Effects of parasites on fish behaviour: a review and evolutionary perspective. *Reviews in Fish Biology and Fisheries*, **10**, 131–165.

Bennett, N.C. and Faulkes, C.G. (2000). *African Mole-Rats: Ecology and Eusociality* Cambridge University Press, Cambridge.

Biron, D.G., Ponton, F., Marche, L. *et al.* (2006). 'Suicide' of crickets harbouring hairworms: a proteomics investigation. *Insect Molecular Biology*, **15**, 731–742.

Boomsma, J.J. and Franks, N.R. (2006). Social insects: from selfish genes to self organisation and beyond. *Trends in Ecology and Evolution*, **21**, 303–308.

Boomsma, J.J., Baer, B., and Heinze, J. (2005). The evolution of male traits in social insects. *Annual Review of Entomology*, **50**, 395–420.

Bourke, A.F.G. and Franks, N.R. (1995). *Social Evolution in Ants*. Princeton University Press, Princeton, NJ.

Camazine, S., Deneubourg, J.L., Franks, N.R., Sneyd, J., Theraulaz, G., and Bonabeau, E. (2001). *Self-Organization in Biological Systems*. Princeton University Press, Princeton, NJ.

Carney, W.P. (1969). Behavioral and morphological changes in carpenter ants harbouring dicrocoeliid metacercariae. *American Midland Naturalist*, **82**, 605–11.

Cervo, R. and Dani, F. (1996) Social parasitism and its evolution in *Polistes*. In: S. Turillazzi and M.J. West-Eberhard (eds), *Natural History and Evolution of Paper-wasps*, pp. 98–112. Oxford University Press, Oxford.

Costa, J.T. (2006) *The Other Insect Societies*. Harvard University Press, Cambridge, MA.

Cremer, S., Armitage, S.A.O., and Schmid-Hempel, P. (2007). Social immunity. *Current Biology*, **17**, R693–R702.

Crespi, B.J. and Cho, J.C. (eds). (1997). *The Evolution of Social Behaviour in Insects and Arachnids*. Cambridge University Press, Cambridge.

Crespi, B.J. and Yanega, D. (1995). The definition of eusociality. *Behavioral Ecology*, **6**, 109–115.

Dawkins, R. (1976). *The Selfish Gene*. Oxford University Press, Oxford.

Dawkins, R. (1982). *The Extended Phenotype*. W.H. Freeman, Oxford.

Dawkins, R. (1990). Parasites, desiderata lists and the paradox of the organism. *Parasitology*, **100**, S63–S73.

Dawkins, R. (2004). Extended phenotype—but not too extended. A reply to Laland, Turner and Jablonka. *Biology and Philosophy*, **19**, 377–396.

Dawkins, R. and Krebs, J.R. (1979). Arms races between and within species. *Proceedings of the Royal Society of London Series B: Biological Sciences*, **205**, 489–511.

Epsky, N.D. and Capinera, J.L. (1988). Efficacy of the entomogenous nematode *Steinernema feltiae* against a subterranean termite, *Reticulitermes tibialis* (Isoptera, Rhinotermitidae). *Journal of Economic Entomology*, **81**, 1313–1317.

Flack, J.C., Girvan, M., de Waal, F.B.M., and Krakauer, D.C. (2006). Policing stabilizes construction of social niches in primates. *Nature*, **439**, 426–429.

Foster, K.R. and Ratnieks, F.L.W. (2005). A new eusocial vertebrate? *Trends in Ecology and Evolution*, **20**, 363–364.

Foster, K.R., Wenseleers, T., and Ratnieks, F.L.W. (2006). Kin selection is the key to altruism. *Trends in Ecology and Evolution*, **21**, 57–60.

Franks, N.R. and Richardson, T. (2006). Teaching in tandem-running ants. *Nature*, **439**, 153–153.

Fujiyuki, T., Takeuchi, H., Ono, M. *et al.* (2004). Novel insect picorna-like virus identified in the brains of aggressive worker honeybees. *Journal of Virology*, **78**, 1093–1100.

Gardner, J., Marsack, P., Trueman, J., Calcott, B., and Heinsohn, R. (2007). Story-telling: an essential part of science. *Trends in Ecology and Evolution*, **22**, 510–510.

Garnier, S., Gautrais, J., and Theraulaz, G. (2007). The biological principles of swarm intelligence. *Swarm Intelligence*, **1**, 3–31.

Gould, S.J. and Lewontin, R.C. (1979). Spandrels of San-Marco and the Panglossian paradigm—a critique of the adaptationist program. *Proceedings of the Royal Society of London Series B: Biological Sciences*, **205**, 581–598.

Greene, M.J. and Gordon, D.M. (2003). Social insects—cuticular hydrocarbons inform task decisions. *Nature*, **423**, 32.

Hamilton, W.D. (1963). Evolution of altruistic behavior. *The American Naturalist*, **97**, 354–356.

Hamilton, W.D. (1964a). The genetical evolution of social behaviour. I. *Journal of Theoretical Biology*, **7**, 1–16.

Hamilton, W.D. (1964b). The genetical evolution of social behaviour. II. *Journal of Theoretical Biology*, **7**, 17–52.

Hansell, M. (2004). *Animal Architecture*. Oxford University Press, Oxford.

Hansell, M.H. (1996). Wasps make nests: nests make conditions. In: S. Turillazzi and M.J. West-Eberhard (eds), *Natural History and Evolution of Paper-wasps*, pp. 272–289. Oxford University Press, Oxford.

Hölldobler, B. (1979). Territories of the African weaver ant (Oecophylla longinoda (Latreille)). A field study. *Zeitschrift für Tierpsychologie—Journal of Comparative Ethology*, **51**, 201–213.

Hölldobler, B. (1981). Foraging and spatiotemporal territories in the honey ant *Myrmecocystus mimicus* Wheeler (Hymenoptera: Formicidae). *Behavioural Ecology and Sociobiology*, **9**, 301–314.

Hölldobler, B. (1995). The chemistry of social regulation: multicomponent signals in ant societies. *Proceedings of the National Academy of Sciences of the USA*, **92**, 19–22.

Hölldobler, B. and Wilson, E.O. (1990). *The Ants*. Harvard University Press, Cambridge, MA.

Hoover, J.P. and Robinson, S.K. (2007). Retaliatory mafia behavior by a parasitic cowbird favors host acceptance of parasitic eggs. *Proceedings of the National Academy of Sciences of the USA*, **104**, 4479–4483.

Hughes, D.P. (2005). Parasitic manipulation: a social context. *Behavioural Processes*, **68**, 263–266.

Hughes, D.P. and Kathirithamby, J. (2005). Cost of strepsipteran macroparasitism for immature wasps: does sociality modulate virulence? *Oikos*, **110**, 428–434.

Hughes, D.P., Moya-Raygoza, G., and Kathirithamby, J. (2003). The first record among Dolichoderinae (Formicidae) of parasitism by Strepsiptera. *Insectes Sociaux*, **50**, 148–150.

Hughes, D.P., Kathirithamby, J., and Beani, L. (2004a). Prevalence of the parasite Strepsiptera in adult *Polistes* wasps: field collections and literature overview. *Ethology, Ecology and Evolution*, **16**, 363–375.

Hughes, D.P., Kathirithamby, J., Turillazzi, S., and Beani, L. (2004b). Social wasps desert the colony and aggregate outside if parasitized: parasite manipulation? *Behavioral Ecology*, **15**, 1037–1043.

Jablonka, E. (2004). From replicators to heritably varying phenotypic traits: the extended phenotype revisited. *Biology and Philosophy*, **19**, 353–375.

Jeanne, R.L. (1979). Construction and utilization of multiple combs in *Polistes canadensis* in relation to the biology of a predaceous moth. *Behavioral Ecology and Sociobiology*, **4**, 293–310.

Kastberger, G. and Sharma, D.K. (2000). The predator-prey interaction between blue-bearded bee eaters (*Nyctyornis athertoni* Jardine and Selby 1830) and giant honeybees (*Apis dorsata* Fabricius 1798). *Apidologie*, **31**, 727–736.

Kathirithamby, J. and Hughes, D.P. (2005). Description and biological notes of the first species of *Xenos* (Strepsiptera : Stylopidae) parasitic in *Polistes carnifex* F. (Hymenoptera : Vespidae) in Mexico. *Zootaxa*, 35–45.

Keller, L. and Genoud, M. (1997). Extraordinary lifespans in ants: a test of evolutionary theories of ageing. *Nature*, **389**, 958–960.

Kilner, R.M. and Davies, N.B. (1999). How selfish is a cuckoo chick? *Animal Behaviour*, **58**, 797–808.

King, L.E., Douglas-Hamilton, I., and Vollrath, F. (2007). African elephants run from the sound of disturbed bees. *Current Biology*, **17**, R832–R833.

Laland, K.N. (2004). Extending the extended phenotype. *Biology and Philosophy*, **19**, 313–325.

Laland, K.N. and Sterelny, K. (2006). Seven reasons (not) to neglect niche construction. *Evolution*, **60**, 1751–1762.

McAuliffe, K. and Whitehead, H. (2005). Eusociality, menopause and information in matrilineal whales. *Trends in Ecology and Evolution*, **20**, 650–650.

Maeyama, T., Terayama, M., and Matsumoto, T. (1994). The abnormal behavior of *Colobopsis* sp. (Hymenoptera: Formicidae) parasitized by *Mermis* (Nematoda) in Papua New Guinea. *Sociobiology*, **24**, 115–119.

Martin, S.J., Beekman, M., Wossler, T.C., and Ratnieks, F.L.W. (2002). Parasitic Cape honeybee workers, *Apis mellifera capensis*, evade policing. *Nature*, **415**, 163–165.

Matsuura, M. (1999). Size and composition of swarming colonies in *Provespa anomala* (Hymenoptera, Vespidae), a nocturnal social wasp. *Insectes Sociaux*, **46**, 219–223.

Mayr, E. (1997). The objects of selection. *Proceedings of the National Academy of Sciences of the USA*, **94**, 2091–2094.

Miura, O., Kuris, A.M., Torchin, M.E., Hechinger, R.F., and Chiba, S. (2006). Parasites alter host phenotype and may create a new ecological niche for snail hosts. *Proceedings of the Royal Society B: Biological Sciences*, **273**, 1323–1328.

Mock, D.W. and Parker, G.A. (1997). *The Evolution of Sibling Rivalry*. Oxford University Press, Oxford.

Monnin, T., Ratnieks, F.L.W., Jones, G.R., and Beard, R. (2002). Pretender punishment induced by chemical signalling in a queenless ant. *Nature*, **419**, 61–65.

Moore, J. (2002). *Parasites and the Behavior of Animals*. Oxford University Press, Oxford.

Moore, J., Adamo, S., and Thomas, F. (2005). Manipulation: expansion of the paradigm. *Behavioural Processes*, **68**, 283–287.

Muller, C.B. and Schmid-Hempel, P. (1993). Exploitation of cold temperature as defense against parasitoids in bumblebees. *Nature*, **363**, 65–67.

Odling-Smee, F.J., Laland, K.N., and Feldman, M.W. (2003). *Niche Construction: the Neglected Process in Evolution*. Princeton University Press, Princeton, NJ.

Oster, G. (1976). Modelling social insect populations I. Ergonomics of foraging and population growth in Bumblebees. *The American Naturalist*, **110**, 215–245.

Oster, G.F. and Wilson, E.O. (1978) *Caste and Ecology in the Social Insects*. Princeton University Press, Princeton, NJ.

Pardi, L. (1948). Dominance order in *Polistes* wasps. *Physiological Zoology*, **21**, 1–13.

Pierce, N.E. (1995). Predatory and parasitic Lepidoptera: carnivores living on plants. *Journal of the Lepidopterists Society*, **49**, 412–453.

Pierce, N.E., Braby, M.F., Heath, A. *et al.* (2002). The ecology and evolution of ant association in the Lycaenidae (Lepidoptera). *Annual Review of Entomology*, **47**, 733–771.

Pigliucci, M. and Kaplan, J. (2000). The fall and rise of Dr Pangloss: adaptationism and the Spandrels paper 20 years later. *Trends in Ecology and Evolution*, **15**, 66–70.

Ponton, F., Biron, D.G., Moore, J., Moller, A.P., and Thomas, F. (2006). Facultative virulence: a strategy to manipulate host behaviour? *Behavioural Processes*, **72**, 1–5.

Poulin, R. (1994). The evolution of parasite manipulation of host behavior: a theoretical analysis. *Parasitology*, **109**, S109–S118.

Poulin, R. (1998) *The Evolutionary Ecology of Parasites*. Chapman & Hall, London.

Poulin, R. (2000). Manipulation of host behaviour by parasites: a weakening paradigm? *Proceedings of the Royal Society Series B: Biological Sciences*, **267**, 787–792.

Poulin, R. (2007). *Evolutionary Ecology of Parasites*, 2nd edn. Chapman & Hall, London.

Pratt, S.C., Mallon, E.B., Sumpter, D.J.T., and Franks, N.R. (2002). Quorum sensing, recruitment, and collective decision-making during colony emigration by the ant *Leptothorax albipennis*. *Behavioral Ecology and Sociobiology*, **52**, 117–127.

Queller, D.C. and Strassmann, J.E. (2002). The many selves of social insects. *Science*, **296**, 311–313.

Ratnieks, F.L.W. (1988). Reproductive harmony via mutual policing by workers in eusocial Hymenoptera. *The American Naturalist*, **132**, 217–236.

Rau, P. (1933). *Jungle Bees and Wasps of Barro Colorado Island*. Published by the author.

Read, A.F. (1994). The evolution of virulence. *Trends in Microbiology*, **2**, 73–76.

Reeve, H.K. and Keller, L. (1999). Burying the units-of-selection debate and unearthing the crucial new issues. In: L. Keller (ed.), *Levels of Selection in Evolution*. Princeton University Press, Princeton, NJ.

Reeve, H.K., Peters, J.M., Nonacs, P., and Starks, P.T. (1998). Dispersal of first 'workers' in social wasps: causes and implications of an alternative reproductive strategy. *Proceedings of the National Academy of Sciences of the USA*, **95**, 13737–13742.

Richardson, T.O., Houston, A.I., and Franks, N.R. (2007). Teaching with evaluation in ants. *Current Biology*, **17**, 1520–1526.

Ross, K.G. and Matthews, R.W. (1991). *The Social Biology of Wasps*. Cornell University Press, Ithaca, NY.

Salt, G. (1927). The effects of stylopization on aculeate Hymenoptera. *Journal of Experimental Zoology*, **48**, 223–331.

Salt, G. (1931). A further study on effects of stylopization. *Journal of Experimental Zoology*, **59**, 133–166.

Schmid-Hempel, P. (1998). *Parasites in Social Insects*. Princeton University Press, Princeton, NJ.

Schmid-Hempel, P. and Schmid-Hempel, R. (1984). Life duration and turnover of foragers in the ant *Cataglyphis bicolor* (Hymenoptera, Formicidae). *Insectes Sociaux*, **31**, 345–360.

Schmid-Hempel, P. and Schmid-Hempel, R. (1990). Endoparasitic larvae of conopid flies alter pollination behavior of bumblebees. *Naturwissenschaften*, **77**, 450–452.

Schmid-Hempel, P. and Stauffer, H.P. (1998). Parasites and flower choice of bumblebees. *Animal Behaviour*, **55**, 819–825.

Schmid-Hempel, R. and Muller, C.B. (1991). Do parasitized bumblebees forage for their colony. *Animal Behaviour*, **41**, 910–912.

Schultz, T.R., Mueller, U.G., Currie, C.R., and Rehner, S.A. (2005). Reciprocal illumination: a comparison of agriculture in humans and in fungus-growing ants. In: F.E. Vega and M. Blackwell (eds), *Insect–Fungal Associations*, pp. 149–190. Oxford University Press, New York.

Seeley, T.D. (1995). *The Wisdom of the Hive*. Harvard University Press, Cambridge, MA.

Seeley, T.D. and Buhrman, S.C. (1999). Group decision making in swarms of honey bees. *Behavioral Ecology and Sociobiology*, **45**, 19–31.

Seeley, T.D., Visscher, P.K., and Passino, K.M. (2006). Group decision making in honey bee swarms. *American Scientist*, **94**, 220–229.

Segerstråle, U. (2007). An eye on the core: Dawkins and sociobiology. In: A. Grafen and M. Ridley (eds), *Richard Dawkins: How a Scientist Changed the Way we Think*. Oxford University Press, Oxford.

Sherman, P.W., Seeley, T.D., and Reeve, H.K. (1988). Parasites, pathogens, and polyandry in social Hymenoptera. *The American Naturalist*, **131**, 602–610.

Shykoff, J.A. and Schmid-Hempel, P. (1991). Incidences and effects of four parasites in populations of bumble bees in Switzerland. *Apidologie*, **22**, 117–125.

Strambi, C., Strambi, A., and Augier, A. (1982). Protein levels in the haemolymph of the wasp *Polistes gallicus* L. at the beginning of imaginal life and during overwintering. Action of the strepsipteran parasite *Xenos vesparum* Rossi. *Experentia*, **38**, 1189–1191.

Strassmann, J.E., Zhu, Y., and Queller, D.C. (2000). Altruism and social cheating in the social amoeba *Dictyostelium discoideum*. *Nature*, **408**, 965–967.

Sumpter, D.J.T. (2006). The principles of collective animal behaviour. *Philosophical Transactions of the Royal Society B: Biological Sciences*, **361**, 5–22.

Thomas, F., Schmidt-Rhaesa, A., Martin, G., Manu, C., Durand, P., and Renaud, F. (2002). Do hairworms (Nematomorpha) manipulate the water seeking behaviour of their terrestrial hosts? *Journal of Evolutionary Biology*, **15**, 356–361.

Thomas, F., Adamo, S., and Moore, J. (2005). Parasitic manipulation: where are we and where should we go? *Behavioural Processes*, **68**, 185–199.

Tibbetts, E.A. (2002). Visual signals of individual identity in the wasp *Polistes fuscatus*. *Proceedings of the Royal Society Series B: Biological Sciences*, **269**, 1423–1428.

Trabalon, M., Plateaux, L., Peru, L., Bagneres, A.G., and Hartmann, N. (2000). Modification of morphological characters and cuticular compounds in worker ants *Leptothorax nylanderi* induced by endoparasites *Anomotaenia brevis*. *Journal of Insect Physiology*, **46**, 169–178.

Turillazzi, S. (1991). The Stenogastrinae. In: K.G. Ross and R.W. Matthews (eds), *The Social Biology of the Wasps*. Cornell University Press, Ithaca, NY.

Turillazzi, S. and West-Eberhard, M.J. (eds) (1996). *Natural History and Evolution of Paper-wasps*. Oxford University Press, Oxford.

Visscher, P.K. (2007). Group decision making in nest-site selection among social insects. *Annual Review of Entomology*, **52**, 255–275.

Von Frisch, K. (1968). Role of dances in recruiting bees to familiar sites. *Animal Behaviour*, **16**, 531–533.

Von Frisch, K. (1974). *Animal Architecture*. Harcourt Brace Jovanovich, New York.

Wcislo, W.T. (1999). Transvestism hypothesis: a cross-sex source of morphological variation for the evolution of parasitism among sweat bees (Hymenoptera: Halictidae)? *Annals of the Entomological Society of America*, **92**, 239–242.

Wenseleers, T. and Ratnieks, F.L.W. (2006). Enforced altruism in insect societies. *Nature*, **444**, 50.

West-Eberhard, M.J. (1969). The social biology of Polistine wasps. *Miscellaneous Publications of the Museum of Zoology, University of Michigan*, **140**, 1–101.

West-Eberhard, M.J. (2003). *Developmental Plasticity and Evolution*. Oxford University Press, New York.

Wheeler, W.M. (1911). The ant-colony as an organism. *Journal of Morphology*, **22**, 307–325.

Wilson, E.O. (1959). Source and possible nature of the odor trail of fire ants. *Science*, **129**, 643–644.

Wilson, E.O. (1971). *The Insect Societies*. Harvard University Press, Cambridge, MA.

Wilson, E.O. (1975). *Sociobiology. The New Synthesis* Belknap Press, Cambridge, MA.

Wilson, E.O. (1983). Caste and division of labor in leaf-cutter ants (Hymenoptera, Formicidae, Atta).3. Ergonomic resiliency in foraging by Atta-Cephalotes. *Behavioral Ecology and Sociobiology*, **14**, 47–54.

Wilson, E.O. (1985). The sociogenesis of insect colonies. *Science*, **228**, 1489–1495.

Wilson, E.O. (2000). *Sociobiology. The New Synthesis 25th Anniversary Edition*. Belknap Press, Cambridge, MA.

Wilson, E.O. (2005). Kin selection as the key to altruism: its rise and fall. *Social Research*, **72**, 159–166.

Wilson, E.O. and Hölldobler, B. (2005). Eusociality: origin and consequences. *Proceedings of the National Academy of Sciences of the USA*, **102**, 13367–13371.

Winfree, R. (1999). Cuckoos, cowbirds and the persistence of broad parasitism. *Trends in Ecology and Evolution*, **14**, 338–343.

Witte, V., Janssen, R., Eppenstein, A., and Maschwitz, U. (2002). *Allopeas myrmekophilos* (Gastropoda, Pulmonata), the first myrmecophilous mollusc living in colonies of the ponerine army ant *Leptogenys distinguenda* (Formicidae, Ponerinae). *Insectes Sociaux*, **49**, 301–305.

Zahavi, A. (1979). Parasitism and nest predation in parasitic cuckoos. *The American Naturalist*, **113**, 157–159.

CHAPTER 11

Synergy in social communication

David J. T. Sumpter and Åke Brännström

11.1 Introduction

Synergy is where a group of individuals co-ordinate to achieve a task that would not be possible were they to act alone. In social insects: the combs inside a honey bee nest (Camazine 1991), the foraging trails or labyrinth of underground tunnels of many ant species (Buhl *et al.* 2004; Franks *et al.* 1991; Theraulaz *et al.* 2003), and the thermoregulatory termite mound (Korb 2003) are made possible only by the interaction of thousands or millions of individuals. A similar observation can be made about our own human society. Without effective division of labour, specialization, and massively co-ordinated effort we would be unable to build bridges, create transport networks or organize the complex economic activity that characterizes modern society. On a smaller scale, wolves, lions and other predators hunt in groups which allows them to tackle larger prey (Packer and Ruttan 1988), while prey may be able to use their numbers to confuse predators (Treherne and Foster 1981). Although not all of these activities require communication, in many cases animals use physical, visual and chemical signals in order to co-ordinate their activities. By communicating with each other the group becomes more than the sum of its parts.

The search for food and other resources provides many examples where communicating individuals can become more than the sum of their parts. Animals often live in environments where resources are distributed in difficult to locate patches which exist only temporarily. In such an environment, a single individual has a very low rate of finding a resource patch if it searches independently. When large numbers of individuals search at the same time, however, the probability that one of them finds one of the patches is considerably larger. If these finds can be effectively communicated between individuals then the rate at which individuals find food increases. Instead of the number of individuals finding resources increasing linearly with group size, it can increase nonlinearly. The overall success of the group becomes more than the sum of each individual's success.

While it may appear that communication improves per capita success in foraging for resources, a number of challenging questions remain. As is often the case in understanding social interactions these questions are both mechanistic and functional (Krebs and Davis 1993; West *et al.* 2007). The mechanistic questions lie in studying different communication mechanisms, such as **pheromone** trails in ants and termites, dances as performed by honeybees and other **signals** and **cues**, and trying to understand how they contribute to group success. This is the question of how the effects of social communication add up. Answering this question is difficult because communication necessarily involves feedback loops, whereby information about the location of resources flows between individuals. These feedback loops can lead to complex relationships between the type of communication mechanism, the environment the animals live in and the per capita success in this environment. The first half of this chapter will review studies where these feedback loops have been identified and discuss how we have, with the help of mathematical models, come to better understand their consequences.

The functional questions are about the conditions under which communication about resources can evolve. While the per capita success of a group might increase with the number of individuals in it, there is not always an 'incentive' on the part of an individual to communicate. If an individual does not communicate the location of a resource then it can monopolize it. Thus while there is an incentive for the average group member to have resource finds communicated, there is not necessarily an incentive for the finder to make that communication. In other words, even if per capita success would increase if all individuals engaged in social communication, this does not imply that it should evolve. There are a number of ways around this dilemma and these are discussed in the second half of this chapter. As the title of this chapter suggests, a particular emphasis is put on synergy: where non-linear increase in foraging success at the group level can lead to incentive at an individual level to cooperate and communicate about resources.

While this chapter will be divided into rather clear halves, first the mechanisms of communication and then the functional consequences, an emphasis will be put on the importance of recognizing the relationship between mechanism and function. How the effects of social communication add up has a strong role in determining why this communication evolved (Sumpter 2006). Indeed, many functional problems in understanding communication may be simply solved by a good mechanistic understanding of the role communication has on per capita foraging success (see Chapter 6 for related point regarding functional and mechanistic approaches). For example, a cornerstone of social evolution is **Hamilton's rule** which provides an inequality between relatedness and the costs and benefits of co-operation (Hamilton 1964). A common criticism, not of Hamilton's rule but of the way it might be applied in practice, is that too much emphasis is put on calculating relatedness (r) and not enough on costs (c) and benefits (b) (Korb and Heinze 2004). A mechanistic approach that recognizes that b and c are not constants, but instead non-linear functions of group size enables us to come to a better understanding of why different forms of social communication evolve in different types of groups. It allows us to bypass much of the abstract, difficult and often unproductive discussion about which level of selection operates and instead concentrate on making empirically testable predictions.

11.2 Mechanisms

Social communication about resources usually occurs between animals which live in a communal nest or den. Communal living provides a good opportunity for transfer of information about the location and quality of resources. Individuals that have found food possess information about where it is located. If successfully communicated, this information can be used by nestmates who can then locate the same food source. This communication can be either through cues, which are "a feature of the world that can be used by the receiver as a guide to future action" (Maynard Smith and Harper 2005) or signals which are "an act or structure that alters the behaviour of another organism, which evolved because of that effect, and which is effective because the receiver's response has also evolved" (ibid) (See also Chapter 1). In this chapter we consider mainly evolved signals. In **eusocial** insects, sophisticated signals have evolved to actively communicate food discoveries, but signals are also found in, for example, cliff swallows as well as other species of birds and mammals.

11.2.1 Ant pheromone trails

Many species of ants deposit pheromone signals marking the route from food to nest (Hölldobler and Wilson 1990; Wilson 1971) (see Chapter 5 and 7 for related discussion on pheromones). These trails allow other ants within the nest and those already exploring to find the food source. As more ants find the pheromone trail and, as a consequence, the food source, a positive feedback loop is started. The pheromone trail is strengthened and after a short time a steady trail of ants is established between food and nest. Pheromone trails are formed purely on the basis of local information. They are started by a single individual or a small group of ants responding to the presence of food and they are reinforced by ants that encounter and successfully follow the trail.

A striking aspect of pheromone trail recruitment is that it needs a minimum number of participants in order to function. A study of the foraging efficiency of different sized Pharaoh's ant colonies revealed that the number of ants arriving at a feeder placed a short distance from the nest increased as a non-linear function of colony size (Beekman *et al.* 2001). Figure 11.1a shows how foraging success changes with colony size for these ants. Below a critical colony size the foragers functioned independently of one another. When the feeder was presented to small colonies there was no increase in the number of ants visiting the feeder. Above the critical size however there was a sudden switch to effective foraging by the ants, and the number of ants visiting the feeder increased with colony size. Similar results, although with a smaller number of colony sizes have also been observed in Argentine ants (Halley and Burd 2004).

Why does foraging success increase non-linearly with colony size? We can answer this question by thinking about what happens when an ant in a colony of a particular size finds food. If an ant in a small colony finds a food source a long way from the nest, then by the time another ant passes over the place she left pheromone trail, the pheromone will probably have evaporated. In this case, the trail doesn't help other ants find the food. For large colonies of ants, however, it is more likely that an ant will find the pheromone trail before it evaporates, follow it and thus reinforce it. Beekman *et al.* (2001) formalised this argument in a differential equation model of the mechanisms underlying pheromone trail laying. Their model is given in Box 11.1. The key assumptions of the model are that the rate of individual ants joining a trail is an increasing function of the number of ants on the trail, while the rate of individuals losing the trail is a decreasing function of group size. In this model a *bifurcation* occurs whereby at a critical colony size foraging trails become effective (Figure 11.2).

While the effect of colony size on foraging success has been investigated in these short time scale experiments, less is known about the long term fitness consequences of using pheromone trails. In general, the study of per capita 'lifetime' productivity of ant societies has been focused on the early stages of colony foundation, where increases in productivity are usually attributed to co-operative building and defense than to foraging success. One of these studies that could relate to information transfer looked at brood raids by fire ants on other nearby ant colonies. Adams and Tschinkel (1995) found that nests consisting of multiple queens produced more workers and then had an increased success during raids on other colonies. Further work is required in this area, especially to link foraging efficiency to colony growth and fitness measures.

Pheromone trails act not only to inform nestmates where food is located, but can also be used to find the shortest path to it. For example, Beckers *et al.* (1992) presented starved colonies of the ant *Lasius niger* with two alternative bridges of different lengths between food and nest. In these experiments, individual ants made little or no comparison of the two bridges, instead the slightly longer trip time means that pheromone builds up slower on the longer bridge. Thus when trail following ants make the choice between two bridges they detect a higher concentration of pheromone on one of the bridges, i.e. the shorter one (Beckers *et al.* 1993). The shorter bridge is chosen with a higher probability by the follower ants and when these ants return home they further reinforce the shortest path.

Theoretical studies have shown that effective group-level decisions about which of several paths to food is shortest also require a minimum number of ants (Nicolis and Deneubourg 1999; Nicolis *et al.* 2003). Once an equilibrium flow of ants has been reached on the bridge, ants in small colonies will choose bridges in proportion to the respective lengths of the bridges, while ants in large colonies will be strongly biased towards the shortest bridge. Thus although in both cases the ants are capable of choosing the shortest path, ants in large colonies focus their workforce more strongly on the shorter bridge. Similar results apply when ants are choosing between two food sources of different quality, ants are more likely to lay a pheromone trail to a better quality food source (Sumpter and Beekman 2003). These studies suggest that larger colonies are likely to be able to build more efficient pheromone trail networks, leading to better quality food sources. Per capita foraging success will thus increase with group size.

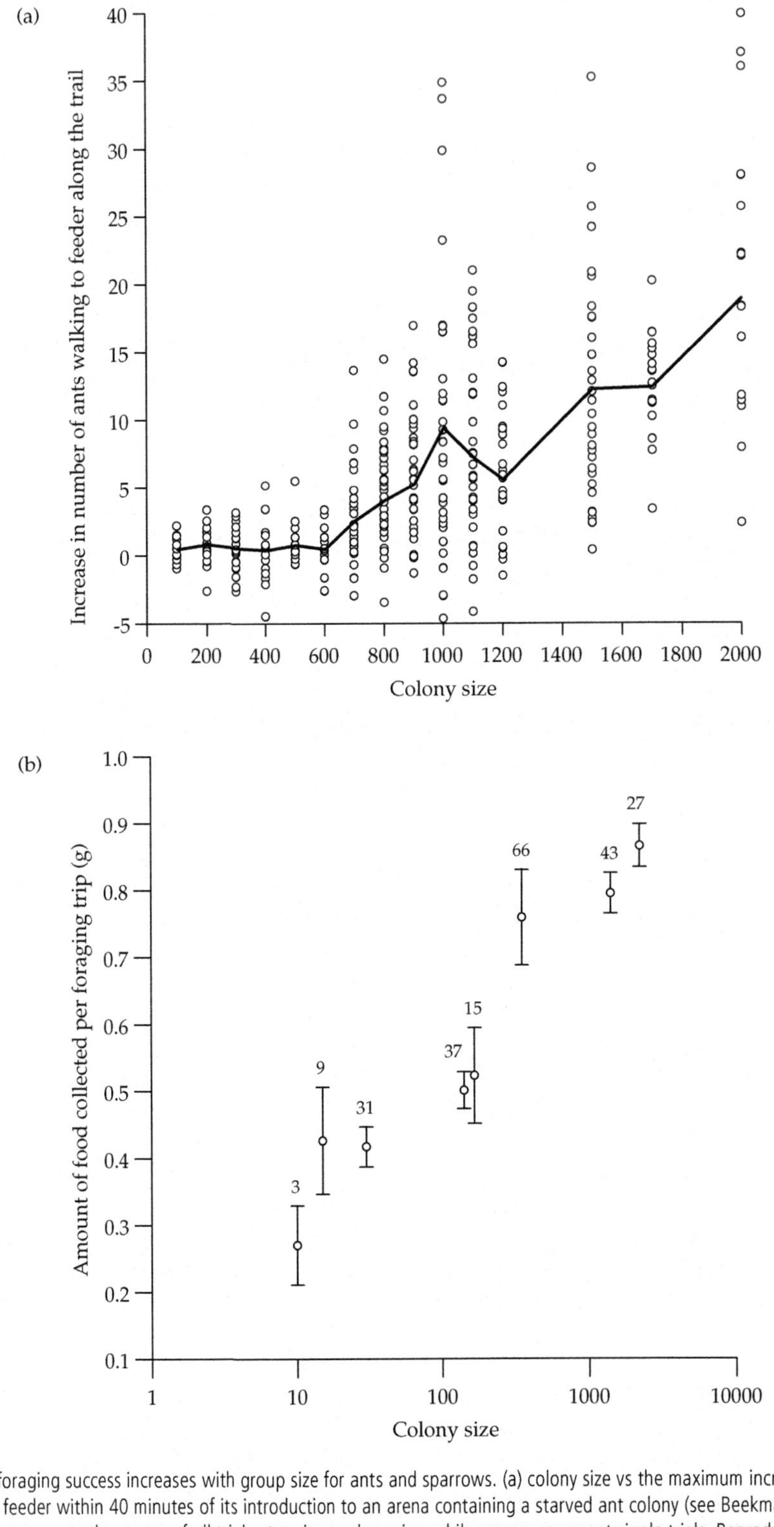

Figure 11.1 How foraging success increases with group size for ants and sparrows. (a) colony size vs the maximum increase in the number of ants walking to a feeder within 40 minutes of its introduction to an arena containing a starved ant colony (see Beekman *et al.*, 2001 for details). The solid line connects the means of all trials at a given colony size, while crosses represent single trials. Reproduced from Beekman *et al.*, 2001. (b) group size vs per capita foraging success. The figure shows how the amount of food collected per parent cliff swallow increases with the size of colony the swallow lives in. Reproduced from Brown and Brown 1996.

Box 11.1 Mechanistic models of foraging

General framework

Here we give a simple mechanistic model, based on a framework provided by Sumpter and Pratt (2003), of how a group of communicating foragers find a food source. We assume a group of n individuals and let x be the number of these individuals that are informed about where the food source is. Thus x is a measure of group productivity, since it is the number of individuals successfully foraging at any point, and x/n is the per capita success since it is the proportion of informed individuals in the group.

We assume that social communication increases the rate at which individuals find a food source. We denote the rate per individual of finding the feeder with $f(x)$ which is an increasing function of x. This function f differs depending on the recruitment mechanism used by the modeled species. We denote the rate per individual of losing the food source as $g(x)$ which again differs dependent on the mechanism by which individuals lose the feeder. We thus denote the rate of change of the number individuals going to the feeder as

$$\frac{dx}{dt} = f(x)(n-x) - g(x)x$$

where t denotes time. Solving

$$f(x_*)(n-x_*) = g(x_*)x_*$$

for x_* gives the equilibrium, or long term, number of individuals going to the feeder. It is by finding how this equilibrium changes with n that we determine how foraging efficiency changes with group size.

Pheromone trail communication

Beekman *et al.* (2001) use a version of this model to look at pheromone foraging ants. In this case, x denotes the number of ants on a pheromone trail to a feeder. They set

$$f(x) = \alpha + \beta x$$

where α is the rate at which individuals find the food source in the absence of a trail and βx models the fact that the number of ants finding the food source increases with the number leaving a trail to it. They further set

$$g(x) = \frac{s}{K+x}$$

which is a decreasing function of ants on the trail.

The equilibrium number of individuals on the trail to the food source is given by

$$\alpha(n-x_*) + \beta x(n-x_*) = \frac{sx_*}{K+x_*} \qquad \text{(A.1)}$$

Figures 11.2a and 11.2b shows how x_* changes with n for two different values of α. When random finds of the feeder are frequent (Figure 11.2a) then there is a unique stable equilibrium. When random finds of the feeder are infrequent (Figure 11.2b) there is an intermediate range of values of n at which there are two stable equilibrium. In both cases the number of individuals visiting the feeder, i.e. foraging success, is a non-linear increasing function of the number of foragers. Figures 11.2c and 11.2d shows that in both cases there is a range of group sizes at which per capita foraging success also increases with group size.

Dance communication

Beekman *et al.* (2007) propose that with dance communication the rate at which bees find a feeder can be written as

$$f(x) = \alpha(1-a/A)^{px} + \beta(1-(1-a/A))^{px}$$

where A is the total area of the honey bee colonies dance floor, a is the area covered by a single dancing bee and p is the proportion of time that a foraging bee spends dancing during a round trip to the feeder. The expression $1-(1-a/A)^{px}$ is the probability that an uninformed bee can find a dancing bee to follow (Beekman *et al.* 2007). α and, respectively, β are the rate

continues

Box 11.1 *continued*

at which bees which could and, respectively, could not find a dance to follow find the food source. The rate at which honey bees retire from following is independent of the behaviour of other foragers, so that $g(x)=\lambda$.

The equilibrium number of bees going to the food source is thus given by

$$\left(\alpha(1-a/A)^{px}+\beta(1-(1-a/A))^{px}\right)(n-x)=\lambda x \qquad \text{(A.2)}$$

Figure 11.3a shows how x_* changes with n for some typical parameter values. In this case there is only one stable equilibrium for the number of foragers going to the food source. The foraging success is a non-linear increasing function of the number of foragers, although not as dramatic as that for the ants. Figure 11.3b shows that there is a range of group sizes at which per capita foraging success also increases with group size.

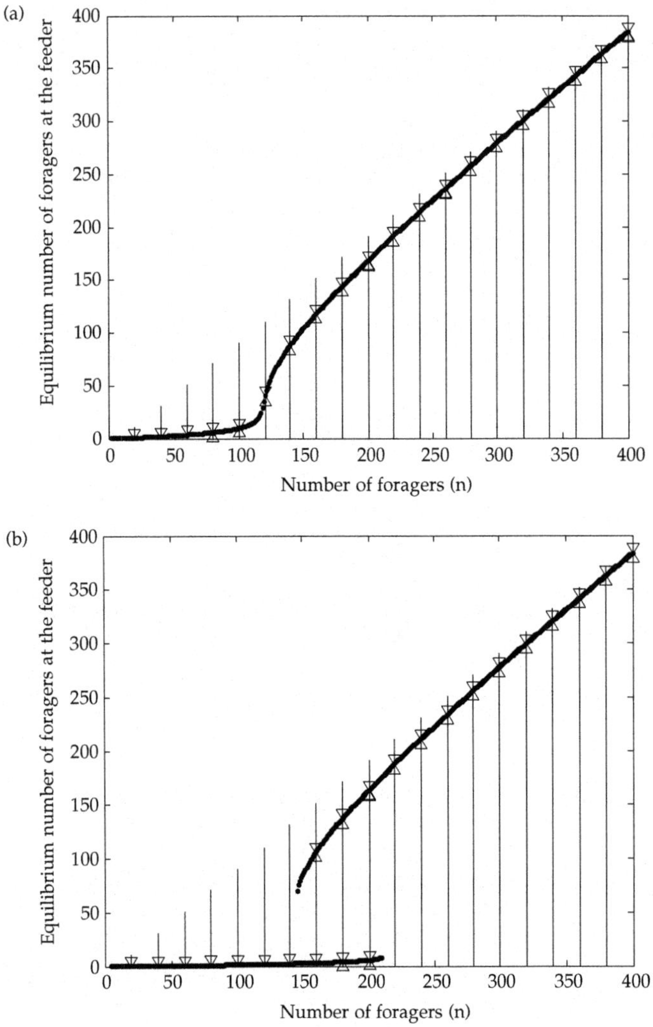

Figure 11.2 *continues*

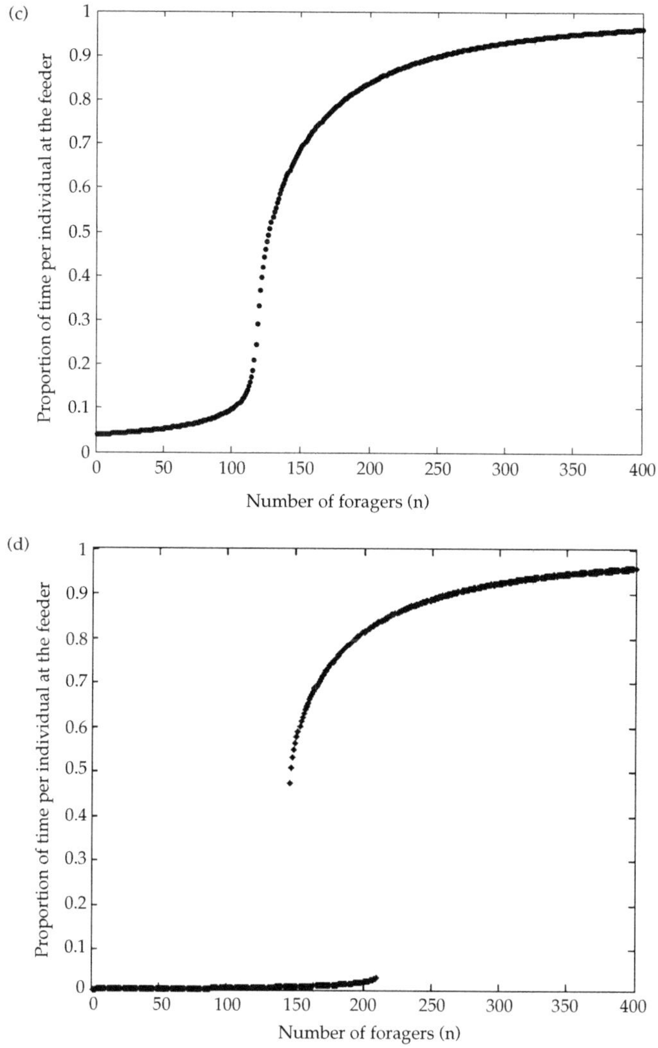

Figure 11.2 Prediction of model given by equation A.1 (see Box 11.1 for details) of how the number of foragers using a trail to food changes with number of foraging ants in the colony in two different cases (a) when random finds of the food source are common ($\alpha = 0.004$) and (b) when random finds are rare ($\alpha = 0.001$). The black lines give the predicted stable equilibrium for number of foragers visiting the feeder. The arrows indicate which equilibrium occurs given different initial number of ants at the feeder. The per individual proportion of time on the trail is shown for (c) frequent and (d) infrequent finds. Note that the dark lines in (c) and (d) are simply the lines on (a) and (b) divided by the number of foragers. Other parameter values are $\beta = 0.00015$, $s = 1$ and $K = 10$. See Sumpter and Pratt (2003) for details of the analysis of this model.

11.2.2 Honey bee dances

One of the most comprehensively studied forms of social communication about resources is the waggle dance of the honey bee (Seeley 1995; von Frisch 1967). Waggle dances are performed by honey bee foragers that have found highly rewarding nectar or pollen sources, and inform nestmates about the location of these resources. Uninformed bees in the hive follow a dance and then fly in the direction of and for the distance encoded by the dance, after which they search locally using odour and visual cues (Riley *et al.* 2005). Usually a recruited bee will fail to find the advertised food site, but by repeatedly returning to the dance floor and following further dances she will eventually find and return with food (Seeley and Towne 1992). Since recruited bees may later perform dances themselves, the waggle dance, like pheromone trails, acts as a positive feedback mechanism through which information about food is transferred.

Theoretical predictions about how foraging success changes with colony size are different for dance communication than for pheromone recruitment.

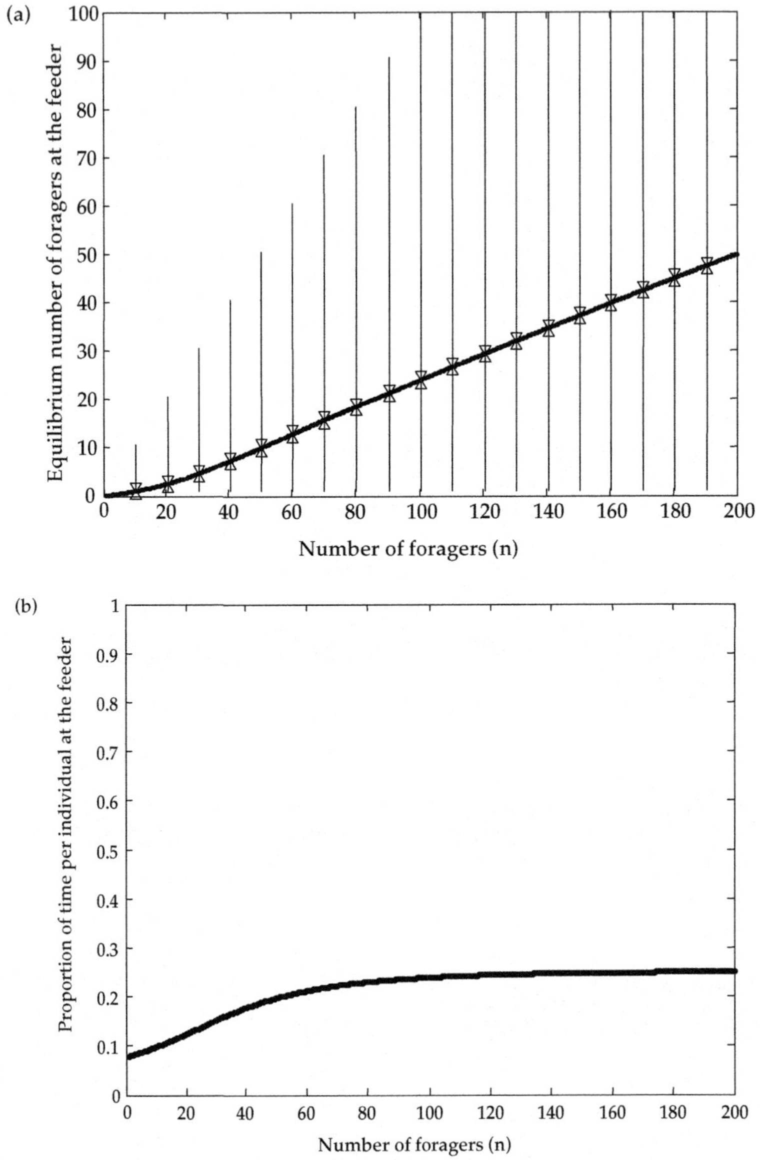

Figure 11.3 (a) Prediction of model given by equation A.2 (see Box 11.1 for details) of how the number of foragers visiting a feeder changes with number of foraging bees in the colony in two different cases. The black lines give the predicted stable equilibrium for number of foragers visiting the feeder. The arrows indicate which equilibrium occurs given different initial number of ants at the feeder. (b) The per individual proportion of time spent at the feeder.

The difference between the two forms of recruitment arises from the functions with govern rates of recruitment and retirement from a food source. The rate at which bees are recruited to food is, like the pheromone recruiting ants, an increasing function of the number dancing for the food. However, unlike the ants, the decision by a bee to retire from a food source is made independently of the number of other bees foraging and purely based on the quality of the source (Seeley 1995). Thus rather than retirement rate per individual decreasing with the number of individuals at the food source it is a constant, independent of the number of other foragers. Box 11.1 investigates how independent retirement affects foraging success. As with the ants, foraging success is a non-linear function of colony size although it is less dramatically so than it is with the ants (Figure 11.3). Very small colonies lose out because potential foragers have difficulty locating dances to follow. This result is also obtained in more detailed individual based models of honey bee foraging (Dornhaus *et al.* 2006).

Rather than concentrating on the role of colony size, experimental work on the efficiency of the honey bee dance has looked at the environmental conditions under which dance communication provides a fitness advantage (see Chapter 10 for discussion of another environmental factor, parasites, and their role in collective behaviour). Perhaps surprisingly, given the interest in the evolution of this communication mechanism, the waggle dance does not always result in more efficient foraging by the bees. Sherman and Visscher (2002) found that the dance only provided an advantage, in terms of colony weight gain, during winter months when food was scarce. For the majority of the year, honey bees in hives within which all dance information was disorientated gained weight at a rate not significantly different than control colonies in which dances could be followed as usual. Dornhaus and Chittka (2004) found that in temperate habitats, similar to those in Sherman and Visschers' experiment and in which food is relatively abundant, dance communication again offered no improvement in foraging efficiency. In tropical habitats, however, where food is more tightly clustered, they found that dance disorientated colonies performed worse than control colonies. The dance appears to play an important role in colony survival in circumstances where food is difficult to find and/or highly clustered. Beekman and Lew (2008) used a model to show that if a dancing and no-dancing colonies of bees compete locally for resources, the dancing bees can rapidly monopolise high quality food sources.

One study that did look at the role of colony size in honey bee foraging was conducted by Beekman *et al.* (2004). They looked at how the number of patches visited by bees differed between large and small colonies. Although the number of foragers in these colonies was proportional to colony size, there was little difference in the number of distinct foraging sites danced for by small and large colonies. In other words, foraging was less focused on particular sites in small colonies. One interpretation of these results is that small colonies are unable to focus their foraging on particular sites because there are insufficient dances to efficiently communicate food finds. As a result, the bees in small colonies forage independently of one another and do not benefit from communication about profitable food sources. This hypothesis is supported by experiments in which the number of dances performed within the hive was manipulated (Beekman *et al.* 2007). When there were fewer dances to follow bees scouted independently of each other.

11.2.3 Birds and rodents

Recruitment signals are not limited to social insects. For example, Norway rats deposit odour trails from where they find food back to the nest (Galef and Buckley 1996) (see also Chapter 6). By attracting nestmates, these trails spread information about widely scattered, ephemeral food sources. Naked mole rats also leave odour trails on finding food, make chirping noises during their return trip and display the collected food for nestmates (Judd and Sherman 1996). There is evidence for a weak form of positive feedback with follower naked mole rats vocalising when they find food, but with a lower probability than the initial discoverer. Recruited mole rats appear to look for the trail left by a specific individual, suggesting that recruitment to a particular food source is proportional to the number of recruiting individuals. This would suggest that the improvement in foraging efficiency with group size is more likely to be similar to that predicted for the honey bees (Figure 11.3) than the ants (Figure 11.2).

Cliff swallows nest in colonies at which there is communication about the location of food (Brown and Brown 1996). At the nest this communication is primarily cue-based, with birds that successfully return with food being followed by their nestmates. There is no evidence that successful birds actively advertise food finds at this stage, i.e. communication is cue-based, but there is also no evidence that they disguise these finds. Clear evidence of signalling between birds is however seen in the form of food calling at mobile insect swarms. The signalling birds can track the swarm while being able to make return journeys between the colony and the insects. Brown *et al.* (1991) suggest that cliff swallow signals about food location occur only when the insects upon which they feed are spatially clustered. Indeed, in other contexts sparrows only signal the location of food when it is sufficiently large that it can be shared with other birds (Krebs and Davis, 1993).

When naïve North American ravens were added to communal roosts they followed their informed roost-mates to new feeding sites (Marzluff *et al.* 1996). At the beginning of these flights some birds produce noisy 'kaws' and 'honks', although it is not known whether these are more often produced by informed birds. There is however evidence based on a small number of observations of European ravens that the first birds to be seen at a bait carcass were also those that performed flight displays and vocalisations the evening before and appeared to initiate morning departures from the roost (Wright *et al.* 2003). These observations would suggest that informed ravens actively signal the location of food.

While these examples provide evidence that signaling and social communication exist in a wide range of species, less is known about how the effects of these forms of communication add up. There are however some noteworthy exceptions. For example, it is known that signalling by Norway rats reduces the time it takes individuals to find food (Galef and White 1997), although it is not clear how foraging success changes with group size or the number of trail layers. Foraging success and group size has been studied in cliff swallows. Brown and Brown (1996) found that both the amount of food collected by parent cliff swallow per foraging trip and the frequency of these trips increased with group size. Although each of these factors appeared to increase log linearly with group size (Figure 11.1b), when combined using least squares regression they appeared to add up to a linear increase in per capita success with group size (Brown and Brown 1996).

11.3 Function

Why have animals evolved to communicate with each other about food finds? All of the examples above involve an apparently costly signal between the individual which has found food and those which have not. The cost can be either a direct result of the time or energy expended in making the signal, e.g. in performing a dance or producing pheromone chemicals, or a result of increased competition for the resource signalled for. In order for a costly signal to have evolved there must also be an associated benefit (Maynard Smith and Harper 2005). This benefit must on average be greater than the cost. The key evolutionary question about all systems where we see costly signalling is: what are the benefits of communication?

Such questions do not usually have one simple answer but depend on a whole range of factors. Here, we discuss three types explanations: **inclusive fitness**, synergy and reciprocation. A particular emphasis will be put on synergy, because it arises in situations where foraging success increases as a non-linear function of the number of individuals involved. Indeed, synergistic co-operation requires a good understanding of the mechanisms underlying co-operation.

11.3.1 Co-operation between relatives

The basic idea of inclusive fitness theory as an explanation for the evolution of helping is that genes which promote costly helping towards relatives are likely to be selected for because these relatives have a higher probability than a randomly selected member of the population of carrying the same gene by decent as the helping individual (Hamilton 1964). The gene is passed through the generations not directly but indirectly through these relatives, who because they were helped have a greater reproductive success than individuals without relatives to help them. Thus, in assessing the fitness of a gene we must account for its inclusive fitness, the direct benefit (which may be a negative, i.e. a cost) it gives the individual plus the indirect benefit it gives to relatives (West *et al.* 2006). Social communication about the location of food to relatives can confer indirect fitness benefits to other individuals in the foraging group or to a small group of reproductive individuals. Many social insect species have a high degree of within group relatedness and inclusive fitness is thought to contribute to the evolution of social communication of these species (Bourke and Franks 1995; Foster *et al.* 2006). Indeed, the importance of inclusive fitness is firmly established in evolutionary biology and is the focus of several chapters in this book.

When individuals gain inclusive fitness by co-operating then it is useful to consider what types of communication are best for the group as a whole, rather than for which are best for each individual. The observation that small colonies of ants cannot effectively forage using pheromone trails has interesting consequences for the evolution of communication mechanisms within these colonies. Pheromone trail laying has evolved primarily in ant species which contain large numbers of workers (Beckers *et al.* 1989), with ants that typically live in smaller colonies using tandem running or group recruitment where either another individual or a small group of individuals is directly guided towards the food source found (Hölldobler and Wilson 1990). Furthermore, *Lasius niger* ants change their trail laying behaviour dependent on their colony size (Devigne and Detrain 2002). Ants in small colonies do not leave trails, while those in large colonies do.

11.3.2 Synergy

Consider an animal foraging for food which is located in a difficult to find clump, but with large capacity once it is found. We assume that if a focal individual finds food and communicates its location to its partner then the partner can find the food. Meanwhile the focal individual, which has a limited capacity for how much food it can carry, returns to its nest to feed its offspring. In order for the focal individual to locate the food source again however the partner must also communicate its location. For example, we can imagine a situation where it is difficult to reliably navigate to the food source or that it changes position through time. If the link of communication is broken the food becomes difficult to find again.

To illustrate this idea, we can describe this situation as a two player game where individuals have one of two phenotypes: 'Communicate' or 'Do nothing'. The payoff table for this game is given in Table 11.1. If the focal animal communicates the presence of food it pays a signaling cost C but gets a direct benefit, from for example predator dilution, D. The partner gets benefit B irrespective of whether it signals itself or not, but if it signals it pays cost C. If both signal then both individuals get an extra benefit, E. This extra benefit is acquired because both individuals signal food location and thus are always able to find the food source again after they return to the nest. The term E makes the animals interactions more than the sum of its parts. It arises only when they both communicate.

Assuming that the direct benefit, D, is less than the cost, C, but the extra benefit E is greater than the cost, i.e. $E>C>D$, then there are two evolutionarily stable states for this game. If a focal individual lives in a population where everyone cooperates then it is always better to cooperate, since $B+E-C>B$. However, if the focal individual lives in a population where everyone defects then, because $C>D$, it is better to also defect and avoid a negative payoff. Thus, the two possible evolutionarily stable states are one corresponding to everyone co-operating and another corresponding to

Table 11.1 Payoff table for two player social communication game. The values in the table determine the fitness gained by the focal individual as a function of its own strategy and that of its partner. For a model of synergy we assume that $E>C>D$.

Focal/Partner	'Communicate' (Cooperate)	'Do nothing' (Defect)
'Communicate' (Cooperate)	$B-C+E$	$D-C$
'Do nothing' (Defect)	B	0

everyone defecting. Which evolutionarily stable state the population evolves to depends on the initial conditions. If the population initially contains more than C/E co-operators then evolution will lead to full co-operation, otherwise evolution will lead to full defection.

A counter-intuitive prediction of this model is that costly behaviour can evolve even when the focal individual gets no benefit from co-operating when interacting with other individuals that defect. This point is not always given full consideration when discussing the evolution of costly signals. For example, cliff swallows call to signal the location of insect swarms thus paying a, probably small, cost but provide nearby foraging partners a positive benefit in finding food. Brown *et al.* (1991) suggest, quite correctly, that swallows may have evolved call signalling because "even if other birds do not also call, the caller could benefit through local enhancement simply by watching the nearby group members as some of them track the subsequent movement of the prey". If this is the case, then there may be no cost to interacting with a defector, i.e. $D > C$, and full co-operation always evolves even in a population of defectors. However, our model suggests that a direct benefit of signalling is not a requirement for the evolution of food calling. Rather, the model predicts that provided there is an extra benefit when both birds call that is greater than the cost of calling, then co-operation can evolve independent of any direct benefits in the absence of calling, i.e. $B < C$. It is plausible that such extra benefits exist for cliff swallows. Groups that contain individuals that always signal can continuously track the movement of insect swarms. When interacting with a co-operator the focal individual gets the additional benefit, $E > C$, of being able to refind its own discovery. Defection would reduce both the focal and the partner birds' ability to find food.

The last paragraph takes a two player game and suggests it may apply to multi-player interactions. Swallows don't forage in pairs but rather in large groups. Under what circumstances can synergistic co-operation persist in larger groups? Box 11.2 describes a continuous strategy game with group size N in which each individual can make an investment p_i in communicating. This investment incurs a cost, but group productivity increases as a function of the total communication by the group. This productivity is shared equally between individuals, so benefit to an individual increases as productivity divided by number of group members. Co-operative communication is evolutionarily stable for large groups in this model provided that group productivity increases as at least the square of the group size, or equivalently provided that the benefit per individual increases at least linearly with group size.

Clearly group productivity cannot increase indefinitely with group size. Once the group is so large that any exploitable food sources are quickly exhausted, then it is no longer beneficial to signal about its location. By the time signalling individuals return to the food source it will have been consumed by all the other group members. Figure 11.5 shows how the evolutionarily stable states change with group size with a productivity function which first increases non-linearly then only linearly with group size. For small groups there is an evolutionarily stable state where all individuals cooperate, but as the group becomes too large this becomes unstable and all defect is the only evolutionarily stable state. The important message here is that when resources are sufficiently large but difficult to find then co-operation can evolve.

Cliff swallows nest in colonies of genetically unrelated individuals and inclusive fitness plays little or no role in the evolution of their foraging behaviour. In Figure 11.1b we saw how per capita delivery of food to the nest changes as a function of group size for cliff swallows. Although it is not entirely clear whether this increase is equal or greater to linear, as would be required under the model in Box 11.2, the fact that an increase is seen suggests that the swallows interactions could be synergistic. The evolution of these signals, and others seen in rats and other birds, is intimately linked with positive feedback. Signalling by a focal individual improves other group members' chances of discovering food and since these group members are also signallers then this then improves the chance of rediscovering the same food or finding other nearby sources. The positive feedback continues and group productivity increases as more than the sum of the group's parts.

Box 11.2 Synergy in groups of size N

Consider a population that on each generation randomly aggregates in isolated groups of size *N*. Each individual can choose to invest an amount $p_i \in [0, 1]$ in a co-operative behaviour. The benefit to each individual, $g(\sum_{j=1}^{N} p_j)/N$, is assumed to be a function of the overall productivity of the group members, *g*, divided by the total number of group members. We assume that this function is the same for all group members. Thus the payoff for an individual *i* is

$$\frac{g\left(N, \sum_{j=1}^{N} p_j\right)}{N} - p_i c$$

where *c* is the cost of the co-operative behaviour. This model is an example of a structured-deme model (Nunney 1985; Wilson 1983). Here, we further assume that productivity increases as some power α of the level of co-operation of, i.e. $g(P) = bP^{\alpha}$.

We now follow the method outlined by Doebeli *et al.* (2004). Assume that all individuals have the same strategy *q* apart from a mutant with strategy *p*. The selection gradient is then

$$D(q) = \frac{\partial}{\partial p}\left(\frac{b}{N}(p + (N-1)q)^{\alpha} - cp\right)\Bigg|_{p=q} = \frac{b\alpha}{N}(Nq)^{\alpha-1} - c$$

Since we insist that investment is between 0 (defection) and 1 (co-operation), we can evaluate the selection gradient at these two extremes in order to see whether they are stable strategies. Evaluating $D(0) = -c$ tells us that the all defect is an evolutionarily stable state. Similarly, $D(1) = b\alpha N^{\alpha-2} - c$ tells us that the all cooperate is also evolutionarily stable, provided $b\alpha N^{\alpha-2} > c$. When all cooperate is stable there exists, although we don't determine it explicitly here, a single steady state q_* between these two extremes which is not convergent stable. This steady state acts as a repellor: when initially $q > q_*$ then $q \to 1$ and when initially $q < q_*$ then $q \to 0$. Qualitatively, the situation is the same as in the two player discrete game discussed in the text: both all cooperate and all defect are evolutionarily stable.

The condition for synergistic co-operation in this model is $b\alpha N^{\alpha-2} > c$. If $\alpha < 2$ then as group size increases the cost an individual is willing to pay in co-operating decreases. For example, when $\alpha = 1$ we recover $b/N > c$. If $\alpha \geq 2$, however, then as group size increases the cost an individual is willing to pay tends toward a positive but finite limit. In particular, when $\alpha = 2$, co-operation is stable if $2b > c$ independent of *N*. Figure 11.4 shows how the steady states change with group size for $\alpha = 3$.

Figure 11.5 shows similar analysis for $g(P) = T^2 \frac{P^3}{T^2 + P^2}$. This productivity function initially grows cubically, but when group size exceeds *T* the growth becomes more linear. For large *P* growth is purely linear. Here there are three different parameter regimes. For very small group sizes all individuals evolve to invest nothing ($p = 0$) in co-operation, but as group size increases the strategy of full investment ($p = 1$) becomes stable. At intermediate group sizes the full investment becomes unstable and a compromise of partial investment becomes stable. As group size increases still further all communication becomes evolutionarily unstable and $p = 0$ is the only evolutionarily stable state.

Synergism is likely to combine with inclusive fitness in promoting co-operation. Several authors have argued that because relatedness within social insect colonies is lower than first predicted, inclusive fitness may have a less important role in co-operation than once supposed (Korb and Heinze 2004; Wilson and Holldobler 2005). It is here that the observation that signalling in foraging increases per capita foraging success with group size plays an important role. In terms of Hamilton's rule, synergism leads to an increase in benefits and thus a lower requirement for within colony relatedness for the evolution of co-operation. For naked mole rats, the relevance of high within

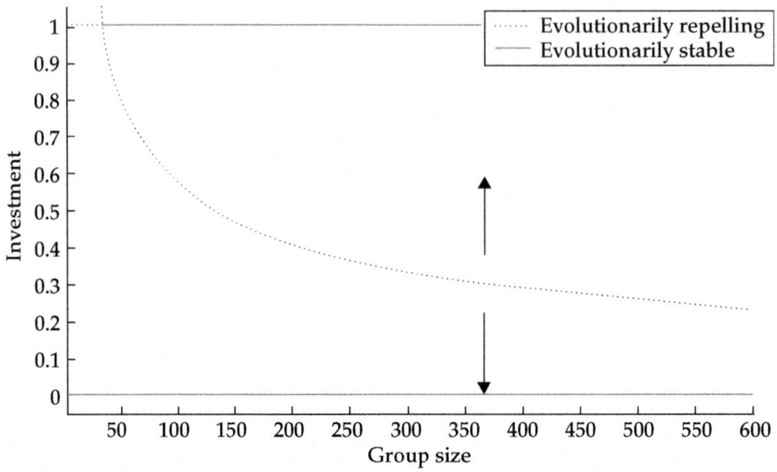

Figure 11.4 Model of synergy described in Box 11.2 with productivity that increases with the cube of group size, i.e. $g(P)=bP^3$. Bifurcation plots showing the location and stability of interior singular strategies and boundary points as a function of group size *N*. We choose $c/b=100$ so that for very small groups there is no benefit to co-operation, i.e. $p=0$ is the only stable strategy. With increasing group size a repelling interior singular strategy emerges and both no investment (p=0) and maximal investment (p=1) are locally stable strategies. Arrows indicate for which initial investment in co-operation these strategies will evolve.

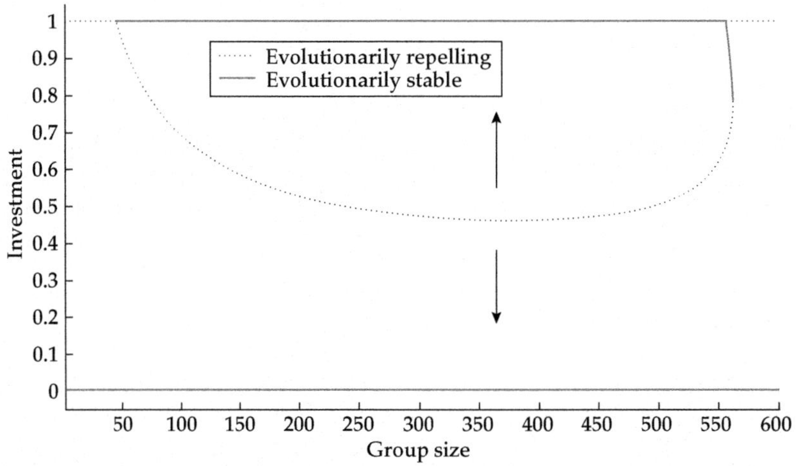

Figure 11.5 Model of synergy described in Box 11.2 with group productivity that first increases with the cube of group size but later saturates to linear increase, i.e. $g(P) = bT^2 \frac{P^3}{T^2 + P^2}$. Parameters are $c/b=5$ and T=40. Bifurcation plot showing the location and stability of interior singular strategies and boundary points as a function of group size *N*. As in Figure 11.5 with increasing group size a repelling interior singular strategy emerges and both no investment ($p=0$) and maximal investment ($p=1$) are locally stable strategies. In this case however as group sizes increases further $p=1$ becomes unstable and a strategy corresponding to an intermediate investment in communication becomes stable. As group size increases still further the intermediate investment strategy disappears and $p=0$ is the only stable state. The arrows from points indicate for which initial investment in co-operation the various stable strategies will evolve.

group relatedness (Reeve *et al.* 1990) has been questioned because the degree of competition between relatives has not been measured (Griffin and West 2002). With or without competition for resources, synergism whereby co-operation increases the amount of available resources could lead to the evolution of signalling during foraging. Further empirical tests of the foraging performance of different sized colonies are needed to clarify the relationship between synergism and relatedness for these species.

Synergy plays a role at all levels of biology. While we have used animal groups to illustrate these ideas, it is perhaps at the level of microbiology that synergy plays its most important role (Maynard Smith and Szathmáry 1995). In Chapter 2, Diggle *et al.* give an overview of quorum signaling and sensing performed by bacteria. Here we see a wealth of examples where group productivity is a non-linear function of the local density of co-operators. For example, the pathogenic bacterium *Pseudomonas aeruginosa* produces costly siderphores which act to release iron from the host organism (Harrison and Buckling 2005). The production of siderphores is only effective when they are produced in sufficient quantities: siderphore production below a threshold level will have little effect in releasing iron but above this threshold iron will be effectively released promoting growth of the bacteria (A. Buckling, personal communication). As in ant foraging, a minimum threshold of signalers is required for the co-ordinated action to succeed. We would predict that if typical group sizes were close to that of the minimum threshold of signalers then full co-operation can evolve without invoking inclusive fitness arguments. As with social insects, there may be a degree of between individual relatedness within hosts that further promotes co-operation. Synergy and inclusive fitness effects can interact to promote co-operation.

11.3.3 Reciprocal interactions

Of the examples of social communication in foraging given in the previous section, the one for which it is hardest to provide a functional explanation based on either synergy or inclusive fitness is the flight displays by ravens (Wright *et al.* 2003). These displays lead non-relatives directly to food items that have only a limited capacity and do not suffer from a large predation risk. For these birds, groups may be sufficiently small that repeated interactions, either in terms of direct reciprocation or indirect reputation building, could play a role in their evolution. In general, reciprocal interactions can occur when groups are small and repeated interactions are guaranteed (Axelrod and Hamilton 1981; Trivers 1971). This may apply in the case of ravens, although it is also the case that groups are relatively fluid and change membership regularly. What is needed here is data on the probability of repeated interaction and the benefits and costs of communicating.

11.4 From mechanism to function

We have emphasized synergy as an important explanation of the evolution of co-operative signaling in animal groups. While most evolutionary biologists are on some level aware that synergy can play a role in the evolution of co-operation between large numbers of unrelated individuals, it is seldom stressed when considering social communication. For example, much of the work on **central place foraging** starts from the view that signaling about food cannot evolve unless there is always a direct benefit, such as anti-predator vigilance, from travelling to food in groups (Dall 2002; Richner and Heeb 1996). The argument is that if there is not a direct benefit to costly signaling then a 'do nothing' strategy can invade the population. In terms of the model in Table 11.1, in order for communication to evolve we must have $D>C$. In this case, a single mutant co-operator can invade a population of defectors.

While the 'direct benefits' argument is correct, it can obscure the fact that co-operation can still evolve even when there is no direct benefit to co-operation for a mutant in a population of defectors, i.e. $D<C$. In terms of two player games this is when $E>C$ and there is some form of extra or synergistic benefit to co-operation. In the model in Box 11.2, costly communication can be an evolutionarily stable strategy provided that per capita productivity increases at least linearly with group size. In this case, 'do nothing' will not benefit from failing to communicate because their breaking the communication link will lead to a decrease in success not only for other group members but also for themselves. As more individuals participate in communication the greater their effectiveness of the groups' actions. Such an argument gives a strong justification for the information centre hypothesis for central place foraging first proposed by Zahavi (1971).

Synergy has important consequences for how we conduct experiments to test the evolution of co-operation. West *et al* (2006) propose that a first step are experiments looking at growth rates of isolated groups of 'communicator' or 'do nothing' individuals with growth rates of mixtures of the two types. Consider two such experiments we might perform on bacteria or animal groups. The first experiment establishes that a 'do nothing' mutant can invade a resident population of communicating individuals. In a second experiment a 'communicate' mutant is shown to be unable to invade a population of 'do nothing' residents due to a cost of signalling. The second of these experiments is similar to the that performed by Diggle *et al* (in press) and discussed in Chapter 2 of this book. They found that non-signaling bacteria grow faster in nutrient rich environments than signaling wild types and that mutant non-signalers invade a population of signalers.

Should we necessarily conclude from our hypothetical experimental results that the evolution of costly signaling requires some form of inclusive fitness benefit? The answer is 'no'. To see this, consider the selective pressure in each of these experiments with relation to Figure 11.5, for group sizes slightly larger than that at which all cooperate is a stable strategy. Here there is a stable state corresponding to a mixture of co-operators and defectors, as well as a stable state corresponding to all defect. The results of these experiments are consistent with this model: communication can be selected against in very large populations of 'do nothing' individuals and communication is also selected against in a large population of communicators. However, at an intermediate group size where a substantial number of individuals are already co-operating the selective pressure is for increased and ultimately full co-operation. In such a situation there is no need to invoke inclusive fitness to justify communication. Costly signaling can evolve between non-relatives, although non-signalling individuals will remain when group sizes become very large.

The difficulties arising in interpreting experiments illustrates the intimate connection between mechanisms and function. Without knowing the mechanism of communication and how they depend upon the number of interacting individuals we cannot predict whether or not a particular form of communication will evolve. The prediction of the model in Box 11.2 depends crucially upon the non-linear shape of the benefit function. It is important therefore to measure this benefit function in a wide range of contexts. This has been done in some of the studies of social foraging that we have reviewed here. A number of studies of nest establishment by insects have also shown that per capita success increases with group size. Further work is needed to link together mechanisms, of for example, costly signaling chemicals with growth of bacteria populations. Establishing the form of group benefit functions will help demystify the wealth of co-operative interactions seen at all levels of biological organization.

Summary

Synergy is where the interactions of a group of individuals becomes more than the sum of their parts. In this chapter we review how, through the use of social communication, foraging animals can increase their rate of finding food. We discuss how mechanisms such as pheromone trails, dancing and other signals act to increase group, and thus individual, success. We also discuss how social dilemmas can arise where costly signalling can be exploited by non-signallers. We show that under a range of conditions, specifically when group success increases more than linearly with group size, co-operative signalling can evolve without kin selection or reciprocity. This study serves to emphasise the importance in linking mechanism with function when studying collective behaviour of animals.

Acknowledgements

Thank you to Madeleine Beekman for comments on this chapter.

References

Adams, E.S. and Tschinkel, W.R. (1995) Density-Dependent Competition in Fire Ants—Effects on Colony Survivorship and Size Variation. *Journal of Animal Ecology*, **64**, 315–324.

Axelrod, R. and Hamilton, W.D. (1981) The Evolution of Cooperation. *Science*, **211**, 1390–1396.

Beckers, R., Deneubourg, J.L. and Goss, S. (1992) Trails and U-Turns in the Selection of a Path by the Ant Lasius-Niger. *Journal of Theoretical Biology*, **159**, 397–415.

Beckers, R., Deneubourg, J. L. and Goss, S. (1993) Modulation of Trail Laying in the Ant Lasius-Niger (Hymenoptera, Formicidae) and Its Role in the Collective Selection of a Food Source. *Journal of Insect Behavior*, **6**, 751–759.

Beckers, R., Goss, S., Deneubourg, J.L. and Pasteels, J.M. (1989) Colony size, communication and ant foraging strategy. *Psyche*, **96**, 239–256.

Beekman, M., Gilchrist, A.L., Duncan, M. and Sumpter, D.J.T. (2007). What makes a honeybee scout? *Behavioral Ecology and Sociobiology*, **61**, 985–995.

Beekman, M. and Lew, J.B. (2008) Foraging in honeybees—when does it pay to dance? *Behavioral Ecology*.

Beekman, M., Sumpter, D.J.T. and Ratnieks, F.L.W. (2001) Phase transition between disordered and ordered foraging in Pharaoh's ants. *Proceedings of the National Academy of Sciences of the United States of America*, **98**, 9703–9706.

Beekman, M., Sumpter, D.J.T., Seraphides, N. and Ratnieks, F.L.W. (2004) Comparing foraging behaviour of small and large honey-bee colonies by decoding waggle dances made by foragers. *Functional Ecology*, **18**, 829–835.

Bourke, A.F.G. and Franks, N.R. (1995) *Social Evolution in Ants*. Princeton, Princeton University Press, New Jersey.

Brown, C.R. and Brown, M.B. (1996) *Coloniality in the Cliff Swallow*. The University of Chicago Press, Chicago.

Brown, C.R., Brown, M.B. and Shaffer, M.L. (1991) Food-Sharing Signals Among Socially Foraging Cliff Swallows. *Animal Behaviour*, **42**, 551–564.

Buhl, J., Gautrais, J., Deneubourg, J.L. and Theraulaz, G. (2004) Nest excavation in ants: group size effects on the size and structure of tunneling networks. *Naturwissenschaften*, **91**, 602–606.

Camazine, S. (1991) Self-organizing pattern formation on the combs of honey bee colonies. *Behavioral Ecology and Sociobiology*, **28**, 61–76.

Dall, S.R.X. (2002) Can information sharing explain recruitment to food from communal roosts? *Behavioral Ecology*, **13**, 42–51.

Devigne, C. and Detrain, C. (2002) Collective exploration and area marking in the ant Lasius niger. *Insectes Sociaux*, **49**, 357–362.

Doebeli, M., Hauert, C. and Killingback, T. (2004) The evolutionary origin of cooperators and defectors. *Science*, **306**, 859–862.

Dornhaus, A. and Chittka, L. (2004) Why do honey bees dance? *Behavioral Ecology And Sociobiology*,**55**, 395–401.

Dornhaus, A., Klugl, F., Oechslein, C., Puppe, F. and Chittka, L. (2006) Benefits of recruitment in honey bees: effects of ecology and colony size in an individual-based model. *Behavioral Ecology*, **17**, 336–344.

Foster, K.R., Wenseleers, T. and Ratnieks, F.L.W. (2006) Kin selection is the key to altruism. *Trends In Ecology and Evolution*, **21**, 57–60.

Franks, N.R., Gomez, N., Goss, S. and Deneubourg, J. L. (1991) The Blind Leading the Blind in Army Ant Raid Patterns—Testing a Model of Self-Organization (Hymenoptera, Formicidae). *Journal of Insect Behavior*, **4**, 583–607.

Galef, B.G. and Buckley, L.L. (1996) Use of foraging trails by Norway rats. *Animal Behaviour*, **51**, 765–771.

Galef, B.G. and White, D.J. (1997) Socially acquired information reduces Norway rats' latencies to find food. *Animal Behaviour*, **54**, 705–714.

Griffin, A.S. and West, S.A. (2002) Kin selection: fact and fiction. *Trends in Ecology and Evolution*, **17**, 15–21.

Halley, J.D. and Burd, M. (2004) Nonequilibrium dynamics of social groups: insights from foraging Argentine ants. *Insectes Sociaux*, **51**, 226–231.

Hamilton, W.D. (1964) The genetical evolution of social behaviour. I, II. *Journal of Theoretical Biology*, **7**, 1–52.

Harrison, F. and Buckling, A. (2005) Hypermutability impedes co+operation in pathogenic bacteria. *Current Biology*, **15**, 1968–1971.

Hölldobler, B. and Wilson, E.O. (1990) *The ants*. Belknap Press of Harvard University Press, Cambridge, Massachusetts.

Judd, T.M. and Sherman, P.W. (1996) Naked mole-rats recruit colony mates to food sources. *Animal Behaviour*, **52**, 957–969.

Korb, J. (2003) Thermoregulation and ventilation of termite mounds. *Naturwissenschaften*, **90**, 212–219.

Korb, J. and Heinze, J. (2004) Multilevel selection and social evolution of insect societies. *Naturwissenschaften*, **91**, 291–304.

Krebs, J. R. and Davis, N. B. (1993) *An introduction to behavioural ecology*. Blackwell Science, Oxford.

Marzluff, J. M., Heinrich, B. and Marzluff, C. S. (1996) Raven roosts are mobile information centres. *Animal Behaviour*, **51**, 89–103.

Maynard Smith, J. and Harper, D. (2005) *Animal Signals*. Oxford University Press, Oxford.

Maynard Smith, J. and Szathmáry, E. (1995) *The Major Transitions of Evolution*: W.H. Freeman Spektrum.

Nicolis, S.C. and Deneubourg, J.L. (1999) Emerging patterns and food recruitment in ants: an analytical study. *Journal of Theoretical Biology*, **198**, 575–592.

Nicolis, S. C., Detrain, C., Demolin, D. and Deneubourg, J.L. (2003) Optimality of collective choices: A stochastic approach. *Bulletin of Mathematical Biology*, **65**, 795–808.

Nunney, L. (1985) Group selection, alturism, and structured-deme models. *American Naturalist*, **126**, 212–230.

Packer, C. and Ruttan, L. (1988) The Evolution of Cooperative Hunting. *American Naturalist*, **132**, 159–198.

Reeve, H.K., Westneat, D.F., Noon, W.A., Sherman, P.W. and Aquadro, C.F. (1990) Dna Fingerprinting Reveals High-Levels of Inbreeding in Colonies of The Eusocial Naked Mole-Rat. *Proceedings of The National Academyo Sciences of The United States of America*, **87**, 2496–2500.

Richner, H. and Heeb, P. (1996) Communal life: Honest signaling and the recruitment center hypothesis. *Behavioral Ecology*, **7**, 115–118.

Riley, J.R., Greggers, U., Smith, A.D., Reynolds, D.R. and Menzel, R. (2005) The flight paths of honeybees recruited by the waggle dance. *Nature*, **435**, 205–207.

Seeley, T.D. (1995) *The wisdom of the hive*. Belknap Press of Harvard University Press, Cambridge, Massachusetts.

Seeley, T.D. and Towne, W.F. (1992) Tactics of dance choice in honey bees: do foragers compare dances? *Behavioral Ecology and Sociobiology*, **30**, 59–69.

Sherman, P.W.V. and Visscher P.K. (2002) Honeybee colonies gain fitness through dancing. *Nature (London)*, **419**, 920–922.

Sumpter, D.J.T. (2006) The principles of collective animal behaviour. *Philosophical Transactions of the Royal Society B-Biological Sciences*, **361**, 5–22.

Sumpter, D.J.T. and Beekman, M. (2003) From nonlinearity to optimality: pheromone trail foraging by ants. *Animal Behaviour*, **66**, 273–280.

Sumpter, D.J.T. and Pratt, S.C. (2003) A modelling framework for understanding social insect foraging. *Behavioral Ecology and Sociobiology*, **53**, 131–144.

Theraulaz, G., Gautrais, J., Camazine, S. and Deneubourg, J. L. (2003) The formation of spatial patterns in social insects: from simple behaviours to complex structures. *Philosophical Transactions of The Royal Society of London Series A-Mathematical Physical And Engineering Sciences*, **361**, 1263–1282.

Treherne, J.E. and Foster, W.A. (1981) Group transmissin of predator avoidance behaviour in a marine insect: the Trafalgar effect. *Animal Behaviour*, **28**.

Trivers, R.L. (1971) Evolution of Reciprocal Altruism. *Quarterly Review of Biology*, **46**, 35–57.

von Frisch, K. (1967) *The Dance Language and Orientation of Bees*. The Belknap Press of Harvard University Press, Cambridge, Massachusetts.

West, S.A., Gardner, A. and Griffin, A.S. (2006) Altruism. *Current Biology*, **16**, R482-R483.

West, S.A., Griffin, A.S. and Gardner, A. (2007) Social semantics: altruism, cooperation, mutualism, strong reciprocity and group selection. *Journal of Evolutionary Biology*, **20**, 415–432.

Wilson, D.S. (1983) The group selection controversy: history and current status. *Annual Review of Ecology and Systematics*, **14**, 159–187.

Wilson, E.O. (1971) *The Insect Societies*. Belknap Press of Harvard University Press. Cambridge, Massachusetts.

Wilson, E.O. and Holldobler, B. (2005) Eusociality: Origin and consequences. *Proceedings of The National Academy of Sciences of The United States of America*, **102**, 13367–13371.

Wright, J., Stone, R.E. and Brown, N. (2003) Communal roosts as structured information centres in the raven, Corvus corax. *Journal of Animal Ecology* **72**, 1003–1014.

Zahavi, A. (1971) The function of pre-roost gatherings and communal roosts. *Ibis*, **113**, 106–109.

CHAPTER 12

Conflicting messages: genomic imprinting and internal communication

David Haig

12.1 Introduction

Cell biologists and behavioural ecologists make different implicit assumptions about how **signals** evolve. Cell biologists are usually interested in signals that are transmitted within cells or between the cells of a single body. Signaller and receiver are implicitly assumed to have identical interests. The question of whether signals are credible does not arise because signallers do not have incentives to deceive. In behavioural ecology, on the other hand, signaller and receiver are different individuals, potentially with **conflicting** interests. Receivers must decide whether signals can be trusted. Although behavioural ecologists recognize the possibility of conflicts between individuals, they usually assume that individuals have well-defined, unitary interests.

These two areas of inquiry are, of course, intimately linked. A behavioural signal is usually the external output of a complicated process of signaling among and within cells of the sender. The perception and interpretation of the behavioural signal usually involves an equally complex process of signalling among and within the cells of the receiver. Despite these intimate connections, different kinds of questions are typically asked about signalling within individuals and signalling between individuals. Communication within individuals is usually viewed as a problem in signal engineering. Relevant questions are how to send signals efficiently, how to cope with noise and 'cross-talk' from other signalling pathways, and how to correct errors. Questions of signal efficacy also arise for communication between individuals (Guilford and Dawkins 1991; see also Chapter 3) but behavioural ecologists usually focus on question of signal credibility: (1) can a signal be trusted (see also the handicap principle, Chapter 1), (2) what is the sender's motive, and (3) does the sender have something to hide?

Neither cell biologists nor behavioural ecologists have given much thought to the implications, for signalling theory, of conflicts within the genomes of individual organisms. The possibility of a fractious genome raises interesting questions about how conflicts among different genomic factions could influence communication within cells and among individuals. Is deception possible within an individual? Could different parts of an individual disagree over whether to send a signal to another individual? Would such a signal be sent?

There are many sources of **intragenomic conflict** (Burt and Trivers 2006; see also Chapter 13), but this chapter will focus on just one: antagonism between genes of maternal and paternal origin (Wilkins and Haig 2003; Haig 2004). Section 12.2 gives a theoretical introduction to conflicts that arise within the genome when one individual has different degrees of kinship to another individual's mother and father. This conflict can result in patterns of gene expression that are contingent on an allele's parental origin, so-called genomic imprinting. Section 12.3 discusses two disorders of imprinted genes that exhibit perturbations of social communication among individuals. The remaining sections focus on communication within organisms: section 12.4 considers relationships among genes, proteins, cells, and organisms while section 12.5 illustrates these relationships with the examples of two signalling pathways that respond to information from the external environment. Section 12.6 discusses interactions among imprinted genes.

12.2 Genomic imprinting and kinship

Hamilton (1963) argued that **natural selection** will favour one individual performing a costly action that benefits another individual if

$$k = \frac{1}{r} \tag{12.1}$$

where k is the ratio of benefit (to other) to cost (to self), and r is the beneficiary's coefficient of relatedness to the donor. Roughly speaking, r is the probability that the gene responsible for the behaviour has an identical-by-descent copy in the recipient. For many classes of relatives, this probability differs for genes of maternal and paternal origin. It is a simple matter to define distinct coefficients of matrilineal and patrilineal relatedness. Thus, r_m is the probability that another individual carries an identical-by-descent copy of an egg-derived (madumnal) allele of the focal individual. Similarly, r_p is the probability that the other individual carries an identical-by-descent copy of a sperm-derived (padumnal) allele of the focal individual. Individuals for whom $r_m = r_p$ are *symmetric kin* of the focal individual whereas individuals for whom $r_m \neq r_p$ are *asymmetric kin* of the focal individual.

If the beneficiary is asymmetric kin of the donor, madumnal genes would favour performing the costly action whereas padumnal genes would favour blocking the action if

$$\frac{1}{r_m} < k < \frac{1}{r_p}. \tag{12.2a}$$

Conversely, padumnal genes would favour performing the action whereas madumnal genes would favour blocking the action if

$$\frac{1}{r_m} > k > \frac{1}{r_p}. \tag{12.2b}$$

Consider a warning call emitted by one member of a litter that benefits a littermate but increases the caller's risk of predation. In this example, $r_m = 1/2$ and $r_p = p/2$, where p is the probability of shared paternity. If $k > 2/p$, madumnal and padumnal alleles would both benefit from sending the call. If $k < 2$, neither madumnal nor padumnal alleles would benefit from sending the call. If $2 < k < 2/p$, the call would enhance madumnal **fitness** but reduce padumnal fitness. Thus, an internal conflict exists for intermediate benefit-to-cost ratios. Kinship theory, by itself, does not predict whether the signal will be sent if either (12.2a) or (12.2b) is satisfied. Knowledge of proximate mechanisms is required to know how the conflict will be resolved.

The simplest resolution of the conflict occurs if the relevant genes lack information about their parental origin. In the absence of such cues, a gene is constrained to exhibit the same behaviour when it is transmitted via an egg or via a sperm. An average gene on an autosome is transmitted with equal frequency via eggs and sperm. Therefore, such a gene would be selected to send the warning signal if

$$k > \frac{2}{r_m + r_p} \tag{12.3}$$

which is equivalent to (12.1).

Genomic imprinting refers to **epigenetic** modifications, in either the mother's or the father's germ line, that provide a historical record (in offspring) of whether a gene resided in a male or female body in the previous generation. In the presence of genomic imprinting, genes possess information about their parental origin and can employ conditional strategies that do one thing when maternally derived and a different thing when paternally derived (Haig 1997). The simple strategy described by inequality (12.3) may be evolutionarily unstable if alleles possess information about their parental origin and $r_m \neq r_p$.

Consider a locus at which natural selection has favoured a higher level of gene product when a gene is maternally derived than when the gene is paternally derived. The **evolutionarily stable strategy** (ESS) for an allele at such a locus is to be silent when paternally derived and expressed at the madumnal optimum when maternally derived. The nature of the ESS is reversed at a locus where natural selection favours higher expression when a gene is paternally derived. At such loci, the ESS is to be silent when maternally derived but produce the padumnal optimum when paternally derived. I have called this the loudest-voice-prevails principle (Haig 1997). This principle applies whenever (1) madumnal and padumnal alleles contribute gene products to a common pool and (2) fitness is determined by the size of the pool and not by the relative contributions of the two alleles (see Box 1 of Wilkins and Haig 2003).

The loudest-voice-prevails is a simple form of 'conflict resolution': whichever allele favours the higher amount of gene product produces that amount. Silencing of one of the two alleles at a diploid locus has a number of important consequences, of which I will discuss two. First, alternative alleles at the locus have phenotypic effects when inherited from one sex but are without effect when inherited from the other sex. Therefore, alleles at a madumnally silent locus will be selected solely for their effects on patrilineal fitness whereas alleles at a padumnally silent locus will be selected solely for their effects on matrilineal fitness (Haig 1997, 2000). Second, the loudest-voice-prevails principle reveals a sender's identity to the recipient. If both alleles are transcribed, a recipient of a signal (gene product) has no way of telling whether the sender is a madumnal or padumnal allele. If one of the two potential sources of a signal is reliably silent, then the actual signaller's identity is revealed.

At a single locus, the loudest-voice-prevails principle suggests that whichever allele favours the larger amount 'wins'. However, most organismal outcomes are influenced by many genes. For example, madumnal and padumnal alleles may disagree over how much investment an offspring extracts from its mother. Padumnal alleles at a demand-enhancing locus may produce their favoured amount of a demand enhancer, but these effects may be countered by madumnal alleles at another locus producing their favoured amount of a demand inhibitor (Haig and Graham 1991). A joint ESS is possible for the two loci; with madumnal silencing of the demand enhancer and padumnal silencing of the demand inhibitor (Haig and Wilkins 2000; Wilkins and Haig 2001). The joint ESS is associated with a 'conflict cost', because the same overall level of demand could be achieved with lower production of both enhancer and inhibitor.

The two-locus joint ESS has the form of a stalemate: the marginal cost of an increment of demand enhancer balances any benefit padumnal alleles would gain from increased demand; likewise, the marginal cost of an increment of demand inhibitor balances any benefit madumnal alleles would gain from reduced demand. In general, neither matrilineal nor patrilineal fitness is optimized at a joint ESS. 'Costs' are required to stabilize the joint ESS. If there were no costs, simple models predict an infinite escalation of enhancer and inhibitor. My current interpretation is that these stabilizing costs represent negative **pleiotropic** effects of high levels of demand enhancers and inhibitors (Haig 2006).

The loci that influence an organismal outcome may have more than two sets of interests. Haig (2006) explored interactions among multiple 'factions' with respect to a single organismal trait ('demand'). It was found that the factions tended to align into two 'parties': one favouring increased demand and the other favouring reduced demand. More theoretical work is needed to see whether this result can be generalized to conflicts over multiple traits. (See related discussion regarding conflict within the superorganism in Chapter 10.)

12.3 Prader–Willi and Angelman syndromes

The paucity of spontaneous movement in infants with Prader–Willi syndrome, and their described placid nature, may result in decreased interaction with care-givers.

(Cassidy 1988)

Human chromosome 15q11–q13 is home to a cluster of imprinted genes. Deletion of the padumnal copy of this cluster results in Prader–Willi syndrome whereas deletion of the madumnal copy results in Angelman syndrome. These syndromes have complex phenotypes (Holm *et al.* 1993; Williams *et al.* 2005), but I will focus on abnormalities of communication. My intent in discussing these syndromes is two-fold. First, the phenotypes of Prader–Willi and Angelman syndromes provide evidence that imprinted genes can influence social communication between individuals. Second, the imprinted genes of the cluster provide illustrations of the ways in which internal genetic conflicts can influence signalling within organisms.

Prader–Willi syndrome is caused by the absence of expression of padumnal alleles whereas Angelman syndrome is caused by the absence of expression of madumnal alleles. Therefore, Prader–Willi syndrome is predicted to exaggerate behaviours that, in a normal child, benefit a mother's residual reproductive value, at a cost to the child's expected fitness, and Angelman syndrome is predicted to exaggerate behaviours with the opposite effect (Haig and Wharton 2003).

Newborn infants with Prader–Willi syndrome have an abnormal cry that has been variously described as feeble, weak, squeaky, peculiar, or not sustained (Aughton and Cassidy 1990; Butler 1990; Miller *et al.* 1999; Õiglane-Shlik *et al.* 2006). Such infants have poor suck and are often fed directly through a tube to the stomach (Cassidy 1988). Their voice has hypernasal resonance and abnormally high pitch (Åkefeldt *et al.* 1997). Speech is dysfluent, but without features of typical stuttering (Defloor *et al.* 2000). Morphosyntax (word and sentence structure) is generally poor (van Borsel *et al.* 2007). Repetitive, perseverative speech is a problem behaviour in many older individuals with Prader–Willi syndrome (Dykens and Kasari 1997; Walz and Benson 2002; Jauregi *et al.* 2007).

The weak cry and poor suck of infants with Prader–Willi syndrome suggest that padumnally-expressed genes from 15q11–q13 promote signals that elicit maternal care. However, these phenotypes have not been studied in detail and it is possible that the weak cry and poor suck are secondary effects of the generalized hypotonia and poor respiratory control that is characteristic of Prader–Willi syndrome. Older children with Prader–Willi syndrome develop an insatiable appetite and become massively obese. Haig and Wharton (2003) suggested that the switch from anorexia to hyperphagia, often in the second year of life, may reflect evolutionary conflicts associated with weaning.

Communication is more severely perturbed in Angelman syndrome than in Prader–Willi syndrome. Infants with Angelman syndrome have an abnormal high-pitched cry (Clayton-Smith 1993) and babbling is delayed or absent (Yamada and Volpe 1990; Penner *et al.* 1993). Affected children never learn to talk (Williams *et al.* 1995; Clarke and Marston 2000). This absence of speech appears out of proportion to the underlying level of mental retardation (Penner *et al.* 1993; Pembrey 1996; Alvares and Downing 1998). Children with Angelman syndrome have poor motor imitation skills and most fail to imitate verbal behaviour (Jolleff and Ryan 1993; Penner *et al.* 1993; Duker *et al.* 2002; Didden *et al.* 2004). 'Motor theories' of the evolution of human language posit that language is based on the perception and imitation of gestures of the vocal tract (Gentilucci and Corballis 2006; Galantucci *et al.* 2006; see also Chapter 13). An interesting possibility is that the ataxia and absence of speech of individuals with Angelman syndrome may have a common aetiology with defects in the neural representation of motor actions.

Children with Angelman syndrome use non-verbal communication primarily for making requests and rejecting offers (Didden *et al.* 2004). Such communication usually involves direct manipulation of the other person (pushing a hand away, leading by the hand, touching to gain attention) rather than the use of gesture or pointing (Jolleff and Ryan 1993). Joint attention, joint action, and taking turns are poorly developed (Penner *et al.* 1993).

Angelman syndrome is characterized by positive affect with frequent laughter and smiling (for a comprehensive review see Horsler and Oliver 2006a; Isles *et al.* 2006 emphasize the contrast with the negative affect of Prader–Willi syndrome). Laughter has been described as inappropriate and unprovoked, but careful behavioural studies suggest that laughter is rare in non-social contexts and is particularly pronounced after eye contact (Oliver *et al.* 2002; Horsler and Oliver 2006b). On the other hand, a small study of two young boys with

Angelman syndrome (18 and 42 months old) found no consistent difference in laughing and smiling between social and non-social contexts (Richman *et al.* 2006).

Oliver *et al.* (2007) studied 13 children with Angelman syndrome in a school setting and compared them with a matched group of children with other forms of intellectual disability. The children with Angelman syndrome smiled more than the comparison group, were more likely to reach out to or touch adults before smiling, and their smiles were more effective at eliciting adult smiles in return. This behavioural profile was interpreted as compatible with Brown and Consedine's (2004) suggestion that children with Angelman syndrome exhibit exaggerated expression of behaviours that normally elicit increased maternal investment and minimize the likelihood of rejection.

Autistic-like symptoms are common in Prader–Willi syndrome (Koenig *et al.* 2004; Veltman *et al.* 2005), especially in individuals with maternal uniparental disomy (rather than padumnal deletion) of chromosome 15q11–q13 (Milner *et al.* 2005) (see also discussion of autism in Chapter 13). Moreover, madumnal duplications of 15q11–q13, but not padumnal duplications, are associated with autism (Cook *et al.* 1997). These observations suggest that overexpression of madumnally expressed genes from this region may play a role in the development of autistic symptoms (see Badcock and Crespi 2006 for a general discussion of links between imprinting and autism). The situation is less clear for Angelman syndrome. Some studies suggest that autism is common in Angelman syndrome (Steffenburg *et al.* 1996; Bonati *et al.* 2007) whereas others consider autism to be rare in Angelman syndrome (Veltman *et al.* 2005). These disagreements arise from the difficulty of assessing autistic symptoms against the backdrop of severe mental retardation in Angelman syndrome (Trillingsgaard and Østergaard 2004). Children with Angelman syndrome actively seek social interactions, a characteristic that clearly distinguishes them from typical autism (Williams *et al.* 2001). A group of children with Angelman syndrome who had been classified as autistic exhibited more social interest and responsiveness in dyadic interaction than a comparison group of autistic children without Angelman syndrome, but the two groups resembled each other on tasks requiring more complex forms of social communication (Trillingsgaard and Østergaard 2004).

The causal connections between genes and behaviour are not understood in Prader–Willi and Angelman syndromes. Nevertheless, a brief discussion of what is known about the functions of these genes will provide a sense of the many different ways that imprinted genes could influence signalling and information processing within individual organisms.

The 15q11–q13 imprinted cluster encodes a number of padumnally expressed transcripts whose absence may contribute to the symptoms of Prader–Willi syndrome (Nicholls and Knepper 2001). Attention has focused on *Necdin* and *SNURF-SNRPN*. *Necdin* encodes a protein with diverse effects in brain development that include inhibiting apoptosis of post-mitotic neurons (Kurita *et al.* 2006), promoting differentiation of GABAergic neurons (Kuwajima *et al.* 2006), and enhancing elongation of axons (Lee *et al.* 2005). The *SNURF-SNRPN* transcript encodes two polypeptides—SNURF (function unknown) and SNRPN (a brain-specific splicing factor)—and is host to multiple small nucleolar RNAs (snoRNAs). These snoRNAs reside in the introns of the nascent transcript and are released by splicing (Runte *et al.* 2001). One of these snoRNAs, HBII-52, binds to exon Vb of nascent transcripts of *HTR2C* (the X-linked serotonin 2C receptor) and promotes the inclusion of this exon in mature transcripts (Kishore and Stamm 2006). HBII-52 is also suspected of modifying RNA editing of *HTR2C* mRNA, thus influencing which amino acids are incorporated into the mature receptor (Vitali *et al.* 2005). Hippocampal expression of MBII-52, the murine homolog of HBII-52, is increased during consolidation of fearful memories (Rogelj *et al.* 2003).

Angelman syndrome is associated with the absence (or absence of expression) of a madumnal copy of *UBE3A*, a gene from the 15q11–q13 imprinted cluster. *UBE3A* is biallelically expressed in most tissues, but has preferential madumnal expression in the human brain. UBE3A functions both as a ubiquitin ligase and as a nuclear receptor coactivator. Ubiquitin ligases attach ubiquitin to target proteins, thereby marking them for degradation by the proteasome. Some individuals with Angelman syndrome have mutations in the madumnal copy of *UBE3A*. Many of these mutations inactivate the ubiquitin ligase function of UBE3A but leave the coactivator function intact. Taken together, these data suggest that the major symptoms of Angelman syndrome are caused by the failure to degrade one or more substrates of UBE3A in the brain (Lalande and Calciano 2007). The critical substrates are yet to be identified.

The Prader–Willi/Angelman gene cluster thus illustrates diverse ways in which imprinted genes can influence communication within organisms. Imprinted genes can promote or inhibit the development of particular cell types or the formation of connections among neurons, cause the degradation of a signal molecule (or otherwise block a signalling pathway), and even modify the transcript produced by another gene.

Signaller and receiver are easy to identify when a child smiles to engage his or her mother's attention. But who is the signaller, and who the receiver, when HBII-52 modifies the transcript of *HTRC*, thus altering the properties of a serotonin receptor, thereby modifying a neuron's response to serotonin released by another neuron? The next section attempts to refine this question by considering the relationship between genes and the structures that genes create.

12.4 Limber robots and lumbering genes

Now [genes] swarm in huge colonies, safe inside gigantic lumbering robots, sealed off from the outside world, communicating with it by tortuous indirect routes, manipulating it by remote control.

This famous sentence from *The Selfish Gene* (Dawkins 1976, p. 21) is often interpreted as claiming that organisms are the slaves of their genes. Genes are in control. They, not the organisms they construct, are the actors in an evolutionary drama on an ecological stage. At the same time, the quotation subtly undercuts this view of genetic autonomy. Genes are 'remote' and 'sealed off' from an outside world with which their communication is 'tortuous' and 'indirect'. To what extent is it fruitful to think of communication from a gene-centred rather than an organism-centred point of view? Do genes communicate with each other or is communication primarily an attribute of the robotic structures—proteins, cells, organs, organisms—that genes construct? Answers to these questions are necessary before one can address the implications of intragenomic conflicts for communication within and between organisms.

As humans we construct robots to free ourselves from mundane decision-making and to perform actions that we are unable to achieve without prosthetic aid. In the control of robotic actions, some decisions are reserved for human intervention but others are delegated to the robot. Consider a robot that has been designed and programmed to explore Mars. The robot interacts directly with the Martian environment but its Earth-bound controllers experience Mars only vicariously through the robot's sensors. Not everything sensed by the robot need be communicated to Earth. The robot and its controllers use information about the environment to modify the robot's behaviour. For some decisions, the robot 'consults' its controllers on Earth but other decisions are taken without consultation. At crucial times, the robot is on its own and must respond to its environment more rapidly than signals can be exchanged with Earth (Dennett 1984, p. 55).

One can imagine a spectrum of robot autonomy. At one end of the spectrum, robots are simple mechanical prostheses with all important decisions taken by humans. At the other end of the spectrum, robots are designed by humans but make all of their own decisions. Where do Dawkins' lumbering robots reside on this continuum? Are the robots on a short leash, with all important decisions taken by their genetic controllers, or are the robots fully autonomous, exploring and exploiting their environment without ever consulting their genes? Clearly, the answer will lie somewhere between these rhetorical poles.

'Lumbering' has connotations of awkwardness, but some robots can perform actions with a delicacy and precision beyond the powers of unaided human actors. After all, the aim of robotic design is to produce supple, rather than clumsy, machines (and the same is surely true of the 'designs' of natural selection). An organism is an agile automaton designed by natural selection to function effectively in a complex world. Simple automata (genes and proteins) interact with each other in complex **networks** to create a hierarchy of higher-level automata (cells, organs, organisms). Simple automata have only a small number of possible states, but as the number of simple automata that constitute a larger system increases, so does the number of possible states of the system, in a combinatorial explosion. As a result, higher-level automata can express more flexible behaviour, and can possess more sophisticated information about their environment, than can lower-level automata. Genes may be less nimble than the robots they construct.

Genes and proteins are both 'simple' components of organisms. Why should we privilege genes and ask whether they are in control? A gene is, after all, no less mindless than a protein. Genes, however, are a very special kind of automaton. They are, to use Dawkins' term, *replicators*. Chemical changes to their structure are transmitted to their descendants, if descendants they have. Genes are thus the evolutionary repositories of the heritable information used to construct organisms. Although genes are the *constructors* of organisms, it does not necessarily follow that genes are also the *controllers* of organisms. To what extent are genes involved in the moment-to-moment decisions that control the behaviour of organisms? And, to what extent are they merely interested bystanders?

An automaton 'detects' a property of the environment (including 'signals' sent by other automata) when this property causes a change in the automaton's state. 'Communication' takes place when one automaton (the sender) causes a change in state of another automaton (the receiver). A receiver detects a sender's change of state, either directly, by physical contact with the sender, or indirectly, by the detection of a change in the environment induced by the sender's change of state. Genes and proteins are automata that can potentially fill the roles of sender and receiver but so too can cells and other higher-level automata.

The principal way that a gene interacts with the world is by the production of transcripts, some of which are translated into proteins. Many proteins have multiple states and function as simple automata. The factors that induce changes of state are what a protein 'knows' about the world. A protein communicates with another protein when it induces a change in the other protein's state. Each protein's repertoire of functional states may include different conformations (e.g. induced by binding of a ligand or by a change in pH), as well as chemical modifications of its structure (e.g. editing of its transcript by HBII-52, ubiquitination by UBE3A, phosphorylation by a kinase, cleavage by a protease).

Genes construct protein automata. But, genes are themselves automata. A gene's states are determined by the binding of transcription factors and other proteins, by interactions with RNAs and with other DNA sequences, and by epigenetic modifications such as cytosine methylation. These factors determine where and when the gene is expressed. A gene may possess information about past environments as well as information about the current environment. An imprinted gene, for example, 'remembers' whether it was present in a male or female body in the previous generation. A common criticism of the **selfish-gene** approach is that it assigns too much agency to genes. A gene is not a homunculus fully aware of everything that is happening to the organism and making plans accordingly. However, one can also err on the side of underestimating the strategic options available to genes.

12.5 Light and cold

Rhodopsin is expressed in rod cells of the retina but is not expressed in other cell types. Thus, *Rhodopsin* 'senses' when it resides in a rod cell and uses this datum to switch between active and silent states. In its active state, *Rhodopsin* produces an mRNA that is translated at the endoplasmic reticulum to produce a protein (opsin) that forms a covalent linkage with the chromophore 11-*cis*-retinal to form the visual pigment rhodopsin. (In the remainder of this chapter, I will adopt the standard convention that genes are italicized, e.g. *Rhodopsin*, whereas their protein products, often of the same name, are not, e.g. rhodopsin.) The receipt of a photon of an appropriate wavelength causes isomerization of 11-*cis*-retinal to all-*trans*-retinal. This causes a conformational change in rhodopsin that is propagated to other proteins, triggering a complex biochemical cascade (Okada *et al.* 2001; Ridge *et al.* 2003) that culminates in hyperpolarization of the plasma membrane and discharge of the rod cell (a higher-level automaton).

Two important points can be made about this example. First, the photon is detected by the protein, not by the gene. *Rhodopsin* neither detects photons nor signals their presence. Rather, the gene directs the construction of protein automata that detect the presence of photons and signal their presence to other protein automata. Second, a *single* state of the gene produces an automaton that switches among *multiple* states in response to 'environmental' cues. A gene can produce a protein with a 'behavioural flexibility' that the gene itself lacks. There is no simple relation between the number of states of a gene and the number of states of the automata it constructs.

Consider a quintessential example of social communication in which rhodopsin plays a part. Multitudinous photons are received by the rod and cone cells of an infant's retina. The pattern of discharge of retinal cells initiates complex processes in the infant's brain that result in the recognition of the infant's mother and the coordination of a motor response (a smile). By an equally complex process, the baby's smile is detected by its mother and elicits a smile in return. The entire chain of events—from receipt of photons at the baby's retina to the contraction of muscles in its mother's face—probably takes place *without the causal intervention of changes of gene state*.

An exchange of smiles is possible because countless gene copies specify the production of innumerable protein automata in an untold number of higher-level automata (nerve cells and muscle fibres). These cellular automata are organized into two very high-level automata (mother and child) who are able to respond to each other's facial gestures. The limber robots communicate without consulting their lumbering genes. The development and maintenance of organism-level automata clearly involve coordinated changes in gene state but the exchange of smiles is too rapid for transcription and translation to play a role. Higher-level automata acquire and act on information that is unavailable to lower-level automata. No gene in the infant's genome perceives its mother's face.

What is true of the perception and interpretation of photons is true of most sources of information about an organism's external and internal environment. Another mammalian example reaffirms the substantial autonomy of protein (and cellular) automata from direct genetic control. Information about external and internal temperature is detected by temperature-sensitive neurons, some located peripherally and others centrally. These inputs are integrated in the hypothalamus and other brain regions to coordinate thermoregulatory responses (Morrison 2004; Romanovsky 2007). One such response is the activation of non-shivering thermogenesis in brown adipose tissue. Uncoupling protein 1 (UCP1) resides in the inner membrane of the mitochondria of brown adipocytes. Activation of UCP1 causes a proton leak that 'uncouples' mitochondrial respiration from oxidative phosphorylation. As a result, organic substrates are 'burned' with the release of heat. The entire chain of events from stimulus (skin-cooling) to response (activation of non-shivering thermogenesis) may take place without the direct intervention of genes.

The efferent arm of this response (from brain to UCP1) involves a complex signalling cascade involving multiple protein automata. Brown adipose tissue is innervated by noradrenergic neurons of the sympathetic nervous system. When these neurons receive appropriate input from the brain they release norepinephrine which binds to β_3-adrenergic receptors (β_3ARs) at the cell surface of brown adipocytes. The cytoplasmic domains of β_3ARs are associated with heterotrimeric G-proteins. Activation of β_3AR by norepinephrine causes the release of the G-protein α-stimulatory subunit (Gαs). Gαs stimulates another protein, adenylyl cyclase, to produce cyclic adenosine monophosphate (cAMP). Via a further series of protein automata, increased cAMP causes activation of UCP1 in the inner mitochondrial membrane (Cannon and Nedergaard 2004; Romanovsky 2007; Nakamura and Morrison 2007).

Although changes of gene state do not play a direct role in the acute response to cold, such changes play important roles in modulating responses over longer timescales. The processes by which cold exposure increases the thermogenic capacity of brown adipocytes are instructive. Noradrenergic signalling via β_3AR not only activates UCP1 (the protein) but also promotes transcription of *Ucp1* (the gene). Noradrenergic neurons also form synapses on pre-adipocytes (adipose stem cells) that express β_1-adrenergic receptors. Cold-induced activation of β_1ARs promotes the recruitment of mature brown adipocytes from pre-adipocytes, a process that involves the transcription of many genes that are inactive in pre-adipocytes. As a result of these changes in gene expression, the inner mitochondrial membranes of brown adipocytes contain more copies of UCP1, and the total number of brown adipocytes is increased, in anticipation of the next cold exposure (Cannon and Nedergaard 2004).

The control of non-shivering thermogenesis is a potential arena of conflict between genes of maternal and paternal origin in species that huddle together for warmth. Heat generation by one individual reduces the heating costs of other individuals in its huddle and creates an evolutionary temptation for free-riding (not paying a fair share of the communal heating bill). If the members of a huddle are asymmetric kin, madumnal and padumnal alleles can disagree over how much heat to contribute to the common good. Specifically, when members of a multiple-paternity litter huddle together, padumnal alleles are predicted to favour a lower set-point for the brown-adipose thermostat than that favoured by madumnal alleles (Haig 2004).

Genomic imprinting influences at least one step in the signalling pathway that activates non-shivering thermogenesis in brown adipocytes. Gαs is one of several protein products of the complex *GNAS* locus (Abramowitz *et al.* 2004). Both alleles of *GNAS* produce Gαs in most tissues of the body, but only the madumnal allele produces Gαs in brown adipose tissue (Yu *et al.* 1998). A second gene product, XLαs ('extra large' αs), is produced by padumnal *GNAS* and antagonizes the effects of Gαs in brown adipose tissue (Plagge *et al.* 2004). Gαs and XLαs mRNAs are transcribed from different *GNAS* promoters, and use alternative first exons, but share their remaining 12 exons. Thus, Gαs is produced by madumnal *GNAS* and promotes non-shivering thermogenesis whereas XLαs is produced by padumnal *GNAS* and inhibits non-shivering thermogenesis. Imprinted genes also influence the recruitment of extra heating units. Two padumnally expressed genes, *Preadipocyte factor-1* and *Necdin*, produce proteins that inhibit the differentiation of pre-adipocytes into brown adipocytes (Tseng *et al.* 2005).

Brown adipocytes are heat-generating automata. Their level of heat production is determined by the combined effects of unimprinted genes, madumnally expressed imprinted genes, and padumnally expressed imprinted genes. The loudest-voice-prevails principle predicts that imprinted genes that increase heat production in multiple-paternity huddles will be madumnally expressed whereas imprinted genes that reduce heat production will be padumnally expressed. However, current theory has little to say about why some genes in a pathway are imprinted while others are not. Why is it that *GNAS* is imprinted in brown adipocytes but *UCP1* is not?

Imprinting can only make a phenotypic difference at loci for which gene dosage matters. If one active allele is just as good as two, then silencing of one allele makes no selective difference. Some effects of Gαs are dosage sensitive. For example, loss of one functional copy of Gαs causes osteodystrophy despite the expression of Gαs transcripts from both alleles in bone (Mantovani *et al.* 2004). The effects of Gαs may be particularly dosage-sensitive because many G protein-coupled receptors activate multiple signalling pathways via alternative G proteins with different α subunits. Thus, the precise stoichiometry of α subunits may determine the balance of signalling among pathways and the nature of the cellular response. For example, the β_3AR of brown adipocytes signals via both Gαs and Gαi (Chaudhry *et al.* 1994). However, it seems unlikely that Gαs is the only dosage-sensitive step in the pathway from the detection of norepinephrine to the activation of non-shivering thermogenesis.

Should a brown adipocyte generate heat only when heat serves the interests of madumnal genes, of padumnal genes, of unimprinted genes, or of something else? Current theory does not predict how an automaton should behave if the automaton's responses are influenced by genes with conflicting interests. I suspect that there is no general answer to this question, and that answers for specific cases will require detailed knowledge of intricate molecular processes within cells. The genome has aspects of a fractious, and poorly informed, committee attempting to set policy, but with most decisions on how to implement policy being taken elsewhere.

12.6 Do imprinted genes talk to each other?

The previous section argued that much intracellular and intercellular communication takes place without the direct intervention of genes. Nevertheless, genes are not merely passive observers. Changes in the state of some genes do cause changes in the state of other genes. This section raises, but does not answer, questions about who imprinted genes talk to and who listens.

Three classes of genes maximize different fitness functions: madumnally expressed genes evolve to maximize matrilineal fitness; padumnally expressed genes evolve to maximize patrilineal fitness; and biallelically expressed genes evolve to maximize the sum of matrilineal and patrilineal fitness (Haig 2000, 2006). Does communication among genes occur at random with respect to class membership, or do madumnally expressed genes preferentially talk to other madumnally expressed genes (and likewise for padumnally expressed genes)? If madumnal genes talk to padumnal genes, can they negotiate a compromise to reduce conflict costs (Haig 2003)?

Recent studies provide evidence for tight interactions among some imprinted genes. *ZAC1* is a padumnally expressed gene that encodes the transcription factor ZAC1. Among the targets of ZAC1 is a differentially methylated region of *LIT1*. ZAC1 binds to the unmethylated (padumnal) copy of *LIT1* and induces transcription of padumnal *LIT1* (Arima *et al.* 2005). *LIT1*'s transcript is a non-coding RNA that causes the silencing of closely linked, protein-coding genes (Horike *et al.* 2000; Mancini-DiNardo *et al.* 2006; in mice, these genes are *Ascl2*, *Kcnq1*, *Cdkn1c*, *Slc22a18*, *Phlda2*, and *Osbp15*). The methylated (madumnal) copy of *LIT1* does not bind ZAC1. Therefore, *LIT1* is expressed solely from its padumnal allele whereas several closely linked genes are expressed solely from their madumnal alleles.

ZAC1 is a key node in a network of imprinted genes controlling embryonic growth (Varrault *et al.* 2006). Expression of *ZAC1* caused increased mRNA levels of madumnally expressed genes as well as padumnally expressed genes. In addition, ZAC1 has been shown to bind to a shared endodermal enhancer of two oppositely imprinted genes: *Igf2* (a padumnally expressed gene) and *H19* (a madumnally expressed gene). Unfortunately, the study did not show whether ZAC1 bound to the madumnal enhancer, the padumnal enhancer, or both.

ZAC1 and *LIT1* provide an example of a particularly tight interaction between imprinted genes. In this example, ZAC1, the protein product of a padumnally expressed gene on one chromosome, binds to the padumnal copy of a DMR on another chromosome, activating the transcription of the padumnal copy of *LIT1* and causing the suppression of the padumnal copies of loci closely linked to *LIT1*. The 'message' from *ZAC1* to *LIT1* is evolutionarily 'trustworthy' because sender and receiver both reside on padumnal chromosomes and thus have congruent interests.

Expression of *ZAC1* also modulates the expression of madumnally expressed genes (Varrault *et al.* 2006), although it has not been determined whether these effects are direct or indirect. The interests of sender and receiver need not coincide when a padumnally expressed gene sends a message to a madumnally expressed gene. Natural selection acting on the sender may be favouring different outcomes from natural selection acting on the receiver. Under some circumstances, this conflict might result in an 'evolutionary chase' in which the receiver attempts to avoid 'accepting' a message from the sender or in which a sender attempts to avoid its message being intercepted by the wrong recipient. Below, I will consider an example where a chase has been sought but not found.

Insulin-like growth factor 2 (*IGF2*) is a padumnally expressed gene that promotes fetal growth. Its protein product, IGF-II, binds to two receptors (IGF1R and IGF2R). IGF1R mediates the growth-promoting effects of IGF-II. IGF2R is a decoy receptor that binds IGF-II and transports it to lysosomes for degradation (Filson *et al.* 1993). In most eutherian mammals, but not primates, *IGF2R* is padumnally silent (Killian *et al.* 2001). Thus, a padumnal gene produces a growth factor (IGF-II) that is degraded by the product (IGF2R) of a madumnal gene (Haig and Graham 1991). This can be considered a simple form of deception: *Igf2* sends a signal to IGF1R, but the message is intercepted by IGF2R before it reaches the intended recipient. In this example, there is no transfer of information between *IGF2* and *IGF2R* (the respective genes). Rather, the message from *IGF2* is intercepted by a protein produced by *IGF2R*.

McVean and Hurst (1997) argued that *IGF2* and *IGF2R* should show evidence of rapid antagonistic co-evolution if the theory of intragenomic conflict were correct. However, comparative sequence analysis found no evidence of an evolutionary chase. Therefore, McVean and Hurst considered this to be a failed prediction of the theory. A key assumption of their prediction was that IGF-II has sufficient degrees of evolutionary freedom to continually evade IGF2R.

No crystal structures are currently available for IGF-II bound to IGF2R. Our imperfect understanding of how IGF-II binds to its receptors is based on the analysis of mutations. Some observations suggest that natural selection might be able to eliminate binding to IGF2R without compromising IGF-II's interaction with IGF1R. On the molecular surface of IGF-II, critical residues for binding to IGF2R do not overlap with critical residues for binding to IGF1R. Moreover, mutations that abolish binding to IGF2R have mild, albeit negative, effects on binding to IGF1R (Delaine *et al.* 2007). On the other hand, IGF-II is a small and highly conserved protein. IGF2R-binding sites on IGF-II partially overlap with binding sites for an important family of IGF-binding proteins (Terasawa *et al.* 1994). Therefore, it is possible that IGF-II's structure is sufficiently constrained that it is a 'sitting duck', a tethered target against which *IGF2R* can perfect its aim.

12.7 Concluding remarks

Both communication *among* organisms and communication *within* organisms can be influenced by internal genetic conflicts. Most of this communication does not take place among genes but rather among the robotic structures produced by genes. Genes usually have little direct information about the environment but genes collaborate to construct high-level automata that are well informed.

If an automaton's structure and responses are influenced by multiple genes with conflicting interests, then the automaton's responses need not conform to the evolutionary interests of any particular gene. This is clearly true at the level of organisms but it may also be true at the level of individual proteins. The serotonin 2C receptor (HTR2C) provides a particularly clear example. The amino acid sequence of the receptor that is inserted in a neural membrane is jointly determined by an unimprinted X-linked gene (*HTR2C*) and by a padumnally expressed snoRNA (HBII-52) encoded in the Prader–Willi region of chromosome 15. Thus, the function of HTR2C is subject to conflicting selective forces.

The indeterminacy of the genetic interests of organisms may be a bitter pill for evolutionary theorists wishing to predict organismal behaviour or for behavioural ecologists wishing to test the theorists' predictions in the field. How an organism is expected to behave, given internal conflicts, requires input from cell biology, but from a cell biology informed by concepts of **inclusive fitness** and of evolutionarily stable strategies, and a cell biology attuned to the trade-offs that whole organisms face in a complex ecology.

Evolutionary theory makes explicit predictions about the selective forces that have acted on genes subject to genomic imprinting. Specifically, the expression of imprinted genes is predicted to have fitness consequences for asymmetric kin. Therefore, the detailed analysis of behaviour in individuals with inactivation, or overexpression, of imprinted genes will provide useful clues about how natural selection has shaped kin interactions in the past and on what have been the most salient trade-offs between an individual's fitness and that of its relatives. For example, the observation that imprinted genes have multiple roles in the differentiation and function of brown fat have made me appreciate the individual costs and kin benefits associated with thermogenesis in small mammals and birds.

Behavioural ecologists have traditionally studied 'wild-type' individuals, because these are the genotypes that have survived the filter of natural selection. However, a behavioural ecology of the genetically perturbed has great potential for illuminating evolutionary questions, whether these perturbations result from naturally occurring variation, such as human genetic disorders, or from deliberately induced mutations, such as the 'knock-outs' of modern mouse genetics (see mouse genetic and scent communication in Chapter 6). For example, mouse litters can now be produced in which some members of the litter do not pay their fair share of heating costs because of the genetic ablation of non-shivering thermogenesis. A behavioural analysis of how warm pups respond to cool pups in their midst would provide insight into how cooperative heat production has been maintained in the face of 'free-riders'.

This chapter is an attempt by one organism-level automaton to communicate to other organism-level automata. Each of us has a set of genetic controllers 'back on Earth' trying to pull the strings, but we make most decisions on our own. Our pains and pleasures are the sticks and carrots that genes use to influence our decisions in pursuit of their ends. Genes do not care for us, know little about our world, and cannot agree among themselves. We should respect their suggestions, but not too much. If this chapter serves my genes' interests, it will be by a very indirect route. Our genes' purposes are not our own.

Summary

Behavioural ecology and cell biology both use the language of communication and signalling. In cell biology, signaller and receiver are cells of a single body or molecules encoded by a single genome and are implicitly assumed to have identical evolutionary interests. The signaller does not have an incentive to deceive. In behavioural ecology, signaller and receiver are different genetic individuals, with possibly conflicting evolutionary interests. Signallers may have an incentive to deceive, so receivers must decide whether signals can be trusted. However, we now know that conflicts within genomes are possible. This raises questions about how internal conflicts influence signalling between and within individual organisms. These questions are explored using the example of genomic imprinting.

Acknowledgements

Bernie Crespi, Patrizia d'Ettorre, Shantanu Gaur, David Hughes, and Manus Patten have provided thoughtful comments on the manuscript. Peter Godfrey-Smith has constantly challenged me to clarify my ideas on the role of replicators.

References

Abramowitz, J., Grenet, D., Birnbaumer, M., Torres, H.N., and Birnbaumer, L. (2004). XLαs, the extra-long form of the α-subunit of the Gs G protein, is significantly longer than suspected, and so is its companion Alex. *Proceedings of the National Academy of Sciences of the USA*, **101**, 8366–8371.

Åkefeldt, A., Åkefeldt, B., and Gillberg C (1997). Voice, speech and language characteristics of children with Prader–Willi syndrome. *Journal of Intellectual Disability Research*, **41**, 302–311.

Alvares, R.L. and Downing, S.F. (1998). A survey of expressive communication skills in children with Angelman syndrome. *American Journal of Speech-Language Pathology*, **7**, 14–24.

Arima, T., Kamikihara, T., Hayashida, T. *et al.* (2005). *ZAC, LIT1* (*KCN1Q1OT1*) and $p57^{KIP2}$ (*CDKN1C*) are in an imprinted gene network that may play a role in Beckwith–Wiedemann syndrome. *Nucleic Acids Research*, **33**, 2650–2660.

Aughton, D.J. and Cassidy, S.B. (1990). Physical features of Prader–Willi syndrome in neonates. *American Journal of Diseases of Children*, **144**, 1251–1254.

Badcock, C. and Crespi, B. (2006). Imbalanced genomic imprinting in brain development: an evolutionary basis for the aetiology of autism. *Journal of Evolutionary Biology*, **19**, 1007–1032.

Bonati, M.T., Russo, S., Finelli, P. *et al.* (2007). Evaluation of autism traits in Angelman syndrome: a resource to unfold autism genes. *Neurogenetics*, **8**, 169–178.

van Borsel, J., Defloor, T., and Curfs, L.M.G. (2007). Expressive language in persons with Prader–Willi syndrome. *Genetic Counseling*, **18**, 17–28.

Bouillard, F., Ricquier, D., Mory, G., and Thibault, J. (1984). Increased level of mRNA for the uncoupling protein in brown adipose tissue of rats during thermogenesis induced by cold exposure or norepinephrine infusion. *Journal of Biological Chemistry*, **259**, 11583–11586.

Brown, W.M. and Consedine, N.S. (2004). Just how happy is the happy puppet? An emotional signaling and kinship theory perspective on the behavioral phenotype of Angelman syndrome children. *Medical Hypotheses*, **63**, 377–385.

Burt, A. and Trivers, R. (2006). *Genes in Conflict*. Harvard University Press, Cambridge, MA.

Butler, M.G. (1990). Prader–Willi syndrome: current understanding of cause and diagnosis. *American Journal of Medical Genetics*, **35**, 319–332.

Cannon, B. and Nedergaard, J. (2004). Brown adipose tissue: function and physiological significance. *Physiological Reviews*, **84**, 277–359.

Cassidy, S.B. (1988). Management of the problems of infancy: hypotonia, developmental delay, and feeding problems. In: M.L. Caldwell and R.L. Taylor (eds), *Prader–Willi Syndrome. Selected Research and Management Issues*, pp. 43–51. Springer-Verlag, New York.

Chaudhry, A., MacKenzie, R.G., Georgic, L.M., and Granneman, J.G. (1994). Differential interaction of β_1- and β_3-adrenergic receptors with G_i in rat adipocytes. *Cellular Signalling*, **6**, 457–465.

Clarke, D.J. and Marston, G. (2000). Problem behaviors associated with 15q- Angelman syndrome. *American Journal on Mental Retardation*, **105**, 25–31.

Clayton-Smith, J. (1993). Clinical research on Angelman syndrome in the United Kingdom: observations on 82 affected kindreds. *American Journal of Medical Genetics*, **46**, 12–15.

Cook, E.H., Lindgren,V., Leventhal B.L. *et al.* (1997). Autism or atypical autism in maternally but not paternally derived proximal 15q duplication. *American Journal of Human Genetics*, **60**, 928–934.

Dawkins, R. (1976). *The Selfish Gene*. Oxford University Press, Oxford.

Defloor, T., van Borsel, J., and Curfs, L (2000). Speech fluency in Prader–Willi syndrome. *Journal of Fluency Disorders*, **25**, 85–98.

Delaine, C., Alvino, C.L., McNeil, K.A. *et al.* (2007). A novel binding site for the human insulin-like growth factor-II (IGF-II)/mannose 6-phosphate receptor on IGF-II. *Journal of Biological Chemistry*, **282**, 18886–18894.

Dennett, D.C. (1984). *Elbow Room*. MIT Press, Cambridge, MA.

Didden, R., Korzilius, H., Duker, P., and Curfs, L.M.G. (2004). Communicative functioning in individuals with Angelman syndrome: a comparative study. *Disability and Rehabilitation*, **26**, 1263–1267.

Duker, P., van Driel, S., and van de Bercken, J. (2002). Communication profiles of individuals with Down's syndrome, Angelman syndrome and pervasive developmental disorder. *Journal of Intellectual Disability Research*, **46**, 35–40.

Dykens, E.M. and Kasari, C. (1997). Maladaptive behavior in children with Prader–Willi syndrome, Down syndrome, and nonspecific mental retardation. *American Journal on Mental Retardation*, **102**, 228–237.

Filson, A.J., Louvi, A., Efstratiadis, A., and Robertson, E.J. (1993). Rescue of the T-associated maternal effect in mice carrying null mutations in *Igf-2* and *Igf2r*, two reciprocally imprinted genes. *Development*, **118**, 731–736.

Galantucci, B., Fowler, C.A., and Turvey, M.T. (2006). The motor theory of speech perception reviewed. *Psychonomic Bulletin and Review*, **13**, 361–377.

Gentilucci, M. and Corballis, M.C. (2006). From manual gesture to speech: a gradual transition. *Neuroscience and Biobehavioral Reviews*, **30**, 949–960.

Guilford, T. and Dawkins, M.S. (1991). Receiver psychology and the evolution of animal signals. *Animal Behaviour*, **42**, 1–14.

Haig, D. (1997). Parental antagonism, relatedness asymmetries, and genomic imprinting. *Proceedings of the Royal Society B: Biological Sciences*, **264**, 1657–1662.

Haig, D. (2000). The kinship theory of genomic imprinting. *Annual Review of Ecology and Systematics*, **31**, 9–32 [see on-line erratum].

Haig, D. (2003). On intrapersonal reciprocity. *Evolution and Human Behavior*, **24**, 418–425.

Haig, D. (2004). Genomic imprinting and kinship: how good is the evidence? *Annual Review of Genetics*, **38**, 553–585.

Haig, D. (2006). Intragenomic politics. *Cytogenetic and Genome Research*, **113**, 68–74.

Haig, D. and Graham, C. (1991). Genomic imprinting and the strange case of the insulin-like growth factor-II receptor. *Cell*, **64**, 1045–1046.

Haig, D. and Wharton, R. (2003). Prader–Willi syndrome and the evolution of human childhood. *American Journal of Human Biology*, **15**, 320–329.

Haig, D. and Wilkins, J.F. (2000). Genomic imprinting, sibling solidarity, and the logic of collective action. *Philosophical Transactions of the Royal Society B: Biological Sciences*, **355**, 1593–1597.

Hamilton, W.D. (1963). Evolution of altruistic behavior. *American Naturalist*, **97**, 354–356.

Holm, V.A., Cassidy, S.B., Butler, M.G. *et al.* (1993). Prader–Willi syndrome: consensus diagnostic criteria. *Pediatrics*, **91**, 398–402.

Horike, S., Mitsuya, K., Meguro, M. *et al.* (2000). Targeted disruption of the human *LIT1* defines a putative imprinting control element playing an essential role in Beckwith–Wiedemann syndrome. *Human Molecular Genetics*, **9**, 2075–2083.

Horsler, K. and Oliver, C. (2006a). The behavioral phenotype of Angelman syndrome. *Journal of Intellectual Disability Research*, **50**, 33–53.

Horsler, K. and Oliver, C. (2006b). Environmental influences on the behavioral phenotype of Angelman syndrome. *American Journal on Mental Retardation*, **111**, 311–321.

Isles, A.R., Davies, W., and Wilkinson, L.S. (2006). Genomic imprinting and the social brain. *Philosophical Transactions of the Royal Society B: Biological Sciences*, **361**, 2229–2237.

Jauregi, J., Arias, C., Vegas, O. *et al.* (2007). A neuropsychological assessment of frontal cognitive function in Prader–Willi syndrome. *Journal of Intellectual Disability Research*, **51**, 350–365.

Jolleff, N. and Ryan, M.M. (1993). Communication development in Angelman's syndrome. *Archives of Disease in Childhood*, **69**, 148–150.

Killian, J.K., Nolan, C.M., Wylie, A.A. *et al.* (2001). Divergent evolution in *M6P/IGF2R* imprinting from the Jurassic to the Quaternary. *Human Molecular Genetics*, **10**, 1721–1728.

Kishore, S. and Stamm, S. (2006). The snoRNA HBII-52 regulates alternative splicing of the serotonin receptor 2C. *Science*, **311**, 230–232.

Koenig, K., Klin, A., and Schultz, R. (2004). Deficits in social attribution ability in Prader–Willi syndrome. *Journal of Autism and Developmental Disorders*, **34**, 573–582.

Kurita, M., Kuwajima, T., Nishimura, I., and Yoshikawa, K. (2006). Necdin downregulates Cdc2 expression to attenuate neuronal apoptosis. *Journal of Neuroscience*, **26**, 12003–12013.

Kuwajima, T., Nishimura, I., and Yoshikawa, K. (2006). Necdin promotes GABAergic neuron differentiation in cooperation with Dlx homeodomain proteins. *Journal of Neuroscience*, **26**, 5383–5392.

Lalande, M. and Calciano, M.A. (2007). Molecular epigenetics of Angelman syndrome. *Cellular and Molecular Life Sciences*, **64**, 947–960.

Lee, S., Walker, C.L., Karten, B. *et al.* (2005). Essential role for the Prader–Willi syndrome protein necdin in axonal outgrowth. *Human Molecular Genetics*, **14**, 627–637.

McVean, G.T. and Hurst, L.D. (1997). Molecular evolution of imprinted genes: no evidence for antagonistic coevolution. *Proceedings of the Royal Society B: Biological Sciences*, **264**, 739–746.

Mancini-DiNardo, D., Steele, S.J.S., Levorse, J.M., Ingram, R.S., and Tilghman, S.M. (2006). Elongation of the *Kcnq1ot1* transcript is required for genomic imprinting of neighboring genes. *Genes and Development*, **20**, 1268–1282.

Mantovani, G., Bondioni, S., Locatelli, M. *et al.* (2004) Biallelic expression of the Gsα gene in human bone and adipose tissue. *Journal of Clinical Endocrinology and Metabolism*, **89**, 6316–6319.

Miller, S.P., Riley, P., and Shevell, M.I. (1999). The neonatal presentation of Prader–Willi syndrome revisited. *Journal of Pediatrics*, **134**, 226–228.

Milner, K.M., Craig, E.E., Thompson, R.J. *et al.* (2005). Prader–Willi syndrome: intellectual abilities and behavioural features by genetic subtype. *Journal of Child Psychology and Psychiatry*, **46**, 1089–1096.

Morrison, S.F. (2004). Central pathways controlling brown adipose tissue thermogenesis. *News in Physiological Sciences*, **19**, 67–74.

Nakamura, K. and Morrison, S.F. (2007). Central efferent pathways mediating skin-cooling evoked sympathetic thermogenesis in brown adipose tissue. *American Journal of Physiology*, **292**, R127–R136.

Nicholls, R.D. and Knepper, J.L. (2001). Genome organization, function, and imprinting in Prader–Willi and Angelman syndromes. *Annual Review of Genomics and Human Genetics*, **2**, 153–175.

Õiglane-Shlik, E., Zordania, R., Varendi, H. *et al.* (2006). The neonatal phenotype of Prader–Willi syndrome. *American Journal of Medical Genetics*, **140A**, 1241–1244.

Okada, T., Ernst, O.P., Palczewski, K., and Hofmann, K.P. (2001). Activation of rhodopsin: new insights from structural and biochemical studies. *Trends in Biochemical Sciences*, **26**, 318–324.

Oliver, C., Demetriades, L., and Hall, S. (2002). Effects of environmental events on smiling and laughing behavior in Angelman syndrome. *American Journal on Mental Retardation*, **107**, 194–200.

Oliver, C., Horsler, K., Berg, K., Bellamy, G., Dick, K., and Griffiths, E. (2007). Genomic imprinting and the expression of affect in Angelman syndrome: what's in the smile? *Journal of Child Psychology and Psychiatry*, **48**, 571–579.

Pembrey, M. (1996). Imprinting and transgenerational modulation of gene expression; human growth as a model. *Acta Geneticae Medicae et Gemellologiae*, **45**, 111–125.

Penner, K.A., Johnston, J., Faircloth, B.H., Irish, P., and Williams, C.A. (1993). Communication, cognition, and social interaction in the Angelman syndrome. *American Journal of Medical Genetics*, **46**, 34–39.

Plagge, A., Gordon, E., Dean, W. *et al.* (2004). The imprinted signaling protein XLαs is required for postnatal adaptation to feeding. *Nature Genetics*, **36**, 818–826.

Richman, D.M., Gernat, E., and Teichman, H. (2006). Effects of social stimuli on laughing and smiling in young children with Angelman syndrome. *American Journal on Mental Retardation*, **111**, 442–446.

Ridge, K.D., Abulaev, N.G., Sousa, M., and Palczewski, K. (2003). Photo transduction: crystal clear. *Trends in Biochemical Sciences*, **28**, 479–487.

Rogelj, B., Hartmann, C.E.A., Yeo, C.H., Hunt, S.P., and Giese, K.P. (2003). Contextual fear conditioning regulates the expression of brain-specific small nucleolar RNAs in hippocampus. *European Journal of Neuroscience*, **18**, 3089–3096.

Romanovsky, A.A. (2007). Thermoregulation: some concepts have changed. Functional architecture of the thermoregulatory system. *American Journal of Physiology*, **292**, R37–R46.

Runte, M., Hüttenhofer, A., Groß, S., Kiefmann, M., Horsthemke, B., and Buiting K (2001). The IC-*SNURF-SNRPN* transcript serves as a host for multiple small nucleolar RNA species and as an antisense RNA for *UBE3A*. *Human Molecular Genetics*, **10**, 2687–2700.

Steffenburg, S., Gillberg, C.L., Steffenburg ,U., and Kyllerman, M. (1996). Autism in Angelman syndrome: a population-based study. *Pediatric Neurology*, **14**, 131–136.

Terasawa, H., Kohda, D., Hatanaka, H. *et al.* (1994). Solution structure of human insulin-like growth factor II; recognition sites for receptors and binding proteins. *EMBO Journal*, **13**, 5590–5597.

Trillingsgaard, A. and Østergaard, J.R. (2004). Autism in Angelman syndrome: an exploration of comorbidity. *Autism*, **8**, 163–174.

Tseng, Y.H., Butte, A.J., Kokkotou, E. *et al.* (2005). Prediction of preadipocyte differentiation by gene expression reveals role of insulin receptor substrates and necdin. *Nature Cell Biology*, **7**, 601–545.

Varrault, A., Gueydan, C., Delalbre, A. *et al.* (2006). Zac1 regulates an imprinted gene network critically involved in the control of embryonic growth. *Developmental Cell*, **11**, 711–722.

Veltman, M.W.M., Craig, E.E., and Bolton, P.F. (2005). Autism spectrum disorders in Prader–Willi and Angelman syndromes: a systematic review. *Psychiatric Genetics*, **15**, 243–254.

Vitali, P., Basyuk, E., Le Meur, E. *et al.* (2005). ADAR2-mediated editing of RNA substrates in the nucleolus is inhibited by C/D small nucleolar RNAs. *Journal of Cell Biology*, **169**, 745–753.

Walz, N.C. and Benson, B.A. (2002). Behavioral phenotypes in children with Down syndrome, Prader–Willi syndrome, or Angelman syndrome. *Journal of Developmental and Physical Disabilities*, **14**, 307–321.

Wilkins, J.F. and Haig, D. (2001). Genomic imprinting of two antagonistic loci. *Proceedings of the Royal Society B: Biological Sciences*, **268**, 1861–1867.

Wilkins, J.F. and Haig, D. (2003). What good is genomic imprinting: the function of parent-specific gene expression. *Nature Reviews Genetics*, **4**, 359–368.

Williams, C.A., Angelman, H., Clayton-Smith, J. *et al.* (1995). Angelman syndrome: consensus for diagnostic criteria. *American Journal of Medical Genetics*, **56**, 237–238.

Williams, C.A., Lossie, A., and Driscoll, D. (2001). Angelman syndrome: mimicking conditions and phenotypes. *American Journal of Medical Genetics*, **101**, 59–64.

Williams, C.A., Beaudet, A.L., Clayton-Smith, J. *et al.* (2005). Angelman syndrome 2005: updated consensus for diagnostic criteria. *American Journal of Medical Genetics*, **140A**, 413–418.

Yamada, K.A. and Volpe, J.J. (1990). Angelman's syndrome in infancy. *Developmental Medicine and Child Neurology*, **32**, 1005–1011.

Yu, S., Yu, D., Lee, E. *et al.* (1998). Variable and tissue-specific hormone resistance in heterotrimeric G_s protein α-subunit ($G_s\alpha$) knockout mice is due to tissue-specific imprinting of the $G_s\alpha$ gene. *Proceedings of the National Academy of Sciences of the USA*, **95**, 8715–8720.

CHAPTER 13

Language unbound: genomic conflict and psychosis in the origin of modern humans

Bernard J. Crespi

Words are the physicians of the mind diseased.

Aeschylus, *Prometheus Bound*

13.1 Introduction

The question of the origin of modern humans has inspired more scenarios, stories, and research than perhaps any other in biology and the humanities. In one of the first such stories, Plato describes how Epimetheus the Titan distributed abilities to each kind of animal, but used them up before reaching humans. His brother Prometheus, seeing that humans had nothing enabling them to survive, stole technology and fire, and knowledge and philosophy, from the gods and bequeathed these skills and abilities upon them. As eating from the tree of knowledge prompted the Biblical God to banish Adam and Eve, so Zeus punished Prometheus by binding him to a rock, and so humans have paid dearly for their gifts of cognition ever since they were bestowed.

In this chapter I will seek to bring the Prometheus myth and metaphor up to date, with a focus on language, the gift most uniquely human (see Chapter 14 for a related discussion of human language). Analysis of the evolution of human language brings together three of the greatest unknowns in biology: the brain, the genome, and the evolution of modern humans. It has thus generated a vast literature, a verbiage so extensive that it tends to obscure the paucity of facts. Moreover, the facts that do exist reside in diverse, specialized disciplines from genetics to phylogeny, palaeontology, anatomy, neuroscience, psychology, psychiatry, and linguistics. My goal is to integrate across these disciplines using the only tool that unites them, evolutionary biology. Language, and humans, are social, so my conceptual monkey-wrench of choice for such a construction is theory of the evolution of social behaviour (Alexander 1980, 1987), the only science that addresses how human sociality, and its genetic underpinnings, change under **Darwinian selection**.

I begin with a few basics about the brain, language, and how natural selection works at different levels from genes to groups. Next, I explain how the brain and language can be studied using the three main approaches for analysing the adaptive significance of traits: functional design, measurement of selection, and the comparative method. I then apply the comparative method to a new form of diversity: autistic and psychotic–affective spectrum conditions, the main generalized 'mutations' of human sociality and language. Our goal here is to understand how human language and communication have evolved by analysing how these adaptive systems can be perturbed. The nature of such perturbations provides insights into our cognitive and emotional architecture, just as mutations in a single gene provide insight into its functions in physiology and development.

Virtually all previous studies of language evolution have focused on cooperative and beneficial aspects of human communication, such as

Box 13.1 Glossary

Asperger syndrome: Idiopathic (with unknown cause) autistic condition that involves specific deficits in social reciprocity but no language delay or mental retardation.

Autism: Idiopathic condition defined by deficits in language, communication, and social reciprocity, and by the presence of restricted interests and repetitive behaviour.

Autistic spectrum conditions: Autism and Asperger syndrome, which grade into normality, as well as genomically based neurological conditions that involve high rates of autism, such as Fragile X syndrome, Rett syndrome, Angelman syndrome and other conditions in Table 13.2.

Broca's area: Region of the left frontal lobe of the brain that is specialized for speech production.

Corpus callosum: large bundle of nerve fibres that connects the left and right hemispheres of the brain.

Dyslexia: Learning disability manifested as impairments in reading.

Genomic conflict: Presence in the same genome of genes that maximize their replication via different, conflicting effects on growth, development, and behaviour (e.g. conflict between Y-linked genes and autosomes over offspring sex ratio, or between paternally imprinted and maternally imprinted genes over growth).

Genomic imprinting: Silencing of a gene in an individual depending upon whether the gene was inherited from the father or the mother. According to Haig's kinship theory of imprinting, paternally silenced (maternally expressed) genes are expected to restrict the 'selfish' interests of offspring, and paternally expressed genes are expected to enhance such interests.

Hyperlexia: Spontaneous, precocious mastery of single-word reading, which often also involves impairments in comprehension of the meaning of written material.

Klinefelter syndrome: Syndrome due to one or more extra X chromosomes in males, usually XXY. This condition involves poor verbal skills, spared visual-spatial skills, and a high incidence of psychotic-affective spectrum conditions.

Positive selection: Selection 'for' specific alleles or haplotypes (contiguous blocks of DNA with the same alleles at polymorphic sites), as indicated by high rates of amino acid substitution or by the presence of haplotypes that are unexpectedly large and have thus recently risen to a relatively high frequency in a population.

Psychosis: Mental state characterized by loss of contact with objective reality, which often involves paranoid or grandiose delusions, hallucinations, or disorganized thinking. Psychosis is common in schizophrenia and not uncommon in bipolar disorder and major depression.

Psychotic–affective spectrum conditions: A suite of genetically related and phenotypically related idiopathic psychiatric conditions that includes schizophrenia, schizotypal personality disorder, bipolar disorder, major depression, anxiety disorders, and panic attacks, as well as genomic conditions such as Klinefelter syndrome, velocardiofacial syndrome, Prader–Willi syndrome and other conditions in Table 13.3. 'Psychotic' refers to cognitive (thought) symptoms, and 'affective' refers to mood (emotional) symptoms.

Schizophrenia: Set of related psychiatric disorders characterized by psychosis or dysregulated affect, such as 'flat' (lack of) affect or affect incongruent with environmental conditions.

Social brain: Distributed, integrated neural systems for the acquisition and processing of social information; also refers to the idea that the human brain evolved in the context of strong selection from the fitness-mediating effects of complex social interactions.

Turner syndrome: Syndrome due to full or partial loss of an X chromosome in females, such that females are mainly XO. This syndrome involves good verbal skills but impaired visual–spatial skills, and a high incidence of autism in females with the intact X inherited from their mother.

Wernicke's area: Area of the left hemisphere involved in the comprehension of spoken language.

coordination of activities, pedagogy, or impressing a potential mate with syntactic and emotive prowess. This perspective is incomplete, because human social interaction is always permeated by complex mixtures of cooperation and **conflict**, which follow inevitably from asymmetries in genetic relatedness (Alexander 1980, 1987; Haig 2006a). In the fourth section of this chapter, I thus explain the potential roles of conflicts—especially **genomic conflicts**—in the evolution of language (see also Chapter 12 for further details on genomic conflict and Box 13.1 for a glossary of terms used in this chapter). Finally, I end the chapter by linking evolutionary genomics with psychiatry and neuroscience, to develop a perspective for understanding the selective pressures involved in the origin of modern humans. We will also unbind Prometheus, and discuss new approaches to free humans from the disorders of our evolutionary legacy.

13.2 What, and where, is human linguistic communication?

Human linguistic communication involves the activation of both the left and right hemispheres of the brain, a complex system of over 100 muscles for articulation, plus a suite of ancillary movements involving manual gestures and changes in facial-expression around the eyes and mouth (Galantucci *et al.* 2006; Lieberman 2007). Crow (2004) and Mitchell and Crow (2005) have described a simple model of the brain as a 'four-chambered organ' in how it processes and produces language via the activation of heteromodal association cortex, the 'thinking' regions of the neocortex that integrate sensory data and motor feedback with thought and memory (Fig. 13.1).

The left hemisphere harbours Broca's area, the locus of encoding and producing speech, which translates 'thoughts' or 'inner speech' into linear strings of neural commands to move specific muscles in specific ways. Also on the left, nearer the back, is Wernicke's area, most simply described as the region of the brain for decoding the literal denotations of speech by others.

On the right, we have a posterior region 'for' establishing the meanings of heard speech and accompanying movements—that is, the prosody and pragmatics (non-literal meanings) of language that are inferred more or less automatically from tone, inflection, and other clues that overlay literal word-for-word interpretation. Here lies irony, sarcasm, humour—and much of emotion. The right frontal region is an analogue of Broca's area, but for generating the thoughts and intentions that precede speech. In this area, discourse plans are generated via some form of spatial recombination of components, and then transferred across the corpus callosum.

The two hemispheres thus process forms of information more or less separately, but they must still regulate unitary behaviour.

For instance, a husband may tell his wife that he is going to be working late. Her left brain hears that he is going to be 'working late' and accepts that on face value. Her right hemisphere, however, hears the melody of his voice, notes the changes in his face and body language as he

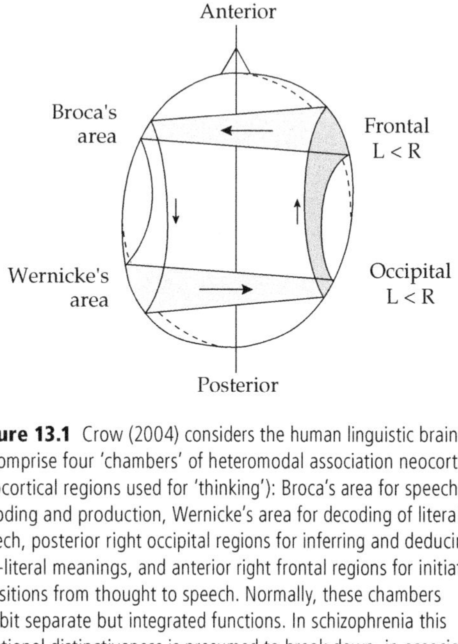

Figure 13.1 Crow (2004) considers the human linguistic brain to comprise four 'chambers' of heteromodal association neocortex (neocortical regions used for 'thinking'): Broca's area for speech encoding and production, Wernicke's area for decoding of literal speech, posterior right occipital regions for inferring and deducing non-literal meanings, and anterior right frontal regions for initiating transitions from thought to speech. Normally, these chambers exhibit separate but integrated functions. In schizophrenia this functional distinctiveness is presumed to break down, in association with neurodevelopmentally-reduced levels of cerebral asymmetry, to produce the disordering of language that characterizes psychosis. This model of the brain in language was developed via consideration of the 'first-rank' symptoms of schizophrenia, and the neuroanatomical changes wrought by the evolution of modern humans.

talks and decides that he is up to something that does not involve work. How she reacts will in turn depend upon which half of her brain prevails as well as on her past experience with her husband and his late night sojourns. In any case, she is in conflict. (Joseph 1992)

Like any discussion involving the brain, this is all a considerable oversimplification. I refer the reader to Cook (2002) and Mitchell and Crow (2005) for subtleties and evidence regarding differential linguistic functions of the two hemispheres. Our main message so far is that left-brain language areas are usually relatively specialized as a system for the rapid, temporal, linear functions of encoding and decoding, while the right-brain areas specialize in processing of spatial, multidimensional information involving emotions, intentions, metaphors, meanings, and one of their external manifestations, sociality. This conceptual, neurological model of language functions has been applied to the components of linguistic discourse by Cook (2002) (Table 13.1).

The actual neurological mechanisms used in processing and producing language appear simpler than one might think. Thus, Ivry and Robertson (1998) provide the rather woolly concepts such as 'coding' and 'meaning' with a solid neurophysiological basis, in showing how the left hemisphere is relatively specialized for higher-frequency, more-local forms of information processing than is the right. Similarly, the well-supported 'motor theory of language' posits that we decode speech in part by activating the same pre-motor neural circuits that we would use to make the very sounds we hear—a reversible sound-to-neuron translation system (Galantucci *et al.* 2006). Perception and production are entwined more generally in the human mirror neuron systems, whereby we interpret hand movements, sounds, and facial expression via activation of the pre-motor neural pathways that we would use to generate them ourselves (Iacoboni and Dapretto 2006). Effective human discourse thus relies on forms of social–emotional resonance, mediated by the left and right brains working in concert. Such mechanisms take on vital importance in understanding how language could have evolved step by step (e.g. Arbib 2005), and in understanding how sociality and language can go awry in human neurodevelopment.

13.3 How, why, and where does language evolve?

We have described human language and its neural instantiations. Now let us wrap this all together, call it 'a trait', and see how it should evolve, from first principles. I am interested not in imagined prehistorical sequences, but in how basic social-evolution theory can help to explain the functions of language, simple or complex, in human interaction. Towards this end we will take a series of small

Table 13.1 Cook (2002) describes how the left and right hemispheres of the brain are more or less specialized for mediating different components of language, from its smallest parts to its largest, conversation or discourse

Level of linguistic complexity	Hemisphere of the brain	
	Left	Right
Phoneme	Auditory segmentation	Intonational decoding
Word	Denotation	Connotation
	Close associations	Distant associations
Noun-adjective phrase	Literal meanings	Metaphorical meanings
Sentence	Literal meanings	Emotional implications
Paragraph	Explicit event-by-event meanings	Implicit meanings
Discourse	Sequential	Contextual

steps down an evolutionary garden path:

1. Humans are expected to behave so as to maximize their **inclusive fitness**, barring errors or rapid environmental change.

This is simply how natural selection works, on all organisms, all the time. Alexander (1980, 1989) expands upon the caveats involved.

2. The best way to maximize one's inclusive fitness is often to alter the behaviour of other humans, given that humans are extremely social, interdependent animals that exhibit pervasive confluences and conflicts of interest.

My inclusive fitness is not my brother's inclusive fitness, nor my mother's or children's—we are family but related only by one-half for autosomal genes, the bulk of our genomes. Worse, non-relatives are motivated to mutual aid only by reciprocity and larger-scale common interests. Human interaction and human history are thus litanies of shifting conflict and cooperation, between and within individuals, families, cultures, and other groups, over resources historically linked to reproduction.

How do we alter the behaviour of others? There are only three ways: 'persuasion' (negotiating and providing mutualistic benefits), 'coercion' (imposing costs on others, or threatening to do so), and force (taking control of others' behaviour away, or threatening such action) (Brown *et al.* 1997). These methods are used from playground to boudoir to battlefield, and their deployment as alternatives depends critically on the presence and form of asymmetries in physical power, resources, and information.

3. One of the best ways to alter the behaviour and thought of other humans, compared to other modalities or actions, is via language and its facial-gestural trappings.

Language offers us the ability to convince, persuade, or coerce other humans with logic, to use emotional prosody for persuasion or coercion, and to lie. Indeed, I would suggest that language and emotion evolved in large part for verbally 'manipulating', in a more or less non-pejorative sense, the thoughts and behaviour of others. Talk may be energetically cheap, but it can be very powerful and thus very expensive, or profitable, socially. Such large gains and losses may be possible because language is the medium of information, information is power, and power in the control of resources and other humans is the most general and flexible of all avenues to reproductive success.

Studies focusing on the human 'social brain' and 'Machiavellian intelligence' have addressed the niceties of selection for social skills in primate and human evolution (Whiten and Byrne 1997), but they have only begun to interface with studies of language evolution. The common currencies between these fields are neurological, developmental, and genetic. For example, the superior temporal gyrus subserves both language processing and social cognition (Bigler *et al.* 2007), social skills and language develop in concert throughout early childhood (Bloom 2004), and psychiatric disorders of the social brain virtually always involve genetically based alterations of language (DeLisi 2001; Seung 2007). Is there recent and ongoing selection in human evolution for social skills, and for aspects of language? As discussed below, we must uncover the genes involved to find out for sure—and we can.

4. Humans are expected to be unaware of, and deny, that their behaviour is **selfish**, nepotistic, or manipulative of others. As a result, much of our most fitness-salient thought, language, and behaviour should be unconscious, repressed, projected, rationalized, deluded, automatic, or self-deceptive—with denial or the social emotions of shame, embarrassment, and admitted guilt quickly deflecting any suggested or actual culpability.

Alexander (1989) made this essential point, which goes to the psychoanalytic core of human consciousness and behaviour. We thus admire but eschew Machiavelli, we esteem altruistic humans who perform heroically for stranger or nation—and we venerate the various gods who sacrifice themselves for us. We also make moral decisions, but cannot provide them any coherent justification because they 'are not open to conscious introspection' (Hauser *et al.* 2007). Civilized behaviour may thus result from repression of Freud's 'sex instinct', modernized in terms of maximizing inclusive fitness.

Robust empirical analysis of such psychological tendencies as delusion and denial is fiendishly difficult, but modern neuroscience can tap the unconscious and offer clues (Trivers 2000; Stein *et al.* 2006). For example, some patients with right (but not left) hemisphere strokes, leading to left-side paralysis, will vehemently deny their obvious inability to move their left arm, offering instead rationalizations such as fatigue (Ramachandran 1996). This and other evidence suggests that the voluble left hemisphere serves as a cognitive 'spin-doctor' that maintains (self-serving) conceptual and world-view consistency, while the mute right hemisphere serves us as 'anomaly detector' or 'devil's advocate', prompting cognitive change should the weight of evidence contrary to the left-hemisphere's beliefs become too great (Ramachandran 1996). The relationship between left and right hemisphere is also indicated by their severance: cutting the corpus callosum (to control intractable epilepsy) results in complete loss of speech for days, weeks, or months in most patients, but right-hemisphere damage does not cause loss of speech. The implication is that the left hemisphere normally awaits cognitive input from the right hemisphere before initiating speech (Cook 2002), as also suggested by Crow's model of the four-chambered brain. To the extent that consciousness (whatever that is) is associated with language and speech production, it is predominantly a left-hemisphere function—but this extent remains quite unknown, and the right-hemisphere also mediates perceptions that we would consider as conscious (Joseph 1992).

The garden path has led us back to the brain. If you politely followed the entire route, you might agree that the lateralized social and linguistic brain is an astoundingly complex parallel processor designed to maximize inclusive fitness, without being aware, or admitting, that it does precisely this. Now, maximizing inclusive fitness is an inherently social enterprise, so we must discuss next the contexts of social and language evolution, the arenas of conflict and cooperation that generate variation in the reproduction of alleles and their bearers. There are three such arenas: within family, within and between group, and within the individual.

13.2.1 Within-family conflicts

Children develop in the womb nourished from their mother-invading placenta, then from breast, hand, and crying or babbling mouth. Their linguistic minds develop mainly through interactions with their mother and other family members, via the simple, exaggerated language of 'motherese' and pretend, scenario-building play with self, mother, toys, and peers (Vygotsky 1962; Bloom 2004; Falk 2004). Childhood is also the main arena for two forms of social strife: **parent–offspring conflict** and conflict that involves **genomic imprinting**. Put most simply, the child's non-imprinted autosomal genes, and the child's paternally expressed imprinted genes, have been selected for expression and activity that provides more in the way of developmental-reproductive resources to the child than the mother's genes, or the child's maternally expressed imprinted genes, have been selected to provide (Haig 2006a; see also Chapter 12). To the extent that language mediates the transfer of resources within families, it should be a key weapon in both forms of conflict (Crespi 2007). Indeed, according to Vygotsky (1962):

> During this stage, the child and adult are constantly issuing instructions or asking questions of each other so that, for the child, the whole process of speaking becomes bound up with attempts by the two parties to control each other's actions.

For children, suckling, crying, cooing, babbling, persuasive requests, charm, and smiles, and coercive tantrums, arguments, and refusals, stock the social armoury (Badcock 1989; Isles et al. 2006; Locke 2006). Mutual dependency and coincident fitness interests temper these battles and reduce associated costs, making many conflicts subtle unless development is perturbed.

13.2.2 Conflicts within and between groups

For our growing child, within- and between-group conflicts come into play with sexual and social maturity. **Sexual selection** and **sexual conflicts** within local groups suffuse adolescence and young adulthood; the former has been postulated as a driving force in the evolution of language (Locke and Bogin 2006), and both processes should contribute to the well-documented sex differences in verbal abilities, with females superior.

Alexander (1989) describes evidence for the pervasiveness of group against group conflicts in human evolution, Bowles (2006) lends population-genetic rigour to the efficacy of this level of selection in humans, and Lahti and Weinstein (2005) explain how the tension between within-group cooperation and within-group conflict should shift in relation to the strength of external threats. Many group-level traits in humans, such as religion, local linguistic distinctiveness, and group-competitive team sport, can best be interpreted in the context of this selective arena and level (Alexander 1989; Nettle and Dunbar 1997). To the extent that group against group conflicts have driven the evolution of the human psyche (Alexander 1989) they must also have mediated the evolution of language, perhaps as the most effective possible means of coordinating within-group cooperation under this lethal selective pressure. Group cohesion should also be greatly strengthened by shared delusions, such as religious and nationalistic beliefs of own-group supremacy and righteousness—righteousness raised from the individual level to that of groups and gods.

Hypotheses regarding the roles of sexual selection and sexual conflict in the evolution of language can be evaluated via joint analysis of sex differences in linguistic abilities and the genetic basis of such abilities. Of particular interest is whether language-related traits are X-linked, and how gene expression levels of X-linked genes co-vary with verbal skills. For example, the corpus callosum, which strongly mediates linguistic abilities (e.g. Dougherty *et al.* 2007), is also sexually-dimorphic in humans—and callosal disorders (such as its absence or reduction) show a strong enrichment to the X chromosome (B. J. Crespi, unpublished data), as do genes whose mutations influence general intelligence (Skuse 2005).

13.2.3 Conflict within individuals

So-called individuals are divisible genetically, because they bear sets of genes with different routes for maximizing their frequency in the next generation, via divergent effects on their bearers. These genetic 'factions' (Haig 2006a), which are more or less in conflict depending on their patterns of relatedness and inheritance, include autosomal genes, sex-linked genes, mitochondrial genes, genes in strong linkage disequilibrium, and paternally vs. maternally inherited genes subject to silencing by genomic imprints. Of these, imprinted genes are one of the best understood, and with sex-linked genes they are also most motivated by selection to influence human cognition, emotionality, and language development. Genes are most often imprinted in the placenta, but the brain runs a close second (Isles *et al.* 2006), presumably because these two organs directly mediate the transfer of fitness-limiting resources in **networks** of kin.

The effects of imprinted genes are usually unseen, because they engage in dynamically balanced, phenotypic 'tugs-of-war', as between mother and placenta in fetal growth (e.g. Cattanach *et al.* 2006). In pathology such conflictual systems are revealed, as one party stumbles and loses ground, if not *because* of the tug of war itself, then due to mutation or **epimutation**. The major disorders of the human 'social placenta', including gestational diabetes, fetal growth restriction, hydatidiform moles, and pre-eclampsia, are mediated in considerable part by imbalances in imprinted gene expression (e.g. Oudejans *et al.* 2004). And so, I argue below, are the main disorders of the social brain. But to consider this, we first need some tools with which to dissect the human brain and language, and to uncover the selective pressures under which they evolve.

13.3 How do we study the evolution of language?

Many studies of the 'evolution' of language have used arguments from plausibility, with a loose rubric of descent with modification as their main evolutionary tool. Such weak inference appears difficult to avoid, given that we seek to understand a revolutionary, about 50,000 years past transition in an organ that we do not understand. I will suggest here that recent, converging evidence from three disciplines, neuroscience, genomics, and psychiatry, is poised to loosen, and ultimately remove, this veil of ignorance and speculation.

Tinbergen described four methods for analysing traits in biology, posed as questions: (1) adaptive function and (2) phylogeny are the two evolutionary, ultimate questions, and (3) ontogeny and (4) mechanism, are the two proximate ones. We will address his evolutionary, ultimate questions with three approaches for analysing the adaptive significance of human language and communication: functional design, measurement of selection, and the comparative method.

13.3.1 Functional design

Functional design refers to what a trait or form of a trait is 'for'—how it enhances performance at some task. Especially for complex traits, analyses of functional design benefit greatly from understanding how the trait 'works' and how its components function together.

For human language and human communication, the burgeoning field of brain imaging, especially functional MRI, is telling us how the brain works—for example, that the medial pre-frontal cortex is for theory of mind and empathy, and that the orbitofrontal cortex is for regulating impulse and socially appropriate behaviour (e.g. Saxe 2006). Analyses of activation patterns in normal brains engaging in various tasks are finally illuminating and mapping the mind's former heart of darkness, and telling us that the brain is both highly modular and tightly integrated. As we decided above, it is 'for' maximizing inclusive fitness, and so we find regions like the insula that mediate both visceral disgust and ingroup–outgroup judgements (Harris and Fiske 2006), the fusiform gyrus specialized for recognizing faces (Gobbini and Haxby 2006), the ventromedial pre-frontal cortex for solving moral dilemmas (Koenigs *et al.* 2007), and a suite of interconnected regions from amygdalae to frontal lobes comprising the human 'social brain' (Saxe 2006).

These natural history studies dovetail with older work on the effects of damage to specific brain regions, such as lesions in Broca's or Wernicke's areas causing forms of aphasia, or impairments in understanding emotional prosody of speech following right-hemisphere damage or sectioning of the corpus callosum. What is perhaps most exciting is the nascent integration of genetics with functional imaging, which has shown that brain activation patterns depend on genotype, for such genes as the serotonin transporter *SLC6A4* (Rao *et al.* 2007) and the dysbindin gene *DTNBP1* (Fallgatter *et al.* 2006). Such studies provide a strong, integrative link between genes and brain function—a link that we might prefer to deny or explain away, but that can some day tell us how we became human, once we understand how genes for our brain evolved.

13.3.2 Measurement of selection

Functional design tells us about performance, but fitness is the currency of evolution, especially given the ubiquity of **pleiotropy** and tradeoffs. Fitness variation is the outcome of selection, the statistical relation between a trait and some measure of reproductive success. So how can we possibly measure selection, the driving force of evolution, on perhaps the most complex structure in the universe, with its 40–100 billion neurons each with 10,000–100,000 synapses (Rapoport 1999)? What's worse, the key selective events took place tens of thousands of years ago, and evolutionary psychologists struggle to measure selection even in extant populations.

There is a way, albeit indirect. We find genes 'for' brain size, structure, laterality, language, spatial skills, and mood—genes whose variants can reliably be associated with variation in neurophysiological, neuroanatomical, cognitive, and emotional traits. Then we use genetic-variation data from extant humans, and from other primates, to test for the presence and form of selection on these genes, especially '**positive selection**', the signature of adaptive directional change in nucleotide sequence.

FOXP2 is perhaps the best-known such gene: it was subject to positive selection in the human lineage roughly 50,000 years ago leading to two key amino acid changes (Zhang *et al.* 2002). The gene is highly expressed in language regions of the brain (Vargha-Khadem *et al.* 2005), its extant mutations have been associated with impaired language and articulation of speech, autism, and schizophrenia with auditory hallucinations (see Vargha-Khadem *et al.* 2005; Crespi 2007), and it may be subject to genomic-imprinting effects (Feuk *et al.* 2006). There are other such genes. For example, higher expression of the X-linked, non-inactivated gene *GTPBP6* is strongly associated with lower verbal skills in Klinefelter syndrome subjects (XXY males) (Vawter *et al.* 2007), and this gene has apparently been subject to positive selection in the human lineage (B. Crespi and K. Summers, unpublished). Genetic variation in the *EFHC2* gene, also X-linked, explains over 13% of the variation in recognition of fear from faces—a social-emotional trait—in Turner syndrome (XO) females, and the better-recognizing **haplotype** appears to have been selected for in recent human evolution (Weiss *et al.* 2007). And our functionally imaged genes *SLC6A4* and *DTNBP1* both show strong evidence of recent positive selection in humans (Voight *et al.* 2006); *DTNBP1* also shows associations of some alleles and haplotypes with schizophrenia risk and general intelligence (Zinkstok *et al.* 2007), and *SLC6A4* harbours variants associated with schizophrenia risk (Fan and Sklar 2005), autism (Brune *et al.* 2006), and major depression (Vergne and Nemeroff 2006).

We are just beginning to close the loops between brain function and genetic variation, between genetic variants and positive selection during recent human evolution, and between genes and psychiatric disorders of the social brain. To understand language evolution, we need more of the genes underlying the primary human disorders of language: autism, schizophrenia, specific language impairment, and dyslexia, and genes 'for' lateralization and language ability in non-clinical populations. One such gene has recently been uncovered: haplotypes of the *LRRTM1* gene on chromosome 2 are associated both with schizophrenia risk, and with handedness in dyslexics (Francks *et al.* 2007). This gene is of special interest because it is imprinted, with expression only from the paternal allele, and it has apparently been subject to positive selection in recent human evolution (Voight *et al.* 2006).

Measuring positive selection on human genes, and linking genetic variants to cognitive and emotional phenotypes, both have severe limits. Some selected variants will be virtually fixed in humans, for example the *FOXP2* functional mutations causing major speech and language impairment are found almost exclusively in a single extended family in London. In other cases, signatures of selection may be erased by recombination of extended haplotypes bearing selected alleles. And there are over 30,000 genes in humans, a very large fraction of which are expressed in the brain. So bottom-up from genes to language will take awhile. What is top down?

13.3.3 Comparative method

An eagle ate the liver of Prometheus each day as he stood chained, paying the price for inspiring humanity with the skills and abilities of gods. Humans likewise pay a huge cost in suffering for their evolutionary legacy of complex social and technical cognition. The cost comes due when some combination of genetic, **epigenetic**, and environmental factors causes neurodevelopment to go wrong. 'Going wrong' is a vague and relative term, but precisely how development is perturbed can provide vital **cues** to understanding human cognition and emotion.

Marcus and Rabagliati (2006) discuss how we can use human developmental disorders to understand the nature and origins of language, especially its modularity. By their exposition, impairments of particular aspects of language should correlate with impairments in particular ancestral cognitive structures. For example, studies of autistic children show that humans can learn the meanings of words (or how to converse) either naturally, via social-cognitive mechanisms, or via a general capacity for logic—brute intellectual force—when social cognition is underdeveloped (Grandin 1995; Marcus and Rabagliati 2006).

We can generalize their approach, and consider neurodevelopmental disorders as relatively generalized 'mutations' of the mind—though not so much mutations as naturally assorting, cognitive-emotional variation that grades smoothly into normality. Indeed, both autism and schizophrenia are usually discussed as discrete conditions, but all of their core phenotypes represent just tails on smooth continua of personality and behaviour (Claridge 1997; Happé *et al.* 2006). Such conditions are each also convergent, in that a very wide range of developmental perturbations can result in relatively small, circumscribed sets of psychological traits—their formal psychiatric diagnostic criteria.

We will consider such convergent neurodevelopmental disorders, especially autistic spectrum conditions and what I call psychotic–affective spectrum conditions, as taxa for comparative evolutionary study. We will thus compare them, to uncover just how and why they show particular patterns of similarities and differences. As the comparative method in evolutionary ecology allows us to infer selection, the comparative method in psychiatry should, in theory, reveal aspects of human cognitive architecture, and especially language, built by evolution. These are the outcomes of selection for performance in particular mental domains, and the results of maximizing inclusive fitness, for genes and humans that cooperate and compete.

13.4 The autistic spectrum

Autism is a spectrum of conditions (Table 13.2), all of which involve some combination of impairment in social interaction, language, and communication, as well as repetitive, stereotyped behaviours (Fig. 13.2) (Happé *et al.* 2006). As regards language the variation is extensive, ranging from mutism in roughly 40% of cases of infantile 'Kanner' autism, to well-developed literal verbal skills in Asperger syndrome, though pragmatic, social–emotional verbal skills remain underdeveloped (Seung 2007). Autism is also highly heritable, with a risk to unborn siblings of autistics 25–100 times higher than in the general population, but its heritability is largely a function of component phenotypes, which are only loosely associated (Happé *et al.* 2006).

Autistic spectrum conditions are normally considered in terms of disability, in part because some degree of mental retardation is so common. The relative weaknesses found in autism can indeed be problematic for social functioning, as they centre around 'mentalistic' skills of language, imagination, and emotion used in social interactions—most importantly, skills used in inferring the motivations, intentions, and thoughts of other humans (Box 13.2). Deficits in mentalism are grounded in egocentrism—hence the very term 'autistic', for self-oriented. Such egocentrism applies most closely to Asperger syndrome, a condition characterized by extremely self-centred behaviour and specific reductions in social cooperation and reciprocity (Frith 2004).

In addition to these relative weaknesses, autism also exhibits a pattern of relative cognitive strengths (Box 13.2). These strengths centre around perceptual, spatial, and mechanistic skills, and indeed Baron-Cohen *et al.* (2001) have found that 'autistic conditions are associated with scientific skills' in non-clinical populations, and Wheelwright and Baron-Cohen (2001) reported a familial association of autism with engineering. This is the world of non-human 'things': tools, systems, and non-human animals, where activities and actions are much

Table 13.2 The autistic spectrum encompasses a suite of conditions. These conditions include autism (Kanner autism), and syndromes or conditions that overlap strongly with autism in terms of their phenotypic expressions, for multiple traits, in at least a substantial proportion of subjects

Condition	Selected recent citation
Kanner (infantile) autism	Happé *et al.* 2006
Asperger syndrome	Frith 2004
Rett syndrome	LaSalle *et al.* 2005
Fragile X syndrome	Belmonte *and* Bourgeron 2006
Angelman syndrome	Cohen *et al.* 2005
Tourette's syndrome	Canitano *and* Vivanti 2007
Turner syndrome	Skuse 2005
Smith-Lemli-Opitz syndrome	Cohen *et al.* 2005
Specific language impairment	Conti-Ramsden *et al.* 2006
Hyperlexia	Newman *et al.* 2007

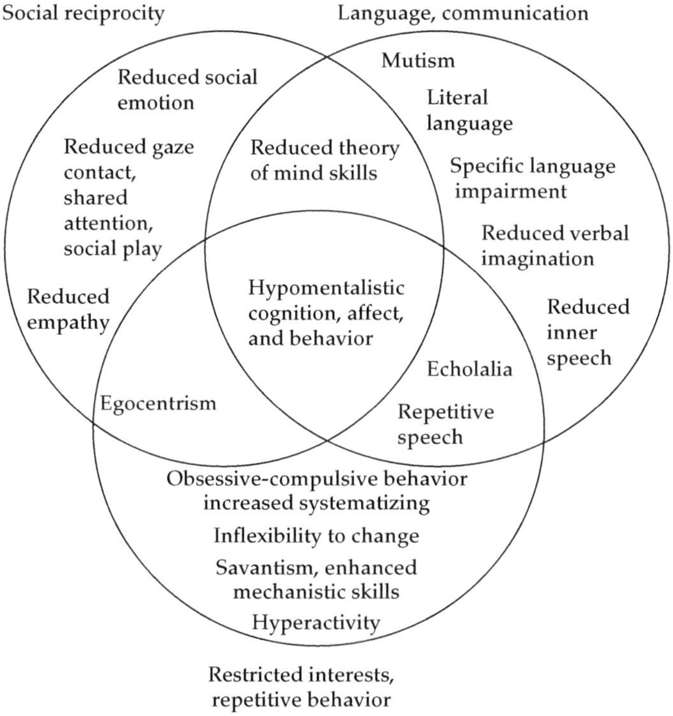

Figure 13.2 The autistic spectrum can be visualized in terms of three suites of traits that partially overlap in their phenotypic expression and genetic underpinnings. These three suites of traits make up the DSM-IV criteria for diagnosis of autism. At the core of these criteria we find a reduction in mentalistic cognition, which is mediated in part by effects on the development of language.

Box 13.2 The autistic spectrum

People with autistic spectrum conditions, especially autism and Asperger syndrome, exhibit relative strengths and weaknesses in aspects of cognition, emotion, and language. See Baron-Cohen (2003, 2006), Mottron *et al.* (2006) and Crespi and Badcock (2008) for details.

Inability to deceive.

Special abilities and savant skills in about 10% of subjects, including calendar calculation, list memory, music memory, 3D drawing, arithmetic computation, perfect pitch, hyperlexia (precocious, untaught high-speed reading).

Relative strengths

Mechanistic cognition involving understanding or engagement of 'systems', 'folk physics' and how things work.

Encoding and decoding skills for language and other tasks.

Some visual–spatial skills involving multidimensional spatial problem-solving, such as block design and embedded figures tests.

Perception of local vs. global features of environment, and 'bottom-up' processing of information.

Relative weaknesses

Mentalistic, theory-of-mind skills, such as interpreting gaze, inferring intentions, sharing attention, and understanding false beliefs.

Pragmatics of language, such as non-literal meanings, metaphors, emotions, humour, irony.

Expression of social emotions such as shame, embarrassment, guilt, contempt.

Executive functioning, central coherence.

Pretend play, imagination, abstraction, inner speech.

more predictable and can often be controlled. This is also a world where language is relatively literal and mechanical, or non-existent. Chen *et al.* (2007) describe how this apparently disparate pattern of enhanced spatial skills, and reduced linguistic and mentalistic skills, may be jointly mediated by the egocentric cognition characteristic of autism and Asperger syndrome.

13.5 How and why is language affected in autism?

In his original description of autistic children, Kanner noted mutism, unresponsiveness to questions and lack of drive to communicate with language or gesture, pronoun reversal (especially discussion of the self in the third person), echolalia (repetition of heard speech), and a linguistic focus on one's own specific, often obsessive interests. These are all deficits in the sociality of language, and they also include the pragmatics of subtle social meaning (Rapin and Dunn 2003). By contrast, literal verbal processing and memory—the mechanical syntax and phonology of language, are relatively preserved, or even enhanced in such skills as hyperlexic reading (Newman *et al.* 2007).

For Asperger syndrome, Ellis and Gunter (1999) and Gunter *et al.* (2002) characterize this general pattern of strengths and deficits, for language and other traits, as indicating relative right-hemisphere impairment and reduced inter-hemispheric connectivity. This inference certainly fits with the pattern of relative social weaknesses in autism, and its underlying cause apparently involves accelerated early brain growth and reversed lateralization in many cases (Flagg *et al.* 2005; Herbert *et al.* 2005), although the actual mechanisms and connections remain unclear. The ultimate result is that autistics tend to use speech primarily as a mechanical tool for serving their self-interest, and they think less in words and inner speech but more in mental pictures (Grandin 1995; Whitehouse *et al.* 2006). Literal and pragmatic speech are thus partly dissociable, as are thinking in words compared to images. In autism, complex language can be acquired, but not through the usual route of Vygotsky's (1962) developmental pathways from external social interactions and relationships to private speech, inner speech, and thought in words. The essence of the autistic spectrum is that the child's assimilation of social interactions, the process that drives this process of enculturation, mental development, and language, is underdeveloped to a greater or lesser degree (Badcock and Crespi 2006).

13.6 The psychotic–affective spectrum

Psychosis is literally a disordering of the psyche, the Greek 'soul'. In schizophrenia, such disordering commonly involves delusions and auditory hallucinations, loss of coherence and logic in thought and discourse, and emotionality ('affect') externally reduced or inappropriate to social context (Tamminga and Holcomb, 2005). Auditory hallucinations, a primary symptom found in over 60% of persons diagnosed with schizophrenia, are also common in bipolar disorder, which involves cycling between manic and depressive states (Baethge *et al.*, 2005), and in major depression. Bipolar disorder and major depression commonly involve other psychotic symptoms such as delusions, as well as symptoms related to dysregulated emotionality (Boks *et al.* 2007a).

Schizophrenia, bipolar disorder, major depression, and related conditions (Table 13.3) thus exhibit broad phenotypic overlap (Fig. 13.3), and they also overlap in their **polygenic** underpinnings (Craddock and Forty, 2006; Blackwood *et al.* 2007). Like the autistic spectrum, psychotic–affective spectrum conditions involve a pattern of relative strengths and weaknesses with regard to cognition, emotion, and aspects of language (Box 13.3).

Most research to date has focused on schizophrenia. Specific symptoms of this condition, such as auditory hallucinations, delusions, and dysregulated affect, are also common in non-clinical settings (Claridge 1997; Bentall 2003), and some schizotypical traits such as belief in supernatural beings and other aspects of 'magical ideation' are taken for granted, and promulgated, in modern society.

Schizophrenia exhibits a lifetime prevalence of about 1% (Tamminga and Holcomb 2005), across virtually all cultures and racial groups, and it is considered unique to humans, in contrast to other major psychiatric conditions which appear to exhibit

Table 13.3 The psychotic–affective spectrum involves a suite of broadly overlapping conditions. The best-known conditions include schizophrenia, bipolar disorder, and major depression. The other conditions overlap strongly with these three in terms of their phenotypic expression, for a substantial proportion of subjects. For example, Klinefelter syndrome, velocardiofacial syndrome, and Prader–Willi syndrome involve notably elevated rates of psychosis, and dyslexia and schizophrenia share a suite of neuroanatomical and cognitive features. Autistic behaviour has been described for velocardiofacial syndrome and Prader–Willi syndrome, but it apparently reflects a personality pre-morbid for schizophrenia or aspects of negative schizotypy (Eliez 2007; Crespi and Badcock 2008) and it is not underlain by autistic-spectrum neurological or physiological traits, or overlap in genetic underpinnings. This spectrum also includes panic disorder, delusional disorder, and anxiety disorders

Condition	Selected recent citation
Schizophrenia	Tamminga and Holcomb 2005
Bipolar disorder	Craddock and Forty 2006
Major depression	Craddock and Forty 2006
Schizotypal personality disorder	Claridge 1997
Klinefelter syndrome	Boks *et al.* 2007b
Velocardiofacial syndrome	Feinstein *et al.* 2002
Prader-Willi syndrome	Soni *et al.* 2007
Metachromatic leukodystrophy	Black *et al.* 2003
Dyslexia	Condray 2005

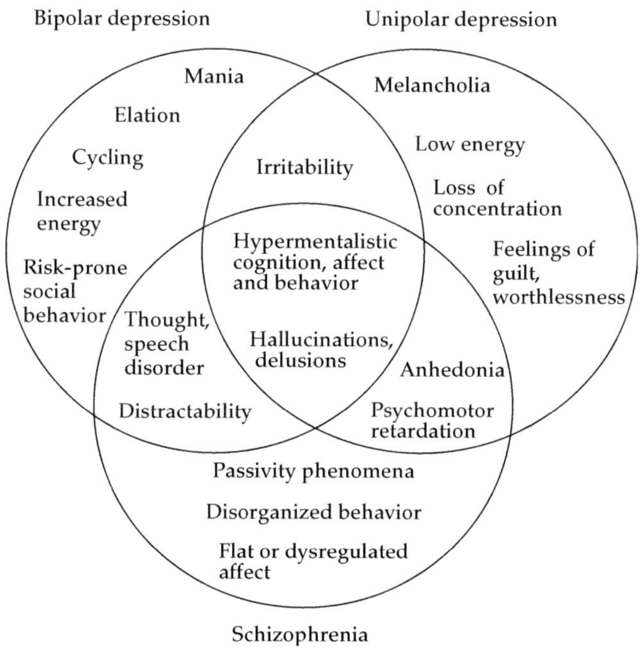

Figure 13.3 The psychotic-affective spectrum can be visualized in terms of three main conditions, schizophrenia, bipolar disorder, and major depression, that exhibit partial overlap in their phenotypic expression and genetic underpinnings. These three conditions have historically been considered as more or less separate, but recent genetic studies, and consideration of intermediate conditions, have demonstrated that they partially share a broad range of features and risk factors. At the core of the three conditions we find hyper-development in aspects of mentalistic cognition and emotion, which is mediated in part by variation in the development of language.

approximate non-human homologues (Crow 1997; Horrobin 1998). This uniqueness derives from the observation that the 'nuclear', or 'first-rank', symptoms of schizophrenia involve language and its relations with thought, and they also involve the most recently evolved and expanded regions and features of the human brain—including strong lateralization of cognitive, emotional, and linguistic

Box 13.3 Psychotic–affective spectrum

Individuals with psychotic-affective spectrum conditions exhibit relative strengths and weaknesses in aspects of cognition, emotion, and language. The strengths are found primarily in individuals exhibiting mild, non-clinical manifestations of these conditions—in the conditions themselves, the 'strengths' are hyperdeveloped and dysfunctional, as shown in parentheses. The evidence regarding strengths is also relatively sparse, because most research on schizophrenia and schizotypy focuses on characterizing deficits in clinical populations with a high incidence of pathology. See Crespi and Badcock (2008) for details, and Kravariti *et al.* (2006) in particular for data on verbal and visual-spatial abilities in schizophrenia and schizotypy.

Relative strengths

Note: pathological over-development shown in parentheses:

Mentalistic cognition involving application of 'folk psychology'.

Perception of global vs. local features of environment, 'top-down' processing.

Sensitivity to gaze, inferring intentions, shared attention, personal agency, deception (over-interpretation of intention, paranoia, erotomania, delusions of conspiracy, megalomania, self-deception).

Pragmatics of language, such as non-literal meanings, metaphors, emotions, humour, irony (misinterpreted language in psychosis, dysregulated or 'flat' affect).

Understanding and expression of social emotions such as shame, embarrassment, guilt, contempt (emotions typically expressed by voices in auditory hallucination, and in depression).

Pretend play, imagination, verbal creativity, inner speech (magical ideation, auditory hallucination, thought insertion, thought disorder, disorganized speech).

Relative weaknesses

Mechanistic cognition; rapid decoding and encoding skills, such as reading.

Some visual–spatial skills involving 2D and 3D spatial problem-solving, which have been characterized as 'trait markers' for schizophrenia.

functions to the left and right hemispheres. One of the most consistent findings in the schizophrenia literature is that structural and functional brain asymmetry is reduced compared to controls, for a variety of cognitive and emotional traits, but most notably for language (e.g. Sommer *et al.* 2001; Spaniel *et al.* 2007).

13.7 How and why is language affected in schizophrenia?

Speech in schizophrenia can be characterized as language expanded pathologically in all possible directions, with the discourse of any schizophrenic individual inhabiting some region of a chaotic linguistic landscape. Symptoms include poverty of speech, continual fast speech, distraction and derailment, incoherence, loss of logic, invention of new words, use of real words in new ways (e.g. 'handshoe' for glove), and choice of words by sound rather than meaning ('clanging') (McKenna and Oh 2005). That said, linguists inform us that schizophrenic speech is only quantitatively, but not qualitatively, different from speech in normal populations (Covington *et al.* 2005). Thought, the self–other distinction, and emotion exhibit comparable bedlam in schizophrenia: for example, thoughts may be removed from one's head, inserted from outside, or broadcast to others, feelings, actions, or thoughts may be controlled by others, and auditory hallucination, the most-common core phenotype of schizophrenia, involves hearing one's thoughts spoken aloud, voices discussing ones-self in the third person, running commentary on one's behaviour, or commands to engage in specific acts (Crow 1997).

What can such unfathomable phenotypes tell us about the evolution of language? Crow (1997, 2004) interprets all of these symptoms in terms of consequences of failure to establish left-hemisphere dominance for speech, such that the four-chambered brain dysfunctions in direction and strength of mental flow. In turn, reduced hemispheric dominance derives from delayed development, especially of the later-maturing left hemisphere, during gestation and childhood. Impaired or reduced left-hemisphere language function in schizophrenia and schizotypy may then result in greater reliance on right-hemisphere processing for some components of thought and language (Fisher *et al.*, 2004; Mohr *et al.*, 2005). A key consequence of such a shift may be more 'coarse' semantic processing, generation of more distant associations between events and thoughts, overestimation of meaningfulness of coincidences, increased magical ideation, and at the extreme, hallucination, delusion, paranoia, and other symptoms of schizophrenia (Claridge 1997; Leonhard and Brugger, 1998; Pizzagalli *et al.*, 2000; Brugger, 2001; Mohr *et al.*, 2005). The hypothesis also provides a simple explanation for the links between creativity and schizotypy as a cognitive style that involves more distant and more novel associations between aspects of thought and language (Gianotti *et al.*, 2001; Brugger, 2001; Barrantes-Vidal, 2004).

The links of imagination and creativity, especially verbal creativity, with the psychotic–affective spectrum (Claridge *et al.*, 1990; Nettle 2001) strongly contrast with the lower levels of pretend play and symbolic creativity in autistics (Blanc *et al.*, 2005), their reduced use of inner speech (Whitehouse *et al.* 2006), and their use of literal rather than figurative or metaphorical language. Indeed, to the extent that thought in words involves play and imagination as social-scenario building (Alexander 1989; Knight 2000), it may be underdeveloped in autism and hyperdeveloped, as well as selectively dysfunctional, in schizophrenia. Whereas in autism the left hemisphere may thus contribute disproportionately to language functions, in schizotypy and schizophrenia we apparently see the reverse—reduced skill with syntax and phonology (DeLisi 2001) but increased contribution of right-hemisphere social–linguistic non-literal meanings and emotion to discourse and thought, even though meanings are misinterpreted through some combination of delusion, rationalization, and confabulation (Arbib and Mundhenk 2005).

We can now revisit Table 13.1 and imagine a continuum between autistic spectrum and psychotic–affective spectrum conditions, mediated in part by lateralized brain structure and function being altered during neurodevelopment in the two possible directions. This is a considerable oversimplification but at least a potentially useful framework, amenable to falsification. Perhaps the most compelling evidence to date is the cognitive similarities between dyslexia (reading impairment) and schizophrenia (Condray 2005), and the virtual restriction of hyperlexia (fast, precocious, untaught reading) to autistics (Newman *et al.* 2007). Convergent evidence also comes from the two main forms of sex-chromosome **aneuploidy** in humans. Turner syndrome (XO females) involves well-developed literal verbal skills (including hyperlexic reading), poor visual-spatial skills, and a high incidence of autism—all suggesting relative right-hemisphere weaknesses (Temple and Carney 1996; Crow 1997; Skuse 2005). By contrast, Klinefelter syndrome subjects (XXY males) exhibit poor verbal skills, relatively preserved visual-spatial skills, and a notably high risk of both dyslexia and schizophrenia (Crow 1997; Geschwind *et al.* 2000; Condray 2005; Boks *et al.* 2007b)—consistent with relative left-hemisphere dysfunction, as in schizophrenia itself (Crow 1997, 2004). These findings also fit with the emergence of a cognitive trade-off between verbal skills and spatial skills, when a usually overriding factor for highly variable general intelligence is factored out (Ando *et al.* 2001; Johnson and Bouchard 2007).

Finally, a central phenotype of schizophrenia, auditory verbal hallucination, can also be understood in terms of dysfunctional mentalizing (Box 13.3), which takes us beyond the simple neurological level of impaired self-monitoring of speech. Vygotsky (1962) described a comprehensive theory for the development of human language and thought, whereby both develop from birth through adolescence via external social interactions and relationships becoming internalized in the brain. Language in particular develops from 'private speech' in pre-schoolers—talking out loud to and with one's-self in social dialogues, commentaries, and commands, usually with Teletubbies or equivalent as avatars of social–emotional verbal play. As the child develops, private speech becomes truly private—in the brain alone as inner speech, but with the same forms of social dialogue, commentary, and command. Jones and Fernyhough (2007) point out that auditory hallucinations in schizophrenia exhibit precisely the same manifestations as private speech in young children, thus providing the first coherent explanation for their social forms and contents, and for the subvocalizations that accompany them in schizophrenia. Moreover, Bentall (2003, p. 354) describes how such hallucinations often involve the voices of 'significant family members', and Birchwood *et al.* (2004) describe them as operating 'like external social relationships'. By implication, we all have voices of sorts in our heads, but after early childhood we do not hear them as such—they only emerge as our 'own' thoughts after our minds have developed to full self-consciousness, and after some sort of neural consensus has been reached (Haig 2006b).

This integration of child development with psychopathology dovetails with the highly speculative psycho-historical hypothesis of Jaynes (1976), that the right hemisphere of humans routinely perceived auditory hallucinations, interpreted as voices from gods, during prehistory from about 10,000 to about 1000 years ago, when true self-consciousness evolved. Does child development recapitulate this process? Functional imaging studies of children that test for spontaneous auditory cortex activity during silence (Hunter *et al.* 2006), may provide clues. Jaynes suggested that symptoms of schizophrenia represent vestiges of the bicameral (two-chambered) hallucinating mind. By contrast, the inner speech that fuels hallucination in schizophrenia is reduced in autism (Whitehouse *et al.* 2006), as is the sense of self-consciousness and personal agency (Toichi *et al.* 2002).

Our consideration of autistic and psychotic–affective spectrum conditions suggests that the human mind and language exhibit a psychological architecture that stretches along a continuum from mentalistic and hypersocial to mechanistic and purely egoistical cognition (Badcock 2004)—from a world of people to a world of things, with so-called normality at the centre exhibiting a balance between the two. The autistic and psychotic–affective spectra may thus be considered as complex and diverse but also diametrical conditions, with diametric profiles of cognitive strengths and weaknesses, as described in Boxes 13.2 and 13.3. The aetiologies of these conditions are thus presumably mediated by some partially shared set of genetic, epigenetic, and environmental factors that influence the development of brain and sociality, and can be perturbed in two main directions (Crespi and Badcock 2008; Crespi 2008). We have seen such diametric causes before, underlying the primary dimensions of human genetics and behaviour. They are two: paternal vs. maternal and male vs. female.

13.8 Genomic conflicts

Development is a trajectory maintained in dynamic balance by forces of homeostasis and canalization. In many situations, the trajectories bifurcate, leading to two more or less distinct forms, such as male and female mammals that diverge as embryos under the cascading effects of genes on the Y and X chromosomes (Davies and Wilkinson 2006). In other situations, divergent genomic interests create a dynamic balance where a single course is followed unless development is perturbed. One example is placental development, where a maternal–fetal tug-of-war, mediated in part by imprinted genes, creates conditions for diametric pathologies such as fetal growth restriction vs. overgrowth, as seen in Silver–Russell vs. Beckwith–Wiedemann syndromes (Monk and Moore 2004).

In sexual differentiation, and in genomic imprinting, we see forms of **divergent selection** at work, generating sets of phenotypes related to sex, growth, development—and language. For example, females exhibit a well-documented superiority to males in verbal skills, apparently in part due to their lower degree of lateralization and relatively large corpus callosum; by contrast, males show relative strengths in some visual-spatial skills (Geary 1998). Females are also, on average, more empathic, and males more systematic (Baron-Cohen 2003, 2006), and these differences are only controversial if ignorance tars them with the brush of determinism or ethics—we could always change the environment and find quite different patterns, or no sex differences, even if the genetics remained the same.

It is my main thesis here that the genomic axes of imprinting and sex mediate in some measure the development of autism and psychotic–affective spectrum conditions, and that liability to these conditions evolved with the origin of modern humans, and hypersociality driven by language. We will evaluate these hypotheses with several lines of evidence.

13.8.1 Imprinted-gene conflicts

A role for dysregulated genomic imprinting in autistic and psychotic–affective spectrum conditions is supported by several lines of evidence, including: (1) strong parent of origin effects in the genomic bases of both sets of conditions, (2) high rates of autism in cytogenetic disorders involving imbalance towards paternally expressed imprinted genes, such as Angelman syndrome and Rett syndrome, while the opposite imbalance involves high rates of psychosis, as in Prader–Willi syndrome, and (3) data from genome scan and genetic-association studies that implicate imprinted genes in the development of autism, Rett syndrome, schizophrenia, and bipolar disorder (Badcock and Crespi 2006; Crespi and Badcock 2008; Crespi 2008). Similarly, the high rates of psychotic spectrum conditions in Klinefelter syndrome, and autism in Turner syndrome (where the X is maternally inherited), can be explained under Haig's (2006a) hypothesis that X-chromosome genes are selected for benefits to matrilineal interests, as are maternally expressed imprinted genes on autosomes.

At the phenotypic level, a bias towards paternal-gene expression should result in more 'selfish' phenotypes (especially in interactions with mother), as seen most clearly in Asperger syndrome and 'high-functioning' autism where pathological effects of disrupted development are relatively small. Benefits to mothers and matrilines from psychotic-affective traits are less obvious, but can potentially be understood in terms of small deviations towards enhanced mentalistic skills in 'healthy' positive schizotypy, which can involve higher verbal fluency, increased 'openness' to the environment, and better-developed empathy, **altruism**, and spirituality (Crespi and Badcock 2008)—precisely the traits of a child who will never see the inside of a psychiatry clinic. But the ultimate currency and evidence is reproductive, and there is evidence from six studies for increased fecundity in first-order relatives of schizophrenics, especially on the maternal line (reviewed in Crespi and Badcock 2008). Mothers with more-autistic offspring should tend to have fewer children, due to their increased costs. This prediction is obvious for cases of Kanner autism due to its high level of impairment at an early age, but cases involving high-functioning autism or Asperger syndrome should provide useful tests.

13.8.2 Sexual differentiation and conflict

What of sex? Our second line of evidence derives from Baron-Cohen *et al.* (2005), who have championed the 'extreme male brain' theory of autism, positing that this condition is due predominantly to an overdose of testosterone in the womb. Baron-Cohen has marshalled an impressive body of evidence showing parallels between males and autistics for cognitive traits, such as high systematizing and low empathizing, and some aspects of neuroanatomy such as reduced large-scale connectivity. However, there is an absence of evidence for higher fetal testosterone in autism, and his evidence is also consistent with a paternal bias for imprinted genes, given the similarities (though not identity) of the paternal vs. maternal cognitive axis with that of males vs. females (Badcock and Crespi 2006), such as the higher cost of rearing males (Gibson and Mace 2003). Baron-Cohen (2003, p. 173) also discussed the 'extreme female brain' as exhibiting high empathy and low systematizing ability, but he dismissed its role in psychopathology on the presumption that hyperdeveloped theory of mind skills would be accurate and adaptive rather than pathologically overdeveloped.

Now let us look at sex and imprinting effects together, as they must occur in nature. The so-called 'male brain' appears relatively similar, neuroanatomically and cognitively, to a brain biased towards increased influence of paternally expressed imprinted genes (Crespi and Badcock 2008). Conversely, a 'more female' brain is similar to a brain developing under a relatively strong influence of maternally expressed imprinted genes. Sex ratios in autism are highly male-biased at the 'mild' end of these conditions, but in severe autism the sex ratio is near equality. Similarly, schizophrenia is relatively mild, with a later onset as well, in females than in males. The most severe neurological and cognitive impairments are found, in both conditions, where the direction of genomic-imprinting dysregulation opposes the sex difference: in females with autism, and in males with schizophrenia (Crespi and Badcock 2008). This hypothesis may also help to explain such patterns as the hypogonadism found in males with Klinefelter and Prader–Willi syndromes, the relatively female-like neuroanatomy and hormonal profile of male schizophrenics (Mendrek 2007), a role for imprinting effects in sexual preference (Green and Keverne, 2000; Mustanski *et al.*, 2005), and Freud's (1911) contention that paranoid schizophrenia in males is underlain by repressed homosexuality. Sexual conflict, with alleles differentially favouring one sex over another (Chapman 2006), represents a third force, in addition to sexual differentiation and imprinted-gene conflict, potentially mediating these effects—which we will not understand until we have dissected the genes and mechanisms involved.

13.8.3 The origin of modern humans

We have two human sexes, and two main disordered spectra of the social–linguistic and technical brain—how did we get this way? Our final line of evidence seeks to connect psychosis, autism, and language with the origin of modern humans. The only real connection, aside from untestable speculation, is genetic: what genes made us human, how did they evolve, and how do they relate to disorders of sociality? By my reckoning, there are three main dimensions of recent human evolution: (1) language, and thinking in words and abstract concepts, (2) emotionality, which became enhanced and encephalized as a social tool for maximizing inclusive fitness by subtle persuasion and coercion, and (3) technical skills, forged by systematic causal thinking and fine-motor abilities (Wolpert 2003). Expansion of each of these dimensions presumably created novel scope for forms of psychiatric dysregulation, but only the first is uniquely human, emerging from Crow's four-chambered brain and potentiating schizophrenia.

Now, a considerable suite of genes are known or suspected to underlie schizophrenia: have they been subject to recent positive selection, with schizophrenia as a maladaptive by-product? The short answer is, apparently, yes—many genes that influence the risk of developing schizophrenia show signatures of recent positive selection in the human lineage, including *DTNBP1*, *FOXP2*, and *MCPH1* (Crespi 2006; Voight *et al.* 2006; Lencz *et al.* 2007) and data from the first-generation human HapMap shows an enriched **signal** of selection for schizophrenia genes (Crespi *et al.* 2007). The long and real answer must address the question of how schizophrenia coevolved with human cognition, emotionality, and language. This answer awaits studies that deeply integrate genomics with neuroscience and psychiatry, in the context of evolutionary theory.

13.9 Conclusions

Myth does not mean something untrue, but a concentration of truths.

Doris Lessing, *African Laughter*

Hercules rescued Prometheus from his bonds, during the course of his 12 labours. Jesus likewise rose from the dead and ascended to heaven, suggesting that altruism has its own special and personal rewards. We poor humans will only be free from soul-wrenching autistic and psychotic–affective disorders once we have dissected their evolutionary-genetic and epigenetic bases, and developed pre-natal tests and preventatives. During this labour, we should also uncover genes underlying the evolution of language, intelligence, emotion, and technical skills, and elucidate how their variants are subject to tradeoffs, pleiotropic effects, and dysregulation.

I have argued here that an important cause of disordered language, cognition, and emotion is conflict, expressed at multiple levels from different human groups, to families, to mother and child, and to genes that harbour divergent interests within individuals. At each of these levels, the nexus of conflict within and between groups of people or genes is divergent avenues of maximizing inclusive fitness, which lead to exceptionally strong selection, tugs of war, and imbalances of power (Alexander 1989). Balancing this conflict are the confluences of interest that emerge from genic cooperation, mother's love for child, and love of God—who, like our circle of kin, created us in body and psyche and promises immortality, and who we serve to give life its meaning. In the beginning was the Word, and the Word was God—as are we, modern humans.

Summary

Human social communication is impaired in the two primary disorders of the 'social brain', autism and schizophrenia. I describe a new hypothesis for the role of language in the evolution and development of autism and schizophrenia: that the cores of these two conditions are disordered social communication, with dysregulated social–linguistic development mediated in part by extremes of bias in maternal vs. paternal imprinted gene expression, resulting in extreme mentalistic vs. extreme mechanistic cognition.

I evaluate this hypothesis via (1) tests of the molecular evolution of genes underlying schizophrenia and language, (2) analyses of how the human social brain has evolved, and (3) evaluation of the roles of genomic conflicts in human development, enculturation, and communication, via study of the genetic and epigenetic underpinnings of social-brain disorders. These convergent lines of evidence from evolutionary theory and neurogenomics support the hypothesis that psychosis represents the 'illness that made us human'.

Acknowledgements

I am grateful to David Hughes and Patrizia d'Ettorre for inviting me to participate in this volume, to Christopher Badcock, and David Haig for comments and discussions, and to the Canada Council for the Arts and NSERC for financial support. I also acknowledge an unpublished work by Richard D. Alexander on 'The concept of God and the meaning of life', for insights into the parallels between inclusive fitness and religion.

References

Alexander, R.D. (1980). *Darwinism and Human Affairs.* University of Washington Press, Seattle and London.

Alexander, R.D. (1987). *The Biology of Moral Systems.* Aldine de Gruyter, New York.

Alexander, R. (1989). Evolution of the human psyche. In: P. Mellars and P. Stringer (eds), *The Human Revolution: Behavioral and Biological Perspectives on the Origins of Modern Humans*, pp. 455–513. Edinburgh University Press, Edinburgh.

Ando, J., Ono, Y., and Wright, M.J. (2001). Genetic structure of spatial and verbal working memory. *Behavioral Genetics*, **31**, 615–624.

Arbib, M.A. (2005). From monkey-like action recognition to human language: an evolutionary framework for neurolinguistics. *Behavioral and Brain Sciences*, **28**, 105–124.

Arbib, M.A and Mundhenk, T.N. (2005). Schizophrenia and the mirror system: an essay. *Neuropsychologia*, **43**, 268–280.

Badcock, C. (1989) Sociobiology, psychoanalysis and a possible evolutionary dimension to an oral phase. *Ethology and Sociobiology*, **10**, 408–409.

Badcock, C. (2004). Mentalism and mechanism: the twin modes of human cognition. In: C. Crawford and C. Salmon (eds), *Human Nature and Social Values: Implications of Evolutionary Psychology for Public Policy*, pp. 99–116. Lawrence Erlbaum Associates, Mahwah, NJ.

Badcock, C. and Crespi, B. (2006). Imbalanced genomic imprinting in brain development: an evolutionary basis for the aetiology of autism. *Journal of Evolutionary Biology*, **19**, 1007–1032.

Baethge, C., Baldessarini, R.J., Freudenthal, K., Streeruwitz, A., Bauer, M., and Bschor, T. (2005). Hallucinations in bipolar disorder: characteristics and comparison to unipolar depression and schizophrenia. *Bipolar Disorder*, **7**, 136–145.

Baron-Cohen, S. (2003). *The Essential Difference: the Truth about the Male and Female Brain.* Basic Books, New York.

Baron-Cohen, S. (2006). Two new theories of autism: hyper-systemising and assortative mating. *Archives of Disease in Childhood*, **91**, 2–5.

Baron-Cohen, S., Knickmeyer, R.C., and Belmonte, M.K. (2005). Sex differences in the brain: implications for explaining autism. *Science*, **310**, 819–823.

Baron-Cohen, S., Wheelwright, S., Skinner, R., Martin, J., and Clubley, E. (2001). The autism-spectrum quotient (AQ): evidence from Asperger syndrome/high-functioning autism, males and females, scientists and mathematicians. *Journal of Autism and Developmental Disorders*, **31**, 5–17.

Barrantes-Vidal, N. (2004). Creativity and madness revisited from current psychological perspectives. *Journal of Consciousness Studies*, **11**, 58–78.

Belmonte, M.K. and Bourgeron, T. (2006). Fragile X syndrome and autism at the intersection of genetic and neural networks. *Nature Neuroscience*, **9**, 1221–1225.

Bentall, R.P. (2003) *Madness Explained: Psychosis and Human Nature*. Allen Lane, London.

Bigler, E.D., Mortensen, S., Neeley, E.S. *et al.* (2007). Superior temporal gyrus, language function, and autism. *Developmental Neuropsychology*, **31**, 217–238.

Birchwood, M., Gilbert, P., Gilbert, J. *et al.* (2004). Interpersonal and role-related schema influence the relationship with the dominant 'voice' in schizophrenia: a comparison of three models. *Psychological Medicine*, **34**, 1571–1580.

Black, D.N, Taber, K.H. and Hurley, R.A. (2003). Metachromatic leukodystrophy: a model for the study of psychosis. *Journal of Neuropsychiatry and Clinical Neuroscience*, **15**, 289–293.

Blackwood, D.H., Pickard, B.J., Thomson, P.A., Evans, K.L., Porteous, D.J., and Muir, W.J. (2007). Are some genetic risk factors common to schizophrenia, bipolar disorder and depression? Evidence from DISC1, GRIK4 and NRG1. *Neurotoxicity Research*, **11**, 73–83.

Blanc, R., Adrien, J.L., Roux, S., and Barthelemy, C. (2005). Dysregulation of pretend play and communication development in children with autism. *Autism*, **9**, 229–245.

Bloom, P. (2004). *Descartes' Baby: How the Science of Child Development Explains What Makes Us Human*. Basic Books, New York.

Boks, M.P., Leask, S., Vermunt, J.K., and Kahn, R.S. (2007a). The structure of psychosis revisited: the role of mood symptoms. *Schizophrenia Research*, **93**, 178–185.

Boks, M.P., de Vette, M.H., Sommer, I.E. *et al.* (2007b). Psychiatric morbidity and X-chromosomal origin in a Klinefelter sample. *Schizophrenia Research*, **93**, 399–402.

Bowles, S. (2006). Group competition, reproductive leveling, and the evolution of human altruism. *Science*, **314**, 1569–1572.

Brown, W.D., Crespi, B.J. and Choe, J.C. (1997). Sexual conflict and the evolution of mating systems. In: J. Choe and B.J. Crespi (eds), *Evolution of Mating Systems in Insects and Arachnids*, pp. 352–77. Cambridge University Press, Cambridge.

Brugger, P. (2001). From haunted brain to haunted science: a cognitive neuroscience view of paranormal and pseudoscientific thought. In: J. Houran and R. Lange (eds), *Hauntings and Poltergeists: Multidisciplinary Perspectives*, pp. 195–213. McFarland, Jefferson, NC.

Brune, C.W., Kim, S.J., Salt, J., Leventhal, B.L., Lord, C., and Cook, E.H., Jr (2006). 5-HTTLPR genotype-specific phenotype in children and adolescents with autism. *American Journal of Psychiatry*, **163**, 2148–2156.

Canitano, R. and Vivanti, G. (2007). Tics and Tourette syndrome in autism spectrum disorders. *Autism*, **11**, 19–28.

Cattanach, B.M., Beechey, C.V., and Peters, J. (2006). Interactions between imprinting effects: summary and review. *Cytogenetic and Genome Research*, **113**, 17–23.

Chapman, T. (2006). Evolutionary conflicts of interest between males and females. *Current Biology*, **16**, 744–754.

Chen, F., Planche, P., Lemonnier, E., and Lazartigues, A. (2007). How could language interact with visuospatial performance in autism? *Medical Hypotheses*, **69**, 695–697.

Claridge, G. (ed.) (1997) *Schizotypy: Implications for Illness and Health*. Oxford University Press, Oxford.

Claridge, G., Pryor, R., and Watkins, G. (1990). *Sounds from the Bell Jar: Ten Psychotic Authors*. Macmillan, London.

Cohen, D., Pichard, N., Tordjman, S. *et al.* (2005). Specific genetic disorders and autism: clinical contribution towards their identification. *Journal of Autism and Developmental Disorders*, **35**, 103–116.

Condray, R. (2005). Language disorder in schizophrenia as a developmental learning disorder. *Schizophrenia Research*, **73**, 5–20.

Conti-Ramsden, G., Simkin, Z., and Botting, N. (2006). The prevalence of autistic spectrum disorders in adolescents with a history of specific language impairment (SLI). *Journal of Child Psychology and Psychiatry*, **47**, 621–628.

Cook, N.D. (2002). Bihemispheric language: how the two hemispheres collaborate in the processing of language. In: T.J. Crow (ed.), *The Speciation of Modern Homo sapiens*, Proceedings of the British Academy, Volume 106, pp. 169–194. Oxford University Press, Oxford.

Covington, M.A., He, C., Brown, C. *et al.* (2005). Schizophrenia and the structure of language: the linguist's view. *Schizophrenia Research*, **77**, 85–98.

Craddock, N. and Forty, L. (2006). Genetics of affective (mood) disorders. *European Journal of Human Genetics*, **14**, 660–668.

Crespi, B. (2006). The natural selection of psychosis. *Behavioral and Brain Sciences*, **29**, 410–411.

Crespi, B.J. (2007). Sly FOXP2: genomic conflict in the evolution of language. *Trends in Ecology and Evolution*, **22**, 174–175.

Crespi, B. (2008). Genomic imprinting in the evolution and development of psychosis. *Biological Reviews* (under revision).

Crespi, B. and Badcock, C. (2008). Psychosis and autism as diametrical disorders of the social brain. *Behavioral and Brain Sciences* (in press).

Crespi, B., Summers, K., and Dorus, S. (2007). Adaptive evolution of genes underlying schizophrenia. *Proceedings of the Royal Society Series B: Biological Sciences*, **274**, 2801–2810.

Crow, T.J. (1997). Is schizophrenia the price that *Homo sapiens* pays for language? *Schizophrenia Research*, **28**, 127–141.

Crow, T.J. (2004). Auditory hallucinations as primary disorders of syntax: an evolutionary theory of the origins of language. *Cognitive Neuropsychiatry*, **9**, 125–145.

Davies, W. and Wilkinson, L.S. (2006). It is not all hormones: alternative explanations for sexual differentiation of the brain. *Brain Research*, **1126**, 36–45.

DeLisi, L.E. (2001). Speech disorder in schizophrenia: review of the literature and exploration of its relation to the uniquely human capacity for language. *Schizophrenia Bulletin*, **2**, 481–496.

Dougherty, R.F., Ben-Shachar, M., Deutsch, G.K., Hernandez, A., Fox, G.R., and Wandell, B.A. (2007). Temporal-callosal pathway diffusivity predicts phonological skills in children. *Proceedings of the National Academy of Sciences of the USA*, **104**, 8556–8661.

Eliez, S. (2007). Autism in children with 22q11.2 deletion syndrome. *Journal of American Academic Child and Adolescent Psychiatry*, **46**, 433–434.

Ellis, H.D. and Gunter, H.L. (1999). Asperger syndrome: a simple matter of white matter? *Trends in Cognitive Science*, **3**, 192–200.

Fallgatter, A.J., Herrmann, M.J., Hohoff, C.*et al.* (2006). DTNBP1 (dysbindin) gene variants modulate prefrontal brain function in healthy individuals. *Neuropsychopharmacology*, **31**, 2002–2010.

Falk, D. (2004). Prelinguistic evolution in early hominins: whence motherese? *Behavioral and Brain Sciences*, **27**, 491–503.

Fan, J.B. and Sklar, P. (2005). Meta-analysis reveals association between serotonin transporter gene STin2 VNTR polymorphism and schizophrenia. *Molecular Psychiatry*, **10**, 928–938.

Feinstein, C., Eliez, S., Blasey, C., and Reiss, A.L. (2002). Psychiatric disorders and behavioral problems in children with velocardiofacial syndrome: usefulness as phenotypic indicators of schizophrenia risk. *Biological Psychiatry*, **51**, 312–318.

Feuk, L., Kalervo, A., Lipsanen-Nyman, M. *et al.* (2006). Absence of a paternally inherited FOXP2 gene in developmental verbal dyspraxia. *American Journal of Human Genetics*, **79**, 965–972.

Fisher, J.E., Mohanty, A., Herrington, J.D., Koven, N.S., Miller, G.A., and Heller, W. (2004). Neuropsychological evidence for dimensional schizotypy: implications for creativity and psychopathology. *Journal of Research in Personality*, **38**, 24–31.

Flagg, E.J., Cardy, J.E., Roberts, W., and Roberts, T.P. (2005). Language lateralization development in children with autism: insights from the late field magnetoencephalogram. *Neuroscience Letters*, **386**, 82–87.

Francks, C., Maegawa, S., Lauren, J. *et al.* (2007). LRRTM1 on chromosome 2p12 is a maternally suppressed gene that is associated paternally with handedness and schizophrenia. *Molecular Psychiatry*, **12**, 1129–1139.

Freud, S. (1911) Psycho-analytic notes on an autobiographical account of a case of paranoia. In: J. Strachey, A. Freud, A. Strachey, and A. Tyson (eds), *The Standard Edition of the Complete Psychological Works of Sigmund Freud*, Vol. 12, pp. 1–82. The Hogarth Press and The Institute of Psychoanalysis, London.

Frith, U. (2004). Emanuel Miller lecture: confusions and controversies about Asperger syndrome. *Journal of Child Psychology and Psychiatry*, **45**, 672–686.

Galantucci, B., Fowler, C.A., and Turvey, M.T. (2006). The motor theory of speech perception reviewed. *Psychonomic Bulletin and Review*, **13**, 361–377.

Geary, D.C. (1998). *Male, Female: the Evolution of Human Sex Differences*. American Psychological Association, Washington, DC.

Geschwind, D.H., Boone, K.B., Miller, B.L., and Swerdloff, R.S. (2000). Neurobehavioral phenotype of Klinefelter syndrome. *Mental Retardation and Developmental Disabilities Research Reviews*, **6**, 107–116.

Gianotti, L.R., Mohr, C., Pizzagalli, D., Lehmann, D., and Brugger, P. (2001). Associative processing and paranormal belief. *Psychiatry and Clinical Neurosciences*, **55**, 595–603.

Gibson, M.A. and Mace, R. (2003). Strong mothers bear more sons in rural Ethiopia. *Proceedings. Biological Sciences*, **270**(Suppl. 1), S108–S109.

Gobbini, M.I. and Haxby, J.V. (2006). Neural response to the visual familiarity of faces. *Brain Research Bulletin*, **71**, 76–82.

Grandin, T. (1995). *Thinking in Pictures: and Other Reports from My Life with Autism*. Vintage Books, New York.

Green, R. and Keverne, E.B. (2000). The disparate maternal aunt-uncle ratio in male transsexuals: an explanation invoking genomic imprinting. *Journal of Theoretical Biology*, **202**, 55–63.

Gunter, H.L., Ghaziuddin, M., and Ellis, H.D. (2002). Asperger syndrome: tests of right hemisphere functioning and interhemispheric communication. *Journal of Autism and Developmental Disorders*, **32**, 263–281.

Haig, D. (2006a). Intragenomic politics. *Cytogenetic and Genome Research*, **113**, 68–74.

Haig, D. (2006b). Intrapersonal conflict. In: M. Jones and A. Fabian (eds), *Conflict*, pp. 8–22. Cambridge University Press, Cambridge.

Happé, F., Ronald, A., and Plomin, R. (2006). Time to give up on a single explanation for autism. *Nature Neuroscience*, **9**, 1218–1220.

Harris, L.T. and Fiske, S.T. (2006). Dehumanizing the lowest of the low: neuroimaging responses to extreme out-groups. *Psychological Science*, **17**, 847–853.

Hauser, M., Cushman, F., Young, L., Kang-Xing, J., and Mikhail, J. (2007). A dissociation between moral judgments and justifications. *Mind and Language*, **22**, 1–21.

Herbert, M.R., Ziegler, D.A., Deutsch, C.K. *et al.* (2005). Brain asymmetries in autism and developmental language disorder: a nested whole-brain analysis. *Brain*, **128**, 213–226.

Horrobin, D.F. (1998). Schizophrenia: the illness that made us human. *Medical Hypotheses*, **50**, 269–288.

Hunter, M.D., Eickhoff, S.B., Miller, T.W., Farrow, T.F., Wilkinson, I.D., and Woodruff, P.W. (2006). Neural activity in speech-sensitive auditory cortex during silence. *Proceedings of the National Academy of Sciences of the USA*, **103**, 189–194.

Iacoboni, M. and Dapretto, M. (2006). The mirror neuron system and the consequences of its dysfunction. *Nature Reviews Neuroscience*, **7**, 942–951.

Isles, A.R., Davies, W., and Wilkinson, L.S. (2006). Genomic imprinting and the social brain. *Philosophic Transactions of the Royal Society B: Biological Sciences*, **361**, 2229–2237.

Ivry, R.B. and Robertson, L.C. (1998). *The Two Sides of Perception*. MIT Press, Cambridge, MA.

Jaynes, J. (1976). *The Origin of Consciousness in the Breakdown of the Bicameral Mind*. Houghton Mifflin, Boston, MA.

Johnson, W. and Bouchard, T.J. (2007). Sex differences in mental abilities: 'g' masks the dimensions on which they lie. *Intelligence*, **35**, 23–39.

Jones, S.R. and Fernyhough, C. (2007). Thought as action: inner speech, self-monitoring, and auditory verbal hallucinations. *Consciousness and Cognition*, **16**, 391–399.

Joseph, R. (1992). *The Right Brain and the Unconscious*. Plenum Press, New York.

Koenigs, M., Young, L., Adolphs, R. *et al.* (2007). Damage to the prefrontal cortex increases utilitarian moral judgements. *Nature*, **446**, 908–911.

Knight, C. (2000). Play as precursor of phonology and syntax. In: C. Knight, M. Studdert-Kennedy, and J.R. Hurford (eds), *The Evolutionary Emergence of Language*, pp. 99–119. Cambridge University Press, Cambridge.

Kravariti, E., Toulopoulou, T., Mapua-Filbey, F. *et al* (2006). Intellectual asymmetry and genetic liability in first-degree relatives of probands with schizophrenia. *British Journal of Psychiatry*, **188**, 186–187.

Lahti, D.C. and Weinstein, B.S. (2005). The better angels of our nature: group stability and the evolution of moral tension. *Evolution and Human Behavior*, **26**, 47–63.

LaSalle J.M., Hogart, A., and Thatcher, K.N. (2005) Rett syndrome: a Rosetta stone for understanding the molecular pathogenesis of autism. *International Review of Neurobiology*, **71**, 131–165.

Lencz, T., Burdick, K.E., DeRosse, P. *et al.* (2007). MCPH1 is associated with risk for illness and cognitive ability in patients with schizophrenia. *Biological Psychiatry*, **61**, S190S.

Leonhard, D. and Brugger, P. (1998). Creative, paranormal, and delusional thought: a consequence of right hemisphere semantic activation? *Neuropsychiatry, Neuropsychology, and Behavioral Neurology*, **11**, 177–183.

Lieberman, P. (2007). The evolution of human speech—its anatomical and neural bases. *Current Anthropology*, **48**, 39–66.

Locke, J.L. (2006) Parental selection of vocal behavior: crying, cooing, babbling and the evolution of language. *Human Nature*, **17**, 155–168.

Locke, J.L. and Bogin, B. (2006). Language and life history: a new perspective on the development and evolution of human language. *Behavioral and Brain Sciences*, **29**, 259–280.

McKenna, P. and Oh, T. (2005). *Schizophrenic Speech: Making Sense of Bathroots and Ponds that Fall in Doorways*. Cambridge University Press, New York.

Marcus, G. and Rabagliati, H. (2006). What developmental disorders can tell us about the nature and origins of language. *Nature Neurosciences*, **9**, 1226–1229.

Mendrek, A. (2007). Reversal of normal cerebral sexual dimorphism in schizophrenia: evidence and speculations. *Medical Hypotheses*, **69**, 896–902.

Mitchell, R.L. and Crow, T.J. (2005). Right hemisphere language functions and schizophrenia: the forgotten hemisphere? *Brain*, **128**, 963–978.

Mohr, C., Krummenacher, P., Landis, T., Sandor, P.S., Fathi, M., and Brugger, P. (2005). Psychometric schizotypy modulates levodopa effects on lateralized lexical decision performance. *Journal of Psychiatric Research*, **39**, 241–250.

Monk, D. and Moore, G.E. (2004). Intrauterine growth restriction—genetic causes and consequences. *Seminars in Fetal and Neonatal Medicine*, **9**, 371–378.

Mottron, L., Dawson, M., Soulieres, I., Hubert, B., and Burack, J. (2006). Enhanced perceptual functioning in autism: an update, and eight principles of autistic perception. *Journal of Autism and Developmental Disorders*, **36**, 27–43.

Mustanski, B.S., DuPree, M.G., Nievergelt, C.M., Bocklandt, S., Schork, N.J., and Hamer, D.H. (2005). A genomewide scan of male sexual orientation. *Human Genetics*, **116**, 272–278.

Nettle, D. (2001). *Strong Imagination: Madness, Creativity and Human Nature*. Oxford University Press, Oxford.

Nettle, D. and Dunbar, R. (1997). Social markers and the evolution of reciprocal exchange. *Current Anthropology*, **38**, 93–99.

Newman, T.M., Macomber, D., Naples, A.J., Babitz, T., Volkmar, F., and Grigorenko, E.L. (2007). Hyperlexia in children with autism spectrum disorders. *Journal of Autism and Developmental Disorders*, **37**, 760–774.

Oudejans, C.B., Mulders, J., Lachmeijer, A.M. *et al.* (2004). The parent-of-origin effect of 10q22 in pre-eclamptic females coincides with two regions clustered for genes with down-regulated expression in androgenetic placentas. *Molecular Human Reproduction*, **10**, 589–598.

Pizzagalli, D. Lehmann, D. Gianotti, L. *et al.* (2000). Brain electric correlates of strong belief in paranormal phenomena: Intracerebral EEG source and regional omega complexity analyses. *Psychiatry Research*, **100**, 139–154.

Ramachandran, V.S. (1996). The evolutionary biology of self-deception, laughter, dreaming and depression: some clues from anosognosia. *Medical Hypotheses*, **47**, 347–362.

Rao, H., Gillihan, S.J., Wang, J. *et al.* (2007). Genetic variation in serotonin transporter alters resting brain function in healthy individuals. *Biological Psychiatry*, **62**, 600–606.

Rapin, I. and Dunn, M. (2003). Update on the language disorders of individuals on the autistic spectrum. *Brain and Development*, **25**, 166–172.

Rapoport, S.I. (1999). How did the human brain evolve? A proposal based on new evidence from in vivo brain imaging during attention and ideation. *Brain Research Bulletin*, **50**, 149–165.

Saxe, R. (2006). Uniquely human social cognition. *Current Opinion in Neurobiology*, **16**, 235–239.

Seung, H.K. (2007). Linguistic characteristics of individuals with high functioning autism and Asperger syndrome. *Clinical Linguistics and Phonetics*, **21**, 247–259.

Skuse, D.H. (2005). X-linked genes and mental functioning. *Human Molecular Genetics*, **14**(Review Issue 1), R27–R32.

Sommer, I.E., Ramsey, N.F., and Kahn, R.S. (2001). Language lateralization in schizophrenia, an fMRI study. *Schizophrenia Research*, **52**, 57–67.

Soni, S., Whittington, J., Holland, A.J. *et al.* (2007). The course and outcome of psychiatric illness in people with Prader–Willi syndrome: implications for management and treatment. *Journal of Intellectual Disability Research*, **51**, 32–42.

Spaniel, F., Tintera, J., Hajek, T. *et al.* (2007). Language lateralization in monozygotic twins discordant and concordant for schizophrenia. A functional MRI pilot study. *European Psychiatry*, **22**, 319–322.

Stein, D.J., Solms, M., and van Honk, J. (2006). The cognitive-affective neuroscience of the unconscious. *CNS Spectrums*, **11**, 580–583.

Tamminga, C.A. and Holcomb, H.H. (2005). Phenotype of schizophrenia: a review and formulation. *Molecular Psychiatry*, **10**, 27–39.

Temple, C.M. and Carney, R. (1996). Reading skills in children with Turner's syndrome: an analysis of hyperlexia. *Cortex*, **32**, 335–345.

Toichi, M., Kamio, Y., Okada, T. *et al.* (2002). A lack of self-consciousness in autism. *American Journal of Psychiatry*, **159**, 1422–1424.

Trivers, R. (2000). The elements of a scientific theory of self-deception. *Annals of the New York Academy of Sciences*, **907**, 114–131.

Vargha-Khadem, F., Gadian, D.G., Copp, A., and Mishkin, M. (2005). FOXP2 and the neuroanatomy of speech and language. *Nature Reviews Neuroscience*, **6**, 131–138.

Vawter, M.P., Harvey, P.D., and DeLisi, L.E. (2007). Dysregulation of X-linked gene expression in Klinefelter's syndrome and association with verbal cognition. *American Journal of Medical Genetics. Part B, Neuropsychiatric Genetics*, **144**, 728–734.

Vergne, D.E. and Nemeroff, C.B. (2006). The interaction of serotonin transporter gene polymorphisms and early adverse life events on vulnerability for major depression. *Current Psychiatry Reports*, **8**, 452–457.

Voight, B.F., Kudaravalli, S., Wen, X., and Pritchard, J.K. (2006). A map of recent positive selection in the human genome. *PLoS Biology*, **4**, e72.

Vygotsky, L.S. (1962). *Thought and Language*. MIT Press, Cambridge, MA.

Weiss, L.A., Purcell, S., Waggoner, S. *et al.* (2007). Identification of EFHC2 as a quantitative trait locus for fear recognition in Turner syndrome. *Human Molecular Genetics*, **16**, 107–113.

Wheelwright, S. and Baron-Cohen, S. (2001). The link between autism and skills such as engineering, maths, physics and computing: a reply to Jarrold and Routh. *Autism*, **5**, 223–227.

Whitehouse, A.J., Maybery, M.T., and Durkin, K. (2006). Inner speech impairments in autism. *Journal of Child Psychology and Psychiatry*, **47**, 857–865.

Whiten, A. and Byrne, R.W. (eds) (1997). *Machiavellian Intelligence II: Extensions and Evaluations*. Cambridge University Press, Cambridge.

Wolpert, L. (2003) Causal belief and the origins of technology. *Philosophical Transactions of the Royal Society A: Mathematical, Physical and Engineering Sciences*, **361**, 1709–1719.

Zhang, J., Webb, D.M., and Podlaha, O. (2002). Accelerated protein evolution and origins of human-specific features: Foxp2 as an example. *Genetics*, **162**, 1825–1835.

Zinkstok, J.R., de Wilde, O., van Amelsvoort, T.A., Tanck, M.W., Baas, F., and Linszen, D.H. (2007). Association between the DTNBP1 gene and intelligence: a case-control study in young patients with schizophrenia and related disorders and unaffected siblings. *Behavioral and Brain Functions*, **3**, 19.

CHAPTER 14

The evolution of human communication and language

James R. Hurford

14.1 Introduction

Human language stands out in a number of ways from the topics of almost all of the other chapters in this book. Although every communication system can claim in some way to be unique, human language is spectacularly unique in its complexity and expressive power.

Complexity is hard to measure, but a clue is given by the fact that *The Cambridge Grammar of the English Language* (Huddleston & Pullum, 2002), which is just a description of Modern Standard English, weighs in at over 1700 pages. The headings of the first half-dozen descriptive chapters, out of eighteen, are: The verb, The clause: complements, Nouns and noun phrases, Adjectives and adverbs, Prepositions and preposition phrases, and The clause: adjuncts. No non-human communication system demands anything like this degree of detail to describe it. And English is just one of over 6,000 human languages, all of comparable complexity.

As for expressive power, this is also hard to measure. We can't see far into the minds of nonhuman animals to know what exactly they can communicate with each other, but it seems a fair bet that any factual information, and any affective content[1] that can be conveyed by an animal communication system can also be conveyed, or at least satisfactorily paraphrased, in any human language. We can, we believe, concisely summarize the information given by honeybee waggle dances, by chaffinch territorial songs, and by vervet monkey alarm calls. We are not yet sure exactly what is conveyed by whale songs, but a reasonable default hypothesis would seem to be that they convey messages of the same expressive power as complex birdsongs. We may be wrong about this, but the current belief is thus that any human language is capable of communicating the sum total of all that any animal species can communicate, and more. More, because we alone, as far as we know, can tell each other about fictional or abstract objects, and about events far distant in time and space.

In the bulk of this chapter, I will list and discuss some of the most important differences and similarities between human languages and nonhuman communication systems, with an evolutionary perspective, in particular drawing on results from comparative psychology pertaining to our closest relatives, the non-human primates (see related discussion on language in Chapter 13).

14.2 Diversity

We must first make the vital distinction between Language, the biologically given universal human capacity, and languages, such as English, Swahili, Cantonese, Dyirbal and Navajo, which are culturally developed systems enabled by the biological capacity. Noone speaks Language; Language (with a capital L) is not a language. This contrasts with animal communication systems. True, different chaffinch, and other songbird, dialects exist, but their range is far less than that among human languages.

[1] Items in this chapter superscripted by '1' are terms routinely used by linguists about language, and are explained in a glossary at the end of this chapter.

By the usual count, there are over 6,000 different human languages. Defining the difference between a language and a dialect is ultimately not possible, but a rough criterion is that different languages are mutually unintelligible, whereas there is some degree of mutual intelligibility among different dialects of the same language. By this criterion, there are in fact several different Chinese languages, of which Mandarin and Cantonese are two, but Norwegian and Swedish actually count as the same language. In the past, many other languages existed, but are now extinct. It seems likely that the peak number of languages spoken by humans occurred some time in the last few millennia, when the earth was as yet sparsely populated by small groups of humans living in relative isolation. Now, languages are being lost, and we are in an age of mass linguistic extinction, with predictions that about half the world's languages will die out in the next century. The great diversity of human languages is made possible by the fact that they are learned, rather than biologically transmitted from generation to generation via the DNA. The fact that languages are learned is not, however, sufficient to account for their great diversity.

The diversity of biological species arises through accumulated genetic copying errors, geographical isolation, and selective adaptation to new niches. Copying errors in learning and geographical isolation are also responsible for the great diversity of human languages. As early humans spread out over the planet, their group languages accumulated changes which were not constrained by any need to communicate with the groups they had left behind, and these languages struck out on their own. But adaptation to new niches is not a factor affecting the diversity of languages, aside from the relatively simple matter of vocabulary—languages of African pygmies have no unborrowed word for snow. In matters of grammatical structure and structure of their sound systems, there is no correlation between languages and the physical environments of their speakers.

A factor permitting the diversification of languages is the 'arbitrariness of the sign'[1]. A rose by any other name would smell as sweet. So long as people tacitly agree to use the same words for the same things, a language works. In grammar, as long as languages put the words in the same order to convey who did what to whom, and use the same conventional inflections to convey such details as the timing of the event reported and the speaker's attitude to it, the language works. So languages are fairly free to evolve different grammars and sound systems within the limits imposed by the communicative needs of the group.

A contentious issue within linguistics is the degree to which language learning, and therefore linguistic diversity, is further constrained by biology. One can certainly imagine crazy languages that would be impossible to learn. A frequently given example is a language which expressed questions by completely inverting the word order of the corresponding statements. In such a language, you would ask the way to the station by saying *Station the to way the me tell can you?*. The strain on short-term memory in computing how to express questions in such a language rules it out as a possible natural language. The contention within linguistics arises because there may be some such constraints on language-learnability which are not attributable to nonlinguistic factors such as short-term memory, but operate only in the specific domain of language. Here is a candidate for such a Language-specific constraint. Beware, like all such examples, it involves 'thinking the impossible', something that linguists are skilled at. Consider the following pair of sentences.

The man built the house. The house fell down.

We can make a single sentence, expressing the same information, thus:

The house <u>that the man built</u> fell down.

So far, so good. The compressed sentence was formed by making a relative clause (underlined) out of the first sentence, and attaching it to the shared noun phrase, *the house*. The original right-hand sentence *The house fell down* is wrapped around the outside of this underlined relative clause. Now let's try it again, with this last complex sentence as one of the inputs to the process:

The house that the man built fell down. The man escaped.

Here again, there is a shared noun phrase, *the man*. So in principle, it ought to be possible to use the same relative clause-forming process. If we do, we get:

**The man that the house that built fell down escaped.*

This is an impossible sentence, as indicated by the linguist's conventional asterisk. And generally, across languages, we find that sentences like this, and their analogues, adjusting for the differences between languages, such as word order, are also not well-formed. The interesting question is 'Why?' Is it due to a constraint specific to Language, a putative 'Law of Language', that sentences such as this do not occur in languages? Or is this fact due to a more general constraint on processing any kind of serial information, linguistic or otherwise? Both opinions are held in the field, probably with a swing under way to the general non-domain-specific explanation. The original discoverer of this family of constraints, known as 'Syntactic Island Constraints' was J.R.Ross (Ross, 1967, 1986), in the early heyday of the generative linguistics[1] movement, whose goal was, in part, to discover facts peculiar to the human Language faculty. The alternative view that such constraints arise from general constraints on learning any sequential behaviour has been argued by Morten Christiansen, among others (Christiansen *et al.*, 2002, Christiansen & Ellefson, 2002). Note that there are dozens of similar examples, a fact which underlines the great complexity of human languages, as compared with animal communication systems, where considerations of such abstractness and complexity do not arise.

Linguistic inheritance is both vertical, as when children more or less faithfully acquire the languages of their parents, and horizontal, as when languages mix and borrow each other's words and constructions. Branching family tree diagrams are still popular in historical linguistics. But such genealogies are misleading. According to a common classification, English is a Germanic language (along with German, Dutch, Icelandic, Danish and others), while French is a Romance language (along with Romanian, Portuguese, Italian, Spanish and others). But such always-diverging, never-converging tree diagrams distort the extent to which languages have influenced each other across language family boundaries. English and French share a lot of similar vocabulary and grammar, the result of the Norman Conquest of England in 1066 and centuries of contact ever since. Many languages are of such hybrid types, owing their structure to multiple sources. The possibility of such hybridization adds to the overall diversity of languages.

Summarizing the factors contributing to linguistic diversity, (1) the fact that languages are learned, rather than coded into the genes, (2) the arbitrariness of the sign, and (3) the prevalence of horizontal transmission allow for great diversity, but this is significantly constrained by (4) biological factors such as memory and processing limitations, which may or may not be specific to the Language domain.

14.3 Learning

Human languages are learned. Many animals are also adept at learning. But they can't learn human languages. Kanzi has learned about 500 words, but this seems to be close to his limit. And Kanzi is a highly human enculturated bonobo[1]. If animals can learn, why can't they learn languages? What is different about languages? A major factor is the arbitrariness of the Sign. A chimpanzee can learn to use a tool to reach a banana. In this case the function of the tool is transparently mechanical. Physical laws govern the interaction of the tool and the banana. A word, it can be argued, is also a kind of tool. If I want a banana, just saying 'banana' may be enough, if my hearer is cooperative, to get me the banana. But the word *banana* has no physically causal relation to the outcome. In this sense, words are magic; just using words, if the hearers are cooperative, makes physical things happen. Apes have a good understanding of physical cause and effect with everyday objects, and can learn practical tasks. But the arbitrary nature of human symbols is a far greater challenge to learning, because it's not obvious how they work.

To begin to understand, as human infants do, that the noises made by conspecifics carry some informative message, there needs to be, in the child, a presumption of their relevance to its life (Sperber & Wilson,1986). This is the idea that uttering a sentence such as *It's late* not only conveys the information that something is late, but also that the speaker intends the hearer to know that something is late. So an infant on hearing an utterance in a human language knows that the speaker intends the hearer to know something. Count the instances of intentional verbs[1] in this last sentence (*knows, intends, know*), and we see three embedded levels of intentionality[1]. (Intentionality is 'aboutness', some mental attitude to an external state of affairs.) Halliday's (1975) provocative title for his book was *Learning how to mean*. Taken literally, this might suggest a *tabula rasa*[1] in the child, in which even an understanding that utterances mean something has to be learnt. What we see in humans, as opposed to non-humans, is a developmental process whereby this understanding emerges well within the first year of life. It is developmentally programmed into normal early human maturation, in normal circumstances, as opposed to being strictly learned. The human child is pre-disposed to understand that utterances mean something. Play, babbling and imitation are aids to achieving this understanding.

A useful distinction has been made between learning by emulation[2] and learning by imitation. Emulation involves achieving the same goal as was observed, but not necessarily by the same means. For example, if a chimpanzee sees me push a door open with my foot, it may learn to push the door open with its hand; this is emulation. But if the chimp slavishly follows my actual method of opening the door, using its foot, that would be learning by imitation. Whiten *et al.* (2004), in a survey of ape learning, conclude that there is more emulation than pure imitation in apes, and both kinds of copying occur much more readily when the demonstrator is a human trainer than spontaneously among apes themselves. None of the work surveyed, however, involves the copying of clearly communicative behaviours. In some sense, emulation is more intelligent; it gets the job done. Human children are natural imitators. They imitate for no apparent reason, as shown by Meltzoff's (1988) well-known experiment, in which very young babies imitated the facial gestures of adults. Non-human apes are very poor at vocal imitation, but human children are expert at it. The babbling stage in babies is like play among animals, in that it seems to have no immediate purpose. Play-fighting in young animals is plausibly accounted for in evolutionary terms in that it refines motor skills which will be useful later in life. Likewise, both babbling and vocal imitation are practice for use of language later in life. Human babies, unlike other apes, have a natural disposition to engage in these activities, whose payoff only materializes long afterwards.

Children's natural dispositions to imitate and to participate in group activities are signs of a kind of sociability special to humans, a topic to which we will return later in this essay.

14.4 Complexity

We have already touched on the great complexity of languages, compared to any non-human communication system. How does this complexity relate to biology? In linguistic theorizing, two opposing tendencies are felt. On the one hand, people are struck by the universality of such complex facts as the Island Constraints mentioned above, and a host of other universal tendencies involving intricate facts about the reference of pronouns, the scope of quantificational words such as *all* and *each*, the varying semantic effects of verbs such as *promise* and *persuade* and adjectives such as *easy* and *eager*[3] and so forth. Children learn such abstract facts with no overt tuition and from examples which by no means logically determine the conclusions the children come to. The children's more or less faithful learning is evidenced by their later usage in general conformity with other members of the community. This led linguists in the 1970s and 1980s to hypothesize a rich innate structure guiding the acquisition process, consisting of a set of distinct innate modules such as the X-Bar module, the Binding module, the Theta module, and the Case module (Chomsky, 1986). It is not necessary here to explain the purported content of these modules; note that each is a set of propositions determining some independent, but interlocking, aspect of the linguistic knowledge that the child

[2] The term 'emulation' is due to Wood (1988).

[3] Compare *I promised John to go* with *I persuaded John to go*. Who, in each case, is to do the going? Also compare *John is easy to please* with *John is eager to please*. Who, in each case, is understood as doing the pleasing?

will ultimately acquire. Thus syntactic theory, at this stage, responded to the complexity of languages by postulating a complex biological endowment specifically devoted to Language.

On the other hand, of course, there was always Occam's Razor, the normal scientific pressure to adopt theories which are as simple as possible. The postulated richness of the innate language acquisition mechanism was a biological embarrassment, as each of these modules presumably had to be coded somehow into the genome. Further, their interacting[4] nature in modern languages made it necessary, but difficult, to imagine stages in the evolution of the modern Language faculty when some of these modules were present and others had not yet emerged. Certainly it is possible to imagine such undeveloped versions of the modern Language faculty, but it adds to the strain on credibility of the whole story, in an already speculative field. This kind of gradual evolution of the Language faculty was proposed by Pinker & Bloom (1990), in a landmark article arguing the proposition that the most obvious explanation for the complexity of natural language is that it evolved by **Darwinian natural selection**. To many, this had seemed obvious, but it is a sign of the intellectual climate within the dominant paradigm in linguistics in the late 20th century that it needed arguing at all.

The simpler a theory of the innate Language faculty could be made to look, the more it appealed to biologists. Quite recently, a theoretical move has been made toward an extremely simple account of the human language faculty; this is known as the Minimalist Program[1] (Chomsky, 1995), proposing that the language faculty is nothing more than a facility to recursively merge lexical structures (precisely specified dictionary entries) to form larger structures such as phrases and sentences. It is stressed that this is a 'program' rather than a 'theory', and its empirical and predictive delivery remains also minimal; one is reminded of String Theory in physics.

Linguists have become persuaded in the last decade or so that, while human language is clearly spectacularly unique among animal communications systems, its biological foundations rest in a combination of many factors, many of which are not specific to Language. Enhanced memory for large numbers of arbitrary associations is one such factor. Humans typically have vocabularies of about 50,000 items. Kanzi, the best-performing non-human primate in this regard, has mastered about one hundredth of that number[5], and is not expected to learn significantly more. It was always recognized that the lexical component of a language necessitated rote-learning, and this made it the least theoretically interesting component of a language. A recent movement in the theory of grammar, known as Construction Grammar[1,6], suggests that there is in fact no principled distinction between lexical items and grammatical constructions. Grammatical constructions are like lexical items with variables in them, permitting the insertion of a more or less wide range of permitted other constituent items. An example of such a construction in English is *The + COMPARATIVE + CLAUSE + the + COMPARATIVE + CLAUSE*, as for instance in *The more you eat, the fatter you get* or *The bigger they are, the harder they fall*. This theory places less emphasis on economy of statement in the grammar, and recognizes that the representation of language in the brain may be somewhat redundant, and uneconomical, taking advantage of humans' undoubtedly great powers of memorization.

The human Language faculty, rather than being a richly structured independent module of the mind, is a mosaic of many factors which have come together in a unique combination in humans. Many of these factors can be observed, often in a less powerful form, in other animals. The recognition that this is so is seen in a distinction made by Hauser *et al.* (2002) between the 'faculty of language in the broad sense' (FLB[1]) and the 'faculty of language in the narrow sense' (FLN[1]). FLN is whatever is unique to human language; Hauser *et al.* (2002) suggest that this may, at most, be limited to the human capacity for recursion[1], the execution of a computation of a certain type during the execution

[4] These hypothesized modules of the Language faculty are 'interacting' in roughly the same sense as subsystems of physical organisms, such as the respiratory system and the circulation system, work together.

[5] Kanzi uses a lexigram board, with abstract symbols that he points to, as a substitute for uttering spoken words.

[6] Goldberg (1995), Croft (2001), Fillmore (2003), Culicover (2005).

of a similar computation at a higher level. For example, to grasp what *John's father's brother's neighbour's cat* refers to, you have to identify the referent of the subpart *John's father's brother's neighbour,* and to understand that, you have to identify the referent of *John's father's brother,* and so on. Hauser *et al.* leave it open whether such a capacity for recursion can be found in any non-human animals. If it can be, then the human faculty for language in the narrow sense is, in their view, actually empty, leaving us with a picture of FLB as a mosaic of factors all of which can be found in some form or other outside of the domain of human Language. One candidate for recursion in animals is navigation; figuring out how to get from A to B might involve recursively embedded processes. The technical definition of recursion, and how to recognize whether it is in play in a specific animal activity, is not, however satisfactorily pinned down, and there is room for argument about the use of recursion in animal activities. The radical view that human Language may have no unique individual properties is controversial, and I will review below other candidates for a categorical difference between human Language and animal communication systems.

14.5 Compositionality

The example of how we parse a recursive structure such as *John's father's brother's neighbour's cat* highlights another feature of human language that is not found in any animal communication system that we know of (with one odd exception). The Principle of Compositionality[1] states that 'The meaning of the whole is a function of the meanings of the parts, and the way they are structured together'. You understand the meaning of a whole sentence because you know the meanings of the individual words, and you know the contribution the grammar makes to this understanding. This is how you know that *Mary kissed Bill* means something different from *Bill kissed Hillary.* While many complex animal calls are combinatorial, that is, they are made up of several reusable subunits strung together, none is compositional in this sense. The songs of gibbons are sequences of units which occur in other contexts, and can therefore be identified as independent subunits, but there is no sense in which the meaning of the whole gibbon song is understood as any combination of the meanings of these subunits. In this sense, the term 'song' is appropriate, as such complex animal calls are more like music than human language, which expresses semantic content through the application of compositionality.

The 'odd exception' mentioned above is the waggle dance of honeybees, which has two meaningful components, the speed of the dance and its orientation on the honeycomb. The speed conveys the approximate distance of food from the hive, and the orientation conveys the angle, relative to the sun's position, at which the food is to be found. Both distance and angle are necessary to specify the food's location, and the meanings of the two aspects of the dance combine to compose this information. The dance behaviour of honeybees is, however, completely specified in their genes. The example shows how utterly different mechanisms may achieve a communicative effect. The human mechanism of learning languages is clearly more versatile in allowing the possibility of conveying a wide range of different messages, in adaptation to a complex and changing world.

The nearest to an example of compositionality in the communication of an animal closely related to humans is described by Klaus Zuberbühler (Zuberbühler, 2002, 2005). As any kind of precursor to the compositionality in human language, the example is problematic, as it involves the responses of one species, Diana monkeys, to the alarm calls of another species, Campbell's monkeys. Campbell's monkeys have different unitary alarm calls for leopards and eagles. Diana monkeys interpret these appropriately. Occasionally a Campbell's monkey utters a 'boom' about thirty seconds before such an alarm call, and the Diana monkeys then react with less panic than to 'boom-less' alarm calls. "...adding 'booms' before the alarm call series of a Campbell's monkey created a structurally more complex utterance with a different meaning than that of alarm calls alone." (Zuberbühler, 2005:279). Zuberbühler himself is frank about the limitations of non-human primates: "there is no evidence that they are able to invent and incorporate new call types into their repertoires or to combine calls creatively to produce novel meanings" (Zuberbühler, 2005:281)

14.6 Double articulation

The contrast between songs consisting of identifiable subunits and truly compositional **signals** such as human sentences brings out another unique characteristic of human languages, their so-called 'double articulation'[1,7]; at the phonological level, the expressions of human languages are combinatorial but not compositional. That is to say that the signals consist of systematically reusable subunits which themselves carry no meaning. The separate phonemes[1] making up a word have no meaning. The word *cat* consists of three phonemes {/k/ + /a/ + /t/}, but the meaning of the word is not derived from the meanings of the phonemes, because they have no meanings. At the morphosyntactic level, the word *cat* does have a meaning, which contributes, for example, to the overall meaning of a sentence such as *The cat sat on the mat.* So languages are structured in two layers, a semantically compositional morphosyntactic layer, and a phonological layer which is merely combinatorial. All human languages have this property.

Double articulation clearly contributes to the massive expressive power of human languages. New meaningful words can be invented by simply combining phonemes from a given set. The phoneme inventories of languages vary in size from a mere dozen to over a hundred. Obviously, languages with fewer phonemes at their disposal tend to have longer words. The combinatorial power afforded by a phonological layer of structure provides languages with their vocabularies of tens of thousands of meaningful words.

The evolution of communication systems with this feature of double articulation is thought-provoking. It clearly has a biological aspect. Humans must have the facility for combining elementary sounds from a small fixed set in highly flexible and productive ways. But the fixed sets of phonemes vary widely from language to language, so these are not biologically fixed, although the articulatory apparatus within which they are defined is a matter of biological endowment. Recent work by Zuidema & de Boer (in press), using computer modelling, shows a process of self-organization at work. They start with utterances which are not recognizably made up of discrete elements, but are merely random walks ('trajectories') through articulatory space. To imagine such a trajectory, try to make some 'inarticulate' vocal sound, moving your jaw, lips and tongue around while sporadically vibrating your vocal cords, and altering the airflow through nose and mouth; avoid visiting known phonemes of your language. By a process of imitative transmission through a population of agents, with copying errors, and a requirement that the separate trajectories should not collapse with each other, the set of such trajectories through articulatory space gradually settled down to a more systematically organized set. In this evolved system, the same starting and end points were used by many different trajectories.

This can be envisaged in visual terms. Imagine a square with random scribbles on it. The only constraint is that each scribble is a single continuous line. Agents are required to copy these lines and pass their copies on to a successive generation, keeping the same overall number of lines. What happens, over time, is that a more systematically organized set of lines emerges, which start and end at the various corners of the square. At the beginning of the simulation, the corners of the square had no special status in the formation of the lines/scribbles. At the end of the simulation, there is convergence on a system of lines which re-use a small set of starting and end points, and move economically between them. See Figure 14.1. This suggests that, even though humans are biologically capable of making 'inarticulate' vocal gestures, and of attaching some meaning to them, what happens over time in the continuous trade of such gestures is a self-organizing process by which 'articulate', jointed speech, re-using a small set of focal points, emerges.

14.7 Self-organization

The example of the emergence of combinatorial phonology introduces what may be a potent and pervasive force in the evolution of languages, in their grammars as well as in their sound systems. The investigation of such self-organizing processes

[7] Also sometimes called 'duality of patterning'—the terms are equivalent.

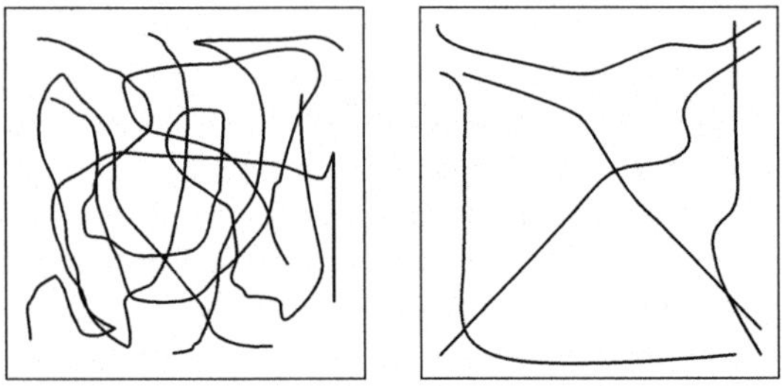

Figure 14.1 Self-organization in articulatory space. The left-hand box contains five randomly scribbled lines, schematically representing random gestural trajectories in articulatory space. The right-hand box shows five trajectories approximately optimized for simplicity and distinctiveness from each other. After Zuidema and de Boer (in press).

in the context of language evolution is relatively new. It is a distinct process from natural selection, but entirely compatible with it. In an early pioneering work on self-organization, Thompson (1961) tended to depict self-organization ('laws of growth and form') and natural selection as mutually exclusive alternatives. More recently, Oudeyer (2006) gives a clear discussion of the relationship between natural selection and self-organization. Self-organization can affect both organic phenomena (e.g. snail shells) and non-organic phenomena (e.g. snow crystals). Self-organization narrows the search space of possibilities from which natural selection selects.

In the evolution of language, the most promising cases of self-organization arise through the interaction of users of a language over historical time. The self-organized object which emerges is not a physical object like a snail shell or a snow crystal, but the language itself, an abstraction over the common behaviours of the speakers of the language. However, a physical, non-linguistic example may help. Consider an informal well-worn footpath diagonally crossing a field. The path was not deliberately designed by any one person, but is simply the end product of hundreds of people taking the shortest route across the field. In the case of language, repeated usage over generations, with idealized copying of the observed patterns by new learners, results in features of language which were not the invention of any one person, and further, were not closely dictated by the genes. In other words, each individual involved in the process could have behaved in a variety of ways, as far as any direct pressure from the genes is concerned. But the accumulation of hundreds of tiny unconscious facultative acts led to the language concerned being the way it is.

Here is an example. A salient feature of very many languages is a correlation between frequency and irregularity. For example, in English, the most common verbs (*be, have, do, make, go,* etc.) are all irregular. This correlation between frequency and irregularity in languages comes about through the repeated action of several processes. One process is the tendency to slur or phonetically erode high frequency words. This erosion creates irregularities. It is well known that children are somewhat resistant to irregularities, tending naturally to regularize even irregular verbs. Thus children learning English go through a stage where they use **goed* instead of *went* and **comed* instead of *came*. In the case of the most frequent irregular verbs, however, the irregular usage in the environment overwhelms the child's natural disposition to regularize, with the result that irregular forms persist in the language, just in the more frequent forms. For less frequent forms, the child is not presented with enough evidence to overrule its natural tendency to regularize, and less frequent forms are mostly regular. This process has been modelled computationally by Kirby (2001).

Such self-organizational processes have been dubbed 'phenomena of the third kind' by Keller (1994). In his taxonomy, phenomena of the first kind are natural phenomena, like oceans and volcanoes; phenomena of the second kind are human artefacts, deliberately made, like telephones and the written constitutions of nations; phenomena of the third kind are the outcome of human action, but not deliberately made by any single, or even collective, conscious decision. Keller also invokes Adam Smith's 'Invisible Hand' (Smith, 1786), paraphrased in modern terminology as 'market forces'. Keller argues that much of the evolution of language should be seen as an Invisible Hand, i.e. self-organizing, process. The self-organization of a communication system along the lines illustrated here can only happen in a relatively complex learned system, such as humans have. With such limited systems as the mostly innate 3-way alarm calls of vervet monkeys, there is far less scope, if any, for the accumulation of tiny facultative actions determining the historical course of the system.

14.8 Stimulus-freedom

There is a considerable difference of degree between humans and non-humans in the extent to which their mental processes are immediate reactions to their environment. Humans can recall, and muse about, specific events from long in the past, and can plan complex series of actions far in the future. One can find the tiny seeds of stimulus-freedom in animals closely related to humans. In object-displacement experiments, for example, a desirable object is hidden from an animal's view, but the animal still seems to know it is there, and searches for it. Thus the animal has a mental representation of an object not currently perceived. Dogs are good at this. Panzee, a chimpanzee, could remember after a night's sleep where food had been hidden the day before (Menzel, 2005). It is often claimed (Tulving, 1972, 2005; Suddendorf & Corballis, 1997, in press) that only humans have episodic memory[1], a recall of specific events, as opposed to non-time-indexed knowledge of some state of affairs (which may result from having observed some event in the past). Experimentally sorting out the difference between recall of events and knowledge of resulting facts is problematic. In recognition of this difficulty, experimenters have attributed 'episodic-like' memory to animals such as scrub jays, which show evidence of remembering what food they hid, and where, and how long ago (Clayton & Dickinson, 1998; Clayton et al, 2001).

The evidence from animals who show some slight signs of episodic memory means that such memory is not absolutely dependent on the prior evolution of language. Certainly in humans, episodic memories are aided by public language. There was presumably some co-evolution of the faculty of Language and a capacity for episodic memory. In humans, the earliest memories of specific lifetime events are typically from roughly around 2 years of age, when syntactic language begins to develop. This suggests some interdependence between episodic memory and language.

Animals can plan future actions, to some degree. Mulcahy & Call (2006) report on experiments in which bonobos and orangutans collected and hoarded appropriate tools for tasks as far ahead as 14 hours before the task was carried out. They comment that "These findings suggest that the precursor skills for planning for the future evolved in great apes before 14 million years ago, when all extant great ape species shared a common ancestor." (p. 1038)

There is a symmetrical relationship between planning and memory, to the extent that planning has sometimes been classified as 'prospective memory' (Meacham & Singer, 1977). In one experiment (Cook *et al.* 1983), rats searching a 12-arm radial maze for food were taken out while still searching and replaced in the maze later. They showed similar accuracy of recall in relation to (1) number of arms already previously searched and (2) number of arms not yet searched and therefore remaining to search. This indicates an overlap of the mechanisms of retrospective and prospective memory. Such memories are stored in the animal's brain and are not dependent on its current perceptions. Humans, however, have much longer retrospective and prospective memories than non-humans. Suddendorf & Corballis (1997) write of the 'unconstrained mental time travel' of humans.

Humans can think about absent things. This stimulus-freedom of human mental processes is reflected, naturally, in our communication systems. We can talk about absent things, and in fact this is the norm for human communication. We constantly bring to mind distant events or possible future events, and talk about them. The structure of modern languages makes this possible, but this is probably a case where language structure has evolved to meet the need to express such 'time-travelling' thought, rather than the structure of language actually enabling such time-travelling thoughts in the first place. A simple story, probably partly correct is: first the private thought capacity, then a communication system adapted to make the private thoughts public.

The relation between language and thought is a hot philosophical issue. Most comparative psychologists, and a growing number of philosophers, are willing to concede some thoughts and concepts to non-humans. But clearly there are thoughts that can only be attained with the help of language. Examples are the concepts expressed by words and phrases such as *Tuesday, unicorn, ninety-three, zero, generosity* and *legal.* Examples such as these rely on verbal definitions made possible by the productive generative capacity of languages. Given compositionality (as discussed above), it is possible to arrive deductively at meanings not previously entertained by the mind. For example, given the concepts expressed by *white, horse, single, horn* and *forehead,* compositionality allows one to deduce what the expression *white horse with a single horn in its forehead* should mean, even though we are never likely to experience such a beast. Presumably such thoughts are permanently denied to non-humans.

Once fictions can be expressed, and thus shared between people, they can become potent cultural forces, defining group identity. A commitment to the proposition that Jesus Christ is the son of God is what centrally divides devout Christians from devout Muslims. Thus, beside the obvious practicality of generative language, for transmitting real-world information, enabling us to build spaceships that reach the Moon, generative language provides for the construction and sharing of rich structures not corresponding to any perceptible reality, defining complex cultures.

14.9 Interpersonal function

The vast potential of languages for describing the real world, and fictitious worlds, in detail, should not lead us to ignore the fact that making descriptive statements about a world must have a social purpose. Austin (1962) wrote of the 'descriptive fallacy', the idea that the point of language is to describe a world. He famously stressed that when we use any language at all, we are thereby doing something, carrying out some social act. Much animal communication carries this purely social force, and it is inappropriate to paraphrase such signals in declarative terms. For example, a threat signal is just doing the threat, or it just is the threat. Translating an animal threat signal into a human declarative sentence, such as *If you don't back off, I will attack you* may be useful for our purposes, but there is no reason to suppose that any such complex thought goes through the mind of the threatener or the threatened animal. A tiny number of human utterances have only this bare 'illocutionary' force. For example, *Hello* just does greeting; it has no declarative content, and doesn't describe any state of affairs. The vast majority of human utterances have some social purpose, an intended impact on a hearer[8], in addition to whatever descriptive content they may have. For example, *It's raining*, in addition to describing the current weather, will always in normal circumstances be said with some intended effect on a hearer, such as to warn them to put on a raincoat, or to prove that your prediction was right, or to jokingly complain about the local climate. As this dyadic 'doing things to each other' function is basic to both animal and human communication, it seems likely that this is a remote evolutionary foundation of human language, and that the vast referential, descriptive, triadic power of language came later. The set of social acts which can be carried out using language exceeds the range of things that non-human animals can do to each other with symbolic signals, such as threat or submission gestures. All of the acts that can be carried out by non-humans can also be carried out using words (see related discussion in Chapter 1). Thus a threat can be made by purely nonverbal means, e.g. by shaking ones fist in a person's face; and it can also be carried out in the calmest of ways with words, with little emotion, by saying *If you move, I'll shoot.* Social acts between animals are mostly dyadic, only involving the sender and the receiver of the signal (see detailed discussion in Chapter 3). Humans can overlay their social acts with descriptive content, as in the previous example, which refers to actions such as moving and shooting. This vastly increases the subtlety and fine-grained detail of things that humans can do to each other, using language.

Some things that humans can do to each other, using language, can only be done with language, or at least in a language-defined context. Thus,

[8] An exception may be private soliloquizing, praying, or talking to oneself. It seems likely that these are uniquely human activities which evolved on top of a prior purely social communicative form of language. Chomsky (1980) is in a minority in holding that such talking to oneself may be the main function of human language.

promising, for example, requires some understanding of what is promised, which can only be expressed with words. True, I can effectively promise something merely by nodding, but in such a case what I am thereby committed to has previously been spelled out in language. Another class of uniquely human communicative acts is those where a social fact or convention is made to exist solely by using an appropriate verbal formula, as in examples like *I name this ship the Mary Rose* or *I hereby declare you man and wife* or *Ego te absolvo*. Such acts are, of course, impossible in the non-human world.

14.10 Mindreading, manipulation and cooperation

Encounters between animals can be either adversarial or mutually beneficial. In both cases, it is advantageous to an animal to be able to predict and influence the actions of the other. Predicting events can involve various degrees of intentionality. Predicting that a falling rock will land near you involves no understanding of the mental processes of another organism. Predicting that a lion skulking nearby will chase you may, or may not, involve attributing some attitude to the lion. A zebra may simply have a built-in avoidance response to nearby skulking lions, just as some people may possibly have built-in, or epigenetically easily-triggered, arachnophobia. But some ability to 'mind-read' accurately the intentions of competitors, predators and prey would clearly be advantageous to any animal (Krebs & Dawkins, 1984).

Experiments with chimpanzees show that they can tell whether a human experimenter is teasing them or merely being clumsy (Call *et al.* 2004), thus demonstrating a degree of mind-reading. There are also many reports of tactical deception among primates, and Byrne & Corp (2004) also found a correlation between neocortex size and rate of tactical deception. Thus one thing bigger brains is good for is deception, which involves both prediction of the likely actions of another organism and deliberate manipulation to influence them. Hare *et al.* (2000) showed experimentally that "Chimpanzees know what conspecifics do and do not see" (their title). In this experiment, a subordinate chimpanzee could see two food items, and was also in a position to see that a dominant chimpanzee could only see one of these food items. When both animals were released from their positions, the subordinate chimpanzee reliably went for the food item that had been invisible to the dominant.

All work of this kind is centered around the complex question of the extent to which non-humans have a 'Theory of Mind'[1], the ability to know that another organism is just like, and therefore thinks like, oneself. Note that there are two main components here (1) the obvious one, just stated, and (2) what 'oneself' is like. Informal character attributions among people show a tendency to project ones own vices and virtues onto other people. Thus a generous person will tend to assume that other people are generous; and a miserly person will tend to assume that other people are also miserly. Crucially, a naturally uncooperative animal will not be able to read cooperative intentions in another animal, although it may well be able to read competitive intentions in another.

There is experimental evidence that chimpanzees can read competitive intentions in human experimenters but not generous cooperative intentions, when the stimuli presented to the animal are very similar. In one condition, a human made a reaching gesture, with hand spread for grasping, toward a container; in this condition, the observing chimpanzee anticipated the human's reach and got to the container first. In another condition, the gesture was very similar, but with fingers together in a whole-hand pointing gesture, indicating the container. The chimpanzee subject did not 'get the point' of this cooperative pointing gesture. A natural interpretation is that chimpanzees can read the intentions of others, but they do not expect those intentions to be cooperative. Thus, a certain category of others' intentions (the cooperative intentions) remains obscure to them.

14.11 Reference

Communicative acts in the animal world are mostly dyadic, not involving any third entity beside the sender and receiver of the signal. A widespread exception is alarm calls. The alarm calls of vervet monkeys (Cheney & Seyfarth, 1990) are especially well known, but many other species of birds and mammals also have ritualized alarm calls for specific classes of predators, typically with separate signals for aerial and terrestrial predators. Alarm calls are triadic because they involve the sender, the receiver, and the referent of the call. Triadic communication is about something, whereas dyadic communication is not. Animal alarm calls are largely genetically determined, in both production and reception, with very little room for voluntary control. In young vervets there is some learning of the specific class of aerial objects for whom it is appropriate to make the eagle alarm call. And there is also an audience effect, with mothers being more likely to make an alarm call when their own offspring are nearby. Since both the stimulus-to-call behaviour and the call-to-response evasive behaviour (e.g. climbing a tree when hearing the leopard call) are strongly specified in the animal's genes, the question arises whether the animals are 'referring', in anything like a human sense, to the predator. It could be the case, for instance, that natural selection has acted in parallel to favour two independent but mutually adaptive behaviours: (1) Bark when seeing a leopard, and (2) Climb a tree when hearing a bark. If this were the case, there would be no human-like sense in which the animal's alarm call means, or brings to mind, the appropriate class of predators.

Klaus Zuberbühler has described experiments which can be naturally interpreted to suggest that animal alarm calls do in fact bring the concept of the appropriate predator to mind, at least for a short period. Zuberbühler *et al.* (1999) worked with Diana monkeys of the African forest who have distinct calls for leopards and eagles. Female monkeys both give spontaneous alarm calls on sensing a predator and respond to alarm calls from males by repeating the call. Beside recording the alarm calls, the researchers also recorded characteristic noises associated with the two predators, such as the growl of a leopard and an eagle's shriek. Next, they played back three different kinds of pairs of stimuli, where the stimuli in each pair were separated by an interval of five minutes silence. On hearing first an eagle alarm call, then (after five minutes) the shriek of an eagle, female monkeys showed less sign of alarm (giving fewer repeat calls) than after hearing, for example, an eagle alarm call followed by the growl of a leopard. The logic is this. If, on hearing an eagle alarm call you are prepared to be wary of an eagle in the area, you are less disturbed by hearing the actual eagle shriek. The eagle shriek merely confirms what the earlier alarm call told you. But if you hear an eagle alarm and then hear a leopard growl, the growl is new information, telling you about a kind of predator that you hadn't been made aware of by the previous call. The researchers did, of course, try out all the necessary control conditions to consolidate this conclusion. The conclusion is that the alarm calls do not merely trigger the relevant evasive action, with no representation of the specific source of danger being kept in the head; the Diana monkeys, on hearing a leopard alarm call, keep the idea of a leopard in their minds for at least five minutes; and likewise with the eagle alarm call. This behaviour meets the criteria set by Marler *et al.* (1992) for calls being 'functionally referential'. It seems likely that similar results would be obtained with all species with small inventories of predator-specific alarm calls.

The term *reference* is used by animal researchers with less care than by most linguists. There are two senses that need to be distinguished. In the discussion above of alarm calls, the question is whether some class of calls, such as a vervet's bark, has a referential meaning in roughly the same way that the English word *leopard* has. Of course translation even from one human language to another is seldom, if ever, perfect, so we should not expect to have a perfect English translation of what the vervet's bark means. But the idea is that the vervets have a (very limited) code, shared by the whole community, according to which 'bark' means what we humans roughly translate as *leopard*. When a vervet hears the bark, it brings a certain concept to mind. Certainly, the vervets' concepts are only protoversions of ours, because they cannot expound on the nature of leopards, and presumably do not ever mutely reflect dispassionately on the nature of leopards. But nevertheless, we may see in such alarm calls a skeletal version of our own shared codes (vocabularies), by which reference to classes of objects and actions is conventionalized to arbitrarily related signals. Putting it anthropomorphically, a vervet's bark denotes something like the class of leopards. Such denotation[1] is one sense of the term *reference*, and in this sense it is the signals themselves, the proto-words, that refer.

The other sense in which *reference* is used does not involve a shared code, conventionally mapping a set of signals onto corresponding classes of objects and actions. In this sense, it is the individual users themselves who do the referring. Here is an example. Third person personal pronouns, such as *he, she, it, this* and *that* can be used with variable reference; what they point out on any occasion depends on the circumstances at the time. If I write here *That's good*, you, as a reader remote in time and space, don't know what I am referring to, and nothing in the word *that* gives you any clue about what class of objects the referent of *that* might belong to. Words such as these are known to linguists as deictic[1], or pointing, words. The word is used in a particular context to draw attention to some particular thing, and on other occasions to draw attention to totally different things. All that is conventionalized about such words is that they are used for pointing to things in the context of the current discourse. What exactly they point to is left to the pragmatic inference of the observer. Pointing with the index finger (or in some cultures the lower lip) is a non-linguistic analogue of the use of deictic words. What precursors can be found in non-human behaviour for this kind of pointing behaviour, whereby an animal draws the attention of another animal to some specific object in their shared context? In the wild, none; in captivity, some, but only with their human keepers. In thirty years of observing chimpanzees in the wild, Jane Goodall (Goodall, 1986) never observed a chimpanzee point to an object with a view to drawing the attention of some other chimpanzee to it. Primates in the wild just don't point to things. In this sense, they don't refer to specific things. This kind of reference is totally absent from non-human communication in the wild. In captivity, chimpanzees and other primates learn to communicate their needs to human keepers by pointing, but the circumstances are limited to the fulfilment of current desires, as when a chimpanzee points to some food item that he wants to be given. Even in captivity, apes do not point to things just to share information about some interesting property they may have.

The absence of pointing in primate communication in the wild highlights the absence of a human level of cooperation.

14.12 Conclusion

There is indeed a wide gap between human language and non-human communication, in the various ways I have surveyed here. The difference cannot be attributed to any single factor. Apes are different from us in many qualitative ways. It seems most likely that at some time a critical combination of factors arose in our ancestors, which gave rise to the rapid expansion of the Language faculty, in its many facets, and a concomitant diversification and enrichment of individual languages and cultures. Exactly what the components of that critical combination were is still to be discerned, and it is not clear what further evidence we may be able to call upon.

Summary

Human languages are far more complex than any animal communication system. Furthermore, they are learned, rather than innate, a fact which partially accounts for their great diversity. Human languages are semantically compositional, generating new meaningful combinations as functions of the meanings of their elementary parts (words). This is unlike any known animal communication system (except the limited waggle dance of honeybees). Humans can use language to describe and refer to objects and events in the far distant past and the far distant future, another feature which distinguishes language from animal communication systems. The complexity of languages arises partly from self-organization through cultural transmission over many generations of users. The human willingness **altruistically** to impart information is also unique.

References

Austin, J.L. (1962). *How to Do Things with Words*. Harvard University Press, Cambridge, MA.

Byrne, R.W. and Corp, N. (2004). Neocortex size predicts deception rate in primates. *Proceedings of the Royal Society of London, Series* B, published online.

Call, J., B. Hare, M. Carpenter, and Tomasello, M. (2004). 'Unwilling' versus 'unable': chimpanzees' understanding of human intentional action. *Developmental Science*, **7**, 488–498.

Cheney, D. and Seyfarth, R. (1990). *How Monkeys See the World: Inside the Mind of Another Species*. University of Chicago Press, Chicago, IL.

Chomsky, N. (1980). *Rules and Representations*. Basil Blackwell, London

Chomsky, N. (1986). *Knowledge of Language*. Praeger, New York.

Chomsky, N. (1995). *The Minimalist Program*. Current Studies in Linguistics 28. MIT Press, Cambridge, MA.

Christiansen, M.H., Dale, R., Ellefson, M.R. and Conway, C. (2002). *The role of sequential learning in language evolution: Computational and experimental studies*, In A Cangelosi and D Parisi, eds. *Simulating the evolution of language*, pp. 165–187. Springer Verlag, London.

Christiansen, M.H. and Ellefson, M.R. (2002). Linguistic adaptation without linguistic constraints: The role of sequential learning in language evolution. In A Wray, ed. *The Transition to Language*, pp. 335–358. Oxford University Press, Oxford

Clayton, N.S. and Dickinson A. (1998). Episodic-like memory during cache recovery by scrub jays. *Nature*, **395**, 272–274.

Clayton, N.S., Griffiths, D., Emery, N. and Dickinson A. (2001). Elements of episodic-like memory in animals. *Philosophical Transactions of the Royal Society of London B*, **356**, 1483–1491.

Cook, R., Brown, M., and Riley, D. (1983). Flexible memory processing by rats: Use of prospective and retrospective information. *Journal of Experimental Psychology. Animal Behavior Processes*, **11**, 453–469.

Croft, W. (2001). *Radical Construction Grammar: Syntactic Theory in Typological Perspective*. Oxford University Press, Oxford.

Culicover, P. and R. Jackendoff (2005). *Simpler Syntax*. Oxford University Press, Oxford.

Fillmore, C., Kay, P., Michaelis, L.A. and Sag, I. *Construction Grammar*. University of Chicago Press, Chicago.

Goldberg, A. (1995). *Constructions: A Construction Grammar Approach to Argument Structure*. University of Chicago Press, Chicago.

Goodall, J. (1986). *The Chimpanzees of Gombe: Patterns of Behavior*. Harvard University Press, Cambridge, MA.

Halliday, M.A.K. (1975). *Learning how to mean: Explorations in the Development of Language*. Edward Arnold, London.

Hare, B., Call, J., Agnetta, B. and Tomasello, M. (2000). Chimpanzees know what conspecifics do and do not see. *Animal Behaviour*, **59**, 771–785.

Hauser, M.D., Chomsky, N. and Fitch, W.T. (2002). The faculty of language: What is it, who has it, and how did it evolve? *Science*, **298**, 1569–1579.

Huddleston, R. and G.K. Pullum (2002). *The Cambridge Grammar of the English Language*. Cambridge: Cambridge University Press.

Keller, R. (1994). *On Language Change: the Invisible Hand in Language*. London: Routledge. (Translation and expansion of *Sprachwandel: Von der unsichtbaren Hand in der Sprache*. Tübingen: Francke.).

Kirby, S. (2001). Spontaneous evolution of linguistic structure: an iterated learning model of the emergence of regularity and irregularity. *IEEE Journal of Evolutionary Computation*, **5**, 102–110.

Krebs, J. R. and R. Dawkins (1984). Animal signals: mind-reading and manipulation. In JR Krebs and N B Davies, eds. *Behavioural Ecology: an Evolutionary Approach*. Blackwell Scientific Publications, Oxford.

Marler, P., Evans, C.S. and Hauser, M.D. (1992). Animal signals? reference, motivation or both? In H Papoucek, U Jürgens and M Papoucek, eds. *Nonverbal Vocal Communication: Comparative and Developmental Approaches*, pp. 66–86. Cambridge University Press, Cambridge.

Meacham, J.A. and Singer, J. (1977). Incentive effects in prospective remembering. *Journal of Psychology*, **97**, 191–197.

Meltzoff, A. (1988). Infant imitation after a one-week delay: long-term memory for novel acts and multiple stimuli. *Developmental Psychology*, **24**, 470–476.

Menzel, C. (2005). Progress in the study of chimpanzee recall and episodic memory. In HS Terrace and J Metcalfe, eds. *The Missing Link in Cognition: Origins of Self-Reflective Consciousness*, pp. 188–224. Oxford University Press, Oxford.

Mulcahy, N.J. and Call, J. (2006). Apes save tools for future use. *Science*, **312**, 1038–1040.

Oudeyer, P.-Y. (2006). *Self-Organization in the Evolution of Speech*. Oxford University Press, Oxford. (Translated from French *L'auto-organisation de la parole* by James R. Hurford).

Pinker, S. and Bloom, P. (1990). Natural language and natural selection. *Behavioral and Brain Sciences*, **13**, 707–784.

Ross, J. R. (1967). *Constraints on Variables in Syntax*. Ph.D. thesis, MIT. (Published as Ross, 1986.).

Ross, J. R. (1986). *Infinite Syntax!* Ablex Publishing Company, Norwood, New Jersey.

Smith, A. (1786). *An Inquiry into the Nature and Causes of the Wealth of Nations: in three volumes*. Fifth edition, A Strahan and T Cadell, London.

Sperber, D. and D. Wilson (1986). *Relevance : Communication and Cognition*. Blackwell, Oxford.

Suddendorf, T. and Corballis M.C. (1997). Mental time travel and the evolution of the human mind. *Genetic, Social, and General Psychology Monograph*, **123**, 133–167.

Suddendorf, T. and Corballis, M.C. The evolution of foresight: What is mental time travel and is it unique to humans? *Behavioral and Brain Sciences*, in press.

Thompson, D. (1961). *On Growth and Form*. JT Bonner, ed. Cambridge University Press, Cambridge. Abridged edition,

Tulving, E. (1972). Episodic and semantic memory. In E Tulving and W Donaldson, eds. *Organization of Memory*, pp. 381–403. Academic Press, New York.

Tulving, E. (2005). Episodic memory and autonoesis: uniquely human? In H. S. Terrace and J. Metcalfe (Eds.), *The Missing Link in Cognition: Origins of Self-Reflective Consciousness*, pp. 3–56. Oxford University Press, Oxford.

Whiten, A., Horner, V, Litchfield, C. and Marshall-Pescini, S. (2004). How do apes ape? *Learning and Behaviour*, **32**, 36–52.

Wood, D. (1988). *How Children Think and Learn*. Basil Blackwell, London.

Zuberbühler, K. (2002). A syntactic rule in forest monkey communication. *Animal Behaviour*, **63**, 293–299.

Zuberbühler, K. (2005). Linguistic prerequisites in the primate lineage. In M Tallerman, ed. *Language Origins: Perspectives on Evolution*, pp. 263–282. Oxford University Press, Oxford.

Zuberbühler, K., Cheney, D.L. and Seyfarth R.M. (1999). Conceptual semantics in a non-human primate. *Journal of Comparative Psychology*, **113**, 33–42.

Zuidema, W. and de Boer, B. The evolution of combinatorial phonology. *Journal of Phonetics*, in press.

Glossary

Affective content: Part of the overall meaning of an utterance which conveys the emotions or attitudes of the speaker, such as anger or sarcasm.

Arbitrariness of the sign: The non-iconic and non-causal relationship between a word and its meaning. Onomatopeic words such as *cuckoo* and *miaow* are rare exceptions to the general arbitrariness of the sign.

Bonobo: *Pan paniscus*, or pygmy chimpanzee, a separate species from the common chimpanzee, *Pan troglodytes*.

Compositionality: The principle whereby the meaning of a linguistic expression (such as a phrase or sentence) is a function of the meanings of its parts (e.g. the words), and the way they are structured together by the grammar of the language concerned.

Construction Grammar: A cluster of theories in syntax, emphasizing the similarity between simple lexical items (words) and more complex constructions. Such theories place more reliance on human memory than other theories, more driven by parsimony of representation. Construction Grammar focusses on the whole spread of possible expressions in a language, including irregular and idiosyncratic ones. Representative works are Goldberg (1995), Croft (2001), Fillmore (2003) and Culicover (2005). Construction Grammar is sometimes said to be opposed to generative linguistics, but both are concerned with the same goal, a clear characterization of humans' impressive syntactic abilities.

Deictic: A deictic, or indexical, word has no constant reference in the external world, being applied to whatever fits in the particular context of a conversation. For example, the English deictic pronouns *I* and *me* do not refer constantly to any particular person, such as the Queen of England or the Dalai Lama; if the Dalai Lama happens to say "I am happy", then on that instance of use, the pronoun *I* refers to the Dalai Lama.

Denotation: The constant relationship between a word and what it picks out. Thus *cat*, for instance, denotes the (fuzzy) set of cats in the world, or alternatively, the concept of what is a cat shared by the language community.

Double articulation: Alternatively known as Duality of Patterning. This is the organization

continues

Glossary *continued*

of language structure into two distinct layers, a phonological layer assembling meaningless units, such as phonemes, into larger sequences such as syllables, and a morphosyntactic layer assembling meaningful units, such as stems and affixes, into larger sequences such as phrases and sentences. All known human languages, including manually signed languages of the deaf, have double articulation.

Episodic memory: An ability to recall specific events which have taken place earlier, as opposed to a memory for constant facts. Amnesics suffer from loss of episodic memory.

FLB: The faculty of Language, in the broad sense, incorporating all that is necessary to the working of language, such as control over the output mechanism (e.g. the vocal tract), memory, understanding of the intentions of other people, and so on. The concept is due to Hauser et al. (2002).

FLN: The faculty of Language, in the narrow sense, including only that which is distinctive of human language. If some property can be found in animal communication, or in human cognition outside of language, then by definition it does not belong in FLN. The concept is due to Hauser *et al.* (2002).

Generative linguistics: An approach to the study of language, pioneered by Chomsky and largely associated with his ideas. In the early days (roughly 1955–1970) it was characterized by emphasis on explicit statement of formal rules, and 'generative' was taken to mean 'explicit and rigorously formulated'. More recently, generative linguistics has become associated with a cluster of claims about the innateness of the human faculty of language and its distinctness from other cognitive domains.

Illocutionary force: The socially conventional act embodied in the performance of an utterance. Common illocutions are thanking, apologizing, congratulating, and greeting. Every utterance in a natural context has some illocutionary force, i.e. is intended in some way to do something to the hearer.

Intentionality: 'Aboutness', some mental attitude to an external state of affairs.

Intentional verb: A verb such as *believe* or *desire*, which relates to some mental state, where the mental state often involves a representation of some state of affairs in the external world.

Minimalist Program: The most recent incarnation of the generative linguistics research programme, summarized by Chomsky (1995). Here the emphasis is on seeing how little theoretical apparatus needs to be postulated to account for the complexity of languages. The Minimalist Program attempts to subsume organizational principles of language previously thought to be independent under a single abstract operation on the basic lexical elements of languages.

Phoneme: A meaningfully distinctive phonetic segment of a language. A phoneme does not have a meaning in itself, but it is capable of signalling a difference in meaning. Thus English /b/ and /p/ are distinct English phonemes, because *bat* and *pat* mean different things in English. In Arabic, by contrast, there is no corresponding difference between these two sounds, and a [p] sound, if it is uttered, is not taken to distinguish a word from a word with a [b] sound; thus in Arabic there is no phoneme /p/. (In linguistics, 'raw' sounds are transcribed between square brackets, and the phonemes of a language are transcribed between oblique slashes.)

Recursion: A property of computational operations which 'call themselves'. A classical formulation of the mathematical notion of a factorial is recursive in this way. The factorial of 1 is 1, and the factorial of any other number is the product of that number and the factorial of the number immediately below it.

Tabula rasa: A blank slate, the expression used to describe the most extreme form of empiricism, claiming that literally all knowledge comes from experience. But any slate, even a blank one, has some properties determining what can, and cannot, be 'written' on it by experience.

Theory of Mind: Put simply, the ability to understand that another organism is an agent just like oneself. Severe autistics lack a Theory of mind.

CHAPTER 15

Why teach? The evolutionary origins and ecological consequences of costly information transfer

Livio Riboli-Sasco, Sam Brown, and François Taddei

15.1 Introduction

Teaching is part of daily life for us as researchers, professors, or students. Teaching is defined in Wiktionary as 'To pass on knowledge', to deliver skills to someone else (see also Box 15.1). Teaching is, among humans, very often a profession, and people are paid to teach. The salary is one of the most obvious benefits for teachers of performing this activity, but it might not be the only one and is probably not the reason for most transfer of knowledge that occurred in our evolutionary past. On the other hand teaching is costly in time, in energy, and benefits pupils, or at least should benefit them if they acquire relevant skills.

In modern societies, organizations have been set up to rule this activity (schools, universities, etc.). One cannot understand our human societies, our activities, and the state of the art of our technologies without teaching. Teaching in this common 'professional' human sense is a very specific behaviour that has received little attention from biologists, and even less from an evolutionary perspective. It is simple to observe that teaching exists in humans far beyond the stereotypical institutional confines of the classroom. For example it is quite easy to observe that young children aged 6 to 10 can teach each other their recent scientific discoveries during class breaks rather than play (Riboli-Sasco *et al.* 2005). But is teaching confined to complex human societies or can we find teaching elsewhere?

Caro and Hauser (1992) gathered observational and experimental data suggesting the existence of 'teaching' in non-human animals. They propose a simple operational definition of teaching, without any demanding cognitive prerequisites such as an attribution of mental states, or sensitivities

Box 15.1 Definitions of teaching

1. Teaching for evolutionary biology: transfer of information, costly for the sender and beneficial for the receiver, that can be further re-sent by the receiver. The information is not lost by the sender when sending it.

2. Teaching as defined by Caro and Hauser (1992): An individual actor A can be said to teach if it modifies its behaviour only in the presence of a naïve observer B at some cost or at least without obtaining an immediate benefit for itself. A's behaviour thereby encourages or punishes B's behaviour, or provides B with experience, or sets an example for B. As a result, B acquires knowledge or learns a skill earlier in life or more rapidly or efficiently than it might otherwise do, or that it would not learn at all.

to a pupil's skills. In their review, they argue that narrow definitions of teaching, derived from adult–infant interactions, have prevented us so far from observing teaching in non-human species. We aim to enlarge on Caro and Hauser's (1992) study by asking evolutionary questions concerning the origins and maintenance of teaching in humans and elsewhere. Understanding how teaching evolved is then likely to be important for comprehending many aspects of evolution of humans and other species. Sticking to a very simple definition of teaching, following from Caro and Hauser (1992), we view teaching as a social transfer of information that imposes direct costs on the sender and potential benefits to the recipient. Teaching can then be studied within the framework of evolutionary biology through the analysis of cost and benefits of teaching behaviours on individual **fitness**.

15.2 From social learning to teaching: infectious transfers of information

Acquiring information from the environment allows behavioural adaptation to the conditions present in this environment. This behavioural plasticity can have a clear selective advantage as it allows a much quicker adaptation to fast changing environments than genetic adaptation (Meyers and Bull 2002). Individuals can also acquire information from each other. Social learning, defined as acquiring information from the behaviour performed by other individuals in their **social network**, has received much attention (Jablonka and Avital 2000). This acquisition mainly happens through observation and imitation. Innovations can appear at different steps of this process by errors in imitation or by the emergence of a new behaviour. Social learning then allows innovations to flow among individuals and across generations. Classical examples include Japanese macaques learning to wash and salt potatoes, and of birds in the United Kingdom opening milk bottles (Reader and Laland 2003).

In most of the social learning literature there is active acquisition but no active sending of information. Teaching can then be defined as a modified, costly behaviour resulting in the sending of information, in the direction of the learner. Caro and Hauser (1992) reviewed anecdotal and quantitative data showing modification in behaviour that has no direct benefit for the 'teacher' but allows the 'learner' to acquire knowledge or perform a skill earlier in life or more quickly and efficiently than it might otherwise do. We consider these behavioural modifications as 'sending of information', information being considered in a broad sense.

While social learning exhibits a clear advantage for the individual performing it, teaching adds a cost to the teacher. These costs may be difficult to measure empirically: the efficiency of prey catching behaviours can be reduced when training youngsters, as for example in meerkats (Thornton and McAuliffe 2006) or more time is spent to perform a single action, as for example in ants (Franks and Richardson 2006)—two cases where researchers used the term 'teaching' to describe these behaviours. Conceptually, one can imagine a continuum between costly teaching and a cost-free transfer of information without benefit for the sender. While costly to the teacher, the active transfer of information can induce an exponential dynamic to the spreading of information within the social network, as every sender can send information to many receivers who then can also send it to multiple further receivers, similar to 'infectious' dynamics (Fig. 15.1). Data exist on the temporal and

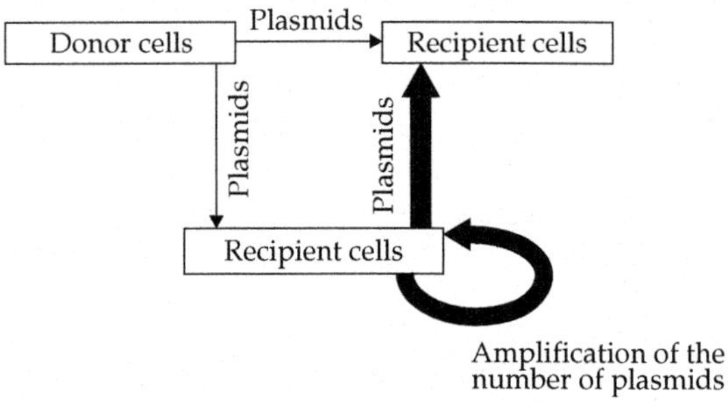

Figure 15.1 Model for the amplification effect, adapted from Dionisio *et al.* (2002) *Genetics* **162**, 1525–1532. Consider three bacterial populations, X, Y, and Z, living in the same habitat. Suppose that Y cells bear a conjugative plasmid. The arrows represent the conjugation events: larger arrows represent higher efficiency of the plasmid transfer. The plasmid from Y is going to infect both X and Z plasmid-free cells. If the conjugation rate among X cells is high, the plasmid number among them will amplify. Following this, plasmids from X cells will be massively transferred to plasmid-free Z cells. Copyright © 2007 by the Genetics Society of America

geographical propagation of milk bottle opening behaviour in birds in the United Kingdom. However, this behaviour seems to have spread through social learning and not through teaching (Hinde and Fisher 1951). If similar data on teaching were to be made available it would be very interesting to compare dynamical properties.

Situations of rapid and exponential behavioural transmission have been observed in humans. For example, 'flash mobilization' or 'flash mob' are aggregation behaviours that spread through text messages on mobile phones that can lead to thousands of persons meeting on a square, within a few hours of the first message being sent (http://en.wikipedia.org/wiki/Flash_mob). Interestingly, this new class of behaviour is itself spreading by different routes including the Web and this chapter. Actually it is clear that new technologies of communication including writing, printing, TV, Internet, etc. play an important role in allowing these ideas to be taught differently in general with a much higher rate and accuracy of transmission.

Humans are not the only species likely to be subject to autocatalytic information exchange. If we consider DNA as information (Maynard Smith 1999), then transfer of plasmids (mobile genetic elements) between bacteria is also transfer of information (see also Chapter 16). Such transfers require specific activity from the donor and the cost for the donor is inherently linked to the ability to transfer (Dionisio *et al.* 2002). Using a population dynamical model of producer (cooperator) and non-producer bacterial lineages mediated by plasmid transfer, Smith (2001) argued that the infectious exchange of information inducing cooperative traits has important selective benefits via the maintenance of the production of group beneficial traits, for instance secreted virulence factors. More precisely, for virulence factors secreted extracellularly by bacteria, **selection** within hosts may favour mutant strains of the pathogen that do not produce the virulence factor themselves but still benefit from factors produced by other members of the pathogen population within a host (Brown *et al.* 2002) (for related issues see Chapters 1 and 2). When this occurs, infectious transmission among bacteria may favour pathogen strains that can reintroduce functional copies of virulence-factor genes into cheaters via horizontal transfer, forcing them to produce the virulence factor (Smith 2001). Smith (2001) focused on the maintenance of microbial social traits, but we can enlarge his scope to teaching in the broad sense defined above, that is transfer of information that offers new skills to the recipient of this information.

In the following section we present a generic version of Smith's (2001) model, focusing on the teaching of any cooperative behaviour. In the discussion we will consider a number of complications and implications that follow from this theoretical first step.

15.3 A simple model of the evolutionary ecology of information transfer

We briefly analyse a simple population dynamical model to explore the competitive fate of individuals that invest variably in cooperation and in information transfer (modified from Smith's 2001 model of bacterial social behaviour mediated by mobile genetic elements). Specifically, we track the fate of two distinct classes of individuals: 'teachers' and 'students'. Each class of individual is characterized by a potentially distinct level of cooperation (variable levels of investment in, and reward from, a shared public good), and potentially distinct abilities to influence the cooperative behaviours of others and to be influenced by others. Labelling these classes T and S (for Teachers, and Students, respectively, $N = T + S$, for total population), the joint dynamics of our model are summarized as:

$$
\begin{array}{lcccc}
 & (1) & (2) & (3) & (4) \\
\mathrm{d}T/\mathrm{d}t = & T(r_T[N] & -x & = d \quad T/N) & +aST \\
\mathrm{d}S/\mathrm{d}T = & S(r_S[N] & & = b \quad T/N) & -aST
\end{array}
\qquad (15.1)
$$

The model terms are organized by columns of related terms, thus column (1) gathers the demographic terms, capturing underlying birth and death processes, column (2) captures the costs of investment in cooperation (environmental change), column (3) captures the benefits derived from cooperators (with potentially different degrees of award differentiated by $d > b$), and column (4) captures the infectious process of teaching, whereby students gain information from teachers, and become teachers themselves. Following Smith's (2001) inspiration, we will begin with a simple demographic appropriate for plasmid-bearing bacteria, allowing a vertical transmission of T and S states ($r_T[N] = r_S[N] = (1-N/k)$ with k being the carrying capacity). However, the following results would also apply for any case where $r_T[N] = r_S[N]$, e.g. change in a single generation, $r_T[N] = r_S[N] = 0$.

If we consider a world where only T have the ability to induce an environmental change without being able to change the behavior of others ($a = 0$), the above model simplifies to

$$dT/dt = T(r[N] - x + d\,T/N) \quad (15.2a)$$

$$dS/dt = S(r[N] + b\,T/N) \quad (15.2b)$$

where T are a cooperative class (paying x to create a social benefit T/N), and S are both non-cooperative and non-teachable but have a social benefit which comes from T's cooperation. When there is no differential access to the public good generated by the innovation ($b = d$), we have a classic 'tragedy of the commons', a multiplayer generalization of the prisoner's dilemma where non-cooperation is always favoured in the absence of population structure (Frank 1998; Nowak 2006). Accordingly, the above model has a single stable equilibrium of S at their carrying capacity k (the carrying capacity defined by details of $r[N]$) and $T = 0$. The cooperative trait is lost. If, in contrast, there is a differential access to the public good generated by the cooperators, in particular with the cooperators gaining more from their public good than others ($d > b$), then a population of pure cooperators can be locally stable if $d-x > b$. However, no matter how great the difference between $d-x$ and b, a population of pure non-cooperators S is also locally stable (thus the dynamics are bistable; in game-theoretic terms we have a stag-hunt game; empirical examples include chemical weapon production in bacteria (Brown *et al.* 2006). Thus even with a superior adaptation to the benefits they create, cooperators cannot invade from being rare (see related synergy concept, Chapter 11).

The condition $a = 0$ studied here corresponds to many possible scenarios. One of the more particular scenarios is the case where the cooperative T lineage can teach (they have the capacity to exchange information) but the S lineage cannot learn (see also parasitism impairing learning, Chapter 10). Therefore there is no transfer of the cooperative behaviour from T to S and thus there is

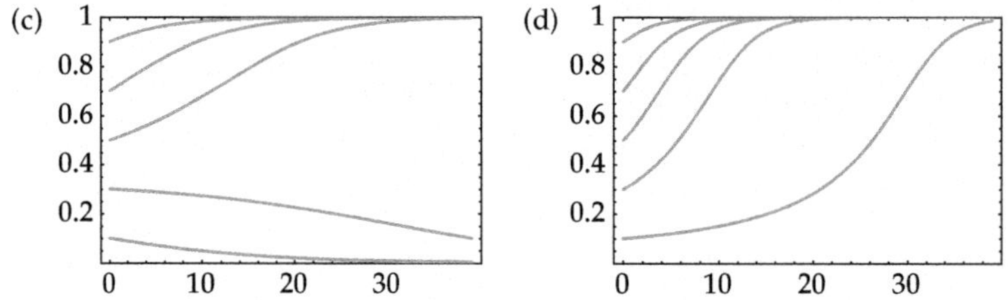

Figure 15.2 Population dynamics of cooperative teachers and students. Lines represent the frequency of cooperative teachers T through time. When there is no differential access to the public good ($b = d = 1$), cooperators cannot be maintained in the no teaching ($a = 0$) conditions (a). With teaching ($a = 10^{-7}$), cooperators dominate, even when starting from rare (b). In the condition when cooperators gain more from their public good than others ($b = 1$, $d = 1.3$, that is $d >$ b and $d-x > b$), without teaching ($a = 0$), we recover classic public goods results, where cooperators can be maintained if they not rare at the beginning (c). With teaching ($a = 10^{-7}$), cooperators dominate, even when starting from rare (d). Other parameters are $k = 10^6$, $x = 0.1$.

no teaching. It is important to stress here that teaching does not only depend on capacities of sending information, but also on the (possibly co-evolved) capacity of receiving information. So in certain circumstances ($a = 0$; $d > b + x$, T initially sufficiently common), the lineage with teaching capacity, T, can dominate and exclude S, without any teaching taking place (Fig. 15.2c). In this case, the information transfer ability that therefore increases merely hitch-hikes on the differential adaptation to the modified environment ($d > b$).

Now we consider a world where teachers are able to influence ignorant students ($a > 0$; teachers can teach and students can learn), returning us to the full model described above in model (15.1):

$$dT/dt = T(r[N] - x + d\,T/N + aS) \qquad (15.3a)$$

$$dS/dt = S(r[N] + b\,T/N - aT). \qquad (15.3b)$$

Given an equivalent access to the public good generated by teachers ($d = b$), teachers alone become stable when $ak > x$ (k being the carrying capacity of students alone), irrespective of their initial frequency. Note this advantage to teachers is nothing to do with the public good they produce (the above result holds even for negative d and b, i.e. if 'teachers' are destroying the shared environment), being driven purely by the infectiousness of the teacher state (Brown *et al.* 2006). Thus infectious transfer can rescue cooperation (Smith 2001). Given a differential access to the public good generated by teachers (with $d > b$), then teachers alone become more readily stable (when $ak + d > b + x$).

The models (with and without infectious transfer of socially beneficial information) highlight the potentially enormous social gains from a process of infectious transfer of elements coding for cooperative traits (Smith 2001). However, this benefit alone does not guarantee maintenance or 'evolvability', as we shall now discuss.

15.4 Discussion

15.4.1 Infectious transfers with no cost

The model (15.1) outlined above is a simple way of exploring some of the questions we stated at the beginning of this chapter. It illustrates the potentially large benefits of establishing a teaching facility in order to guard against social cheats, yet it leaves many open questions and remains to be tested. The second set of equations (15.2) illustrates that if locally dominant, a producer lineage, that is a cooperator contributing by producing a public good, can exclude non-producers in the absence of infectious transfer ($a = 0$). However, the gain to the producing lineage is not intrinsically linked to any assumed ability to send and receive information; rather this ability if it exists would merely hitch-hike on the additional adaptation the lineage carries to the newly engineered environment.

The question then becomes, when does it pay an individual to increase transmission of information? If we assume some low positive baseline a' of information transfer between cooperators and non-cooperators (i.e. due to social learning), then cooperators can overtake non-cooperators from any initial frequency if $a'k > x$, as illustrated above. However, we have no a priori reason to assume the a' is positive, it is quite reasonable to assume in contrast that the net direction of infectious transfer consists of cooperators (paying their social cost x) copying non cooperators (without cost x), leading to the potential for non-cooperators teaching non-cooperation (for instance through the transfer of **selfish** plasmids).

Within human populations an analogy can be made between these 'selfish plasmids' and 'selfish **memes**' as defined by Dawkins (1976). Numerous examples of memes with 'bad' consequences for human life, that is reducing fitness, have been postulated: for instance 'suicide' and 'dangerous games' memes have been reported to spread among young people, leading to epidemics of suicides or of behaviours that increase the death rate of those that play games such as 'the choking game' (Urkin and Merrick 2006). These costly and infectious behaviours lead to a series of questions for future research. Is teaching always an adaptation? Does teaching information transfer in bacteria rely only on the 'selfishness' of some plasmids that spread among their cytoplasm? Given the parallel we drew between bacterial plasmid transfer and teaching one could ask similarly: does teaching in humans rely only on the 'selfishness' of some ideas that spread among our minds? Below we will briefly outline many unanswered questions concerning the (co)evolution of the transmitted 'lesson' content, and teachers and learners strategies (for alternative approaches see prestige effect in Chapter 1 and the synergy effect in Chapter 11, and regarding the adaptive value of psychotic–effective spectrum behaviours in humans see Chapter 13).

15.4.2 Infectious transfer with cost

Another important assumption in the above formal model is that transfer of information is cost free (or at least has a fixed cost subsumed in the cost of cooperation *x*). This is certainly not true in the general case and the model should therefore be expanded by taking into account this cost and the nature of this cost (e.g. what if the cost of teaching is dependent on the number of individuals that are being taught? Then a rare teacher surrounded by ignorants may face a large cost of teaching). Adding either an additional constitutive cost (construction of teacher phenotype) with or without variations in efficiency of teaching/quality of teachers or a variable socially imposed cost (amount of teaching performed) will affect the advantages to infectious cooperation presented in Smith (2001).

Furthermore it is only when taking into account this kind of cost that we reach our initial strict definition of teaching as a cooperative dilemma. Given teaching has direct costs and yields benefits to others, teaching can be linked to other altruistic behaviours, and is therefore potentially explicable by standard explanations of altruistic behaviours. **Kin selection** is one of the basic explanations for the emergence of **altruism** (Frank 1998), and therefore potentially for the emergence of teaching. Teaching to offspring or any genetically related individual is costly but, following **Hamilton's rule** can be beneficial if the pupils are sufficiently closely related to the teacher. The fact that in animals such as ants and meerkats teaching has been documented among kin suggests that kin selection can play a role in the selection of teaching. Yet teaching has also broader 'social' extensions; bacterial plasmid transfer and human teaching can be observed among non-kin. It is quite possible that teaching first evolved among kin and was later on extended to non-kin via direct reciprocity or reputation effects in the case of humans. Teachers could indeed benefit from teaching by other classical mechanisms such as direct and indirect reciprocity. Direct reciprocity corresponds to a situation where the teacher is, in one way or in another, directly 'paid back' by the pupils. This would be the case for instance if in turn the pupil were to teach something to the teacher. Indirect reciprocity is defined as cooperating or paying back those who did beneficial actions to others, but not necessarily directly to you. Indirect reciprocity relies on mechanisms of observation and recognition of those who did beneficial actions to others (Nowak 2006), allowing teachers to build a positive reputation.

The potential for 'teaching' information transfer within bacteria is interesting as it shows us that teaching can evolve without any reciprocity and reputation or mental state. Furthermore, as plasmid transfer can occur among very different species and even across kingdoms, classical kin selection cannot be the main driver of such a costly transfer of information. We have proposed above that information transfer to non-kin could evolve either as a selfish behaviour of the transmitted elements (e.g. plasmids) or as a way to recruit helpers in a cooperative setting if the transferred information codes for a cooperative trait such as the production of a secreted enzyme. We therefore suggest that the spread of collective action sustained by plasmid transfer in bacteria or teaching in humans can be a powerful mechanism to modify the environment, and thus be selected via its indirect environmental effect.

15.4.3 Teach to collectively alter one's environment

The notion of the extended phenotype developed by Dawkins (1982) to describe behaviours such as beavers building dams has been broadened through the concept of niche construction; any process in which an organism alters the environment. It has been recently argued that niche construction has an important role in evolution (Odling-Smee *et al.* 2003). Niche construction emphasizes feedback loops on **natural selection**: species can change the environment and thus the selective pressures they face. Organisms inherit two legacies from their ancestors, namely genes and a modified environment (Odling-Smee *et al.* 2003). Obviously many environmental dimensions can be affected by organisms and the niche construction framework has suffered from the broadness of its claims (Keller 2003). It is useful to distinguish modifications that are cost free for the individual and that do not affect it (e.g. photosynthesis modifying the oxygen concentration of the atmosphere on geological timescale) from those that are costly and which would affect the organisms performing such behaviour (e.g. a beaver building its dam) (see related arguments in Chapter 16). Interestingly, in the latter case the modification of the shared environment can present significant social dilemmas (Brown and Taddei 2007). Cooperating individuals might benefit from altering their environment but may also suffer from the competition of cheaters benefiting from this alteration without contributing to the costly action performed. Because of its infectious properties, teaching behaviours that lead to cooperative environmental change may further accelerate the feedback loops between alteration of the environment and natural selection.

A very common example shows us how niche construction and cooperation can be linked. Building a bacterial biofilm results sometimes (but perhaps not always; Xavier and Foster 2007) in a cooperative dilemma (see Chapter 2). Constituents of the biofilm are secreted by bacteria and biofilm formation is enhanced by natural conjugative plasmids, as conjugation involves firstly bacterial cohesion through pili formation, coded by transmitted genetic information, and secondly adhesion factors secreted after pili have recessed (Ghigo 2001). Non-producers or 'cheaters' can take advantage of the biofilm, for example gaining better access to oxygen at a liquid–air interface (a classic, or 'first order' cooperative dilemma) (Spiers *et al.* 2003). In this situation cooperators could be maintained for example if they were able to transfer plasmids bearing the genes enforcing recipient bacteria to contribute to building the biofilm, they could turn 'cheaters' into cooperators (Smith 2001) (see Chapter 1 for an alternative interpretation based on direct benefit). Interestingly the probability of transferring or receiving plasmids is very polymorphic among natural isolates (Dionisio *et al.* 2002). Variability in the transfer rate of plasmids can be explained by the cost to the sender of plasmid transfer (leading to a second-order cooperative dilemma). This second-order dilemma can be also stated as: when you are already a cooperator are you going in addition to teach and turn cheaters into cooperators or not? Questions for future research include: do second-order cooperators (teachers) bear a higher cost, because of double cooperation, or can they be selected because of complex interactions between the two levels?

15.4.4 Teach or punish?

There has been much recent focus on punishment, considered by some authors (West and Gardner 2004; Hauert 2007) as a second-order cooperative dilemma within humans, as punishers contribute to maintain first-order cooperation by punishing non-cooperators and as punishment is costly and its benefits are shared by everybody in the group when cooperation is enforced. Teaching could open new perspectives to reconsider the potential and non-exclusive role of second-order cooperative dilemmas, in particular where field data do not support the punishment hypothesis (Boyd and Mathew 2007). The studies on punishment encourage us to think about the diversity of 'phenotypes' among teachers, as there can be different kind of punishers (unconditional punisher/conditional/rational/random). The role of the diversity of 'teaching' behaviours needs further investigation. We suggest that a similar classification (unconditional/conditional/rational/random) could be used. In humans, studies have also shown that 'beautiful' teachers get higher ratings from students than less beautiful ones (Hamermesh and Parker 2005). This leads us to the question of the correlation between teaching activities and other kinds of social **signals**. A link can also be made to studies on **quorum sensing** (mechanisms that allow bacteria to sense their cell density). In some bacteria the transfer of information can be under the control of quorum sensing processes, taking place only when bacteria sense a sufficient density of potential recipients (Miller and Bassler 2001). In humans, is a piece of information worth being published in a news paper only when it will interest a sufficient quorum of readers?

15.4.5 Theoretical and experimental perspectives

Experiments should be developed on bacterial model systems that are easy to constrain and control and where individual parameters can be measured experimentally and used as parameters for a model describing the global dynamics. Indeed, such a framework linking modification of the environment and infectious transfer had already been developed when we were able to show that bacteria can use viruses that they carry within their genome to trigger epidemics among their competitors, improving the competitive environment of the virus-carrying lineage (Brown *et al.* 2006). Moreover, we hope to see new experiments on humans where the question of the evolutionary origin of cooperative behaviours has already received much attention, but where information transfer seems to have received much less attention. These two systems are extremely different, but as we have shown in this chapter they share similar features as in both cases one can observe costly information transfer and cooperation. Field studies and experiments on other systems would be useful to see what is the number of species that are able to teach, in particular those who teach when facing cooperative tasks and whether information transfer is needed to perform these tasks. Concerning theoretical models, we should go further in understanding the effect of the cost for the receiver and the sender in the information transfer process. Moreover, information could evolve, it could be honest or dishonest, faithfully transferred or not, aimed at recruiting cooperators or manipulating competitors (Brown and Johnstone 2001). Clearly, evolution of the ability to receive information could be counter-selected in this latter context. Yet, given the potential benefits of information exchange, an alternative would be to exchange information only with those that are trusted, leading to the question of the evolution of green beard dynamics (Jansen and van Baalen 2006).

Clearly, this field is only in its infancy and the 'why teach' question needs further investigation. The current models are still very primitive and do not take into account most of the complexity of human teaching. It draws links to many other questions related to communication among social organisms developed in this book. It would benefit from an interdisciplinary approach collecting data from psychology, science of education, ethology, ethnology, or web science. Such an interdisciplinary approach would help to inform us about the conditions in which costly transfer of information is performed in humans and in animals. Knowing what is taught to whom and by whom would help to develop more elaborate models and rigorous experiments.

Given the amount of time that we spend teaching or being taught, understanding the evolution of our ability to transfer information is not only interesting but could also be useful as a basis for improving our education systems. Furthermore, given the speed at which our communication technologies are developing, such an approach can contribute to more efficient use of the potential of Web 2.0, where everyone can start producing and transferring information, and Web 3.0, where 3D virtual worlds are developed allowing non-verbal transfer of information between avatars to happen.

Summary

If teaching is omnipresent in our knowledge societies, we know little about its evolutionary origins and we can hardly predict the outcome of today's ever faster speed of information transfer made possible by the emergence of information and communication technologies used in wiki, e-mail, or Web 2.0. To explore these issues, we reformulate the 'why teach' question by: why should an individual invest resources in transmission of information to another individual? A qualitative difference between teaching and other forms of altruism associated with material exchanges is that information copy number increases during teaching, allowing information to spread autocatalytically. We introduce models where such autocatalytic transfer of information can modify the behaviours of individuals and thus impact upon their production of public good, altering the shared environment. We discuss the evolutionary causes and ecological consequences of such dynamical processes that can be observed in organisms as diverse as bacteria and humans.

Acknowledgements

We thank our collaborators Julien Capelle, Sandrine Adiba, Bernard Hennion and Sophie Heinrich, as well as David P. Hughes for criticism and suggestions in the course of this work. We are grateful to Orange for its support to the 'Centre de Recherche Interdisciplinaire' through its 'Chaire' program.

References

Boyd, R. and Mathew, S. (2007). A narrow road to cooperation. *Science*, **316**, 1858–1859.

Brown, S.P., Hochberg, M.E., and Grenfell, B.T. (2002). Does multiple infection select for raised virulence? *Trends in Microbiology*, **9**, 401–405.

Brown, S.P. and Johnstone, R.A. (2001). Cooperation in the dark: signaling and collective action in quorum-sensing bacteria. Proceedings of the Royal Society B: Biological Sciences, **268**, 961–967.

Brown, S.P. and Taddei, F. (2007). The durability of public goods changes the dynamics and nature of social dilemmas. *PLoS ONE*, **7**, e593.

Brown, S.P., Le Chat, L., De Paepe, M., and Taddei, F. (2006). Ecology of microbial invasions: amplification allows virus carriers to invade more rapidly when rare. *Current Biology*, **20**, 2048–2052.

Caro, T.M. and Hauser, M.D. (1992). Is there teaching in non-human animals? *Quarterly Review of Biology* **67**, 151–174.

Dawkins, R. (1976). *The Selfish Gene*. Oxford University Press, Oxford.

Dawkins R. (1982). *The Extended Phenotype*. Oxford University Press, Oxford.

Dionisio, F., Matic, I., Radman, M., Rodrigues, O.R., and Taddei, F. (2002). Plasmids spread very fast in heterogeneous bacterial communities. *Genetics*, **162**, 1525–1532.

Frank, S. (1998). *Foundations of Social Evolution*. Princeton University Press, Princeton, NJ.

Franks, N.R. and Richardson, T. (2006). Teaching in tandem running ants. *Nature*, **439**, 153.

Ghigo, J.M. (2001). Natural conjugative plasmids induce biofilm development. *Nature*, **412**, 452–456.

Hager, H. and Hammerstein, P. (2006). Game theory and human evolution: a critique of some recent interpretation of experimental games. *Theoretical Population Biology*, **69**, 339–348.

Hamermesh, D.S. and Parker, A. (2005). Beauty in the classroom: instructors' pulchritude and putative pedagogical productivity. *Economics of Education Review*, **24**, 369–376.

Hauert, C. (2007). Via freedom to coercion: the emergence of costly punishment. *Science*, **316**, 1905–1907.

Hinde, R.A. and Fisher, J. (1951). Further observations on the opening of milk bottles by birds. *British Birds*, **44**, 392–396.

Jablonka, E. and Avital, E. (2000). *Animal Traditions: Behavioural Inheritance in Evolution*. Cambridge University Press, Cambridge.

Jansen, V.A. and van Baalen, M. (2006). Altruism through beard chromodynamics. *Nature*, **440**, 663–666.

Keller, L. (2003). Changing the world. *Nature*, **425**, 769–770.

Maynard Smith, J. (1999) The idea of information in biology. *Quarterly Review of Biology*, **74**, 395–400.

Meyers, L.A. and Bull, J.J. (2002). Fighting change with change: adaptive variation in an uncertain world. *Trends in Ecology and Evolution*, **17**, 551–557.

Miller, M.B. and Bassler, B.L. (2001). Quorum sensing in bacteria. *Annual Review of Microbiology*, **55**, 165–199.

Nowak M.A. (2006). Five rules for the evolution of cooperation. Science, **314**, 1560–1563.

Odling-Smee, F.J., Laland, K.N., and Feldman, M.W. (2003). *Niche Construction: the Neglected Process in Evolution*. Princeton University Press, Princeton, NJ.

Reader, S.M. and Laland, K.N. (2003). *Animal Innovation*. Oxford University Press, Oxford.

Riboli-Sasco, L., Eastes, R.-E., Pellaud, F., Capelle, J., and Sabuncu, E. (2005). L'effet récré, ou l'application à la didactique des théories évolutives de la propagation des idées. *Actes des XXVIIèmes JIES* (http://www.cognition.ens.fr/traces/ressources/articles/effet_recre.pdf).

Smith, J. (2001). The social evolution of bacterial pathogenesis. Proceedings of the Royal Society B: Biological Sciences, **268**, 61–69.

Spiers, A., Bohannon, J., Gehrig, S., and Rainey, P. (2003). Biofilm formation at the air–liquid interface by the *Pseudomonas fluorescens* SBW25 wrinkly spreader requires an acetylated form of cellulose. *Molecular Microbiology*, **50**, 15–27.

Thornton, A. and McAuliffe, K. (2006). Teaching in wild meerkats. *Science*, **313**, 227–229.

Urkin J. and Merrick, J. (2006). The choking game or suffocation roulette adolescence. *International Journal of Adolescent Medicine and Health*, **18**, 207–208.

West, S.A. and Gardner, A. (2004). Cooperation and punishment, especially in humans. *The American Naturalist*, **164**, 753–764.

Xavier, J. and Foster, K. (2007). Cooperation and conflict in microbial biofilms. *Proceedings of the National Academy of Sciences of the USA*, **104**, 876–881.

CHAPTER 16

Grades of communication

Ronald de Sousa

16.1 Introduction

The chapters in this volume illustrate the astounding variety and pervasiveness of communication in the living world. From individual alleles through bacteria and social species to human speech, communication is everywhere. The means of communication also include all channels, chemical, electrical, visual, tactile, auditory, and olfactory, whereby organisms acquire any sort of information at all. For some purposes, one might want to limit the relevant sense of 'communication' to transactions involving conspecifics: so far, when we write articles and books, we intend them only for human readers. But that is only a special case: we also need to take account of cases where individuals of one species affect the behaviour of another in such a way as to affect their own reproductive success and thereby their own genome. In such cases, the individuals of the second species are part of what Dawkins has called 'extended phenotypes' of the first (see also Chapter 10). Typical examples involve parasites that manipulate their hosts, or predator and prey that have co-evolved in the course of an arms race. But a concept may lose some of its usefulness if its application is excessively broad. True extended phenotypes must be carefully distinguished from the endless variety of mere effects that individuals of one species can have on another, without being reflected in the former's genome (Dawkins 1982, 2004). We need to be able to say what, in the interactions of cells, organs, or individuals, is *not* communication. What exactly, then, do all those phenomena have in common which may legitimately fall under the concept of 'communication'?

Let me start with the obvious. Where there is communication, there must be a sender *S*, communicating something, *X*, to a targeted receiver, *R*. This simple truism raises several questions: first, about the potential partners *R* and *S*; second, about the nature of *X*; and third, about the constraints on and the point of the whole process.

On the first question, our initial intuition might be that both *S* and *R* must be individuals. They need to be sufficiently separate for one to be in possession of information that the other lacks. The concept of a biological individual is a complex and elusive one, however (de Sousa 2005), and it is soon apparent that *S* and *R* don't have to be individuals in any strong sense. As several of the chapters in the present volume make clear, the entities engaged in communication can be part of a kind of 'superorganism', such as colonies of ants or bees (see also Chapters 2, 7, and 10) and a surprising number of cases of *intra-organismic* **conflict** and cooperation testify to the fact that parts of a single organism, or even of a single gene (see Chapters 12 and 13), can communicate among themselves to cooperate or compete.

On the second question, we can start by assuming that what *X* stands for at its most general is *information*; but the meaning of that word is in need of elucidation. What has come to be known as 'Shannon information' (see Box 16.1) affords a useful measure of *quantity* of information (Rheingold 2000, Chapter 6), but it is, as we shall see, notoriously inadequate for providing an assessment of its *quality* in terms of what communicators might be interested in.

As to the third questions, we should ask whether information counts as communication any time some X is transmitted from S to R, or whether further constraints should be imposed if the concept is to be of any interest. To understand the point of communication we should require, for example, that the participants in a process of communication derive some benefit from the process. But if so, must benefit accrue to both the sender and the receiver, or only to the sender? And what further conditions apply?

Once the issue of benefit is raised, the issues of reliability, error, and deception loom large. Language, in a famous quip attributed to Talleyrand, was given to the human race in order to enable us to conceal our thoughts. As if to echo that saying, Crespi (Chapter 13) notes that we use language to 'manipulate the thoughts of others'. And it is obvious from surveying instances of non-human mimicry in nature that it is not only humans whose messages are not invariably veridical. Intelligence, it is sometimes said, is an arsenal of weapons in an arms race, an essentially Machiavellian tool (Dunbar 1993), the real point of which is rarely the simple conveying of information, but rather the manipulation of others' responses.

This last observation challenges us to understand how, in the case of organisms to which we are not tempted to ascribe conscious intentionality, we might make sense of the application of the idea of 'manipulation': is it a mere metaphor or does it have literal application? What suggests itself here is that we need to allow for different grades of **signalling**, corresponding to different positions on a continuum of degrees of intentionality. At one end there may be a kind of signal that we can, without metaphor, ascribe to unicellular organisms, or even to their component parts. At the other end, there will be the fully fledged intentional communication of explicit language. The former offer examples of special purpose *functions*, while the latter have collected, over the long stretch of evolution, increasingly sophisticated devices serving an unbounded variety of individual purposes. Many of the special tricks of language are doubtless unique to it; yet we should not assume that we have lost the non-verbal aspects that marked the communication styles of our simpler ancestors. On the contrary, they may remain to constitute a mainstay of the pragmatics of language, with connotation, innuendo, irony, and other figures of speech and exploitation of context responsible for shifts or elaborations of meaning.

In Section 16.2 I begin with an outline of an influential account of 'non-natural meaning', as it

Box 16.1 Shannon information

A quantitative measure of information introduced by Claude Shannon (1948). Intuitively, if a message is considered as a series of random variables selected from a finite set, the information afforded by each variable is measured by the number of yes/no questions that must be answered to guarantee that the value has been identified. Hence the *Shannon entropy* of a message is the minimum average message length, in binary units or 'bits' (using base-2 logarithms), that must be sent to communicate the true value of the random variable to a recipient.

More formally, the information entropy of a discrete random variable X, that can take on possible values $\{x_1 \ldots x_n\}$ is

$$H(X) = E(I(X)) = \sum_{i=1}^{n} p(x_i)\log_2(1/p(x_i)) = -\sum_{i=1}^{n} p(x_i)\log_2 p(x_i)$$

where $I(X)$ is the information content or self-information of X, which is itself a random variable, and $p(x_i) = \Pr(X = x_i)$ is the probability mass function of X.

Equivalently, the Shannon entropy is a measure of the average information content the recipient is missing when he does not know the value of the random variable.

(Partially adapted from the article on 'Shannon entropy' in Wikipedia, at http://en.wikipedia.org/wiki/Shannon_information).

applies to human intentional communication, followed, in Section 16.3, by a brief summary of the best current philosophical solution to the problem of how to analyse teleology without intention. In Section 16.4, I give a very informal sketch of the notion of Shannon information, and explain why it is both indispensable and insufficient for a satisfactory account of communication. In Section 16.5, I address some of the conditions under which it is useful to describe a given transmission of information as a reliable signal, in terms of the different ways in which the process has a distinctive biological function. Finally, in a short conclusion, Section 16.6, I raise a couple of tentative and speculative questions.

16.2 Natural and non-natural meaning

In an influential 1957 article, the philosopher Paul Grice proposed an analysis of the notion of 'non-natural meaning' (Grice 1989). He started by noting a number of contrasts between (1) 'Those spots meant measles' and (2) 'Those three bells meant that the bus is full'. In the case of (2), but not (1), one can consistently infer that *someone meant to convey* something. One can speak of the *content* of what is conveyed in (2), and place it between quotation marks, as in 'three bells meant "the bus is full"'. But it would be nonsensical to say 'those spots meant "measles"', as we might say 'In French, "varicelle" means "measles"'. Furthermore, in the case of (2) one could go on to say: 'but it was a mistake, as the bus wasn't full', but it would seem odd to add to (1), 'but it was a mistake, as it was not measles'. (The mistake in (1), we might say, is made not by the sender but by the receiver.) Grice referred to the meaning in (1) as 'natural meaning', and to the kind of meaning alluded to in (2) as 'non-natural meaning'. The former involves inferences from perceptions of facts or events to correlated facts or events, including causal antecedents or consequences. The latter, on the other hand, involves an intention to communicate. In human language, that intention is crucial, and in later work by Grice and others the analysis of the role of intention reached truly daunting levels of sophistication and complexity. A representative sample of an intermediate level is the following definition (Grice 1989, pp. 99–100):

'U meant something by uttering x' is true if and only if (for some A and for some r):

(a) U uttered x intending
 (1) A to produce r
 (2) A to think U to intend (1)
 (3) A's fulfillment of (1) to be based on A's fulfillment of (2)

(b) there is no inference-element E such that U uttered x intending both
 (1′) that A's determination of r should rely on E and
 (2′) that A should think U to intend that (1′) be false.

Although further analysis found this characterization inadequate (Schiffer 1972), the example is sufficient to give some idea of the complex nexus of intentions involved in an unadorned case of someone meaning something by an utterance.

Quite obviously this isn't the sort of thing that can be involved when we speak of communication among bacteria, genes, or neurons. Nor can it account for birds or mammals 'signalling' to mates or predators. If we are to speak meaningfully of information and communication among organisms not suspected of being capable of formulating conscious intentions we need to find a way of cashing out what in such contexts can only appear as metaphors.

To do this requires two closely related but importantly different tasks. The first is to explain how teleology can be brought under the aegis of ordinary causality. The second is to show that a concept of teleology thus explicated is adequate to provide a theory of communication, spanning both the most basic types and the more elaborate forms of conscious intentionality.

16.3 Objective teleology

In the past 50 years philosophers of biology have successfully accomplished the first task. The work of Larry Wright (1973), refined and elaborated by many others and especially Ruth Millikan (1984, 1993), has established that the core concept of function or teleology can indeed be explicated in terms of an *aetiological schema*, applicable equally well, with minor adjustments, to the functions of human actions (including communicative acts such as uttering a meaningful sentence), of artefacts, and of biological organs. The aetiological schema is so named after the Greek word for 'cause', and its aim is to reduce teleological notions such as purpose, goal, or function to purely causal notions (Box 16.2). The idea is that the function of an act, object, organ or token of behaviour is distinguished among its many potential effects as the one that *causally explains its presence*.

In the case of an intentionally produced object or behaviour, this explanation typically refers to an intentional agent's pre-existing plan or purpose. In the case of an organism or part of an organism, where the functionality in question cannot be attributed to any conscious intention, biologists tend to be somewhat casual about cashing out talk of function or purpose. This is perfectly sensible, given the heuristic fruitfulness of talk of purpose and design in nature. Given a piece of anatomy or a mode of behaviour, the first thing to ask about it is what it is *for*, even though we are clearly not expected to infer that any actual purpose or design is involved. But talk of 'purpose' or 'design' needs to be cashed out. To effect such a cashing out by reducing it to causal terms is the point of the aetiological analysis.

In regard to non-intentional teleology, the aetiological analysis adverts to the effect causally responsible for the relative reproductive success of some lineages over others. The vertebrate heart, among other effects, both produces rhythmic sounds and circulates the blood. To say that the latter is its function, while the former is not, is to say that present-day vertebrates have hearts as a result of the advantage afforded in ancestral hearts by the circulation of the blood, and not by any advantage conferred by rhythmic sounds. It is important to note that this analysis does not require that everything that serves the interests of a particular organism on a specific occasion must be held to be an adaptation. As Sober (1984) has shown, the factors that confer selective advantage on a type of organism are those that the organism is selected *for*, but there will be many others that happen to go along with that without in themselves conferring a **fitness** advantage. These last are *selected*, but not *selected for*. What was previously selected without being selected for—as well as traits that merely resulted from random genetic drift—can become functional if, in changed circumstances, they have effects that confer new fitness advantage, and thereafter begin to be preserved by **natural selection**.

Box 16.2 Objective teleology and the aetiological analysis of function

There are two principal varieties of teleology: *goal* or *purpose*, and *function*. One can say of a tool that it has a function rather than a goal, but it was with the goal of serving such a function that the tool was designed. A goal, then, will commonly be a certain state of affairs, while a function will more likely be identified with a specific means of achieving that state of affairs. Where intentional actions and artefacts are concerned, functions are relative to the goals and interests of agents, and goals differ from one agent to another. But in biology, there are no agents, and so no real goals. We can, however, identify *replication* as a metaphorical 'goal' of nature. Whether a particular gene or set of genes is or is not more successful than another is a matter of fact which—however difficult it might be to ascertain—is not relative to any agents or interests. We can therefore identify purely objective teleological properties such as functions providing we analyse them in strictly causal terms. That is the point of the aetiological analysis (AA) which explicates the intuition that the function *F* of an element *X* (an organ or part or an organism) can be identified with the specific effect *because* of which *X* currently exists:

(AA) An existing element *X* has the direct proper function *F* if and only if:

1. *X* results from the reproduction of an antecedent element, ancestral *X*;
2. Ancestral *X* effected *F* in the past, in virtue of properties reproduced in *X*.

Such traits are first what S. J. Gould called 'spandrels', and when further shaped by natural selection they become 'exaptations' (Gould and Vrba 1982). Good examples are the signals involved in **sexual selection**, which may begin either as random individual preferences by females, favouring inheritance by their offspring of both (through the male) the character preferred and (through the female) the preference itself. Or else they can begin with a trait actually correlated to fitness, particularly where it is subject to allometric development (Cronin 1991, pp. 183–204). The first, which Helena Cronin calls the 'good taste' variant, exacerbates a trait that did not in itself have a selective advantage before it became the object of female preference. The latter, which Cronin calls the 'good sense' variant, originally functioned as a perceptible indication of the presence of a desirable trait. It therefore acted merely as a piece of useful information guiding mate choice, rather than a fully fledged signal. Once it becomes enhanced by sexual selection, however, it can be seen as a genuine signal.

The aetiological theory is well equipped to explain apparent outcomes of design where no intentionality or even mentality is involved. Many well-known examples of mimicry are of this kind. And so is camouflage, though as we shall see below there is reason to think that camouflage, though functional, should not count as a true signal. The function of mimicry and camouflage is to deceive, although no intention can be ascribed at all. The deceptive traits are merely the outcome of reproductive lineages of organisms among whom a higher degree of resemblance to a poisonous species or resemblance to the background afforded a fitness advantage. In such cases, the mimicry does not depend on any representation of the current situation in the organism in question (see also Chapter 4).

When there is such a representation which responds to changes in the immediate environment, we can speak of a ground level of intentionality. It presupposes no sentience in the usual sense of the word, but only its simpler ancestor, a capacity for detection. Higher levels of intentionality, more plausibly attributed to conscious mental states, are involved in the sort of communication made possible by second- and third-order representation, culminating in the process of human communication Grice describes. Thus children and some animals are said to acquire a 'theory of mind' which allows them to respond to their second-order representation of what is represented in the minds of others. Primatologists have found clear evidence that low-ranking animals can be aware of what can be seen by higher-ranking animals and modify their behaviour accordingly (Cheney and Seyfarth 1990).They have also found apparent cases of active deception which presupposes awareness of what the other would discover if placed in a position to do so (Tomasello and Call 1997). The anthropologist Robin Dunbar has suggested that humans are characterized by a fourth level of intentionality, involving the capacity to mention a reference made by an individual to a reference by a second individual to what a third thinks about a fourth (Dunbar 2004). It's not clear that such an achievement requires, as Dunbar intimates, more conceptual resources than are afforded by a clear capacity to go to a third level of representation. As we are able to speak of the content of another's mind, it would seem that one is *ipso facto* enabled to envisage higher orders of reference. But while that is merely a plausible supposition when what is in question are only non-intentional functional capacities, it is certain that once language is available to codify such representations and embed them in iterable syntactic structures, there is no clear theoretical limit to the number of iterations that become possible (see also Chapter 14).

16.4 Information: quantity and quality

So far the notion of 'information', as the X that gets transmitted from S to R, has been taken for granted. It is time to look at it more closely.

Let us again begin with a truism: what I already know conveys no information. This suggests a first approach to the characterization of information, as *a measure of surprise*. Learning something highly unlikely is maximally surprising and therefore maximally informative. One might be tempted, then, to identify information simply with the inverse of probability: the lower the prior probability of p, the more informative it is to learn that p is true. But there are several problems in the way before one can make this into a usable idea.

The first problem is that there are various ways of mathematizing the raw intuition. One little-noticed problem with the idea that information is inversely proportional to probability is that it appears to entail a paradoxical consequence. From the point of view of purely epistemic rationality, concerned exclusively with information and truth in abstraction from any other values (see Box 16.3), believing any proposition and believing its negation will come out to be equally rational.

Endorsing an improbable belief would, of course, be unlikely to pay off; but that would be offset by the large gain in information that would accrue if we turned out to be right. Rather as a perfectly fair bet has the same expected utility as not betting at all, the case of belief would work like this. Supposing I'm wondering whether to believe X, which has a probability of p. By definition *Not-X* has a probability of $(1-p)$. But if $(1-p)$ also the epistemic value of X and p the epistemic value of ~*X*, then the expected desirability of believing X [EU(BX)] is precisely the same as the expected desirability of believing *Not-X* [EU(B~X)], namely zero, or the desirability of believing neither:

$$\mathrm{EU(BX)} = [p \times (1-p)] + [(1-p) \times -p] = \mathrm{EU(B{\sim}X)} = 0.$$

In the first term, p is X's probability and $(1-p)$ is its information value if true. In the second, the two parameters are simply reversed.

To avoid this unwelcome result, the measure of information generally prescribed for Shannon information is not the inverse of probability, but its logarithm base 2. That is not a merely arbitrary dodge designed to avoid the awkward result just mentioned. The log base 2 can be thought of as the number of questions required to arrive at a solution if one is attempting to identify a single item in a structured set. Thus 20 binary questions will suffice to zero in on a single number between zero and 1 million. Provided that the state space of possibilities is suitably structured, then the number of binary units (or 'bits') of information represented by the identification of that one number in a million is determined by the maximum number of binary partitions needed to home in on it. This gives an *objective quantitative measure* of information.

Furthermore, this conception of information is linked to both the physics of thermodynamics and statistical theory by means of the notion of *entropy*. This can be illustrated in terms of the classical model of an ideal gas. Imagine two containers of equal size, linked by a passage currently blocked by a gate. In the right container there is a volume of gas, while the other is a pure vacuum. Now consider what happens when the gate is opened, from the three different perspectives of physics, statistics, and information theory.

From the thermodynamic point of view, the gas in the right container has been exerting a certain pressure on the gate. According to the classical corpuscular theory of gases, that pressure depends on the temperature of the gas, and is actually equivalent to the vector that represents the mean molecular momentum of the particles that constitute the gas exerting pressure on the gate. The pressure represents usable energy: a piston placed between the containers might use it to effect some work (Fig. 16.1, A). But if we simply open the gate, the gas will

Box 16.3 Epistemic rationality

A rational strategy maximizes the probability of success. But what kind of success is relevant? Practical or economic rationality looks to gains and losses, and in biology those get cashed out in terms of fitness. But in the context of information and knowledge, we can think of success exclusively in terms of the likelihood of being *right*, or *believing truly*. That is the point of view of *epistemic rationality* (Levi 1967).

Practical and epistemic rationality can be thought to conflict. It can arguably be of practical benefit to have a false but encouraging belief in one's strength, health, or ability, or to flee from what is falsely believed to be a predator on the maxim 'better safe than sorry'. For a recent discussion of when it might be practically rational to be epistemically irrational, see Stephens (2001).

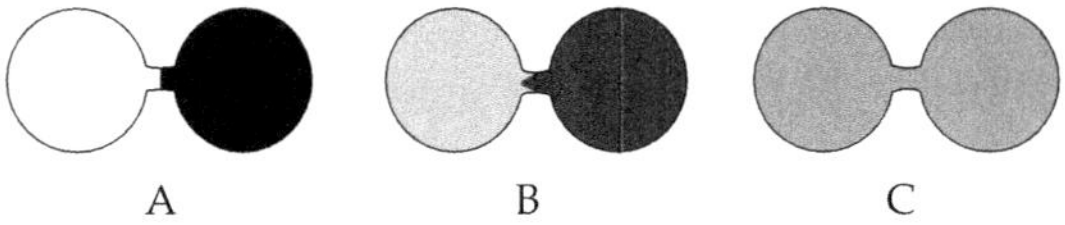

Figure 16.1 Movement of particles of an ideal gas between containers. See text for details.

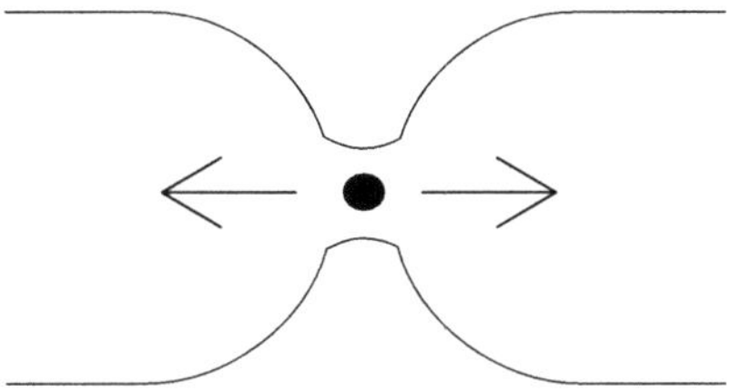

Figure 16.2 Movement of a single particle between containers. See text for details.

gradually diffuse into the second container (Fig. 16.1, B) until the pressure in both containers is equal (Fig. 16.1, C). At which point, in accordance with the second law of thermodynamics, there is no further possibility of using, from within the closed system described, the energy embodied in the motion of the particles of gas. From the physical point of view this final state is known as the state of maximum entropy.

Switching to the statistical point of view, consider a single particle positioned right between the two containers, at the location of the now open gate (Fig. 16.2). Since particles move randomly, it has an equal chance of heading left or right. But since all the particles are originally in the right container, the initial probability of the particle passing from the right into the left container is 0.5, while the probability of a particle going from left to right is zero. As more and more particles end up in the left container, the probability of a random particle going from left to right will increase, in exact proportion to the ratio of particles in the left to those in the right container. At the end of the process, for purely mathematical reasons, the probability of a particle passing from right to left will be precisely equal to the probability of its going the other way. That constitutes the highest point of statistical entropy. This is equivalent to saying that no state of the whole set-up is more probable.

This probabilistic interpretation of entropy provide a handy way to think of the second law, loosely paraphrased as ruling that the passage from order to disorder is always to be expected, as a mere consequence of the mathematical tautology that less probable states are likely to give way to more probable ones. The notion of probability provides the link to the informational point of view, in terms of the basic intuition about surprise with which I started. The state in which all the gas is in the right container (Fig. 16.1, A) is intuitively a state of 'order' as opposed to the state of 'disorder' represented by the fully diffused state of the gas (Fig. 16.1, C). That original state of maximal order, if it were to result from random fluctuations alone, would be highly surprising by dint of its being the least probable outcome of the random motions of the individual molecules constituting the gas. Just as maximal disorder, or maximal entropy, is equivalent to the most probable distribution of particles in an enclosed area, so minimal entropy can be identified with maximal information.

The foregoing considerations bear on communication in two ways. The first concerns the trade-off between the length of a message and what we might call its *informational density*. The second is that further constraints on the notion of information are needed before it can be of practical use in understanding concrete cases of communication between parties that have different interests and different states of prior information.

To understand the trade-off between length and density, note that while Shannon information is conveniently measured by counting binary units, it does not require to be packaged in such units. Implementation in the form of bits is indeed the obvious solution for information processed in computers, in which every basic unit can be regarded as either on or off. The simplicity of the basic vocabulary, which we can then conceive of as being made up of just two elements, 1 and 0, comes at the price of long messages. Thus 999 takes just three elementary signs in the decimal system, but takes up 10 elementary signs in the binary system, where it is represented as 1111100111. Adding the 26 letters of the English alphabet plus a space to the decimal digits 0–9 makes a total of 37 elementary signs, which affords the same amount of information in just two characters, and is, in fact, to an order of magnitude comparable to the number of phonemes in human spoken languages, which ranges between a dozen and a hundred (see also Chapter 14). At the other end of the spectrum from machine language, written Chinese language provides a striking example of a system capable of minimizing message length at the cost of requiring mastery of a very large number of distinct elementary signs. The first Chinese dictionary, commissioned by the Emperor Kang Xi, contained about 47,000 distinct characters, of which a literate but not erudite Chinese might be expected to know some 10,000. There is a trade-off between the memory storage required to distinguish 10,000 characters and the economy this allows in the length of each message. The way this trade-off plays out relates to informational density: the 0 or 1 of machine language represents single bit. A Chinese character, by contrast, compresses 16 bits of information, which is why two 8-bit 'words' or bytes are required to specify it in your word processor. Since an English character takes up just one byte, it is not surprising that a Chinese text invariably takes up much less space on the page than its English translation. On the other hand, the English alphabet can be memorized in an hour, whereas it takes years to master 10,000 Chinese characters. Since oral language is probably a more significant indicator than written language of the constraints under which our brains function, this suggests that the most efficient point at which the trade-off between memory and message length tends to settle requires somewhere between a dozen and a hundred basic elements.

Of course, complete messages are not built directly out of those elements. The power of language largely derives from the way that one or two intermediate levels of structure intervene between the atoms (phonemes) and complete messages (stories, arguments, speeches, poems, etc.), in the form of words, themselves articulated into phrases and sentences.

Here then is one hypothesis that can be derived from these abstract considerations about the notion of information. If a sender is to communicate information to a receiver, energetic factors will be involved, but they will be secondary to a sender and receiver's capacity to discriminate and produce a number of distinct elementary signs, and to their capacity to process signals of a given length and density. Humans naturally 'chunk' information when it involves more than a few elements that need to be held in immediate memory (Miller 1956), and we would find it extremely difficult fluently to read an English sentence transcribed into machine language, with a string of eight ones and zeros replacing each letter of the alphabet and punctuation. The brains and other storage devices of other communicating organisms will necessarily be subject to similar constraints.

Implicit in this discussion has been the assumption that most animal communication is based on a digital system of representation. That implication was carried in the very notion of a signal's density, which presupposes that the information capacity of any single element is finite. A repertoire of possible signals, like the set of phonemes or letters, typically constitutes a finite set, into which any variant along some continuum would be slotted as belonging to one or another of the elements in question. Clearly, however, there are some dimensions of animal signals that may vary in analogue rather than discrete degrees. The orientation and velocity of a bee's dance comes to mind, as representing the continuous factors of direction and distance (Michelson 1993). But it is noteworthy that neither human language nor the 'language' of the genes could function unless they were organized as digital systems. Both genes and language have a comparable density at the lowest level of analysis, comprising four bases (two bits) or 10 to 100 phonemes (four to seven bits) respectively. Both are then chunked into a larger but still relatively small number of 'words' specifying one of some two dozen amino acids in one case and a few hundred thousand words in the other, which in turn become the components of a huge number of possible protein strings or sentences. It would be interesting to know whether these two examples of genes and language, lying as they do about as far from one another on the scale of mentality, exemplify a very general requirement of signalling, favouring digital systems of representation.

I come now to the shortcoming of the notion of information just detailed. Intuitively, Shannon information fails to connect with our intuitive understanding of 'information', because it fails to relate to any of either the sender's or receiver's interests, and so fails to connect with the very idea that led to its characterization in terms of surprise. In short, Shannon information in itself *means* nothing *to* anyone. If I display for you the number 4,987,654,294,997, you are seeing something the prior probability of which was $1/10^{13}$ and so have just acquired 43 bits of information. You should be surprised! Your surprise is damped, however, by the fact that this 'information' is completely meaningless. But what does 'meaningless' mean? In itself, what is transmitted at any given synapse or between any two bacteria is also quite properly 'meaningless'. It is only its role in some larger process that has meaning. We need, therefore, to look more closely at the specific conditions that make it possible and profitable for senders and receivers to exchange information.

16.5 Reliable signals

Any organism equipped with some sort of sensor is capable of acting as an information receiver. But not all information acquired by such an organism should count as a signal. Every cell is capable of some form of tropism, but not all are the effect of signalling. Strategic issues arise only when there is interaction. Even then, the acquisition by one entity of information about another may not constitute communication. This is the case of information acquired by simple perception. If I want to drink, and see water, that's useful information, but it isn't any sort of communication. Perceptual information about other organisms is frequently of that sort. Take, for example, humans' ability to infer information relevant to mate choice from the texture of a person's skin, described by Craig Roberts (see Chapter 9), and contrast it with the human ability to infer socially relevant information from a person's posture and movement. The second appears more likely than the first to be an evolved form of functional signal, in the sense of being in some sense manipulated by the sender, though both may be likely to act as triggers for a certain behaviour. In the terms elaborated by Marc Hauser (1996, p. 24), the former type of signal typically pertains to the sending organism's 'resource holding potential' (RHP): the important point about it is that the RHP cannot, while an optional signal generally can, be manipulated at will to mislead. Thus a larger animal will obviously seem to be larger, which may well be crucial information. A smaller animal can send a misleadingly formidable image of its size only to a very limited extent, if it is able to spread its plumage or its fur, or inflate like a blowfish. But even then, as we shall see in a moment, it isn't clear that what is being sent is a signal in the full sense, rather than something like camouflage, which there is reason to think is not yet quite a signal.

When the issue arises of the relative value of information to a sender and a target of transmission, it raises strategic and economic issues. Some critics of sociobiology—including some biologists—have complained that economic concepts are inappropriate imports in biology, and introduce an ideological bias into our conception of the natural world. Margulis and Sagan, for example, have charged that 'vogue words like "competition", "cooperation", "**mutualism**", "mutual benefit", "energy costs", and "competitive advantage" have been borrowed from human enterprises and forced on science from politics, business, and social thought' (Margulis and Sagan 2002, p. 16). But the mere fact that concepts can bridge two domains says nothing about whether their claim is illegitimate in one or in the other. In fact, as Maynard Smith foresaw, economic and game theoretic concepts have proved to be of enormous importance in evolutionary biology (Maynard Smith 1984, p. 2000). In fact, we could say that economic concepts and game theory actually apply *literally* only to biology. Their application to human choices require psychological assumptions, notably that humans are utility maximizers, which idealize human motivation out of recognition. By contrast, when economic concepts are applied to biology, none of those questions arise assumptions are needed. Fitness provides an objective equivalent of 'interest' or 'advantage' requiring no intermediate psychological hypotheses.

Information is unique among all goods that can be acquired or exchanged, in that it alone does not share the 'zero sum' characteristic of other desirable things such as food, territory, or building materials. Unlike the proverbial cake, I can consume information, and give it to another, and still have it too. The strategic issues raised by information transmission are therefore distinctive in certain ways: when one entity dispenses information, the cost to the former does not amount to the loss of the information in question. But that doesn't mean it is cost-free. On the contrary, in their book-length study of animal signalling, Maynard Smith and Harper (2003) focus principally on the dependence of a signal's reliability on its cost. They first define a signal as follows:

> A signal is any act or structure which [1] alters the behaviour of other organisms, which [2] evolved because of the effect, and which [3] is effective because the receiver's response has also evolved.... the requirement that a signal evolved *because* of its effect on others distinguishes a signal from a 'cue', [defined as] any feature of the world, animate or inanimate, that can be used by an animal as a guide to future action.
>
> Maynard Smith and Harper (2003)

(compare Zahavi in Chapter 1).

The difference between a cue and a signal is illustrated in terms of the following example. Riechert (1978) studied contests between funnel-web spiders, *Agelenopsis aperta*, over web sites. She found that if there was a difference in weight between two spiders of 10% or more, the smaller spider retreated without risking a fight. A spider can perceive its weight relative to that of an opponent because the contests take place on the web. The spiders signal by vibrating the web, transmitting information about their size: a smaller spider can be converted into a winner by attaching a weight to its back. Thus size itself is not a signal by our definition. It did not evolve *because* of its effect on other spiders. However the act of vibrating the web *is* a signal if, as seems plausible, it evolved because of its effect on the behaviour of an opponent through the information it provides about size (Maynard Smith and Harper 2003).

As I understand it, the crucial difference between the vibration of the web in this case and the blowfish's inflated size is that the receiver of the spider's message, but not a predator watching the blowfish, has had to evolve or learn the meaning of what it perceives. Size is size, even where it is deceptive; but vibration is merely a indicator of size. On the other hand, most species of spiders are not equipped with a sufficiently effective visual sense to perceive the other's size, so that the lyriform organ in their feet is the most direct sensory channel available to them to sense the size of anything on the web (David Hughes, private communication). The case cited by Reichert must therefore count as a signal *only if* it is the outcome of an arms race in which spiders have evolved the capacity to amplify their apparent size by vibrating the web in a misleading way, whilst also having refined their capacity to infer real size from vibration.

Following Zahavi and Zahavi (1997), Maynard Smith argues that the costliness of a display can provide a warrant of its reliability. That isn't always the case. Merely being costly and providing information to a receiver does not suffice to make a piece of behaviour count as a signal. *Camouflage* is a case in point, and we can see now why it doesn't count as a fully fledged signal. It is presumably costly to produce, and was selected for. It therefore meets the first two conditions of Maynard Smith's definition. But it fails the third: for the effect that camouflage has on the predator—making the latter less likely to eat it—was not itself selected for. Similarly, if a hare outruns a fox, that may be costly, and the necessary capacity has evolved in the hare. But the insufficient speed of the fox wasn't an evolved response. If, however, the fox has acquired a disposition to economize its own strength by not bothering to undertake the chase, then the hare's speed can count as a signal to the fox. Stotting in gazelles or the alarm calls of vervet monkeys in the presence of predators are costly both in terms of their intrinsic energetic expenditure and owing to the fact that they draw attention to themselves in such a way as to put them at increased risk. And, here again, they will count as genuine signals provided that there is some *evolved* effect on the receiver, causing the latter to save energy by abandoning the chase before it is even begun, as opposed to giving chase in response to the prey's apparent availability.

Once signals evolve to be potentially misleading, the game-theoretical perspective comes into its own. At the ground level, the fact that a signal is not necessarily reliable creates a measure of uncertainty. The frequency of its indicative and misleading occurrences could be assessed by a receiver equipped with the right kind of memory, and there could be different decision outcomes depending on the stakes implicated in different contexts. But where a certain probability of its being misleading gives the receiver the option of ignoring it, an arms race will give rise to second-order signals, in which some additional element is included to warrant reliability. As is obvious from the experience of ordinary conversation, merely asserting 'This is really true!' before reiterating a dubious proposition does not add credibility. Hence Zahavi's 'handicap principle', which applies particularly to predator–prey communication and to mating signals. The handicap principle posits that an animal can warrant the reliability of the signals by which it advertises health and fitness by indulging in an additional and costly display (Zahavi and Zahavi 1997). The cost of the display is itself an additional signal, providing a kind of warranty insofar as it proves that the animal displaying it has strength and resources to spare.

At first sight, this signal serves the prey rather than the predator, since it might deter the latter from pursuit. But Zahavi claims it also serves the predator in saving him a pointless pursuit, as in the case of the fox just mentioned. The situation is worth looking at more closely, however. For there is an important asymmetry between the two. Applying the aetiological analysis of function discussed above, it seems that the handicap serves a *strategy* in the case of the prey, but functions merely as a natural signal or *cue* in the case of the predator. The reason is apparent if we compare two counterfactuals suppositions. For the prey, the point of the handicap lies in the effect it has on the predator. If it did not influence the predator, the handicap would not have been selected. The effect on the predator therefore explains the presence of the handicap in the usual way that the presence of a trait is explained by its function. But from the point of view of the predator, there is nothing more to the signal than its capacity to provide useful information, indicating that the chase is not worth it. So while it may seem to be functional for the predator, it is so only in sense of being informative. The asymmetry resides in the fact that one party gets information from the other while the other actually manipulates the information.

To illustrate the point, consider the difference between the information provided to a human being by another's expression of emotion and that provided by a bruise. From a bruise, I can make the inference that the world affords a blunt object in the vicinity and that the person in whom the bruise is seen has had an encounter with it. From the expression of anger, I can similarly infer that the person in question has suffered some injury. (Let us set aside the further information afforded by the observation that the anger is or is not directed at me, in a way that can't be made sense of in the case of the bruise.) The crucial difference between the bruise and the expression of emotion is that it is reasonable to suppose that the indicative function of the latter, but not that of the former, has been selected for. In other words, the expression of emotion exists *so as to provide information*. A bruise, by contrast, is just an effect of burst capillaries, resulting in blood collecting close enough to the skin to be visible. Because it is visible, it can afford information; but it is unlikely that its visibility was selected for. Most likely, then, a bruise has no evolved signalling function. The bodily manifestations of emotion, by contrast, have functions: they are supposed to tell us something.

16.6 Some remaining empirical and theoretical questions

In contrast to the other chapters in the present book, what I have presented here is highly abstract. Little more, I fear, can be expected from a philosopher, whom a scientist can plausibly regard as a kind of scientific poacher, a free-rider who waits in his proverbial armchair for others to conduct serious research in the lab so that he will have something to talk about. But perhaps I can make amends by suggesting, in conclusion, yet another couple of questions the answer to which is not obviously available to simple cogitation, and to which, as far as I can see, answers have not explicitly been provided by the chapters in the present book. Both bear on the question of the potential practical uses to which the findings of animal communication might be put.

The first question is sparked by a recent report in *New Scientist* about a possible strategy for neutralizing viruses by trapping them in cells that lack the machinery the viruses need to reproduce (Ginsburg 2007). Viruses typically reproduce by hijacking the DNA in the cells they invade for their own reproduction. They respond to particular molecules on the surfaces of cells. These molecules must first act as cues as to the availability of the necessary reproductive machinery inside (though they can be regarded as genuine signals if they result from the fact that cells of that kind have, through a past process of natural selection, been manipulated by the virus). But if these same molecules can be attached to cells that do not contain the machinery in question, they will become *misleading* signals, by which a virus could be lured into a dead end. Since red blood cells have no DNA, a virus trapped in a blood cell will replicate no further. In confirmation of this idea, experiments using genetically modified blood cells bearing glycans on their surface succeeded in trapping a virus into complete extinction *in vitro*, and with notable if incomplete success *in vivo* (Asher *et al.* 2005). Finberg and his colleagues, according to Ginsburg, are now working on synthetic traps that would not require the genetic modification of blood cells. Would this be, in effect, a case of artificial mimicry? That depends on whether glycans (or other similar molecules attaching to the surfaces of cells) already have a function in communication. That isn't known for certain, but it seems possible that glycans on blood cells might indeed function to trap and neutralize viruses. Is this a case of communication or a case of mere camouflage? In its use as a virus trap, the DNA-free cell benefits only the sender; to the receiver it acts as a cue. Yet it can clearly be described as a type of non-intentional level deception. If we are able to use the effect in the manufacture of artificial 'virus traps', it might spark a further step in an arms race. That would happen if it subsequently modifies the behaviour of the virus. And that, in turn, is to be expected, since only mutant versions of the virus that avoid that particular marker will leave copies of themselves. In a case of this sort, perhaps it doesn't matter whether a genuine signal is involved or not. For the purposes of our manipulations, the causal properties of the molecules in question are what counts, not their original functions. But that view may be hasty. For if we are indeed able to classify such cases under the general heading of 'signals', or merely 'camouflage', rather than merely causal mechanisms, might it suggest analogies with other cases of the exploitation of signalling mechanisms that could yield a broader range of applications?

However that may be, the difficulty of deciding whether one is dealing with a genuine signal in these cases may be merely empirical, rather than theoretical. More thorough knowledge and technological refinements will answer the question. But some difficulties entailed by the mere complexity of the systems involved may be more resistant to empirical testing. I have in mind the question of whether the aetiological analysis endorsed above will scale up smoothly to large *networks* of communicating animals. It seems reasonable to assume that it will. But we may not be able to arrive at a precise formula to describe it, just as we can solve the three-body problem only by simulations and not by mathematical equations. We should perhaps also envisage the possibility that additional effects and properties might be emergent in very large networks of communication. Such very large networks have been discussed in the present volume, at both the 'low' and the 'high' ends of the living world. An example of the former is the quorum sensing of bacteria discussed by Diggle *et al.* (Chapter 2). At the other end of the spectrum are the social networks described in the work of Matessi *et al.* (Chapter 3). In such very large networks, it is conceivable that emergent phenomena might arise that are not clearly predictable on the basis of the teleosemantic approach I have endorsed here. Some of Stuart Kauffman's work on self-organization in complex systems (Kauffman 1995) suggests that such systems are subject to surprisingly strong constraints on a purely mathematical level. Similarly, the work of Albert-László Barabási and others leads us to expect certain emergent properties in the structure of very large unplanned networks. The degree of their connectivity, for example, seems 'naturally' to organize itself in conformity with power laws (Barabási 2002). At that level of complexity, we may find the sort of large effects from small causes that are characteristic of chaos. We may then need to use different techniques for understanding and predicting the spread of information in networks.

Summary

Philosophers interested in meaning have tended to look at the extremes of mere causality on the one hand and fully fledged 'non-natural meaning' in human language on the other. But the former (though not simple, as attested by the long and largely vain attempt of philosophers to analyse it) is too simple to count as information, while the complexity of the latter places it far beyond many other forms of genuine communication found in the living world, from bacteria to mammals. Those other forms of communication involve 'Shannon information' but aren't wholly captured by that notion. In this chapter I look at some of the work that biologists have done to construct a coherent concept of information able to span a wide spectrum of communication from such phenomena as 'quorum sensing' among bacteria to sophisticated infra-linguistic signalling in primates.

References

Asher, D.R., Cerny, A.M., and Finberg, R.W. (2005). The erythrocyte viral trap: transgenic expression of viral receptor on erythrocytes attenuates Coxsackievirus B infection. *Proceedings of the National Academy of Sciences of the USA*, **102**, 12897–12902.

Barabási, A-L. (2002). *Linked: The New Science of Networks*. Perseus, Cambridge, MA.

Cheney, D.M. and Seyfarth, R.M. (1990). *How Monkeys See the World*. Chicago University Press, Chicago.

Cronin, H. (1991). *The Ant and the Peacock: Altruism and Sexual Selection from Darwin to Today*. Cambridge University Press, Cambridge.

Dawkins, R. (1982). *The Extended Phenotype: the Gene as Unit of Selection*. Oxford University Press, Oxford.

Dawkins, R. (2004). extended phenotype—but not too extended: a reply to Laland, Turner and Jablonka. *Biology and Philosophy*, **19**, 377–396.

Dunbar, R.I.M. (1993). Co-evolution of neocortex size, group size and language in humans. *Behavioral and Brain Science*, **16**, 681–735.

Dunbar, R.I.M. (2004). *The Human Story: a New History of Human Evolution*. Faber, London.

Ginsburg, J. (2007). Weapons of mass deception. *New Scientist*, **196**, 43–45.

Gould, S.J. and Vrba, E.S. (1982). Exaptation: a missing term in the science of form. *Paleobiology*, **8**, 4–15.

Grice, P. (1989). Meaning. In: *Studies in the Way of Words*, pp. 213–223. Harvard University Press, Cambridge, MA.

Hauser, M.D. (1996). *The Evolution of Communication*. MIT Press, Cambridge, MA.

Kauffman, S. (1995). *At Home in the Universe: the Search for the Laws of Self-Organization and Complexity*. Oxford University Press, Oxford.

Levi, I. (1967). *Gambling with Truth*. Alfred A. Knopf, New York.

Margulis, L. and Sagan, D. (2002). *Acquiring Genomes: a Theory of the Origins of Species*. Basic Books, New York.

Maynard Smith, J. (1984). Game theory and the evolution of behavior. *Behavioral and Brain Sciences*, **7**, 95–126.

Maynard Smith, J. (2000). The concept of information in biology. *Philosophy of Science*, **67**, 177–194.

Maynard Smith, J. and Harper, D. (2003). *Animal Signals*. Oxford University Press, Oxford.

Michelson, A. (1993). The transfer of information in the dance language of honeybees: progress and problems. *Journal of Comparative Physiology A*, **173**, 135–141.

Miller, G.A. (1956). The magical number seven, plus or minus two: some limits on our capacity for processing information. *Psychological Review*, **63**, 81–97.

Millikan, R. (1984). *Language, Thought, and Other Biological Categories*. MIT Press, Cambridge, MA.

Millikan, R.G. (1993). *White Queen Psychology and Other Essays for Alice*. MIT Press, Cambridge, MA.

Rheingold, H. (2000). *Tools for Thought: the History and Future of Mind-Expanding Technology*. MIT Press, Cambridge, MA.

Riechert, S.E. (1978). Games spiders play: behavioral variability in territorial disputes. *Behavioral Ecology and Sociobiology*, **3**, 1–15.

Schiffer, S.R. (1972). *Meaning*. Oxford University Press, New York.

Shannon, C.E. (1948). A mathematical theory of communication. *Bell System Technical Journal*, **27**, 379–423, 623–656 (available online at http://cm.bell-labs.com/cm/ms/what/shannonday/shannon1948.pdf).

Sober, E. (1984). *The Nature of Selection*. MIT Press, Cambridge, MA.

de Sousa, R. (2005). Biological individuality. *Croatian Journal of Philosophy*, **54**, 195–218.

Stephens, C. (2001). When is it selectively advantageous to have true beliefs? Sandwiching the better safe than sorry argument. *Philosophical Studies*, **105**, 161–189.

Tomasello, M. and Call, J. (1997). *Primate Cognition*. Oxford University Press, Oxford.

Wright, L. (1973). Functions. *Philosophical Review*, **82**, 139–168.

Zahavi, A. and Zahavi, A. (1997). *The Handicap Principle: a Missing Piece of Darwin's Puzzle*. Oxford University Press, New York.

Concluding remarks

As befits a concluding essay, we question the progress achieved as a result of this edited volume and reflect on certain recurrent themes running through the multiple contributions. First, the questions: Has this volume delivered on its aim to synthesise the field of social communication? And to what extent is it justifiable to attempt a synthesis given the expansive and multifaceted nature of current communication research? In soliciting contributions and deciding upon the direction of this edited volume we were interested in examining the sociobiology of communication and asking whether common principles have been discovered, what proximate mechanisms can now be studied, and what major questions remain. In recent decades, there have been multiple technological advances that have allowed an ever increasing level of resolution when studying communication. Synthesising while concurrently experiencing a diverging flood of new knowledge therefore proved to be an interesting challenge for us as editors and we hope also for the contributors. Meanwhile in taking an overview of the chapters, we found a number of pervasive themes that seemed worth reflecting upon in some detail.

What was striking to us as editors was the very high level of interdisciplinarity that is a hallmark of each chapter. When we discussed possible contributions we had two principle criteria. We wanted authors that we felt would be able to summarise and discuss an area they were expert in. In addition, we wanted to include a broad range of systems in which communication is studied. We feel that each chapter delivered in these two regards, with the added bonus that the authors also communicated that their topic was a highly interdisciplinary one. So, one of the important take-home messages of this volume is that current research in communication spans multiple disciplines, and that this is necessary to elucidate its functional (why questions) and mechanistic (how questions) foundations.

Knowing that interdisciplinarity exists in current approaches is itself an important and encouraging message, because it suggests that the increasing compartmentalisation of the biological sciences into discrete disciplines does not need to hamper research in communication. That is, an ever more technical language, different interpretations of the same phenomenon, increasingly specialised knowledge of techniques and equipment, and publishing in disjunct journals do not have to preclude a common understanding.

The main principle that stands out from all chapters is that the different levels of selection (e.g. natural, kin, sexual, cultural selection) all have the potential of making communication efficient and signals reliable, but this often requires the suppression of possible elements of corruption. It is clear that economic cost-benefit considerations are behind every form of communication and that this leading principle now seems universally accepted.

Another pervasive theme of this book is the ever greater advances in achieving a proximate level understanding of communication. From the details in the chapters we are now aware, for some model systems at least, how proteins, chemicals and neurons interact to make communication work. In some cases, notably humans, we have even gained the first insight into the genetic basis of communication.

To understand the molecules and genes involved in such complex interactions as a bird displaying ownership of a territory is an important goal in its own right. However, an additional goal of proximate level investigations of communication is to identify if common features exist, whenever natural, sexual and cultural selection have asked for similar solutions when different means were available. Do the pathways leading to bird-bird communication (song using vocal cords, Chapter 3) share similar characteristics to those between rodent-rodent communication (proteins in urine, Chapter 6), insect-insect communication (pheromones, Chapters 5 and 7) and female crickets and humans selecting mates (song produced using the wings, or body posture, Chapters 8 and 9 respectively). Does the evolution of language with its supposedly conflicting origins (Chapter 13) share features with intragenomic conflict (Chapter 12)? Does intragenomic conflict obscure communication in the same way as parasite-associated conflicts within insect societies (Chapter 10)? And, when do those same parasites, by mimicking host signals to hide their identity, use similar or highly divergent proximate pathways (Chapter 4)? Finally, as the paragon of animals, does our own communication, which is mostly shaped by cultural evolution (Chapter 14) have shared characteristics with the "language" of genes (Chapter 16) that is shaped by natural selection? It is too early to decide upon such comparisons but there are encouraging signs that developments towards interdisciplinary synthesis are beginning to happen. Also the use of mathematical modelling illustrates that complex behaviours can be deconstructed to determine which elements of communication systems are crucial and which are secondary. A key example is the phenomenon of collective action, and its associated communication of many individuals at once, that is practiced by insect societies (Chapter 11). Perhaps the time is not far off that such modelling approaches can be united with the many advances in pheromonal and neurological research (Chapter 5, 7). An encouraging glimpse of the potential of such approach can in fact be obtained from this book by contrasting the complementary approaches of Chapter 2 and 15.

The final theme that we want to highlight is the broadness of current studies of communication: in the diversity of taxa that are currently investigated, in the multiple conceptual approaches that are considered, and in the many levels of communication that are being elucidated. We have been fortunate to have persuaded authors to deliver chapters ranging from bacteria, via insects, rodents and birds, to humans and to have them cover the different relevant levels of selection: the gene, the individual, and the group. Kin selection figures prominently, but is also challenged when the handicap principle is applied to 'cooperative' slime molds (Chapter 1). At the same time different levels of phenotype are addressed and leading theories of cooperation, conflict, sexual selection, linguistics and philosophy are evaluated in the context of communication. Chapter authors have discussed communication from bacterial biofilms to human societies, offering a rich palette of examples about bacterial and metazoan cells, proteins in mice urine, nerve clusters in the honeybee brain, manipulative pheromones, bird and insect song, communicating networks, collective actions and cheating in societies, and human language, teaching, internet-use and philosophy.

What we have learned from the different contributions to this book is that interactions among genes, cells, individuals, societies and cultures may indeed be partly obeying the same rules, because the evolutionary principles of communication affect and shape biological (and cultural) systems in a similar way. It was the broadness of approaches coupled with the multiple conceptual paradigms that lead us to choose the painting, *Per ici et ver lá*, by François Géhan: the diversity of forms represented from unfamiliar angles, as the cover for this book. We hope that readers may find other pervasive general themes besides those listed; or that they may merely dip into the book for key chapters of most relevance to their area of expertise, while peeking over the fence of what complementary fields are currently achieving, or use it as a reference source for the current state of the art in communication studies. Whatever the case, we hope that this volume conveys an image of an exciting vanguard that is taking us, as the title of the painting says, from here to there.

David P. Hughes and Patrizia d'Ettorre

Glossary

Altruism: This is not equivalent to the standard English definition. In evolutionary biology, an action performed by individual A that affects individual B is said to be altruistic if it increases the **fitness** of the recipient B, and decreases the fitness of A. Altruism usually involves *cooperation* (see also **mutualism**), the phenomenon of different species, or members of the same species, working together for a common goal. How can a behaviour that is individually costly evolve under a regime of **natural selection**? Richard Dawkins recently summarized four good Darwinian reasons for individuals to be altruistic (*The God Delusion*, p 251, Bantam Press, 2006). 1) Genetic kinship (see **Kin selection** and **Hamilton's rule**); 2) *reciprocity*: the repayment of favours given, and the giving of favours in 'anticipation' of payback. A helps B today, and then B may help A tomorrow; 3) the Darwinian benefit of acquiring a reputation for generosity and kindness, for example in humans; 4) conspicuous altruism as a possible way of buying reliably authentic advertising (cf. *Handicap Principle*).

Aneuploidy: A change in the number of chromosomes so as to have more or fewer chromosomes than an exact multiple of the haploid number. Aneuploidy can potentially cause chromosomal disorders such as Down's syndrome.

Badges of status: Phenotypic traits that animals may use to signal their size and dominance. They are a subset of signals of quality that reveal information about an individual's rank. The classic examples are signals used to settle dominance contests in flocks of birds. Adult male house sparrows possess a distinctive bib of black plumage, the size of which positively correlates with ability to win dominance encounters against other individuals. However, badges of status can be used in other contexts.

Central Place Foraging: The behaviour of a forager that must return to a particular place in order to consume its food, or perhaps to hoard it or feed it to a mate or offspring. Models of central place foraging have been one of the central components of *Optimal Foraging Theory* which argues that because of the key importance of successful foraging to an individual's survival, selection should favour foraging strategies that maximise the rate at which resources can be gathered.

Channel of communication: The medium (e.g. air, water) used to convey information from a *sender* to a *receiver*; it is the environment through which the **signal** (or **cue**) is transmitted. Signals can be transmitted in different **modalities**, e.g. visual, acoustic, chemical, vibrational. *Multi-modal* signals are characterized by the use of more than one sensory modality. The courtship behaviour of *Drosophila melanogaster*, during which both males and females use multi-modal communication involving visual, acoustic, olfactory, gustatory, and tactile signals, provides an example.

Commensalism:—see **Symbiosis**

Conflict: In biology, the situation where individuals, or other units such as groups or genes, have an actual or perceived oppositie interests.

Parent-Offspring Conflict: Is used to signify the evolutionary conflict arising from differences in strategies that maximize the fitness of parents and their offspring. While parents tend to maximize the number of offspring, offspring can increase their fitness by getting a greater share of parental investment, often by competing with their siblings. The theory was proposed by Robert Trivers in 1974 and extends the more general **selfish gen**e and **kin selection** theory.

Sexual Conflict: Occurs when the two sexes have conflicting optimal fitness strategies concerning reproduction, potentially leading to evolutionary arms races between males and females.

Genomic, or intragenomic, conflict: Describes a range of conflicts between different genetic elements within the genome of an individual. It can occur when these elements do not have aligned interests, and opportunities for selfish transmission exist (see also **selfish gene**). In most cases the separate elements of the genome, i.e. genes, chromosomes, do have aligned interests so that cooperation prevails.

Cue: A trait that may provide information although it has not evolved for that purpose (see also **signal**).

Epigenetic: (from Greek *epi-* meaning "upon") Mechanisms causing heritable changes in gene function that are stable over rounds of cell division but do not involve changes in underlying DNA sequences of the organism. Epigenetic processes thus link genotype and phenotype through various actions other than initial gene action or environmental effects (see **phenotypic plasticity**), allowing cells to stably maintain different characteristics despite containing the same genomic material.

Epimutation: Heritable changes in genes which are not due to changes in DNA sequence (See also **Epigenetic**).

Evolutionarily Stable Strategy (ESS): In game theory and evolutionary biology, an ESS is a strategy (e.g. a behaviour) that cannot be invaded or beaten by a mutant performing any alternative strategy once the ESS has been adopted by the majority of individuals in a population.

Eusocial: (from Greek *eu-* meaning 'well' or 'good') The pinnacle of animal social organisation. Eusociality is the special term used to describe the most advanced societies of ants, bees, wasps and termites (eusocial insects) having three main characteristics 1) overlapping generations; 2) reproductive division of labour and 3) cooperative care of young. In recent decades, a few other animals have been shown to approach the eusocial condition: some aphids and thrips, an ambrosia beetle, some shrimps and, among the vertebrates, two species of mole rats.

Fitness: A central concept in evolutionary theory. It describes the capability of an individual of a certain genotype to contribute to future generations, and usually is equal to the proportional representation of copies of the individual's genes in the next generation. If differences in individual genotypes affect fitness, then the frequencies of genotypes will change over generations; as the genotypes with higher fitness become more common. This process is called **natural selection** and leads to *evolution* understood as a change in the population's gene frequencies over time. Fitness can be defined in a number of ways depending on which levels of selection (gene, individual, deme) are involved.

Direct Fitness: Calculated by considering the number of genes resulting from an individual's own reproduction, i.e. the gene copies in that individual's offspring.

Inclusive Fitness: Considers an individual's own reproduction, plus the increase in the the next generation of number of copies of that individual's genes in related individuals. It is calculated using **Hamilton's Rule.**

Genomic Imprinting: A genetic phenomenon by which certain genes are expressed differently depending on which parent they are inherited from. These 'imprinted genes' are either expressed only from the allele inherited from the mother, or in other instances from the allele inherited from the father. Forms of genomic imprinting have been demonstrated in insects, mammals and flowering plants.

Hamilton's Rule: Is an inequality named after W.D. Hamilton who in 1964 published the first formal quantitative treatment of **kin selection** to deal with the evolution of apparently altruistic acts. These are acts that seem to be costly for the performer but beneficial for the receiver. The rule states that genes that cause an individual to behave altruistically will spread in a population if the average number of such genes increases, or, in mathematical terms, that **selection** should favour the evolution of altruism if $r \times B > C$, where C is the cost in fitness to the actor (personal cost to the altruist), r the genetic relatedness between the actor and the recipient (average coefficient of relatedness between altruist and recipient, giving the likelihood that the recipient shares a copy of the 'altruistic' gene identical by descent) and B is the fitness benefit to the recipient (benefit gained thanks to the altruistic act).

Haplotype: The set of alleles borne on one of a pair of homologous chromosomes, or on an un-paired genetic element such as a mitochondrial genome.

Hormone: A chemical substance produced in the body that controls and regulates the activity of certain cells or organs. Thus, a hormone can be a chemical signal that cells use to send messages to other cells (see also **Pheromone**).

Lek: A communal mating arena within which individuals (usually males) hold small territories which contain no resources and are used solely for courtship displays and copulation.

Meme: This term was coined by Richard Dawkins (1976) as analogous to *gene* but referring to non-biological entities (see also **Selfish gene**). Memes (or genes) are *replicators*, entities capable of producing copies of themselves. Thus, for example, a meme could be an idea that is passed on from one human generation to another; i.e. the cultural equivalent of a gene.

Modality:—see **Channel of communication**

Monogyny: A type of mating system. *Polygyny, Monandry, Polyandry, Polygamy* are terms used to describe mating systems, but they are used in rather different ways when applied to social insects or other organisms. For social insects, monogyny means the presence of only one egg-laying queen in the colony, while polygyny is the presence of multiple queens within a single colony. Polyandry in social insects refers to queens that mate with multiple males. Monogamy can be used to describe a colony structure in which a colony is headed by a single queen that has mated with just one male. Males of many social insects die immediately after mating, so that they are represented only by their sperm for most of the life of the colony, and most queens will not re-mate after an initial mating flight. The largest exception to this is the termites, in which a colony is headed by a queen and a king who remate frequently, and where polyandry via serial monogamy has been reported. In other organisms (particularly vertebrates), monogyny is understood too be mating of a male with only one female, while polygyny means mating of a single male with more than one female. Monandry and polyandry are used to describe mating of a single female with only a single, or multiple males, respectively. Monogamy and polygamy are mating systems in which an individual associates with either only one or more than one individual of the opposite sex to reproduce. The 'social' mating system may or may not correspond to the 'genetic' mating system, because *extra pair copulations* may occur.

Mutualism: A relationship where both partners have a net gain (see also **Symbiosis).** Mutualism involves *cooperation*. Cooperation is the behaviour between two or more biological units that increases the **fitness** of each. If each interacting organism gets an immediate net benefit from the interaction, or will pay an immediate net cost for not cooperating (the so-called *boomerang factor*: defecting is detrimental to the defector itself), then cooperation via *by-product mutualism* is expected to be favoured by **natural selection**. If, however, benefits can be gained but no immediate costs paid by defecting, then natural selection is expected to favour defection (as any individual that cooperates against defectors will pay a high cost). Cooperation with delayed pay-offs can be evolutionarily stable (see also **ESS**) if the interacting organisms are related, so that **kin selection** can operate, or if interactions are likely to be repeated, through reciprocal **altruism** or gain in reputation.

Network: A *social network* is the set of conspecifics with which an individual regularly interacts either directly, e.g. through aggression, cooperation, signalling or simple spatial association, or indirectly, e.g. by witnessing interactions among them (bystander). A *communication network* refers to the set of conspecifics that are within signal receiving range of an individual at any particular time.

Parasitism: A relationship where one partner exploits another, the latter suffering a **fitness** loss (see also **Symbiosis** and **Social parasitism)**

Pleiotropy: The influence on multiple phenotypic traits by a single gene. This can, for example, be because the gene codes for a product which is part of multiple pathways.

Phenotypic plasticity: The ability of an organism with a given genotype to express different phenotypes under different environmental conditions.

Pheromone: A chemical that mediates communication between members of the same species, as opposed to *allelomone*, which is a chemical that mediates communication between members of different species. There are two distinct types of pheromones: *releasers* and *primers*. Releaser pheromones initiate immediate behaviuoral responses in an individual upon reception, while primer pheromones cause physiological changes in an individual that ultimately result in a behavioural response.

Polygenic trait: A phenotypic character that is determined by the interaction of two or more genes and their environment. Examples are height, skin colour and autism in humans.

Polygyny:—see **Monogyny**.

Quorum: In standard English it means the minimum number of persons required to reach a decision. In biology, quorum sensing refers to the phenomenon whereby organisms communicate to achieve a consensus to enable a coordinated response. Examples are swarming in bees, house-hunting in ants, and biofilm formation in bacteria. For bacteria, where the term is most commonly used, the accumulation of 'signalling' molecules in the surrounding environment enables a single cell to sense the number of bacteria (cell density), so that the population as a whole can make a coordinated response.

Reproductive skew: It means an uneven sharing of reproduction within a social group. In particular, it refers to the distribution of reproductive output among co-breeding individuals, normally based on dominance rank or a particular phenotypic trait. High reproductive skew implies that few individuals (e.g. highest ranking ones) produce most of the offspring in a generation or season; while low skew implies a more even distribution of reproductive output among individuals.

Selection: The process by which favourable traits that are heritable become more common in successive generations of a population of reproducing organisms, and unfavourable traits become less common. There are many types of selection and concepts associated with the term:

Levels of selection: There is a hierarchy of biological organization from populations, through smaller groups, to individuals and down to genes. Selection can act at different levels of this hierarchy, but its strength may vary.

Natural (Darwinian) selection: Is the category most commonly intended when discussing selection, and its definition is essentially the same as that of selection above, with favourable traits being defined as those that best adapt an organism to its environment (see also **fitness**). The theory of evolution by means of natural selection laid down by Charles Darwin originally referred to the survival and differential reproduction of individuals.

Sexual selection: The process by which heritable traits within one sex that specifically increase the chance of being selected as a mate by the opposite sex (*mate choice*) become more common in successive generations. Such traits may at the same time reduce other components of **fitness**, such as survival. Sexual selection can lead to distinct differences in the appearance of the two sexes within a species, and can potentially lead to exaggerated traits if there is mate choice based upon the absolute size of a trait. For example, the tail of the male peacock is the result of sexual selection.

Individual selection: The process whereby traits become more common in successive generations because of the benefits they provide to individual survival and reproduction. Both **natural** and **sexual selection** can increase the relative representation of an individual's genes in subsequent generations. It is a term that is often used to distinguish the process from **kin selection**.

Kin Selection: The process whereby traits become more common in successive generations because of the benefits they provide to relatives that also possess those traits because of a shared common ancestry. Kin selection may operate through **altruism** according to **Hamilton's Rule**.

Social selection: Occurs whenever the **fitness** of an individual depends, in part, on the phenotype of its social partners. Social selection acts on *interacting phenotypes*; those phenotypes that have reduced or no meaning outside of a social context (such as social dominance or dominance status, courtship, and, of course, traits that function in communication).

Cultural selection: Is **selection** at the cultural level. *Culture* denotes beliefs and attitudes affecting behaviour which can be transmitted from mind to mind by imitation and/or learning. Cultural evolution may be much faster than genetic evolution and *cultural selection theory* investigates phenomena which can spread in a society (see also **Meme**).

Group Selection: Is thought to be the process whereby traits become more common in successive generations because of the benefits they might afford at the level of groups (multiple individuals that may or may not be related).

Directional Selection: Occurs when **natural** or **sexual** selection favours a single phenotype causing the allele frequency to continuously shift in one direction. Therefore the advantageous allele will increase in frequency independently of its dominance relative to other alleles, e.g. causing certain morphological states to become more common within a population.

Divergent Selection: Selection favouring forms that deviate in either direction from the population average, also called *disruptive selection*.

Frequency-dependent selection: Occurs when the fitness of particular genotypes is related to their frequency in the population. Negative and positive frequency-dependent selection denotes the cases where relatively rare and relatively common genotypes, respectively, are at an advantage compared with the other genotypes in the population.

Selfish behaviour: Is a behaviour that reduces the fitness of another individual and is the opposite of **altruistic** and *cooperative* behaviour. Usually selfish behaviour increases the **fitness** of the performer by for example, securing resources, but it can reduce fitness of the performer as well as the receiver, in which case it is called 'spiteful behaviour'.

Selfish Gene: Is the title of a book by Richard Dawkins (1976), which viewed (Darwinian) selection as acting at the level of the gene, rather than the individual or group. This is because it is only gene copies that are transferred between generations. A gene can be thought of as acting selfishly, relative to other genes in the population of individuals or even in the same individual, if it increases the success of its own transmission relative to other genes. The gene usually uses the phenotype to affect transmission, but some genes can transmit without the phenotype and are known as selfish DNA, or selfish genetic elements. The use of the term 'selfish' does not infer any conscious intent on the part of the gene.

Signal: Signals as characters (traits) that evolve in a signaller in order to provide information to a receiver, aiming to change the behaviour of the receiver to the benefit of the signaller (see also **cue**).

Social parasitism: A parasitic relationship in which the **parasite** exploits the common resources of a society rather than directly exploiting any one individual within that society. This can occur within a single species (intraspecific) or between two species (interspecific). The term has been used in human societies to refer to those who exploit the labour and/or capital of others without contributing to either. Biological examples can be found among many social insects, particularly ants, whose colonies can be exploited by a broad range of related and unrelated taxa that feed on the colony food stores or exploit the worker force for their own means. Also brood parasites, such as cuckoo birds, are a form of social parasitism.

Symbiosis: (from Greek *symbios*; *syn-* meaning 'together' + *bios* meaning 'life') is a term meaning living together. It was initially coined by Anton de Bary in 1879 to mean the "living together of unlike organisms". There is no universally agreed definition, and in this book we use it to mean the long-term association of organisms belonging to two or several species. This broad definition includes relationships where one partner has a net loss in fitness (**parasitism**), both have a net gain (**mutualism**) or where one partner gains, but there is no benefit or cost to the other (**commensalism**). Some people only use symbiosis in a narrow sense to refer to mutualistic relationships that are thought to be beneficial to both parties (and thus exclude commensalism and parasitism) (see also **Social parasitism).** The advantage of our broad definition is that it addressed all biological interaction of unilateral exploitation (parasitism) and bilateral exploitation (mutualism) in a single concept.

Index

2D:4D ratio 161, 163, 164
acoustic signals, rapid differentiation 149
Acromyrmex ants
  A. insinuator, sister-taxa social parasitism 67
  inquiline parasites 61
  intraspecific social parasitism 60
across-fibre pattern hypothesis, olfactory coding 122, *123*, 132, 134
action at a distance 172–3, 174
  social parasites 178
active cell death, slime moulds 5–6
N-acylhomoserine lactone (AHL) signal molecules 14, 16, 22, 24–5
  structure *16*
adaptive behavioural manipulation, parasites 177–8
addictive stimuli, use by social parasites 64–5
adoption procedure, *Maculinea* spp. 69
advertising information
  in colonial communities 45
  in family-territorial communities 44
  in lekking communities 46
  in pair-territorial communities 43
Aeromonas spp., QS systems *15*
aetiological analysis *278*
  application to large networks 286
aetiological schema 277–8
affective content **263**
African honeybee (*Apis mellifera scutellata*) 185
age of scent marks 99–102
aggregation, bacterial *12*
agr (accessory gene regulator) 17
Agrobacterium spp., QS systems *15*
agro-predation 59–60
AiiA 25
alarm calls 254, 259–61
alarm pheromones, honeybees 121, 130, 132
allelomones *120*
allo-feeding, babblers *3*
allomones *120*
altruism **291**
  in babblers 2–4, 8
  misdirected 181–2
  in slime moulds 4–6
  social insects 173
    effect of behaviour-modifying parasites 179–83
  in teaching 270
  see also Hamilton's rule
amplification effect *266*
analogue information systems 282
anarchistic workers 58
4,16-androstadien-3-one 92
aneuploidy **291**
Angelman syndrome 212–14, *234*, 241
animal architecture 172, 174
annual cycle, social insect colonies *176*, 182
Anopheles gambiae, olfactory receptors 127
antennae, honeybees *126*
antennal lobe, honeybees *124*, 125, 134
  anatomical differences between castes *128*
  processing
    drones 130–2
    queens 132
    workers 128–30
antennal nerve, honeybees 125
antenno-cerebral tracts, honeybees *124*, 125, 132–3
antibiotic production, bacteria *12*, 14
antibiotic resistance 11
ants
  agro-predation 59–60
  behaviour manipulation by parasites 177
  brood pheromones 64
  colony size, relationship to foraging success 193, *194*
  gestalt odours 60, 62
  inquiline parasites 61
  intraspecific social parasitism 60, 66–7
  Lasius niger, trail laying behaviour 201
  misdirected altruism 181
  multicomponent communication 87–8
  multipurpose pheromones 88
  pheromone trails 192–3, 201
    mechanistic model *195*, *196–7*
  raft-making, fire ants 175
  sensory trap exploitation 64
  sister-taxa social parasitism 61, 67
  slave-making 58, 67–8
  social parasites 70
  teaching behaviour 185
apes
  learning 251, 252
  mind-reading 259
  planning for future actions 257
  pointing behaviour 261
aphrodisin 111
apoptosis 5–6
apparent audiences **34**
  pair-territorial communities 43
appeasement signals 64, 65
  Polyergus rufescens 67–8
Arabian babbler (*Turdoides squamiceps*) *3*, 38
  altruism 2–4, 8
arbitrariness of sign, human language 250, 251, **263**
Archer, D. and Akert, R.M. 162
Asian army ant (*Leptogenys distinguenda*), misdirected altruism 181
Asian honeybees (*Apis dorsata*), bee-curtain formation 174
A-signal, *Myxococcus xanthus* 16
Asperger syndrome **226**, *234*, 241
  language problems 236
  relative strengths and weaknesses *235*
association networks *37*
asymmetric kin 210

attractiveness
  body 160–1
  dynamic versus static stimuli 163–4
  face 158–60
  kinesic information 162–3
  odour 161–2
  voice 161
audience **34**
audience effects **34**
  alarm calls 260
  in colonial communities 5
  in non-territorial communities 42
  in pair-territorial communities 43
auditory hallucinations 236, 239–40
auditory signals
  exploitation by social parasites *62*
  multiple components 89
autism/ autistic spectrum conditions 213, **226**, 232, 233, *234–6*
  contrast with psychotic-affective spectrum conditions 239
  extreme male brain theory 241
  language problems 236
  relative strengths and weaknesses *235*
  role of imprinted genes 241
  sex differences 242
  traits *235*
autoinducers
  autoinducer-2 (Al-2) *16*, 23–4
  bacteria 11, 14, *16*
  diffusion sensing explanation 26–7
autoinducing peptides (AIPs) 22
automata 215–17, 219
axillary secretions, as human pheromones 92

babblers *3*, 38
  altruism 2–4, 8
babbling, babies 252
backup signals 84
bacteria
  public goods production 17
  virus-carrying lineages 272
bacterial communication 11–12, 21, 27
  complexities 24–7
  discovery 14–17
  Gram-negative, QS systems 14–16, *15*
  Gram-positive, signalling molecules 16–17
  interkingdom 24
  interspecies 23–4
  intraspecies 22–3
  see also quorum sensing (QS)
bacterial cross-talk 24
badges of status 35, **291**
  pheromones 89
Baldwin effect 148
'bang-bang' mode of colony production *176*, 182
bark beetles (*Ips paraconfusus*), pheromones 86, 91
barn swallow (*Hirundo rustica*), rapid evolution *143*
beavers, dam building 172, 174
bee-curtains 174
  bees
  cold-seeking behaviour 178
  see also bumblebees; honeybees
begging behaviour, parasitized social insects 181–2
behaviour, human 162
  dynamic versus static stimuli 163–4
  honesty 164–7
  kinesic information 162–3
behavioural plasticity 140, 266
  facilitation of urban colonization 146
  relationship to brain size 146–8
  role in rapid evolution 144–6, 148–9
behaviour manipulation, humans 229
behaviour-modifying parasites, social insects 176–9
  effect on communication channels 179–86
belief 280
benefits of communication 200, 201–2, 205, 276
Bicyclus anynana, sex pheromones 86
bighorn sheep (*Ovis canadensis*), rapid evolution *143*
binary units (bits) of information 280
biofilms 11, *12*, 19, 25–6, 27
  cooperative dilemmas 271
bioluminescence, bacteria *12*, 14, 22
biosurfactant production, bacteria *12*
bipolar disorder 236, *237*
  role of imprinted genes 241
birds, brain size 147
bird song
  adaptation to urban environment *143*, 146, 151
  fairy-wrens 49
  great tits 47–8
blackbirds (*Turdus merula*) 37
  audience effects 43
black grouse (*Tetrao terix*), social system 41
blue-tailed damselfly (*Ischnura elegans*), rapid evolution *143*, 150
blue tits (*Cyanistes caeruleus*) 37
body, attractiveness 160–1
body language, human 162
  dynamic versus static stimuli 163–4
  honesty 164–7
  kinesics 162–3
bombykol, bombykal 83, 84
Bombyx mori, pheromones 83, 84, 119
bonobos **263**
boring by-products of infection 178, 179
boundary activities, social insects 185
brain, language processing *227–8*
brain size, relationship to behavioural plasticity 140, 146–8
'brain-worm' 177
Broca's area **226**, 227, 232
brood parasites 58
  manipulation of host signals 64
brood pheromones
  ants 64
  honeybees 122
brown adipose tissue, role in thermoregulation 216–17
bumblebees
  annual cycle 182
  Bombus terrestris, courtship behaviour 84
  parasite-induced behavioural modification 185
Burkholderia spp.
  B. cepacia, interspecific signalling 24
  QS systems *15*
burrower bug, pheromones *85*
burrowing owls (*Athene cunicularia*), defensive calls 148
burying beetle, pheromones *85*
1-butanol 121
butterflies, sex pheromones 86
2-*sec*-butyl-4,-5-dihydrothiazole 102
bystanders **34**

C4 HSL 22
C57BL/6J inbred mouse 99
cabbage looper moths (*Trichoplusia ni*), pheromones *85*, 86
calcium imaging, honeybees
  antennal lobe 128–*9*

drones 131
workers 129–30
mushroom body 133
camouflage 279, 284
use by social parasites 62, *63*, 70, 71, 179
Polyergus rufescens 68
Campbell's monkey, alarm calls 254
Cape honeybee (*Apis mellifera capensis*) 185
anarchistic workers 58
captive breeding programmes, effects 152
carbapenem (1-carbapen-2-em-3-carboxylic acid) production 14, 22
carotenoids as signals 6
castration, as effect of parasites 180
central place foraging **291**
channels of communication **291**
cheating behaviour 17
in biofilms 271
parasitized social insects 180–1, 184
in quorum sensing 18, 21, 23
chemical camouflage, *Polyergus rufescens* 68
chemical communication 35, 81–2, 91–2
complexity 83–4
multicomponent pheromones *85*, 86–8
multiple messages 84, 86
multipurpose signals 88–9
costs 91
exploitation by social parasites *62*
handicap principle 6, 8
insects as model system 81–2
molecular approaches 98–9
multiple components 90
rodents 97–8, 112–13
individual identity recognition 102–7
kin recognition 108–11
role of volatile scents 107–8
scent mark age 99–102
species-specific signalling 111–12
subordinate status advertisement 102
territorial advertisement 99
see also pheromones
chemical insignificance 61–2
Acromyrmex insinuator 67
Polyergus rufescens 67
childhood, language development 240
childhood conflicts 230
children, imitation 252
chimeric social insects 186
altruistic action disruption 179–81
impact on colony-level decisions 184–5
impact on extra-nidal activities 185
impact on reproductive decision-making 183
misdirected altruism 181–2
chimpanzees
mind-reading 259
pointing behaviour 261
Chinese, written language 282
Chinook salmon (*Oncorhynchus tshawytscha*), rapid evolution *143*
chirality *83*–4
choking game 269
chorus signalling 34
chromatography 98
Chromobacterium violaceum, QS system *15*
'chunking' of information 282
cichlids
sexual selection 151
social structure 38
citral 121, 132
Clethrionomys glareolus (bank vole), urinary proteins *112*
cliff swallows, recruitment signals 199–200, 202
climate change 139, 142, 152
clustering, networks *38*
great tits 48
non-territorial communities 41
cockroaches, pheromones 88–9, 91
coercion 13
Pseudomonas aeruginosa 22
of QS signalling 25
role of language 229
cold-seeking behaviour, parasitized bees 178
collaboration in signalling 1, 7–8
collared flycatcher (*Ficedula albicollis*), rapid evolution *143*
collective action, social insects 173–4, 175, 183–5
colonial communities 38–9, *39*, *40*
information flow 45
colonization
association with rapid evolution 141
facilitation by behavioural plasticity 146
colony fission, social insects 182
colony recognition, social insects 60
colony size
ants, effect on trail laying behaviour 201
and evolutionary stability of co-operative communication 202, *204*
relationship to foraging success 193, *194*, 197–9
cliff swallows 200
colour patches, multiple components 89–90
commensalism 294
communication, nature of 275–7
communication networks 33–4, *37*, 50
changes through lifetime 41
fairy-wrens 48–50
great tits (*Parus major*) 47–8
modelling in pace and time 34–5
communication types, fitness consequences *13*
community level networks *37*–41
competition, as cause of rapid evolution 142
competitive devaluation of signal strength 20
competitive signalling
scent mark age 99–102
subordinate status advertisement 102
territorial advertisement 99
complexity
of human language 249, 251, 252–4
of scent signals 97
complex traits, rapid evolution 139
composite facial images *159*
compositionality, human language 254, **263**
condition-dependence, human mate preference 166
conflict 275, **291**
humans 230–1
genomic 240–2
in information acquisition 43
intragenomic 209–10, 219–20, 227
maternal and paternal genes 210–11
in non-shivering thermogenesis 217
within social insect colonies 175–6
connectivity of nodes *38*
conopid flies, manipulation of host behaviour 178
consciousness 230
conservation efforts, effects 152
consistency, human behavioural traits 165–6

constraints on human language 250–1
constraints to information flow
in family-territorial communities 44–5
in lekking communities 46
in non-territorial communities 42
in pair-territorial communities 43–4
constraints to rapid evolution 143–4, 151
construction behaviour, as defence against parasites 178
Construction Grammar 253, **263**
contemporary evolution 139
see also rapid evolution
context-dependence, human mate preference 166
cooperation
evolutionary stability 17
link to niche construction 271
population dynamic model of information transfer 267–9
cooperative heat production 217, 220
cooperative networks *37*
co-opted signals 148
corpus callosum **226**
cutting of 230
sexual dimorphism 231
costs of chemical communication 91
costs of information flow 41, 266, 270, 272
in colonial communities 45
in family-territorial communities 44–5
in lekking communities 46
in non-territorial communities 42
in pair-territorial communities 43
costs of joint ESS, gene expression 211
costs of response 18–19
costs of signals 18, 284–5
costs of social communication 200, 201–2
cotton aphids (*Aphis gossypii*), pheromones 91
courtship behaviour
Bombus terrestris 84
Drosophila 81, 84
fairy-wrens 49
pair-territorial communities 43
creativity, in psychotic-affective spectrum conditions *238*, 239
crickets, parasite manipulation of behaviour 172
cross-talk, bacterial 24
cryptic species pairs 150–1
C-signal, *Myxococcus xanthus* 16
cuckoo, host behaviour manipulation 64, 172, 178, 179
cuckoo wasps 179
cue range 34–5
cues 2, 13, 33, 192, 284, **291**
in identity recognition 103
to infection status 181, 182
use by cliff swallows 199–200
use by social parasites 56, 57, 61
use within social insect colonies 57
cultural selection **294**
evolution of societies 55
culture, role of language 258
cuticular hydrocarbons
Drosophila 84
recognition of nestmates 87
social insects 60, 71
Acromyrmex spp. 67
chemical camouflage 68
chemical insignificance 61–2
effect of social parasites 65
mimicry 59, 63, *69*, 70
cystic fibrosis
interspecific bacterial signalling 24
quorum sensing 11

dance, as indicator of mate quality 162–3, 165
dance communication, honeybees 197–9, 282
compositionality 254
mechanistic model *195–6*
dark-eyed juncos (*Junco hyenalis*)
behavioural plasticity 144
rapid evolution 142, *143*, *145*
death's head hawk moth (*Acherontia atropos*), honey feeding 59
decay rate of signals 35, 45, 46
deception 259, 279
in social parasitism 61, 62–3, 68, 69, 70, 71, 179–80, 181
defensive behaviour against parasites 178
degradation, of bacterial signalling 24–5
degree of nodes *38*
deictic words 261, **263**
Delisea pulchra, QS disruption 24
delusion 229–30
denial 229–30
denotation **263**
depression 236, *237*
descriptive fallacy, language 258
desertion, parasitized social insects 179, 180
developmental disorders, in study of language evolution 233–4
dialects 250
Diana monkeys 254, 260
diastereomers *83*
DIF (differentiation inducing factor), slime moulds 4–6
diffusible factor (DSF) *16*
diffusion sensing (DS) 26–7
digital information systems 280, 281, 282
digits, 2D:4D ratio 161, 163, 164
dihydroxyphenylalanine (DOPA) 7
3,11-diMeC_{27} 88, *89*
direct fitness **292**
directional selection 139, **294**
direct reciprocity 270
disguises, use by social parasites 61–3
divergent selection 151, 240, **294**
diversity, human language 249–51
double articulation, human language 255, **263–4**
drift, social insect colonies 58–9
drone honeybees
antennal lobe processing 130–2
mushroom body and lateral horn processing 132–3
pheromones 122
Drosophila spp.
courtship behaviour 81, 84, 291
D. melanogaster, olfactory receptors 127
D. mojavensis, courtship song 150
pheromones *85*
D. gryimshawi 91
rapid evolution 141
DTNBP1 232, 233
duality of patterning (double articulation), human language 255, **263–4**
Dufour's gland substances 67–8, 88
dyslexia **226**, *237*, 239

eavesdropping **34**
in colonial communities 45
fairy-wrens 49–50
in family-territorial communities 44
great tits 47–8
in lekking communities 46
in non-territorial communities 42
in pair-territorial communities 43
economic concepts, use in biology 283

efficiency sensing (ES) 27
EFHC2 233
egg-dumpers 58–9
egocentrism, in autistic spectrum conditions 234, 236, 241
electroantennogram (EAG) technique 127
Emery's rule 61, 67
emotion, expression of 285
emulation 252
enantiomers *83*
engineering, association with autism 234, 236
Enteromorpha, effect of AHLs 24
entropy *276*, 280–1
environment
  effect on information acquisition 34–5
  as factor in waggle dance efficiency 199
enzyme degradation, AHLs 25
epigenetic mechanisms **291–2**
epimutation **292**
episodic memory 257, **264**
epistematic rationality *280*
ermine moths (*Yponomeuta* spp.), pheromones 86
Erwinia carotovora, carbapenem production 14, *15*, 22
4-ethyl-2-methoxyphenol 88–9, *90*
ethyl palmitate 122
eusocial insects 82, 173, 192
  see also social insects
eusociality **292**
evolution 209
  of human languages 250
  of multiple signals 84
  of reliable signals 1–2
  of societies 55
  see also rapid evolution
evolutionarily stable strategies (ESS) **292**
  cooperation 35, 202, *203*, *204*, 205–6
  genomic imprinting 211
  quorum sensing 20
evolutionary arms races, in sister-taxa social parasitism 61
evolutionary audiences **34**
  pair-territorial communities 43
evolutionary chases 218
evolutionary problems
  cooperation and communication 17
  quorum sensing 17–19
exaptations 279
exoenzymes *12*
exopolysaccharides, bacterial production *12*
exoproteases *12*
expressive power, human language 249
extended family groups 38
extended phenotype 171–3, 270, 275
  in parasites of social insects 177–9
    effects on communication channels 179–86
  in social insects 173–4, 175
The Extended Phenotype, Richard Dawkins 171, 173
extinction of human languages 250
extra-nidal activities, social insects 185
extra-pair copulation (EPC)
  fairy-wrens 49
  great tits 47, 48
extreme female brain 241
extreme male brain theory, autism 241

facial attractiveness 158–60
faculty of language in the narrow sense (FLN) 253–4, **264**
faculty of language in the broad sense (FLB) 254, **264**
fairy-wrens (*Malurus* spp.) 38
  communication networks 48–50
fallow deer (*Cervus dama*), social system 41
familial imprinting 108
family-territorial communities 38, *39*, *40*
  fairy-wrens 48–50
  information flow 44–5
(*E,E*)-farnesol 121
fear recognition 233
feedback loops 191
  cliff swallows 202
  naked mole rats 199
  in niche construction 270, 271
  pheromone trails 192
  waggle dance 197
'female brain' 242
female preference
  body features 160–1
  evolution 150–1
  facial features 158–60
  odour 161–2
  vocal characteristics 161
femininity, facial 159, *160*
field crickets (*Teleogryllus oceanicus*)
  cryptic species 151
  rapid evolution 142, *143*, 144–6, *145*, 147, 150
fingers, 2D:4D ratio 161, 163, 164
fire ants
  brood raids 193
  raft-making 175
fitness 4, **292**
  correlation with body attractiveness 160
  effect of network links 34
  inclusive 200–1
  in social insect colonies 55
    effect of behaviour-modifying parasites 179
fitness advantage 278
fitness consequences
  of communication types *13*
  of quorum sensing 18–19
'flash mobilization' 267
flatwing morphology, field crickets 144–5, 147
fluctuating asymmetry
  of body 161
  correlation with dancing ability 163, 165
  facial 158
follicle-stimulating hormone (FSH) 7
foraging
  cliff swallows 199–200
  mechanistic models *195–6*
  ravens 200
  social insects 185
    ant pheromone trails 192–3
    waggle dances *195–6*, 197–9
  synergy 191, 192
foraging groups, information flow 41–2
foraging success, relationship to colony size 193, *194*, 197–9
four-chambered brain model *227*, 230
FOXP2 gene 232, 233
fragile X syndrome *234*
framework information theory 57
free-riding 17
  in cooperative heat production 217, 220
  in quorum sensing 18
frequency-dependent selection 150, **294**
'frozen behaviour' 172
fruiting body formation, slime moulds 5–6
functional complexes 144
functional design, human language studies 232
fungus-growing termites, architecture 174

funnel-web spiders (*Agelenopsis aperta*), contests over web sites 284

game theory 283, 284
Gangestad, S.W. *et al.* 163
Gangestad, S.W. and Scheyd, G.J. 162
Gardner, M.G. *et al.* 65
garter snake, pheromones *85*
gas chromatography 98
gases, movement between containers 280–*1*
gel electrophoresis, proteins 98
gender traits, humans 240–1
gene flow, role in rapid evolution 151
generalized signals
  in colonial communities 45
  in non-territorial communities 42
generative linguistics **264**
genes, as units of selection 171–2
genetic architecture, rapid evolution 149–51
genetic assimilation (accommodation) 148
genetic autonomy 214, 215
genetic diversity, social insect colonies 65
genetic drift 141
genetic effects, in social selection 82
genetic heterozygosity, relationship to facial attractiveness 160
genetic studies
  integration with functional brain imaging 232
  social parasitism 71–2
genome sequencing 98
  benefits in protein characterization 99
genomic conflict 209–10, 219–20, **226**, **291**
  human 240–2
  maternal and paternal genes 210–11
  in non-shivering thermogenesis 217
genomic imprinting 210–11, 219, **226**, 230, 231, **292**
  conflicts 241
  influence in non-shivering thermogenesis 217
  interactions of imprinted genes 218–19
  Prader–Willi and Angelman syndromes 211–14
genotype by environment (G × E) interactions 150
geographical variation, social parasitism 71, 72
geranic acid 121
geraniol 121, 132
gestalt odours, ant colonies 60, 62
global sensing, bacteria 24
glomeruli, honeybee antennal lobe 125
  drones 130–2
  queens 132
  workers 128, 130
glycans, mimicry in virus traps 286
GNAS locus 217
goals *278*
good sense variants 279
good taste variants 279
G-protein α-stimulatory subunit (Gαs) 216, 217
grammatical constructions 253
Gram-negative bacteria, QS systems 14–16, *15*, 22
Gram-positive bacteria, signalling molecules 16–17, 22
great tit (*Parus major*) 37
  communication networks 47–8
  rapid evolution *143*
'green beard' recognition tags 60
green tree frogs (*Hyla cinerea*), rapid evolution 142
grey tree frogs (*Hyla versicolor*) 41
group conflicts, humans 230–1
group recruitment, ants 201
group selection 2, **294**
  slime moulds 5
grouse, social systems 41
GTPBP6 233
guest ant (*Formicoxenus nitidulus*), exploitation of sensory traps 64
guppies, rapid evolution 142, *143*, 151

hairiness, female preference 161
Hamilton's rule 19, 192, 203, 270, **292**
handicap principle 1–2, 8, 91, 165, 285
  altruism in babblers 3–4
  in chemical signals 6
  mathematical model 4
  signals within multicellular organisms 7
haplotype **292**
HBII-52 213, 214
healthiness, relationship to facial attractiveness 159, 160
height, female preference 161
helpers, in family-territorial communities 44–5, 49
2-heptanone 121, 130, 132
herbicide resistance 141
heterozygosity, relationship to facial attractiveness 160
1-hexanol 121
hexyl acetate 121
N-(3-oxohexanoyl) homoserine lactone (3-oxo-C6-HSL) 14
honesty, human behavioural attributes 164–7
honeybees
  African honeybee (*Apis mellifera scutellata*) 185
  Asian honeybees (*Apis dorsata*), bee-curtain formation 174
  antennal lobe processing 128
    drones 130–2
    queens 132
    workers 128–30
  Cape honeybee (*Apis mellifera capensis*) 185
    anarchistic workers 58
  dances 197–9, 282
    compositionality 254
    mechanistic model *195–6*
  intraspecific social parasitism 60
  mushroom body and lateral horn processing 132–4
  olfactory coding 122
  olfactory receptors (Ors) 127–8
  olfactory system 122, *124*–5
    sensillae 126–7
  pheromones *85*, 120–1, 134
    brood pheromones 122
    drone pheromones 122
    multipurpose 88
    queen pheromones 121
    worker pheromones 121
  virus-induced behavioural modification 184
horizontal information transmission, non-territorial communities 41
hormones 7, 8, 81, **292**
host-finding, social parasites 56
house finch (*Carpodacus mexicanus*), rapid evolution *143*
house-hunting, social insects 183
3O-C6-HSL 22
3O-C12-HSL 22
HTR2C 213, 214, 219
human activities, as cause of rapid evolution 139, 141, 142, 151–2
human chromosomes, 15q11-q13 cluster 211–14
human conflicts 230–1
human language 227–8, 249, 261, 276
  complexity 252–4

compositionality 254
diversity 249–51
double articulation 255
interpersonal function 258–9
learning 251–2
mind reading, manipulation and cooperation 259
self-organization 255–7
stimulus-freedom 257–8
human language evolution 225, 227, 228–30, 243
role of conflicts 230–1
study methods 231–2
comparative method 233–4
functional design 232
measurement of selection 232–3
human mate quality communication 157–8, 167
behavioural attributes 162
dynamic versus static stimuli 163–4
honesty 164–7
kinesics 162–3
physical traits
body 160–1
face 158–60
odour 161–2
voice 161
human pheromones 92
humans, origin 225, 242
3-hydroxy-2-butanone 88–9, *90*
4-hydroxy-3-methoxyphenylethanol (HVA) 121
olfactory responses 131
9-hydroy-(*E*)-2-decenoic acid (9-HDA) 121
hyperlexia **226**, *234*, 236, 239

identity recognition, scent marks 102–3
individual recognition 106–7
laboratory evidence 103–4
MUP polymorphisms 104
recognition of own versus other scents 104–6
illocutionary force **264**
imitation, human children 252
immune response modulation, *Pseudomonas aeruginosa* 11, 22
impossible sentences 251
imprinted genes 210–11, 219, 231
conflicts 241
influence in non-shivering thermogenesis 217
interactions 218–19
Prader-Willi and Angelman syndromes 211–14
imprinting 108
incentives for social communication 192
inclusive fitness 108, 200–1, 203, 206, **292**
humans 229, 230
social insects 174
indirect genetic effects, social selection 82
indirect reciprocity 270
individual recognition, scent marks 106–7
individuals, genetic conflicts 231
individual selection **293–4**
babblers 3
slime moulds 5, 6
infection status, cues 181, 182
infectious information transmission *266*, 267
population dynamical model 267–9
with cost 270
without cost 269–70
information **34**, 275
quantity and quality 279–83
information acquisition 266
informational density, trade-off with message length 281–2
information centre hypothesis 205
information flow 41
in colonial communities 45
in family-territorial communities 44–5
in lekking communities 46
in non-territorial communities 41–2
in pair-territorial communities 42–4
information requirements
Australian fairy-wrens 49
great tits 47–8
information transfer
with cost 270
population dynamical model 267–9
teaching 265–6
without cost 269–70
see also teaching
inheritance, linguistic 251
innate human language faculty 252–4
innovations 266
inquilines 58, 61, 70
Acromyrmex insinuator 67
Microdon spp. 68
microgyne *Myrmica rubra 66–7*
insecticide resistance 141
insects
chemical communication 81–2, 92
pheromones 119–20
see also ants; honeybees; social insects
insignificance in social parasitism 61–2, *63*, 71
Acromyrmex insinuator 67
insulin-like growth factor 2 (IGF2) 218–19
intentionality 279
in human language 252, **264**
intentional verbs **264**
interdisciplinarity 289
interference with altruism, babblers 3
interkingdom signalling, bacteria 24
interpersonal function, human language 258–9
interspecies communication, bacteria 23–4
intragenomic conflict *see* genomic conflict
intraspecies communication, bacteria 22–3
intraspecific social parasitism 60, 62
microgyne *Myrmica rubra 66–7*
Invisible Hand 256
ipsdienol 86, 91
ipsenol 86, 91
irregular verbs 256
isoamyl acetate 132
isopentyl acetate 121

joint ESS, gene expression 211

kairomones 119, *120*
Kakugo virus 184
Kanner (infantile) autism *234*
see also autism/ autistic spectrum conditions
Kanzi, leaning of human language 251, 253
Kenyon cells 125, 133
keystone individuals *38*
in family-territorial communities 44
in lekking communities 46
kinesic information 162–3, 165
kin recognition 108
laboratory evidence in rodents 108–9
in wild mice 109–11
kin selection 2, 13, 19, 82, 173, 270, **294**
kin selection model, quorum sensing (QS) 20–1, 22, 23

kinship 210
Klinefelter syndrome **226**, 233, *237*, 239, 241, 242
knockout approach, social insect research 186
Koschevnikov gland, honeybees 121

labelled-line hypothesis, olfactory coding 122, *123*, 131, 132, 134
Lactococcus lactis, bacteriocide production 25
language 249, 261, 276
  complexity 252–4
  compositionality 254
  diversity 249–51
  double articulation 255
  interpersonal function 258–9
  learning 251–2
  mindreading, manipulation and cooperation 259
  reference 259–61
  self-organization 255–7
  stimulus-freedom 257–8
language development 240
language evolution 225, 227, 228–30, 243
  role of conflicts 230–1
  study methods 231–2
    comparative method 233–4
    functional design 232
    measurement of selection 232–3
language problems
  in autism 236
  in schizophrenia 238–40
lantibiotics 25
lapwings (*Vanellus vanellus*), social system 38–9
Lasius niger ants, trail laying behaviour 201
las QS system, fitness consequences 19
lateral horn (LH), honeybees *124*, 125, 134
  olfactory processing 133–4
laughter, in Angelman syndrome 212–13
leaf-cutting ants, division of labour 175
learning
  of languages 251–2
  role in evolution 148
learning ability 147
Learning how to mean, M.A.K. Halliday 252
left cerebral hemisphere 227–*8*, 230
  in schizophrenia 239
lekking communities *39*, *40*, 41, 157
  information flow 46
leks **292**
length of message, trade-off with information density 281–2
leptothoracine ants, social parasitism 61
levels of recognition, social insects *87*
levels of selection **293**
lifetime social network model 36–7
linguistic communication, human 227–8
linguistic inheritance 251
linkage disequilibrium 150
links, networks 34, *38*
link strength *38*
  Australian fairy-wrens 49
  in colonial communities 45
  in family-territorial communities 44
  in lekking communities 46
  in non-territorial communities 41
  in pair-territorial communities 42
lipocalins *100*, 111, *112*
liquid chromatography 98
LIT1 218
little owl (*Athene noctua*), begging call 148
lobster cockroach, pheromones *85*
loudest-voice-prevails principle 211, 217
LRRTM1 233
luciferase reaction 14
luteinizing hormone (LH) 7
lux genes 14, 22, 25
LuxS 23
Lyphyra brassolis, social parasitism 57

Machiavellian intelligence 229
macroglomeruli
  honeybee drones 131
  honeybee queens 132
Maculinea spp., social parasitism *63*, 64, 65, 68–70
madumnal alleles 210–11, 218
Mafia hypothesis, social parasitism 179
magnetic resonance imaging (MRI), in human language studies 232
major histocompatibility complex (MHC)
  role in kin recognition 108–9, 110
  as source of individual scents 103, 113
major urinary proteins (MUPs) 99–102, 111–*12*, 113
  mass spectrometry *100*
  polymorphism 104
  role in individual recognition 104–7
  role in kin recognition 109–11
  structure *100*
'male brain' 242
manipulation 276
masculinity, facial 158–9, 160
mass spectrometry, major urinary proteins (MUPs) 99, *100*
mate choice
  advantages of sociality 157
  fairy-wrens 49
  great tits 47
  humans 157
    see also human mate quality communication
  rodents 106
    role of MHC 108–9
    role of MUPs 109–11
mate choice copying, lekking communities 46
maternal genetic effect 82
mating systems 292–3
matrilines, intraspecific social parasitism 60
meaning 283
  natural and non-natural 277
meerkats (*Suricata suricatta*) 38
memes 269, **292**
memory
  episodic 257
  relationship to planning 257
  role in human language 253
  role in social networks 35, 45
menstrual cycle, shifts in female preference *160*, 161, 163, 166
Meriones unguiculatus (Mongolian gerbil), urinary proteins *112*
metachromatic leukodystrophy *237*
methyl-*p*-hydroxybenzoate (HOB) 121
  olfactory responses 131
methyl linolenate 122
2-methylthiazolidine 88–9, *90*
mice
  major urinary proteins (MUPs) 99–102
    polymorphism 104
    role in kin recognition 109–11
  pheromones *85*, 102
  urinary lipocalins 111–*12*
Microdon spp., social parasitism 68

microgyne *Myrmica rubra*, social parasitism *66–7*
Micromys minutus (harvest mouse), urinary proteins *112*
Microtus agrestis (field vole), urinary proteins *112*
microvesicles, production in biofilms 25
mimicry 279
death's-head hawk moth 59
of pheromones 64
in social parasitism 61, 62–3, 68, 69, 70, 71, 179
virus traps 286
mindreading 259
Minimalist Program, human language faculty 253, **264**
mirror neuron systems 228
misdirected altruism, social parasitism 181–2
modality, sexual signals 149
modular theory, human language 252–3
monandry 292
monogamy 292
monogyny **292–3**
moths, sex pheromones 86
motor theory of language 228
mouse lemurs (*Microcebus* spp.), acoustic differences 149
multivariate approach, multicomponent pheromones 87
multicellular organisms, signals within the body 6–7
multicomponent pheromones 83, *85*, 86–8, 91
multicomponent signals 84
multiglomerular projection neurons (mPNs) 125
multimodal signals 81, 84, 91, 291
multiphotonic microscopy 130, 131
multiple chemical messages 84, 86
multiple queens, social insect colonies 60
multipurpose signals, chemical 88–9
multisensory signals 84
mushroom body, honeybees *124*, 125, 134
olfactory processing 132–3
Mus spp.
M. musculus domesticus (house mouse)
MUP polymorphism 104
scent mark characterization 97
urinary proteins 111, *112*
mutants, value in research 186
mutualism **293**
distinction from parasitism 59
mutualistic explanation of cooperation 21
Myrmica spp.
exploitation by *Maculinea* spp. 63, 69–70
M. rubra
intraspecific social parasitism 60
social parasitism *66–7*
sister-taxa social parasitism 61
Myxococcus xanthus
fruiting body formation 14
QS system 16
swarming 11

naked mole rats, recruitment signals 199
Nasanov pheromone, honeybees 121
natural meaning 277
natural selection 1, 82, 172, 210, 278–9, 292, **293**
evolution of societies 55
of human language 253
relationship to self-organization 256
Nauphoeta cinerea, pheromones 88–9, *90*, 91, 92
Necdin 213, 217
Neolamprologus pulcher, social structure 38
nerol 121
nerolic acid 121
nest desertion, parasite-induced 178, 185
nestmate recognition cues 87, 92
networks 33, **34**, 50, **293**
and behavioural ecology of individuals 37–41
information flow 41
levels of *39–40*
see also communication networks; community level networks; social networks
neurodevelopmental disorders, in study of language evolution 233–4
niche construction 270–1
Niphanda fusca, social parasitism 65
nitric oxide (NO), as a signal 7
nodes 34, *38*
2-nonanol 121
non-natural meaning 277
'non-pathfinders', pharaoh ants 88
non-shivering thermogenesis 216–17
non-territorial communities 37, *39*, *40*
information flow 41–2
non-verbal communication 162
in Angelman syndrome 212
dynamic versus static stimuli 163–4
honesty 164–7
kinesics 162–3

object-displacement experiments 257
objective teleology 277–9
Occam's Razor 253
1-octanol 121
octyl acetate 121
odour, attractiveness 161–2
odour cues, social insects 60
odour trails, rodents 199
Oecophylla smaragdina, social parasites 57
olfactory coding, honeybees 122, *123*
olfactory communication 119–20
see also pheromones
olfactory receptor neurons (ORNs), honeybees 124–5
olfactory receptors (Ors), honeybees 127–8
olfactory signals, rapid evolution 149
olfactory system, honeybees 122, *124*–5
antennal lobe processing 128–32
mushroom body and lateral horn processing 132–4
sensillae 126–7
Optimal Foraging Theory 291
orb-weaving spiders 148
Ormia ochracea, as parasite of field crickets 144
oropharyngeal flora (OF) bacteria, interspecific signalling 24
Oudeyer, P.-Y. 256
3-oxo-C6-HSL (*N*-(3-oxohexanoyl) homoserine lactone) 14
9-oxo-(*E*)-2-decenoic acid (9-ODA) 121, 134
olfactory response 127–8, 131, 132

Pachycondyla inversa, pheromones 88, *89*
padumnal alleles 210–11, 218
pair-territorial communities 37–8, *39*, *40*
great tits 47–8
information flow 42–4
PAME (hydroxyl-palmitic acid methyl ester) *16*
Pantoea (Erwinia) stewartii, QS system *15*

paper wasps (*Polistes* spp.), *en masse* dancing 174, 178
paralanguage 162
parallel adaptation, rapid evolution 140–1, 143–4, 150–1
parasites
as cause of rapid evolution 142, 144
manipulation of host behaviour 172
of social insects 175–9
parasitic castration 180
parasitism **293**, 294
distinction from mutualism 59
distinction from predation 59–60
parental investment, pair-territorial communities 43
parent–offspring conflict 230, **291**
Partan, S. and Marler, P. 84
particles, random movement 281
Passer domesticus (house sparrow), social system 39
'pathfinders', pharaoh ants 88
path length, pheromone trails 193
patrilines, social parasitism 60
payoff table, social communication game *201*
peacock's tail, evolution 1
peptide ligands, MHC 103
perennial social insect colonies *176*, 182
pest control, use of pheromones 119
pharaoh ant (*Monomorium pharaonis*), pheromones *85*, 88
phenomena of the third kind 256
phenotype 172
extended *see* extended phenotype
phenotype interactions, social selection 82
phenotypic plasticity 139, **293**
interaction with sexual selection 149
pheromonal parsimony 88
pheromone processing, honeybees 134
antennal lobe 128
drones 130–2
queens 132
workers 128–30
mushroom body and lateral horn 132–4
pheromones 8, 64, 81, 91–2, 98, 119–*20*, **293**
ants 173
chirality 83–4
costs 91
genetic variation 82
honeybees 120–1, 134
brood pheromones 122
drone pheromones 122
EAG responses 127
queen pheromones 121
worker pheromones 121
mice 102
mimicry of 64
multicomponent 83, *85*, 86–8
multiple components 90
multipurpose 88–9
rapid evolution 149
yeast cells 6
pheromone trails, ants 192–3, 201
mechanistic model *195*, *196–7*
Phodopus roborovskii (Roborovsky hamster), urinary proteins *112*
phonemes 255, **264**
phytophagous insects, unpalatability 91
pied flycatchers (*Ficedula hypoleuca*) 43
pigment production, bacteria *12*
pigment symmetry 90
placental disorders 231, 240
planning, relationship to memory 257
plant chemicals, use by insects 91
plasmid conjugation *12*
plasmids, presence of *lux* homologues 25
plasmid transfer *266*, *267*, 270
rate variability 271
pleiotropy **293**
plethodontid receptivity factor (PRF) 149
pointing behaviour 261
polyandry 292
social insects 65
Polyergus rufescens, slave-making 67–8
polygamy 37, 292–3
polygenic traits **293**
polygyny 292
social insects 65
poreplates, honeybees *126*
positive selection **226**
in language evolution 232–3
Prader–Willi syndrome 211–14, *237*, 241, 242
Preadipocyte factor-1 217
predation, distinction from parasitism 59–60
predators, as cause of rapid evolution 142
prediction 259
priming effect, queen mandibular pheromone 121
private speech 240
privatization of information
in colonial communities 45
in family-territorial communities 44
great tits 48
in non-territorial communities 42
in pair-territorial communities 43
probability, relationship to information 279–80
processing 128
producers 41–2
Prometheus myth 225, 233
promising 259
propaganda signals 64
prospective memory 257
protein binding, scent marks *101*–2
protein characterization 98–9
proteins, autonomy 215–17
proteomics 98
Provespa anomala, parasite-induced behavioural modification 184
Pseudomonas spp.
P. aeruginosa
biofilm formation 25, *26*
immune response modulation 11, 22
interspecific signalling 24
QS defective isolates 18, 23
quorum sensing 11, 12, *15*, 16, *18*, 19
siderophore production 205
signal blindness 18, 23, 27
virulence factor production 22
QS systems *15*
pseudomonas quinolone signal (PQS) *16*
psycho-historical hypothesis 240
psychosis **226**
psychotic-affective spectrum conditions **226**, 236–8
relative strengths and weaknesses *238*
role of imprinted genes 241
public goods production, bacteria 17
in biofilms 25
punishment 271
PvdQ 25

quantitative trait loci (QTL) 150
queen honeybees, antennal lobe processing 132
queen-intolerant inquilinism 58
queenless ants (*Dinoponera quadriceps*), action at a distance 174
queen pheromones, honeybees 121

drone responses 131
EAG responses 127
queens, multiple, social insect colonies 60
queen signal, *Pachycondyla inversa* 88, *89*
quiet song 34
quorum quenching 25
quorum sensing (QS) 11–13, 22–3, 27, 183, 205, 271, 286, **293**
in biofilms 25–6
degradation 24
diffusion sensing explanation 26–7
disruption by eukaryotes 24
evolutionary problems 17–19
Gram-negative bacteria *15*
Pseudomonas aeruginosa *18*
kin selection model 19–21
multiple functions of signalling molecules 25
mutualistic explanation 21

raft-making, fire ants 175
rainbow networks model *36*
Ralstonia solanacearum, QS system 16
rapid evolution 139–40, 151–2
of bird song 146
genetic architecture 149–51
parallel adaptation 140–1
precipitating factors 141–2
role of behavioural plasticity 148–9
role of gene flow 151
and sexual selection 142–6
of sexual signals *143*
signal modalities 149
speciation 151
rationality, epistematic *280*
ratios, sex pheromone components 86–7
Rattus norvegicus (Norway rat)
recruitment signals 199, 200
urinary proteins *112*
ravens, recruitment signals 200, 205
receiver bias 64
receivers 275
selection of signals 1, 209
reciprocal altruism 2, 35
reciprocal interactions 205, 270, 291
recruitment signals
ant pheromone trails 192–3
in birds and rodents 199–200, 202
honeybee dances 197–9
recursion 253–4, **264**
red fire ants (*Solenopsis invicta*), behavioural changes 144
red-legged salamanders, pheromones *85*
reference 259–61
refreshment, scent marks 99, 102
relatedness, relationship to cooperation and signalling *20*–1
releaser effect, queen mandibular pheromone 121
reliability of signals 1, 8, 17, 283–5
of human behavioural attributes 164–7
pheromones 91
signals within the body 7
religion 243, 258
replicators, genes as 215
repression 229–30
reproductive decision-making, social insects 182–3
reproductive skew **293**
in family-territorial communities 44
resource holding potential (RHP) 283
great tits 47
resource location, synergy 191, 192
resources, social insects 55–6
retrospective memory 257
Rett syndrome *234*, 241
rewards, use by social parasites 64
rhamnolipid 25
Rhizobium leguminosarum bv. *viciae*, QS system *15*
Rhodobacter sphaeroides, QS system *15*
Rhodopsin/ rhodopsin 215–16
right cerebral hemisphere 227–8, 230
rivals, deterrence of 4
robot autonomy 214
rock sparrows (*Petronia petronia*)
audience effects 43
social system 39
rodents, chemical communication 97–8, 112–13
individual identity recognition 102–7
kin recognition 108–11
odour trails 199
role of volatile scents 107–8
scent mark age 99–102
species-specific signalling 111–12
subordinate status advertisement 102
territorial advertisement 99
roles, networks *38*
roosting groups, information flow 41
ruffs (*Philomachus pugnax*), social system 41
sage grouse (*Centrocercus urophasianus*), social system 41
Salmonella typhimurium, AI-2 signalling 23–4
scarab beetles, pheromone enantiomers 83–4
scent mark age 99–102
scent signals
protein binding *101*–2
rodents 97–8, 112–13
individual identity recognition 102–7
kin recognition 108–11
scent mark age 99–102
species-specific signalling 111–12
subordinate status advertisement 102
territorial advertisement 99
volatile scents 107–8
schizophrenia **226**, *227*, 232, 233, 236–8
language problems 238–40
role of imprinted genes 241
sex differences 242
schizotypal personality disorder *237*
scientific skills, association with autism 234, 236
Sciurus carolinensis (grey squirrel), urinary proteins *112*
scroungers 41–2
scrub-jays (*Aphelocoma coerulescens*) 38
episodic-like memory 257
secondary sexual characteristics
evolution 140
role in signalling 140–1
second-order cooperative dilemmas 271
selection **293–4**
selection measurement, human language evolution 232–3
selective advantage 278
self-assemblages, social insects 175, 183
selfish behaviour **294**
selfish gene 171, 214–15, **294**
The Selfish Gene, Richard Dawkins 214
selfish trait, altruism as 4
self-organization
human language 255–7
social insects 183–5
self-recognition, scent-marking 104–6
semiochemicals *120*
molecular studies 98–9

see also chemical communication; pheromones
senders 275
sensillae, honeybees 124, 126–7
sensory traps, exploitation by social parasites 64–5
sentinel activities, babblers 3, 4
separation methods, semiochemicals 98
Serratia spp., QS systems *15*
sex differences, in autism and schizophrenia 242
sex linkage, role in evolution of sexually dimorphic traits 150
sex pheromones 119
  moths 86
sexual conflict **291**
  role in language evolution 230–1
sexual differentiation, humans 240–1, 241–2
sexual dimorphism
  corpus callosum 231
  facial 159
  urinary lipocalins 111, *112*
sexual selection 82, 279, **293**
  of human male voice 161
  interaction with phenotypic plasticity 149
  and rapid evolution 142–6
  role in language evolution 230–1
sexual signals
  divergence, role in speciation 151
  rapid evolution 140, 141–2, *143*
    constraints 143–4
    genetic architecture 149–51
    signal modalities 149
Shannon entropy *276*
Shannon information 275, *276*, 280, 282–3
sibling rivalry, social insects 182
side-blotched lizard (*Uta stansburiana*), rapid evolution *143*, 150
siderophore production *12*
  Pseudomonas aeruginosa 205
signal blindness, *Pseudomonas aeruginosa* 18, 23, 27
signalling modalities, pair-territorial communities 42
signalling molecules 11
  autoinducer-2 (AI-2) 23–4
  multiple functions 21, 25
  see also *N*-acylhomoserine lactone (AHL) signal molecules
signalling theory 165
signal range 35
signals
  altruistic activities 4
  chemical 6
  definitions 1, 7, 13, 33, 57, 192, 284, **294**
  DIF 4–6
  evolution 1–2, 209
  grades of 276
  learning of 148
  manipulation by social parasites 63–5
  multicomponent 84
  multimodal 84
  multipurpose 88–9
  parasitized social insects 180, 181
  quorum sensing as 12–13
  rapid evolution 139–40, 141–2
    role of behavioural plasticity 148–9
  reliability 283–5
  secondary sexual characteristics 140–1
  use by social insects 56
  within multicellular organisms 6–7
silkworm moths (*Bombyx mori*), pheromones 83, 84, 119
Sinorhizobium meliloti, QS system *15*
sister-taxa social parasitism 61, 70
  Acromyrmex insinuator 67
size, as a signal 283, 284
skin condition, facial 158–9, 160
Slabbekoorn, H. *et al.* 146
slave-makers 58, 61
  effect on host social structure 65
  Polyergus rufescens 67–8
  propoganda signals 64
SLC6A4 232, 233
sliding motility, bacteria *12*
slime moulds, altruism 4–6
smiling, in Angelman syndrome 212–13
snails, trematode parasite infection 185
SNURF-SNRPN 213
soapberry bug (*Jadera hamatoloma*), rapid evolution 139
social animals 33, **34**
social brain **226**, 229, 232
social communication 191–2
  evolution 205–6
  function 200
    inclusive fitness 200–1
    reciprocal interactions 205
    synergy 201–5
  human language 258–9
  mechanisms 192
    ant pheromone trails 192–3
    in birds and rodents 199–200
    honeybee dances 197–9
  mechanistic models of foraging *195–6*
  role of rhodopsin 216
social communication game *201*
social eavesdropping **34**
  see also eavesdropping
social insects 55–*6*
  behaviour-modifying parasites 176–9
    effect on communication channels 179–86
  colonies as superorganisms 174–6
  communication 57–8
  exploitation by social parasites 56–7
  extended phenotype 173–4
  inclusive fitness 201
  levels of recognition *87*
  mating systems 292
  modes of colony development *176*, 182
  mutualistic interactions 59
  see also ants; honeybees; social parasitism
sociality
  insects 81–2
  levels of *39–40*
  reproductive advantages 157
social learning 266–7
social networks 33, 34, *37*, 50, 286, **293**
  and behavioural ecology of individuals 37–41
  changes through lifetime 41
  great tits 48
  modelling in space and time 34–7
social parasitism 55, 70–2, 174, 178, 179–81, **294**
  Acromyrmex insinuator 67
  communication with hosts 57
  distantly related parasites 61
  distinction from mutualism 59
  effect on host social structure 65
  exploitation of host communication *62*
    disguises 61–3
    manipulative signals 63–5
  exploitation of social insects *56–7*
  intraspecific 60
  Maculinea spp. 68–70
  Microdon spp. 68
  microgyne *Myrmica rubra* *66–7*
  Polyergus rufescens 67–8
  sister taxa 61
  types 58–9

social predation 59
social prestige, babblers 4
social recognition, scent marking 107–8
social selection 82, **294**
sociectomy 186
societies, evolution 55
sociobiological theory 171
sockeye salmon (*Onchohynchus nerka*), rapid evolution *143*
Solenopsis invicta, odour cues 60
songs 254, 255
see also bird song
song sparrow (*Melospiza melodia*), rapid evolution *143*
spandrels 279
specialized signals 42
in family-territorial communities 44
speciation 139–40
role of mating signal divergence 151
species-specific signalling, scent marking 111–12
specific language impairment *234*
speech abnormalities, Prader-Willi and Angelman syndromes 212
spiders, contests over web sites 284
stability of networks, non-territorial communities 41
stabilizing selection 139, 140
stag-hunt games 268
Staphylococcus aureus, signalling 17, 22
status communication, social insects 173, 180
stereoisomers *83*
steroids 7
stimulus-freedom, human language 257–8
Streptomyces griseus, streptomycin production 14
strokes 230
structure, networks *38*
structured-deme models *203*
subordinate status advertisement 102
substrate-borne vibrations, exploitation by social parasites *62*
superior temporal gyrus 229
supernormal signals, social parasites 64, 181
superorganisms 174–6
supplementation experiments, cuticular hydrocarbons 87
surprise 279, 283
swallows, colony size, relationship to foraging success *194*
swarming, bacterial 11, *12*, 19
swarming behaviour, *Provespa anomala* 184
swift fox (*Vulpes velox*) 38
swimming motility, bacteria *12*
Syanoeca surinama, cooperative wing beating 174
symbiosis **294**
bacteria *12*
see also mutualism
symmetric kin 210
symmetry
of body 161
correlation with dancing ability 163, 165
facial 158–9
synergy 191, 200, 201–5, 206
mathematical model *203*, *204*
multicomponent pheromones 86
social insects 175, 183
synomones *120*
Syntactic Island Constraints 251

tabula rasa **264**
tactile signals, exploitation by social parasites *62*
talking to oneself 258
tandem running, ants 201
teaching 265–6
areas for future study 272
costs 266
niche construction 270–1
population dynamical model 267–9
with cost 270
without cost 269–70
teaching behaviour, ants 185
teleology 277–9
temporary social parasites 58, 65
termites
architecture 174
social parasitism 71
territorial advertisement, scent marking 99
territory establishment, great tits 47
testosterone
fetal exposure 241
markers of 159, 161
theory of mind 259, **264**, 279
thermodynamics 280–1
thermoregulation 216–17
social insects 183–4
thought, relationship to language 257–8
thought disorders, schizophrenia 238–9
threats, human language 258
time-travelling thought 257
Tourette's syndrome *234*
tragedy of the commons 17, 268
trail markers, ants 88
trematode, parasitism of snails 185
triadic communication 260
trophallaxis *63*, 68, 69, 180
trophobiotic associations, social insects 55
Turner syndrome **226**, 233, *234*, 239, 241

UBE3A 213–14
ubiquitin ligases 213–14
unconditional altruism 4
uncoupling protein 1 (UCP1) 216–17
uniglomerular projection neurons (uPNs) 125
unimodal composite signals 84
universal tendencies, human language 252
unpalatability, phytophagous insects 91
urban colonization, facilitation by behavioural plasticity 146
urbanization, as cause of rapid evolution 141–2
urinary proteins, rodents 111–*12*
see also major urinary proteins (MUPs)

velocardiofacial syndrome *237*
cis-verbenol 86
verbs, irregularity 256
vertical information transmission, non-territorial communities 41
vervet monkeys, alarm calls 259–61
Vibrio anguillarum, interkingdom signalling 24
Vibrio fischeri
bioluminescence 14, 22
interspecific communication 23–4
QS system *15*
victory displays 34
virulence, bacterial *12*
virulence factor production 17, 22
virus-carrying bacterial lineages 272
virus traps 286
visual range 34
visual signals
exploitation by social parasites *62*
role of rhodopsin 216
voice, attractiveness 161

volatile scents, role in individual recognition 107–8
vomeronasal organ, response time 107

waggle dances 197–9, 282
  compositionality 254
  mechanistic model *195–6*
wasps, social parasites 70
Wernicke's area **226**, 227, 232
whitethroats (*Sylvia communis*) 37
  audience effects 43
within-family conflicts, humans 230
wolves (*Canis lupus*) 38
work efficiency, parasitized social insects 179, 180
worker honeybees
  antennal lobe processing 128–30
  mushroom body and lateral horn processing 132, 133
worker pheromones, honeybees 121
worker policing 175, 184

Xanthomonas campestris, QS system 16
XLαs 217
X-linked traits 150
  language-related 231, 233

yeasts, pheromones 6
Yersinia spp., QS systems *15*

ZAC1 218
zero sum characteristic 283

The manufacturer's authorised representative in the EU for product safety is Oxford University Press España S.A. of El Parque Empresarial San Fernando de Henares, Avenida de Castilla, 2 - 28830 Madrid (www.oup.es/en or product.safety@oup.com). OUP España S.A. also acts as importer into Spain of products made by the manufacturer.
Printed and bound by CPI Group (UK) Ltd, Croydon, CR0 4YY
14/09/2026
02216284-0016